Mathematical modeling is the basis of almost all applied mathematics. A 'real-world' problem is dissected and phrased in a mathematical setting, allowing it to be simplified and ultimately solved. This book illustrates how models of continuous processes in a variety of applications can be derived, simplified, and analyzed. Using examples from an impressive array of disciplines, including industrial, environmental, and biological sciences, the presentation emphasizes the uniformity of the approach used by the applied mathematician in these different contexts. It assumes only a basic mathematical grounding in calculus and analysis and provides a wealth of examples for students of mathematics, engineering, and the range of applied sciences.

Mathematical Models in the Applied Sciences

Cambridge Texts in Applied Mathematics

Maximum and Minimum Principles
M.J. Sewell

Solitons
P.G. Drazin and R.S. Johnson

The Kinematics of Mixing
J.M. Ottino

Introduction to Numerical Linear Algebra and Optimisation
Philippe G. Ciarlet

Integral Equations
David Porter and David S.G. Stirling

Perturbation Methods
E.J. Hinch

The Thermomechanics of Plasticity and Fracture
Gerard A. Maugin

Boundary Integral and Singularity Methods for Linearized Viscous Flow
C. Pozrikidis

Nonlinear Systems
P.G. Drazin

Stability, Instability and Chaos
Paul Glendinning

Applied Analysis of the Navier-Stokes Equations
C.R. Doering and J.D. Gibbon

Viscous Flow
H. Ockendon and J.R. Ockendon

Scaling, Self-similarity and Intermediate Asymptotics
G.I. Barenblatt

First Course in Numerical Solution of Differential Equations
A. Iserles

Complex Variables: Introduction and Applications
Mark J. Ablowitz and Athanassios S. Fokas

Mathematical Models in the Applied Sciences
A.C. Fowler

Thinking About Ordinary Differential Equations
Robert E. O'Malley, Jr.

Mathematical Models in the Applied Sciences

A.C. FOWLER
University of Oxford

PUBLISHED BY THE PRESS SYNDICATE OF THE UNIVERSITY OF CAMBRIDGE
The Pitt Building, Trumpington Street, Cambridge CB2 1RP, United Kingdom

CAMBRIDGE UNIVERSITY PRESS
The Edinburgh Building, Cambridge CB2 2RU, United Kingdom
40 West 20th Street, New York, NY 10011-4211, USA
10 Stamford Road, Oakleigh, Melbourne 3166, Australia

First published 1997

Printed in the United States of America

Typeset in Times Roman

Library of Congress Cataloging-in-Publication Data

Fowler, A. C. (Andrew Cadle), 1953–
Mathematical models in the applied sciences / A. C. Fowler.
p. cm. – (Cambridge texts in applied mathematics)
Includes bibliographical references and index.
ISBN 0-521-46140-5 (hardbound). ISBN 0 521 46703 9 (pbk.)
1. Mathematical models. I. Title. II. Series.
QA401.F685 1997
511′.8 – dc21 97-10390
CIP

A catalog record for this book is available from the British Library

ISBN 0 521 46140 5 hardback
ISBN 0 521 46703 9 paperback

To the memory of Alan Tayler

Contents

Preface

The art of applying mathematics to real problems, be they in engineering, geophysics, industry, biology, or any other discipline, is one that is of enormous importance but which is also under relentless pressure: from computational scientists on the one hand and from mathematical analysts on the other. This book is about the middle ground between the two, and in it I hope to demonstrate that mathematical modeling is a subject of enormous potential, and one that has an essential role to play in many areas of modern applied science.

The book is hard, and intentionally so. There are many books on mathematical models that are aimed at a simpler expository level, but what I have endeavored to do here is to provide an insight into a way of thinking, which addresses problems more deeply. The only comparable book that I know of is that by Alan Tayler (1986). The difference between that and this is to some extent the scope of the problem areas and also the detail that I aim to go into. The earlier book by Lin and Segel (1974) is similar in ethos but much simpler.

It is possible to use this book as the basis for a course at graduate level, and indeed it embodies a course I have taught for the past several years at Oxford, in the M.S. program in Mathematical Modeling and Numerical Analysis. In that context, I use (for example) Part three as a rapid refresher/induction for basic models of partial differential equations, and then treat a selection of problems from Part four, two lectures per chapter. If you want to use this book as a course text, then you need to be a smart instructor – that is to say, the text is unforgiving, and some of the problems at the end of each chapter are at the level of research exercises. In the spirit of modeling, they may be more about finding *an* answer rather than finding *the* answer.

In writing this book around the course notes, I have added an introductory four chapters for completeness and reference, and I have also added four chapters at the end, which are more at the level of a research topic; indeed, each of them originated as an M.S. dissertation by former graduate students of mine.

Many people offered comments and advice on various chapters in the text, and I would like to thank them here for the efforts they have made in doing so: Alistair Fitt, Sam Howison, Jeff Dewynne, John Hinch, David Acheson, Hilary Ockendon, Malcolm Hood, John Willis, John Ockendon, Philip Maini, John Tyson, Lynn Van Coller, Donald Ludwig, Sean McElwain, Andrew Lacey, David Scott, Gary Parker, Joe Walder, Chris Huntingford, Chris Aldridge, Paul Emms, Richard Hindmarsh,

Leslie Morland, Kolumban Hutter, Giri Kalamangalam, Chris Noon, John Holden, Guy Kember, and Steve Davis. I hope I haven't forgotten anybody. I also want to thank Ross Mackay, Bernard Hallet, Bill Krantz, and Claude Jaupart for providing copies of original photographs. Pat Black and John Holden were generous in providing photographs, which I was unfortunately unable to use. The image in exercise 16.6 was kindly provided by Michael Manga, who responded promptly and generously to my request. A morning photographic session in Rosie O'Grady's pub in Oxford showed that it is not easy to take pictures of Guinness!

I wish to thank also Don Drew, Richard Alley, Peter Howell, Felix Ng, and Taryn Malcolm for help in various ways; Brenda Willoughby for her usual flawless manuscript production; Lesley Cosier, who did the line drawings rapidly and efficiently; Alan Harvey and Amy Thomas at Cambridge University Press. A long time ago I went for a job interview at Cambridge University Press, and when I got the rejection letter, it said 'We think it more likely that you will write books for us rather than publish them.' Rather observant. Despite all my efforts, there will inevitably be errors, but I hope they are not too terrible.

This book is dedicated to the memory of Alan Tayler. He was my thesis supervisor, and in his relaxed way inspired me and a host of others to carry the torch of mathematical modeling into the murky depths of science and industry.

Oxford, March 1996

Part one

Introduction

1

Mathematical modeling

1.1 What is a model?

Mathematical modeling is a subject that is difficult to teach. It is what applied mathematics (or, to be precise, physical applied mathematics) is all about, and yet there are few texts that approach the subject in a serious way. Partly, this is because one learns it by practice: There are no set rules, and an understanding of the 'right' way to model can only be reached by familiarity with a wealth of examples. That is what this book aims to provide.

A model is a representation of a process. Usually, a *mathematical* model takes the form of a set of equations describing a number of variables, and we distinguish between continuous models, in which the variables vary continuously in space and time, and discrete models, whose variables vary discontinuously. Examples of discrete models are nonlinear recurrence equations for population size in nonoverlapping generations (for example, the well-known logistic equation $x_{n+1} = \lambda x_n(1 - x_n)$) or probability distributions for Markov processes. Another example would be ARMA (auto-regressive moving average) models for the prediction of stochastic time series.

In this book, we are exclusively concerned with continuous models, and in practice that means models formulated as differential equations, both ordinary and partial. Other types of continuous model give rise to integro-differential equations (e.g., age-dependent population growth, nucleation and kinetics of crystal growth) or delay-differential equations (e.g., in ring-cavity lasers and models of cell maturation and growth).

Applied mathematicians have a procedure, almost a philosophy, that they apply when building models. First, there is a phenomenon of interest that one wants to describe or, more importantly, explain. Observations of the phenomenon lead, sometimes after a great deal of effort, to a hypothetical mechanism that can explain the phenomenon. The purpose of a model is then to formulate a description of the mechanism in quantitative terms, and the analysis of the resulting model leads to results that can be tested against the observations. Ideally, the model also leads to predictions which, if verified, lend authenticity to the model. It is important to realize that all models are idealizations and are limited in their applicability. In fact, one usually *aims* to over-simplify; the idea is that if a model is basically right, then it can subsequently be made more complicated, but the analysis of it is facilitated by having treated a simpler version first.

In formulating continuous models, there are three main ways of prescribing governing equations. The classical procedure is to formulate exact conservation laws. The laws of mass, momentum, and energy in fluid mechanics are obvious examples of these. In certain situations, conservation laws involve empiricism. For example, momentum conservation in a turbulent fluid motion may be represented by the friction correlation $\tau_w = f\rho u^2$, where τ_w is the wall shear stress, ρ is density, u is velocity, and f is a friction factor that is determined from experiment. Such 'laws' may depend on the precise physical constitution of the fluid, and may not be uniquely determined. Lastly, there are what might be termed 'hypothetical' laws, based on qualitative reasoning in the absence of precise rules. For example, in the Lotka–Volterra model of interacting predator and prey populations, the death rate of the prey is supposed to be proportional to the product of each population. This is a phenomenological assumption that is plausibly akin to the law of mass action in chemical reactions but which nevertheless has no quantitative basis. In this case, the usefulness of the model is in explaining the mechanism whereby interacting populations can oscillate.

1.2 The procedure of modeling

Problem identification

Mathematical modeling begins with the identification of a problem. There is something we don't understand, a phenomenon that requires explanation, and we begin by trying to identify a plausible mechanism. Sometimes this is quite straightforward. For example, consider the exasperating effect on the motorist of traffic jams on motorways. You drive along until quite suddenly you hit a slow-moving wall of traffic. You then move at a snail's pace until all of a sudden, the road is clear and you can drive freely. You wonder why, if the road is clear, couldn't the drivers ahead get on with it?

How should we model the density of traffic flow? First, we need to define a model context, and we need to know what appropriate variables are. Traffic density (cars per unit length of road) is one such variable and traffic speed is another, and we will want these to be functions of time and distance. We idealize the real situation by supposing that traffic travels in a single lane and also that the variables may be represented as continuous (indeed, differentiable) functions of space and time: This is essentially the *continuum approximation*, familiar in the modeling of fluids and other continua.

To formulate a model, we require laws (often of conservation type) and constitutive relations between variables, which may be based on experiment or empirical reasoning. For example, if the density of cars in a (single) line of traffic is $\rho(x,t)$, where x is distance along the traffic lane and t is time, the conservation of cars requires (subscripts denote partial derivatives)

$$\rho_t + q_x = 0, \tag{1.1}$$

where q is the car flux, that is, $q = \rho v$, and v is the car velocity. A simple phenomenological assumption is that car speed is determined entirely by the density, that is, $v = v(\rho)$, and v decreases as ρ increases. For example, take $v = 1 - \rho$, where car density is measured by its ratio to that of the maximum (bumper to bumper) density.

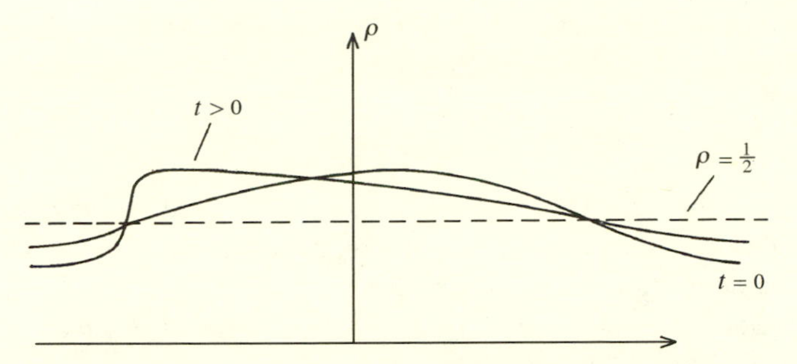

Fig. 1.1. Traffic jam formation: If a uniform density stream of traffic (with $\rho < \frac{1}{2}$) is perturbed so that $\rho_{\max} > \frac{1}{2}$, a jam forms, which is experienced by the oncoming driver as a shock (sudden deceleration) followed by a rarefaction wave (gradual acceleration)

Then $q = \rho(1-\rho)$, and the local wave speed is $q'(\rho) = 1 - 2\rho$. Thus $q' > 0$ for $\rho < 1/2$, $q' < 0$ for $\rho > 1/2$. An arbitrary initial disturbance to a uniform density state $\rho = \rho_0$ evolves as shown in Fig. 1.1 if the maximum value of ρ at $t = 0$, $\rho_{\max}$, is greater than $1/2$. Values of $\rho > 1/2$ move backward and form a shock that moves with speed

$$v_d = 1 - (\rho_+ + \rho_-), \tag{1.2}$$

where $\rho_\pm$ are the values just in front and just behind the shock. Because $v_- > v_d$, cars approach the shock and suffer a drop of speed as they pass through it. Further, because $v > q'$, they pass through the jam and eventually emerge in the undisturbed region again.

Through this simple model, we gain insight. We see that sudden changes in car speed can be associated with passage through a shock wave; the traffic flow is essentially that of a compressible medium. However, we also see that the model has important limitations. For example, one expects traffic jams to form because of, for example, lane closures on motorways, or where traffic lanes merge. Simple models to represent such phenomena can be based on Eq. (1.1) but require more realistic governing equations. Another limitation is that shocks (crashes) are usually avoided, and this may be represented by allowing traffic speed to depend on the spatial derivative of the density, which introduces a crucial diffusive term.

Thus we see the resolution (to some extent) of the problem in terms of a *mechanism*: the dependence of car speed on density and the resultant variation of the kinematic wave speed $q'(\rho)$. More generally, one often seeks a mechanism for the problem as specified. Often, this will be a verbal description only, which one then endeavors to translate into a mathematical formulation. For instance, consider the formation of underground cave systems, particularly in karst regions such as those near Postojna in Slovenia or near Doolin, County Clare, in the west of Ireland. Underground rivers are formed by the dissolution of the alkaline rocks by meteoric[1] groundwater that seeps through the porous fissured rock. There is an obvious mechanism here, because if the porosity (i.e., void space) is larger at some location, then fluid flow will be larger there, and hence also the rate of dissolution. This gives a positive feedback

[1] Meaning derived from the atmosphere, i.e., rainwater, which tends to be slightly acid.

mechanism whereby channels can form through increased amounts of dissolution at greater flow rates.

Model formulation

Once a problem is identified and a mechanism proposed, then one must formulate it mathematically. Often the difficulty lies in the choice of complexity: one wants the ease of a simpler model, but on the other hand one should include every relevant process. Different modelers will differ on what is important, and there is no unique 'right answer.' Formulation involves equations and boundary conditions, and if the problem is a sensible representation of the physics, it will usually (though not always) be well-posed. Mathematical analysts have as their program the establishment of well-posedness of a model, with the view that such results help design and validate suitable numerical solution procedures. In this book, we will not emphasize this approach, but rather will emphasize that of the applied analyst, whose business is to find the actual solutions.

Reduction

Solution of the proposed model now proceeds differently according to the modeler's background. Often (and this is largely the case for engineers and applied scientists) a model is a numerical model, and a solution means a numerical solution. There are two levels of difficulty with this. At the primitive level, direct numerical computations can founder because of ill-posedness or stiffness of the equations. More seriously, computation can limit insight, because of an inability to pose questions properly.

The first difficulty is aided by some pretreatment of the governing equations. It is, for example, often unhelpful to solve problems unless they have been nondimensionalized. When this is done properly (and this forms the focal point of this book), then the presence of small or large dimensionless parameters can be an indicator of singular perturbations, and thus stiffness. In many cases (most, in fact), this numeral inconvenience is an aid to analysis, facilitating the use of perturbation methods that can be used to gain insight into the solutions.

The second difficulty resides in the ability to use big computers to solve problems directly. Here is an example. Computation of convective flow in the Earth's mantle is a problem that realistically depends on a number of different dimensionless parameters: a Rayleigh number Ra, a viscosity number ε, an activation volume number μ, a dissipation number D, an internal heating number H, and so on. One aim of solving the equations representing this flow is to determine the dependence of the dimensionless heat flux, given by the Nusselt number Nu, on these various parameters: $Nu = Nu[Ra, \varepsilon, \mu, \ldots]$. The simplest situation, that of constant viscosity and no internal heating, is well understood both analytically and numerically. However, even with two independent parameters (Ra, ε), results are confusing and inconclusive. The problem is that the kinds of value of relevance to the Earth ($Ra = 10^7$, $\varepsilon = 1/40$, for example) have been, in the past, inaccessible to computer simulation–the asymptotic limits are too severe. One therefore has to extrapolate results at smaller Ra and/or larger ε to more extreme values: but, really, to do this sensibly, one needs a theoretical understanding of the correct limiting behavior, and although this has been done for Ra and ε, it has not for combinations of more parameters.

What happens is that the original question, what is $Nu[Ra, \varepsilon, \ldots]$ when $Ra = 10^7$, $\varepsilon = 1/40$, etc., is replaced by the question, what is $Nu[Ra, \varepsilon, \ldots]$ when, say, $Ra = 10^6$, $\varepsilon = 1/10$; the answer is then extrapolated to the more extreme parameter values. In this situation, the correct extrapolative procedure is to analyze the problem at extreme values and use the numerical results as a test for the predictions: each should complement the other. Successful mathematical modeling needs to combine different approaches, rather than elevate any one in a misplaced ascendancy.

The first thing we wish to do with a continuous model is to nondimensionalize it. It is then possible to identify in a rational way whether different terms are large or small. If the latter, they can in some (but not all) circumstances be ignored. One is thus led to a reduced model, which is a simplification of the original problem but not significantly less accurate. One must keep in mind what the question is. If one seeks not so much a quantitative simulation as a theoretical insight, then it may be judicious to simplify further. For example, detailed simulation of turbulent fluid flow probably requires use of averaged models such as k-ε models, but in some circumstances, Bernoulli's law may be sufficient and even Laplace's equation for the velocity potential![2]

Analysis

Reduction is the process whereby a model is simplified, most often by the neglect of various small terms. In some cases, their neglect leads to what are called *singular perturbations* (for which, see Chapter 4), whose effect can be understood by using the method of *matched asymptotic expansions*. In particular, it is often possible to break down a complicated model into simpler constituent processes, which, for example, operate on different space and time scales. Analytic dissection in this way leads to an overall understanding that is not so simply available through straightforward numerical computations, and indeed it also provides a methodology for simplifying such computations.

In analyzing a model, one is often led through a sequence of similar types of calculation: the existence and nature of steady solutions; their stability and instability, and consequent bifurcations to oscillations and traveling waves; hysteresis and the associated phenomenon of blow-up; secondary instabilities, and the occurrence of chaotic behavior. By studying a variety of models, as we do here, one thus 'learns' modeling by seeing the same sequence of processes carried out on a wide variety of different systems.

Computation

At some judicious point, numerical results need to be obtained. The use of these may be complementary: to obtain quantitative results in parametric regions where analysis is impossible. Or they may be validatory: they provide an independent confirmation

[2] For those unfamiliar with these fluid mechanical terms, further discussion is given in Chapter 6.

of analytic results. Sometimes, scientists think of analysis as confirming numerical results; in reality, one often needs to *design* numerical experiments to complement analytic results: straightforward but unthinking approaches often lead to apparent contradictions where a more carefully designed computation would remove these.

In problems where analytic progress is possible, the eventual problem to be solved numerically is often relatively simple, and we do not dwell on this aspect of modeling in this book; it should, however, be emphasized that it is an important component. Sometimes, indeed, it is *simpler* to solve the original problem numerically *rather* than the simplified model!

Model validation

Ideally, a mathematical model ends by returning to its origin. We look to see whether the model and its analysis explains the phenomenon we are interested in. Does the predicted curve fit the experimental data? Does the predicted stability curve agree with the experimentally determined values? The whole art of mathematical modeling lies in its self-consistency. It is an inexact science that derives its justification from the fact that apparently arbitrary assumptions are seen to work. And ultimately, this is the justification for a model: it helps us to understand an experimental observation. There is no unique or 'correct' model; but there are good models and bad models. The skill of modeling lies in being able to judge which is which.

1.3 Choosing the model

Consider, for example, the problem of modeling the climate. The weather is determined by heat and mass transfer in the atmosphere, which is (more or less) a blanket of air some ten kilometres thick that shrouds the planet (it extends above this *troposphere*, but the processes above the *tropopause* at ten kilometres mainly concern radiative heat exchange and absorption). The basic process of the weather is convection of the atmosphere driven by the differential heat input due to solar radiation between the equator and the poles. This heat imbalance causes a poleward heat flux by various convective processes both in the atmosphere and in the oceans. The vigorous rotation of the Earth causes this slow poleward circulation of the atmosphere to be distorted to a primarily *azimuthal* flow (the zonal wind), which is itself *baroclinically* unstable, and leads to the characteristic large-scale feature of the circulation, a wave-like undulation in the latitude of isobars.

Predicting the weather

Meteorologists predict the weather by writing fluid dynamic equations describing the convective motion of the atmosphere and seeking to solve them numerically. Predictions can be made for periods on the order of days, but it is generally thought that forecasting is impossible beyond about a week because of the chaotic nature of the atmospheric flow. One might infer from this that describing atmospheric dynamics on longer timescales is impossible, but this is not so – it is a question of choosing an appropriate model.

Ice ages

For example, the regular occurrence of ice ages at intervals of about a hundred thousand years can be predicted in some models of climate dynamics. Typically, these models describe the heat and mass transport in atmosphere and oceans by using semiempirical constitutive laws (for the average transport rates), with variable coefficients that depend on the size of the ice sheet cover. The ice sheets wax and wane over thousands of years, so that short term variability on a timescale of days, weeks, years, or even decades is unimportant and can be *averaged.* Thus the short-term unpredictability of the weather is irrelevant to the problem of modeling the longer term evolution of the climate, *provided* the more rapid fluctuations of the various atmospheric transports can be suitably parameterized. The type of model one obtains is very different and can be dramatically simpler. The simplest (energy balance) models are zeroth-order (i.e., algebraic)!

The smaller scale

But equally, we can go to the opposite extreme. One often observes, from aircraft, beautiful roll-like patterns in cloud formations. These are due to local convective processes, and they, too, can be modeled by fluid dynamic equations: However, the space and time scales are much smaller than those involved in weather prediction, and the fundamentally important larger scale effects of rotation are irrelevant on such smaller scales. Therefore, although the relevant model to describe these convective flow patterns is based on fluid equations, it will be rather different to a weather prediction model: For example, phase change effects are important (owing to the presence of water droplets in clouds) but rotation is not.

In summary, modeling is a subjective pursuit, and the nature of the problems dictates the type of model that is relevant. The same process may require different models, depending on the question that is of interest.

1.4 Some examples

In this section, we discuss further some examples of the mathematical models that were mentioned above.

Age-dependent population growth Suppose a population has a size distribution $\phi(a,t)$, where a is age and t is time: $\phi\delta a$ is the number between ages a and $a+\delta a$. The birth rate $b(a)$ depends on age, as does the mortality rate $m(a)$. These may also depend on time through social effects, for example, the baby boom of the sixties. A cohort of individuals age at a constant rate, $\dot{a}=1$, while their numbers decline at a rate $m(a)$, whence $\dot{\phi}=-m\phi$. These are the characteristic equations for the partial differential equation

$$\phi_t+\phi_a=-m\phi, \tag{1.3}$$

known as the Von Foerster equation, and the birth rate appears in the boundary condition

$$\phi(0,t) = \int_0^\infty b(a)\phi(a,t)\,da. \tag{1.4}$$

Further discussion of age-structured population models can be found in the books by Murray (1989) and Hoppensteadt (1975).

Nucleation and kinetics of crystal growth The classic theory of phase change kinetics is given in two papers by Avrami (1939, 1940). Suppose crystals are nucleated and grow from a melt, such that at time t, they occupy a volume $V(t)$ per unit volume. We also define $V'(t)$ to be the volume fraction the crystals *would* have had if different crystals did not meet. We define Y to be the rate of growth of crystal interfaces and I to be the rate of nucleation of new crystals. Both Y and I depend on temperature. In a time interval $(\tau, \tau + \delta\tau)$, $\delta N' = I(\tau)\,\delta\tau$ new crystals are created, and each of these attains a volume $v = a\{\int_\tau^t Y(\theta)\,d\theta\}^3$ at time t, where a is a shape factor ($a = 4\pi/3$ for spheres, $a = 8$ for cubes, for example). It follows that

$$V' = a\int_0^t I(\tau)\left\{\int_\tau^t Y(\theta)\,d\theta\right\}^3 d\tau \tag{1.5}$$

is the *fictive* volume, and

$$\frac{\partial V'}{\partial t} = 3a\int_0^t I(\tau)Y(t)\left\{\int_\tau^t Y(\theta)\,d\theta\right\}^2 d\tau. \tag{1.6}$$

Now we suppose that at time t, the fraction of the fictive crystal surface that lies inside the actual crystals should be V, and thus of the fictive growth $\delta V'$ in the interval δt, only the fraction $1 - V$ contributes to actual growth. Thus $\delta V = (1 - V)\,\delta V'$, so that

$$\frac{\partial V}{\partial t} = 3a(1 - V)\int_0^t I(\tau)Y(t)\left\{\int_\tau^t Y(\theta)\,d\theta\right\}^2 d\tau. \tag{1.7}$$

Recently, this theory (used in the study of solid-solid phase transitions in metallurgy) has been applied to the solidification of magma chambers (Brandeis, Jaupart, and Allegre, 1984).

Delay differential equations These occur in a number of different applications, particularly medical, where the delay may be due to finite maturation time (for example in white blood cell populations (Mackey and Glass, 1977) or in the humoral immune response (Dibrov, Livshits, and Volkenstein, 1977a,b), or to a finite transport time (hence the delay in respiratory control models due to transport of blood gases in the arteries). A common form of such equations is the *delay-recruitment* equation

$$\varepsilon\dot{x} = -x + f(x_1), \tag{1.8}$$

where $x_1 = x(t-1)$, which can be derived in respiratory control models (Fowler, Kalamangalam, and Kember, 1993), blood cell populations, ring cavity lasers (Ikeda and Matsumoto 1987) and population biology (May, 1980; Gurney, Blythe, and Nisbet, 1980). In the last case, the decay term $-x$ is the mortality rate, whereas the delay term is the regeneration rate (or recruitment rate), taken as a nonlinear function of the population size at an earlier time (the delay here is the gestation time).

Lotka–Volterra equations The simplest model of interacting populations was proposed by Volterra (1926). The same model was used by Lotka (1920) to illustrate the phenomenon of undamped oscillations in a model chemical reaction. If x and y are the predator and prey populations, then these equations are

$$\begin{aligned} \dot{x} &= \alpha x y - \beta x, \\ \dot{y} &= \gamma y - \delta x y, \end{aligned} \tag{1.9}$$

representing constant specific birth and mortality rates for prey and predator, respectively. The predators' specific growth rate αy depends on availability of the prey as food source, whereas the prey death rate δx depends on the number of predators. These equations have oscillatory solutions but have the unsatisfactory feature of forming a conservative system, and oscillations of any magnitude are possible. More realistic versions of the model can remove this degeneracy (see Murray 1989).

Traffic flow modeling Modeling traffic flow was used by Whitham and Lighthill as an example of their 'kinematic wave theory.' There is a good discussion in Whitham's (1974) book. In general, shocks will form, and as with shock waves in gas dynamics, this suggests hunting for a physically plausible mechanism that can prevent shock collision. One such mechanism is the realization that if there is a change in density, this also affects drivers' reactions. We might expect that v is lower if the density increases ahead of the driver, thus $\partial v/\partial \rho_x < 0$, and a simple model is then

$$v = 1 - \rho - \delta \rho_x; \tag{1.10}$$

this leads to the nonlinear diffusion equation

$$\rho_t + (1 - 2\rho)\rho_x = \delta(\rho \rho_x)_x, \tag{1.11}$$

which allows a diffusive shock structure of width $O(\delta^{1/2})$.

Formation of cave systems A recent model analyzing the onset of cave systems in limestone regions is that of Groves and Howard (1994). As with many applied problems, one finds similar phenomena in a wide variety of different contexts. For example, meltwater at the surface of a valley glacier finds its way through crevasses to the bed, where it drains along the valley floor through a network of channels. The development of a channeled flow occurs in a similar way to that in limestone, except that melting plays the part of erosion. The basic theory is given by Röthlisberger (1972). Similar erosive/melting instabilities are the cause of channel formation in dendritically solidifying alloys (Copley *et al.*, 1970) and lie at the heart of the erosional formation of river drainage networks (Willgoose, Bras, and Rodriguez-Iturbe, 1991a,b; Smith and Bretherton, 1972; Kramer and Marder, 1992). The web that a particular mathematical model can weave through the different sciences is part of the fun of applying mathematics.

Mantle convection Descriptions of convection of the solid earth's mantle have been reviewed frequently (see for example Turcotte, 1979; Turcotte and Schubert, 1982; Quareni and Yuen, 1988; Hager and Gurnis, 1987). A typical example of the computational approach is given by Quareni and Yuen or Christensen (1984). Examples of how analytic and numerical approaches can be used to complement each other are given by Fowler (1993) and Moresi and Solomatov (1995).

Ice ages The cause of ice ages, which have occurred regularly over the last few million years at intervals of about 100,000 years, is not known for certain, although there is a wealth of hypotheses. Ice ages consist of long periods of cold climate when ice sheets grow in the northern hemisphere and move south below the U.S./Canada border in North America and over the North Sea from Scandinavia in Europe. These great ice sheets may have been several kilometers thick and are associated with a lowering of sea level of the order of 100 meters.

Various ideas have been advanced to explain them. The simplest models are one-dimensional and were proposed by Budyko (1969) and Sellers (1969). These energy-balance climate models are reviewed by North, Calahan, and Coakley (1981). The idea is that the albedo (reflectivity) of the planet increases when significant ice sheets are present. This tends to reduce planetary temperature, and there is thus a positive feedback that can cause ice sheets to grow if the solar insolation (radiation input) is reduced. Following Ghil and Childress (1987), a simple model for the Earth's globally averaged temperature is

$$c\frac{dT}{dt} = Q[1-\alpha(T)] - \sigma g(T)T^4. \tag{1.12}$$

Here, c is the specific heat of the atmosphere. The first term, $R_i = Q[1-\alpha(T)]$ is the absorbed radiation: Q is the solar radiation, and α is the albedo or reflectivity. The latter depends on cloud and ice cover, but its dependence on cloud cover is unknown. Ice is more reflective, and as we associate increased sea and land ice with a decreased T, we take $\alpha = \alpha_l$, $T \leq T_l$, corresponding to an ice-covered Earth, $\alpha = \alpha_u$, $T \geq T_u$, corresponding to an ice-free Earth, with a linearly decreasing α between these values.

The second term, $R_e = \sigma g(T)T^4$, represents the emitted radiation R_e; σT^4 is the black-body radiation, and g is the 'greyness' of the system. This must reflect the greenhouse effect, whereby outgoing infrared radiation is partly absorbed by clouds and trace gases in the atmosphere. Sellers (1969) suggested

$$g(T) = 1 - m\tanh\{(T/T_0)^6\}, \tag{1.13}$$

where $m = 0.5$, $T_0 = 284$ K. Thus as T increases, g decreases, and less radiation is emitted. Equilibrium temperature is determined by intersections of R_i and R_e, and a typical situation is shown in Fig. 1.2. Evidently, multiple equilibria can occur, and it is easily shown that the upper and lower equilibria are stable. In this interpretation, the present climate is represented by the upper equilibrium, whereas the lower one corresponds to climatic conditions during an ice age. The transition between the two is effected by slow variations in the solar radiation Q, which causes $R_i(T)$ to shift up and down (see Exercise 4).

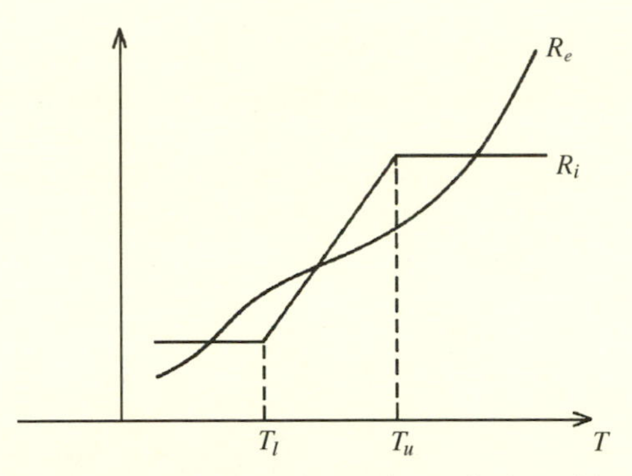

Fig. 1.2. Variation of incoming and outgoing radiation with temperature

In fact, there is some evidence that variation in the solar radiation is the root cause of ice ages. This is the *Milankovitch* theory, which relies on the fact that variations in the Earth's orbital parameters (eccentricity, obliquity, and precession) cause a variation in the solar radiation Q over time of about 5–10%. The supporting evidence is circumstantial–climatic time series data that are proxy measurements of temperature and ice volume have strong spectral frequencies of periods 100 ka (1 ka = a thousand years), 41 ka, 24 ka, and 19 ka. These frequencies are also present in the solar radiation curves, which strongly suggests a causative influence.

However, there is a problem. The hundred thousand year signal is the weakest in the radiation signal, but is dominant in the climatic signal. This suggests that the Earth's climate may act as a forced nonlinear oscillator, with internal feedbacks amplifying the weak forcing. Theories that describe, in a somewhat lumped fashion, such interactive dynamics between ice sheets, oceans and atmosphere, have been discussed by, for example, Weertman (1976), see also Saltzman (1978). The book by Ghil and Childress (1987) has a useful discussion.

Exercises

1. (a) Justify the derivation of the boundary condition (1.4) for age-structured populations.

   (b) Show that the steady size distribution with age of a population satisfying Eqs. (1.3) and (1.4) is given by the linear integral equation

$$\phi(a) = \int_0^\infty G(a, \xi)\phi(\xi)\, d\xi,$$

   where $G(a, \xi)$ should be specified.

   (c) Use the method of characteristics to show that for $t > a$, the solution of Eqs. (1.3) and (1.4) is

$$\phi = \left[\int_0^\infty b(\xi)\phi(\xi, t-a)\, d\xi\right] \exp\left[-\int_0^a m(\eta)\, d\eta\right].$$

   (d) Deduce an approximate equation for ϕ if $b = 0$ for $\xi < t_m$, $b = B$ for $t_m < \xi < t_m + t_b$, and $b = 0$ for $\xi > t_m + t_b$, where t_b is small, and hence show that if $x(t) = \phi(t_m, t)$, then

$$x(t) \approx \delta x(t - t_m),$$

where $\delta = Bt_b \exp[-\int_0^{t_m} m(\eta)\, d\eta]$. Why is this obvious?

2. (i) Show that the kinetic growth model (1.7) can be written in the composite form

$$\begin{aligned} \dot{V} &= (1 - V)Y\xi, \\ \dot{\xi} &= 2Y\eta, \\ \dot{\eta} &= Y\zeta, \\ \dot{\zeta} &= 3aI, \end{aligned} \tag{*}$$

with $\zeta(0) = \eta(0) = \xi(0) = V(0) = 0$. Hence show that if I and Y are constants, the crystal volume fraction is

$$V = 1 - \exp\left[-\frac{t^{*4}}{4!}\right],$$

where $t = t_c t^*$, and the *crystallization time* is

$$t_c = [6aIY^3]^{-1/4}.$$

Why is this not a good model when V is close to 1?

(ii) An alternative model for crystallization replaces I by $(1 - V)I$, on the basis that this is the nucleation rate in the free space. Write down the equivalent to Eq. (*) above, and show that $V = 1 - e^{-\psi}$, where ψ satisfies

$$\frac{d^4\psi}{dt^{*4}} = e^{-\psi}.$$

What is the asymptotic behavior of V as $t^* \to \infty$ in this case?

3. A population of size N grows according to the delay-recruitment model

$$\dot{N} = -\mu N + rN(t - \tau)[N_{\max} - N(t - \tau)]_+,$$

where the parameters represent a specific mortality rate μ, a gestation time τ, and a resource-limited specific growth rate $r(N_{\max} - N)_+$ (note that $z_+ = \max(z, 0)$). By choosing suitable nondimensionalized variables, show that the system can be written in the form

$$\varepsilon\dot{x} = -x + f(x_1),$$

where $f(x) = \alpha x(1 - x)_+$, $\alpha = rN_{\max}/\mu$, $\varepsilon = 1/\mu T$. Deduce that the limit of $\varepsilon \to 0$ corresponds to that of 'large' delay.

The stability of a fixed point x^* ($x^* = f(x^*)$) is determined by the evolution of small perturbations $\propto e^{\lambda t}$ to x^*. Show that stability is determined by solving

$$\varepsilon\lambda = -1 + \mu e^{-\lambda},$$

where $\mu = f'(x^*)$, and x^* is (linearly) stable if and only if $\text{Re}\,\lambda < 0$ for all solutions.

(i) Show that Re $\lambda < 0$ if $|\mu| < 1$.
(ii) Show that if $\mu > 0$, instability occurs at $\mu = 1$ when λ increases through zero.
(iii) Show that if $\mu < 0$, an oscillatory instability first occurs (a Hopf bifurcation) as μ decreases through $\mu_h = \sec\Omega$, where Ω is the root of $\tan\Omega = -\varepsilon\Omega$ between $\pi/2$ and π.
(iv) Show that the curve in the (μ, ε) plane on which this Hopf bifurcation occurs is given by

$$\mu\cos\left[\frac{(\mu^2-1)^{1/2}}{\varepsilon}\right] = 1.$$

4. (i) Show that an equilibrium solution of $\dot{T} = f(T)$ (i.e., where $f(T^*) = 0$) is stable if $f'(T^*) < 0$, and unstable if $f'(T^*) > 0$. Hence show that for the ice age climate model (1.12), as illustrated in Fig. 1.2, the upper and lower equilibria are stable but the middle one is unstable.
(ii) Suppose $c\dot{T} = R_i(T) - R_e(T)$, with $R_i = Q[1-\alpha(T)]$, $R_e = \beta T$, and $\alpha = \alpha_l$ for $T < T_l$, $\alpha = \alpha_u$ for $T > T_u$, with $\alpha_l > \alpha_u$, and α is linear between T_l and T_u. Assuming $\beta < Q(\alpha_l - \alpha_u)/(T_u - T_l)$, show that if $Q < \beta T_u/(1-\alpha_u)$, the unique steady state is stable and corresponds to an ice-covered Earth. Show that if $Q > \beta T_l/(1-\alpha_l)$, the (stable) steady state corresponds to an ice-free Earth. Show also that if Q varies slowly between values $Q_{\max} > \beta T_l/(1-\alpha_l)$ and $Q_{\min} < \beta T_u(1-\alpha_u)$, then the temperature will oscillate between the ice-free and ice-covered values.
5. One thinks of ice ages as 'large' events. Ice sheets kilometers thick grew in North America and Europe, extending for thousands of kilometers, causing huge changes in atmospheric dynamics, surface topography, etc. Yet the Milankovitch theory suggests that this is all caused by a (small) variation in radiation of around 5–10%. Can you reconcile these two disparate observations?

Part two

Methods

2

Nondimensionalization

2.1 Introduction

Confronted with, or having created, a mathematical model of a continuous physical system, which consists of a set of differential equations and associated boundary conditions, the first thing that an applied mathematician will want to do is nondimensionalize the system. Practically useful (as opposed to academically titillating) models are very often far too complicated to analyze rigorously, and the only way forward is some kind of asymptotic reduction based on the idea that certain terms are small and can be neglected.

But what do we mean by small? A little thought shows that the word must be used in a relative manner. A speed of 1 cm s^{-1} is slow for a bullet, but fast for an earthworm. In fact, we can only properly measure the size of a quantity by comparison with another of the same dimension. In this way, terms are evaluated in order of magnitude by means of dimensionless ratios, and the process whereby the terms in an equation are methodically evaluated in this way is called *nondimensionalization*.

The basic process is extremely simple to describe. If a model has a variable u, say, then we nondimensionalize that variable by writing, for example,

$$u = [u]u^*, \tag{2.1}$$

where $[u]$ is the chosen scale and u^* is the corresponding dimensionless variable. In a similar way, we nondimensionalize timescales by writing $t = [t]t^*$, and so on. There is of course nothing sacrosanct about the use of the square bracket or asterisked notation.

Because equations that describe real processes are necessarily dimensionally homogeneous, it follows that the process of nondimensionalization will yield a set of equations, each of whose terms is dimensionless, after division through by the dimension of the equation. It is then possible to compare terms in a meaningful way.

For example, the number of atoms of a radioactive substance is governed (approximately) by the differential equation

$$\frac{dN}{dt} = -\lambda N, \tag{2.2}$$

with an initial condition $N = N_0$ at $t = 0$. Here, λ is the decay constant, having units of $[\text{time}]^{-1}$. Suppose we nondimensionalize as described above, using scales $[N]$ and

$[t]$ (yet to be chosen). Then we get

$$\begin{aligned}\frac{dN^*}{dt^*} &= -\{\lambda[t]\}N^*,\\ N^*(0) &= \{N_0/[N]\}.\end{aligned} \tag{2.3}$$

All the asterisked terms are dimensionless, and the expressions in curly brackets are also; they are called *dimensionless parameters*. There now remains the issue of choosing values for the scales $[N]$ and $[t]$. This process is known as *scaling*.

Scaling

The art of nondimensionalization lies in the choice of scales. There are right ways and wrong ways to do it, and in more complicated problems, the choice of scales can be the most interesting and difficult part of the analysis. The basic principle is that the scales must ultimately be chosen self-consistently by balancing the terms in the equations. Because the purpose is to attain 'properly scaled' equations in which the largest dimensionless parameters are numerically of order one, the simplest choices arise when the scales can be chosen so that all the dimensionless parameters are $O(1)$. For Eq. (2.3), this is effected by choosing

$$[N] = N_0, \qquad [t] = 1/\lambda; \tag{2.4}$$

then $dN^*/dt^* = -N^*$, $N^*(0) = 1$, so $N^* = e^{-t^*}$. Notice that N^* changes by an amount of $O(1)$ in a time t^* of $O(1)$.[1] This provides our rationale. Given no other information, one assumes *a priori* that dimensionless variables and their derivatives are $O(1)$, until we are forced to assume otherwise. It is only when we are then led to inconsistency that the process of *rescaling* may become necessary. The generic situation in which this happens is that where singular perturbation theory is appropriate, and this is considered in Chpater 3.

In general, one cannot choose all the dimensionless parameters to be one, or $O(1)$. It is usually best then to try and choose the largest dimensionless parameters to be equal to one. Usually, this can only be done in one self-consistent manner, which has to be determined by trial and error. We will illustrate this in the next section.

Distinguished limits

Although the point of nondimensionalization is to identify unimportant terms, it is often the case that one is interested in the richest behavior of a model, and this will often occur when a particular choice of scales is selected such that two or more terms are rendered comparable. In this case, we talk of the scaling as a *distinguished limit*. An example is afforded by the Korteweg–de Vries equation for waves of amplitude $\eta(x,t)$ in a channel of depth d (Dodd *et al.*, 1982, page 4; Drazin, 1983, page 3):

$$\frac{\partial\eta}{\partial t} = \frac{3}{2}\sqrt{\frac{g}{d}}\frac{\partial}{\partial x}\left[\frac{2}{3}\alpha\frac{\partial\eta}{\partial x} + \left\{\frac{1}{3}d^3 - (\gamma d/\rho g)\right\}\frac{\partial^3\eta}{\partial x^3}\right]. \tag{2.5}$$

[1] Order notation is described in Chapter 3. Roughly, $O(1)$ means terms of size comparable to unity.

It is clearly possible to choose scales for η, x, and t so that all the terms balance i.e., the dimensionless version of Eq. (2.5) is

$$\zeta_\tau = \zeta_\xi + \zeta\zeta_\xi + \zeta_{\xi\xi\xi}, \tag{2.6}$$

and this is a distinguished limit insofar as it balances the effects of nonlinearity ($\zeta\zeta_\xi$, which tends to cause shock formation) and dispersion ($\zeta_{\xi\xi\xi}$, which causes waves to disperse). As is well known, the balance of these terms allows the existence of solitary waves in the form of particle-like *solitons*.

2.2 Damped pendulum

Motion of a linearly damped pendulum is governed by the equation

$$\begin{aligned} l\ddot{\theta} + k\dot{\theta} + g\sin\theta &= 0, \\ \theta(0) = \theta_0, \quad \dot{\theta}(0) &= \omega_0, \end{aligned} \tag{2.7}$$

where a dot denotes a time derivative. Putting

$$\theta = [\theta]\theta^*, \qquad t = [t]t^*, \tag{2.8}$$

we obtain the equation

$$\left(\frac{l[\theta]}{[t]^2}\right)\ddot{\theta}^* + \left(\frac{k[\theta]}{[t]}\right)\dot{\theta}^* + g\sin\{[\theta]\theta^*\} = 0, \tag{2.9}$$

and

$$\theta^*(0) = \theta_0/[\theta], \quad \dot{\theta}^* = \omega_0[t]/[\theta]. \tag{2.10}$$

Notice that $[\theta]$ is a scale, as θ is dimensionless already. It is usual to omit the asterisks once the dimensionless equations have been obtained. Dividing through to make the equations dimensionless, we thus have

$$\ddot{\theta} + \left\{\frac{k[t]}{l}\right\}\dot{\theta} + \left\{\frac{g[t]^2}{l[\theta]}\right\}\sin\{[\theta]\theta\} = 0, \tag{2.11}$$

together with Eq. (2.10). There are thus five independent dimensionless parameters in Eqs. (2.10) and (2.11), but only two scales, $[\theta]$ and $[t]$. We can therefore eliminate two of the parameters.

For example, because we hope that θ and $\dot{\theta}$ are $O(1)$, an obvious choice is to *balance* the initial conditions by choosing

$$[\theta] = \theta_0, \qquad [t] = \theta_0/\omega_0, \tag{2.12}$$

so that

$$\theta(0) = \dot{\theta}(0) = 1, \tag{2.13}$$

and then

$$\ddot{\theta} + \alpha\dot{\theta} + \beta\sin(\gamma\theta) = 0, \tag{2.14}$$

where

$$\alpha = k\theta_0/\omega_0 l,$$
$$\beta = g\theta_0/\omega_0 l, \tag{2.15}$$
$$\gamma = \theta_0.$$

If α, β, γ are $O(1)$ or smaller, we seem to have a properly scaled problem. However, if any of them is larger, we may not.

As one particular example, suppose that $\alpha \gg 1$, $\beta, \gamma \sim 1$;[2] if we simply divide by α, then the equation is $\dot{\theta}$ = smaller terms, which is inconsistent with the initial conditions. It is better to *rescale*, by choosing a *distinguished limit* in which at least two of the terms in Eq. (2.14) balance. The obvious choice is to bring back the second derivative term, and this can be done by rescaling t, that is,

$$t = \tilde{t}/\alpha. \tag{2.16}$$

We substitute in, and the rescaled equations are then

$$\ddot{\theta} + \dot{\theta} + (\beta/\alpha^2)\sin(\gamma\theta) = 0,$$
$$\theta(0) = 1, \quad \dot{\theta}(0) = 1/\alpha. \tag{2.17}$$

This now provides a sensibly scaled problem as all the coefficients are $\leq O(1)$; moreover the leading order terms give a sensible reduced system, which is in fact a regular perturbation of the full equations.

There is an important principle here: that it is best to choose scales that maximize as many dimensionless parameters as possible. Such distinguished limits allow as much of the dynamics to be included as is feasible and allow the possibility of a solution via perturbation methods.

It would appear that the choice of scales is rather a random procedure. However, where observational values exist, it can be facilitated by assuming that terms in the equations can be described by their observed values. Thus, if a pendulum swinging in air has an amplitude of 45° ($\theta \sim \pi/4$) and oscillates with a period of about a second, then we would guess $[\theta] \sim \pi/4$, $[t] \sim 1$ second, and as is well known, we would then find in Eq. (2.11) that

$$k[t]/l \ll 1, \qquad g[t]^2/l[\theta] \sim 1. \tag{2.18}$$

This would suggest that we *choose*

$$g[t]^2/l[\theta] = 1 \tag{2.19}$$

(and $[\theta] = \theta_0$), so that we would have

$$\ddot{\theta} + \epsilon\dot{\theta} + \sin\gamma\theta = 0, \tag{2.20}$$

where

$$\epsilon = \frac{k[t]}{l} = k\left\{\frac{[\theta]}{gl}\right\}^{1/2}. \tag{2.21}$$

Further variants of this problem are considered in the exercises.

[2] Here the notation $a \sim b$ means a is the same size as b, whereas $a \gg 1$ means a is much larger than one; further explanation is given in Chapter 3.

2.3 Shear flow, heat transport, and convection

Important examples of standard nondimensionalizations and their associated dimensionless parameters are given by the equations of fluid flow, whose solutions will be considered in more detail later. They are, for an incompressible fluid, the Navier–Stokes equations:

$$\boldsymbol{\nabla}.\mathbf{u} = 0, \tag{2.22a}$$

$$\mathbf{u}_t + (\mathbf{u}.\boldsymbol{\nabla})\mathbf{u} = -\frac{1}{\rho}\boldsymbol{\nabla} p + \nu\nabla^2\mathbf{u}, \tag{2.22b}$$

and if temperature is relevant, the heat equation:

$$T_t + \mathbf{u}.\boldsymbol{\nabla} T = \kappa\nabla^2 T. \tag{2.23}$$

Here ρ is the density, p is the pressure, ν is the kinematic viscosity, and κ is the thermal diffusivity.

Shear flow

If we consider a typical case of flow of a uniform stream past a (circular, say) object, then there is a natural velocity scale U (the stream flow far from the object) and length scale l (the linear dimension of the object); from these we construct a *convective timescale* l/U (and this is the norm in problems where advection by a fluid is important). Thus we nondimensionalize by writing

$$\mathbf{u} = U\mathbf{u}^*, \qquad \mathbf{x} = l\mathbf{x}^*, \qquad t = (l/U)t^*; \tag{2.24}$$

it remains to scale the pressure. The Navier–Stokes equation (2.22b) can be thought of as a quadrature for the pressure: it contains a derivative, but no absolute value (the same is true for the temperature T). For this reason, the pressure contains an arbitrary additive pressure constant, which is usually fixed by a boundary condition, for example, the pressure at infinity, and because this additive *ambient* pressure does not affect the solution, it is usual to subtract it off before scaling. Thus we nondimensionalize the pressure by writing

$$p = p_\infty + Pp^*, \tag{2.25}$$

and the pressure scale is chosen by balancing the pressure term with *either* the inertial term $(\mathbf{u}.\boldsymbol{\nabla})\mathbf{u}$ *or* the viscous term $\nu\nabla^2\mathbf{u}$. Choosing the former gives

$$P = \rho U^2, \tag{2.26}$$

and then we find

$$\begin{gathered}\boldsymbol{\nabla}^*.\mathbf{u}^* = 0,\\ \mathbf{u}^*_{t^*} + (\mathbf{u}^*.\boldsymbol{\nabla}^*)\mathbf{u}^* = -\boldsymbol{\nabla}^* p^* + \frac{1}{Re}\nabla^{*2}\mathbf{u}^*.\end{gathered} \tag{2.27}$$

The single dimensionless parameter Re is called the *Reynolds number* and is defined by

$$Re = Ul/\nu, \tag{2.28}$$

and it is (for a given flow geometry) the number that characterizes the flow field.

Heat transport

Considering now the heat equation, we adopt the same scalings. In addition, if there is an ambient temperature T_0, then, just as for the pressure, we subtract this off and write

$$T = T_0 + (\Delta T)T^*, \tag{2.29}$$

where ΔT is an appropriate temperature scale, for example, if the surface temperature of the body is prescribed as $T_s = T_0 + \Delta T$ or if the heat flux is prescribed as $Q = -k\partial T/\partial n \sim k\Delta T/l$. A more general condition we will consider is that of convective cooling from a hot body with temperature T_s on its surface S, that is,

$$-k\frac{\partial T}{\partial n} = h(T_s - T) \quad \text{on } S, \tag{2.30}$$

where h is known as a heat transfer coefficient, and k is the thermal conductivity. We then find that for flow past a body of dimension l,

$$Pe\left[T^*_{t^*} + \mathbf{u}^*.\boldsymbol{\nabla}^* T^*\right] = \nabla^{*2}T^*, \tag{2.31}$$

with

$$\frac{\partial T^*}{\partial n^*} = -\alpha\left(T_s^* - T^*\right) \quad \text{on } S, \tag{2.32}$$

and where

$$Pe = Ul/\kappa, \qquad \alpha = hl/k, \qquad T_s^* = (T_s - T_0)/\Delta T. \tag{2.33}$$

The temperature scale ΔT is not yet chosen in this problem, and we can conveniently choose $T_s^* = 1$, that is,

$$\Delta T = T_s - T_0. \tag{2.34}$$

The quantity Pe is called the Péclet number and measures the importance of heat advection versus that of heat conduction.

The heat equation is well scaled if $Pe < 1$. If $Pe \gg 1$, boundary layers (see Chapter 4) are likely, and a rescaling of T^* may be necessary. If $\alpha \sim 1$, then Eq. (2.32) is well scaled. It is also well scaled if $\alpha \gg 1$, since then $T^* = 1 + O(1/\alpha)$ on S. However, if $\alpha \ll 1$, then Eq. (2.32) suggests a rescaling in which $T^* - T_s^* \sim \alpha$, and then the problem becomes sensibly scaled again. We shall see in Chapter 4 that when $Pe \gg 1$, a more subtle analysis is necessary (and possible: see Exercise 4.7).

Convection

Thermal convection occurs when fluid is heated from below by a prescribed temperature drop ΔT across a layer of thickness d. It is due to the buoyancy of light fluid, that is, that density depends on temperature as

$$\rho = \rho_0[1 - \alpha(T - T_0)]. \tag{2.35}$$

We neglect the variation of ρ except in the buoyancy term in the momentum equation (this is called the Boussinesq approximation and is explained further in Chapter 14):

$$\rho_0[\mathbf{u}_t + (\mathbf{u}.\boldsymbol{\nabla})\mathbf{u}] = -\boldsymbol{\nabla} p + \rho_0\nu\nabla^2\mathbf{u} - \rho g\mathbf{k}, \tag{2.36}$$

where now g is gravity, and $\mathbf{k}$ is a vertically oriented unit vector. The scaling is as before (though using d and not l), but we now subtract off the hydrostatic pressure in writing the pressure, thus

$$p = p_a - \rho_0 g z + P p^*, \tag{2.37}$$

where p_a is atmospheric pressure, for example. The result is set down in the following equations:

$$\begin{aligned} \boldsymbol{\nabla}^*.\mathbf{u}^* &= 0, \\ (\rho_0 U^2/d)\left[\mathbf{u}^*_{t^*} + (\mathbf{u}^*.\boldsymbol{\nabla}^*)\mathbf{u}^*\right] &= -(P/d)\nabla^* p^* + (\rho_0 \nu U/d^2)\nabla^{*2}\mathbf{u}^* + (\rho_0 \alpha \Delta T g)\mathbf{k}, \\ (Ud/\kappa)\left[T^*_{t^*} + \mathbf{u}^*.\boldsymbol{\nabla}^* T^*\right] &= \nabla^{*2} T^*, \end{aligned} \tag{2.38}$$

and where suitable boundary conditions are $O(1)$, for example,

$$\begin{aligned} \mathbf{u}^* &= 0 \quad \text{at } z^* = 0, 1, \\ T^* &= 0 \quad \text{at } z^* = 1, \\ T^* &= 1 \quad \text{at } z^* = 0. \end{aligned} \tag{2.39}$$

Although the choice is somewhat arbitrary, it is common to choose U and P by balancing pressure gradient with viscous terms and heat advection with conduction. Thus

$$U = \kappa/d, \qquad P = \rho_0 \nu \kappa / d^2, \tag{2.40}$$

and the momentum equation takes the form (dropping the asterisks)

$$\frac{1}{Pr}[\mathbf{u}_t + (\mathbf{u}.\boldsymbol{\nabla})\mathbf{u}] = -\boldsymbol{\nabla} p + \nabla^2 \mathbf{u} + Ra\mathbf{k}; \tag{2.41}$$

the parameters are the *Prandtl number* and the *Rayleigh number*:

$$Pr = \nu/\kappa, \qquad Ra = \alpha \Delta T g d^3/\nu\kappa. \tag{2.42}$$

Pr is a property of the fluid alone, whereas Ra is a measure of the thermal forcing ΔT. The convection is determined by these two parameters alone and for a given fluid is determined solely by the Rayleigh number.

2.4 Using numerical estimates: an example from mathematical biology

In 1976, Meinhardt presented a model of phenomenological type, which showed how as few as four main chemical substances could, through their dynamic interaction, lead to solutions that correspond to the formation of various branched structures in morphogenetically developing organisms. He cited the formation of leaf veins and blood vessels as examples. More recently, the model has been applied in a conceptual way to the formation of river networks in drainage basins.

Meinhardt's model is framed in terms of four substances, called the *activator* A, the *inhibitor* H, the *substrate* S, and the *differentiated cell concentration* Y, respectively.

Equations governing these variables are postulated as follows:

$$\begin{aligned}
A_t &= cA^2S/H - \mu A + \rho_0 Y + D_a \nabla^2 A, \\
H_t &= cA^2S - \nu H + \rho_1 Y + D_h \nabla^2 H, \\
S_t &= c_0 - \gamma S - \epsilon Y S + D_s \nabla^2 S, \\
Y_t &= dA - eY + Y^2/(1 + fY^2).
\end{aligned} \tag{2.43}$$

Y is a kind of indicator variable. As can be seen in Fig. 2.1, when A is small, $Y = 0$, but if A becomes sufficiently large, then Y tends to Y^* irreversibly (providing A is positive, as we expect). We associate the transition of Y from 0 to Y^* for high A with the formation of differentiated cells – leaf veins, blood vessels, or the like.

Although the model itself is phenomenological, its interpretation is fairly simple. A is an activator for the differentiation process. It is produced by the differentiated cells themselves and also autocatalytically by the substrate. However, its production is limited by an inhibitor H, which is also produced from the substrate and from differentiated cells. The substrate is consumed by the production of A and H, and consumption is enhanced for differentiated cells but also produced by, (say), metabolic processes.

Our purpose here is to seek a sensible nondimensionalization, based on values chosen by Meinhardt in his numerical simulation. For his Figure 4, he chooses values

$$\begin{aligned}
c &= .004, \\
\mu &= .12, \\
\rho_0 &= .03, \\
D_a &= .02, \\
\nu &= .04, \\
\rho_1 &= .0003, \\
D_h &= .18, \\
c_0 &= .02, \\
\gamma &= .02, \\
\epsilon &= .2, \\
D_s &= .06, \\
d &= .0013, \\
e &= .1, \\
f &= 10,
\end{aligned} \tag{2.44}$$

and we use these as representative in what follows. Meinhardt does not indicate typical amplitudes, nor space or timescales, in his numerical results.

Natural balances for A, S, and H can be determined by balancing the source terms with the decay terms, thus

$$\begin{aligned}
CA^2S/H &\sim \mu A, \\
CA^2S &\sim \nu H, \\
c_0 &\sim \gamma S,
\end{aligned} \tag{2.45}$$

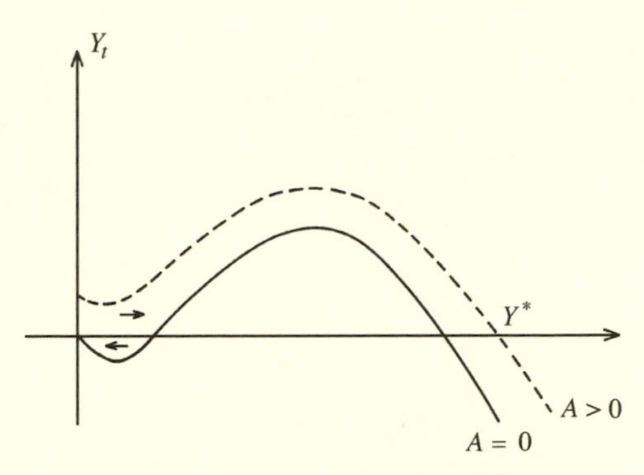

Fig. 2.1. Y_t as a function of Y

whence we define the scales as

$$
\begin{aligned}
[S] &= c_0/\gamma, \\
[A] &= \nu/\mu, \\
[H] &= c[A]^2[S]/\nu = c\nu c_0/\mu^2\gamma,
\end{aligned}
\tag{2.46}
$$

and we nondimensionalize by writing

$$
S = [S]S^*, \qquad A = [A]A^*, \qquad H = [H]H^*; \tag{2.47}
$$

Y is left as it is, in view of Fig. 2.1, which suggests $Y = O(1)$ (since, with $e \ll 1$ and $f \gg 1$, $Y^* \approx 1/ef = 1$ for the values in Eq. (2.44)). Additionally, we choose space and time scales $[x]$ and $[t]$, so

$$
\mathbf{x} = [x]\mathbf{x}^*, \qquad t = [t]t^*, \tag{2.48}
$$

where these remain to be chosen.

Substituting into the equations, we find

$$
\begin{aligned}
(\nu/\mu[t])A^*_{t^*} &= \nu[A^{*2}S^*/H^* - A^*] + \rho_0 Y + (D_a\nu/\mu[x]^2)\nabla^{*2}A^*, \\
([H]/[t])H^*_{t^*} &= \nu[H][A^{*2}S^* - H^*] + \rho_1 Y + (D_h[H]/[x]^2)\nabla^{*2}H^*, \\
(c_0/\gamma[t])S^*_{t^*} &= c_0[1 - S^*] - (\epsilon c_0/\gamma)YS^* + (D_s c_0/\gamma[x]^2)\nabla^{*2}S^*, \\
(1/[t])Y_{t^*} &= (d\nu/\mu)A^* - eY + Y^2/(1 + fY^2).
\end{aligned}
\tag{2.49}
$$

Dividing by the coefficients multiplying the supposedly important source terms, and henceforth omitting the asterisks, we have

$$
\begin{aligned}
(1/\mu[t])A_t &= (A^2S/H - A) + (\rho_0/\nu)Y + (D_a/\mu[x]^2)\nabla^2 A, \\
(1/\nu[t])H_t &= (A^2S - H) + (\rho_1/\nu[H])Y + (D_h/\nu[x]^2)\nabla^2 H, \\
(1/\gamma[t])S_t &= 1 - S - (\epsilon/\gamma)YS + (D_s/\gamma[x]^2)\nabla^2 S;
\end{aligned}
\tag{2.50}
$$

now we choose the time and space scales so that the largest coefficients of time and space derivatives respectively are $O(1)$. Because

$$
1/\mu \sim 8, \qquad 1/\nu \sim 25, \qquad 1/\gamma \sim 50, \tag{2.51}
$$

we choose

$$[t] = 1/\gamma. \tag{2.52}$$

Also, because

$$D_a/\mu \sim 1/6, \qquad D_h/\nu \sim 4.5, \qquad D_s/\gamma \sim 3, \tag{2.53}$$

we choose

$$[x] = (D_h/\nu)^{1/2}. \tag{2.54}$$

Then

$$\begin{aligned}(\gamma/\mu)A_t &= AS^2/H - A + (\rho_0/\nu)Y + (\nu D_a/\mu D_h)\nabla^2 A,\\ (\gamma/\nu)H_t &= A^2 S - H + (\rho_1/\nu[H])Y + \nabla^2 H,\\ S_t &= 1 - S - (\epsilon/\gamma)YS + (\nu D_s/\gamma D_h)\nabla^2 S,\\ Y_t &= (d\nu/\mu\gamma)A + \gamma^{-1}[-eY + Y^2/(1+fY^2)].\end{aligned} \tag{2.55}$$

The parameters are given approximately by

$$\begin{gathered}\gamma/\mu \sim 1/6, \qquad \rho_0/\nu \sim 3/4, \qquad \nu D_a/\mu D_h \sim 1/18,\\ \gamma/\nu \sim 1/2, \qquad \rho_1/\nu[H] \sim .7,\\ \epsilon/\gamma \sim 10, \qquad \nu D_s/\gamma D_h \sim 2/3,\\ d\nu/\mu\gamma \sim .02.\end{gathered} \tag{2.56}$$

Because $e \sim 0.1$, $f \sim 10$, we infer from Fig. 2.1 that with

$$Y = e\eta, \tag{2.57}$$

then

$$\begin{aligned}-eY + Y^2/(1+fY^2) &= e^2 g(\eta),\\ g(\eta) &= -\eta + \eta^2/(1+fe^2\eta^2),\end{aligned} \tag{2.58}$$

and thus

$$e\eta_t = (d\nu/\mu\gamma)A + (e^2/\gamma)g(\eta), \tag{2.59}$$

where

$$e^2/\gamma \sim 1.2. \tag{2.60}$$

We thus have the nondimensional model

$$\begin{aligned}\delta_1 A_t &= A^2 S/H - A + \alpha_1 Y + \epsilon_1 \nabla^2 A,\\ \alpha_2 H_t &= A^2 S - H + \alpha_3 Y + \nabla^2 H,\\ S_t &= 1 - S - \alpha_4 \eta S + \alpha_5 \nabla^2 S,\\ \delta_2 \eta_t &= \epsilon_2 A + g(\eta),\end{aligned} \tag{2.61}$$

where

$$\begin{aligned}
&\delta_1 = \gamma/\mu \sim .16, \qquad \delta_2 = \gamma/e \sim .2, \\
&\alpha_1 = \rho_0/\nu \sim .75, \qquad \alpha_2 = \gamma/\nu \sim .5, \qquad \alpha_3 = \rho_1/\nu[H] \sim .7, \\
&\alpha_4 = \epsilon e/\gamma \sim 1, \qquad \alpha_5 = \nu D_S/\gamma D_h \sim .67, \\
&\epsilon_1 = \nu D_a/\mu D_h \sim .056, \qquad \epsilon_2 = d\nu/\mu e^2 \sim .04.
\end{aligned} \tag{2.62}$$

Of these, α_i are $O(1)$, ϵ_i are small, and δ_i are relatively small. The model appears sensibly scaled and provides the basis for either a numerical solution or a discussion of the likely dynamics.

The model appears sensibly scaled, at least while $\eta \sim O(1)$ ($Y \sim O(e)$), insofar as all the dimensionless parameters are $\leq O(1)$. A preliminary discussion of the dynamics is then as follows. So long as $Y = O(e)$, then $\alpha_1 Y \ll 1$; and because $\epsilon_1 \ll 1$, and if we take δ_1 ($\sim 1/6$) $\ll 1$, then A changes rapidly according to

$$\delta_1 A_t \approx A^2 S/H - A, \tag{2.63}$$

in which S/H is approximately constant. This equation exhibits a threshold phenomenon. If $A < H/S$, then A tends to zero, whereas if $A > H/S$, then A 'blows up,' tending to infinity in finite time. The effect of the small diffusive term is to confine this blowup to a localized spike. If blowup occurs, then H follows A up; also η changes rapidly, and for sufficiently large A, it also grows rapidly until $Y = O(1)$ (see Fig. 2.1).

When $Y = O(1)$, Eq. (2.58) shows that (since $f = 1/e$)

$$eg(\eta) \approx 1 - Y, \tag{2.64}$$

so that

$$\delta_2 Y_t \approx \varepsilon_2 e A + 1 - Y, \tag{2.65}$$

so that (if $A \ll 1/\varepsilon_2 e$), $Y \to 1$ rapidly. S decreases rapidly to $S \approx e/\alpha_4$, and A and H are now described by the approximate equations

$$\begin{aligned}
\delta_1 A_t &\approx \frac{e}{\alpha_4}\frac{A^2}{H} - A + \alpha_1 + \varepsilon_1 \nabla^2 A, \\
\alpha_2 H_t &\approx \frac{e}{\alpha_4} A^2 - H + \alpha_3 + \nabla^2 H,
\end{aligned} \tag{2.66}$$

and if the threshold of A for growth of Y is not too high, we can expect A to be still of $O(1)$ when Eq. (2.66) becomes operative: Then the autocatalytic term proportional to A^2 can be ignored, and

$$\begin{aligned}
\delta_1 A_t &\approx \alpha_1 - A + \varepsilon_1 \nabla^2 A, \\
\alpha_2 H_t &\approx \alpha_3 - H + \nabla^2 H.
\end{aligned} \tag{2.67}$$

Here $A = \alpha_1$, $H = \alpha_3$ are stable steady states. The effect of the diffusion terms is to allow the equilibrated states to propagate as traveling waves, and indeed, this is what numerical simulations indicate. One purpose of scaling the equations is to gain an understanding of the necessary time and space steps that are required to resolve the model calculations numerically. For example, to resolve spatial diffusion of A, we should require $\Delta x \ll \epsilon_1^{1/2}$, for example, $\Delta x < 0.3\epsilon_1^{1/2}$, and similarly $\Delta t \ll \delta_1, \delta_2$, (e.g. $\Delta t < 0.3\gamma/\mu$).

2.5 Notes and references

Nondimensionalization The classic book that devotes a chapter to nondimensionalization is that by Lin and Segel (1974). The description is rather more formal than that given here. As with most of the techniques of mathematical modeling, you can only really learn to do it properly by practice. Of course, *anybody* can nondimensionalize: it is important to realize that there are right and wrong ways to do it.

It is possible to dress up the procedure of nondimensionalization in a rather formal way, and this is discussed in the above text and also in that of Massey (1986). The subject goes by the name of *dimensional analysis*, and we should at least mention Buckingham's Pi theorem, which is this: if n variables $Q_1 \ldots Q_n$ involving r separate dimensional components (usually $r = 3$, these being mass, length, time, i.e. M, L, T) are related by a unique dimensionally consistent function $\psi(Q_1, \ldots, Q_n) = 0$, then one can find $n - r$ dimensionless combinations of Q_i, $\Pi_j(Q_i)$, $j = 1, \ldots, n - r$, such that $\phi(\Pi_1, \ldots, \Pi_{n-r}) = 0$. To illustrate this, consider the problem of determining the pressure drop Δp due to fluid flow in a long, smooth pipe of length l and (circular) diameter d. The parameters that occur in this problem are Δp (of dimension $[ML^{-1}T^{-2}]$), l $[L]$, d $[L]$, mean velocity u $[LT^{-1}]$ (given by the volume flux divided by the cross-sectional area), and the viscosity μ $[ML^{-1}T^{-1}]$ and density ρ $[ML^{-3}]$ of the fluid. There are six variables and three dimensions, so there are three independent dimensionless variables. These can be taken as $\Delta p/\rho u^2$, the aspect ratio l/d, and the Reynolds number $Re = \rho ul/\mu$. Thus the Pi theorem tells us that there is some function such that

$$\Delta p = \rho u^2 f[Re, l/d], \tag{2.68}$$

and through the translation invariance of the flow, we would expect $\Delta p \propto l$ (at least if $l \gg d$), so that

$$\frac{\Delta p}{l} = \frac{\rho u^2}{2d}\phi(Re). \tag{2.69}$$

Measurements of pipe flow show that for $Re < 2000$, $\phi \approx 64/Re$, whereas for $Re > 3000$, ϕ is determined by *Prandtl's law* (see Schlichting, 1979, pp. 597, 598, and 611):

$$\frac{1}{\sqrt{\phi}} = 2\log_{10}\{Re\sqrt{\phi}\} - 0.8. \tag{2.70}$$

Many of the dimensionless numbers (sometimes called dimensionless groups) that occur in various models have attached names: for example, the Reynolds number, the Péclet number, and so on. There is in fact a bewildering variety of these numbers, and Massey's book gives an exhaustive list of them.

Meinhardt's model of cell differentiation Meinhardt's (1976) model for differentiation was applied to developmental biology: the formation of leaf veins, and also blood vessels and the fibers of the nervous system. But as with many mathematical models, the problem of network formation applies to other situations, and the particular problem of river network formation is one such. Willgoose, Bras, and Rodriguez-Iturbe (1989, 1991a) have adapted the Meinhardt model to apply in this situation, with some success. The 'veins' are channels in which streams flow, and the 'substrate' is the hillslope topography. The hillslope evolves through erosion, and the differentiation between stream and hillslope is enabled by erosive instability (deeper film flows erode more, leading to the formation of local deep channels). Models such as this and others are currently of much interest to geomorphologists, hydrologists, and physicists.

Exercises

1. For the damped pendulum equation

$$\ddot{\theta} + \alpha\dot{\theta} + \beta \sin \gamma\theta = 0, \quad \theta(0) = 0, \ \dot{\theta}(0) = 1,$$

find suitable rescalings for the following cases:

(i) $\alpha, \gamma = O(1), \quad \beta \gg 1$;
(ii) $\alpha, \beta = O(1), \quad \gamma \gg 1$;
(iii) $\alpha \sim \beta\gamma \sim 1/\gamma \gg 1$.

2. Show that if $Re \ll 1$ in Eq. (2.27), a suitable problem can be found by rescaling $p \sim 1/Re$, $t \sim Re$. Hence derive the approximate equations for unsteady *Stokes flow*:

$$\begin{aligned} \boldsymbol{\nabla}.\mathbf{u} &= 0, \\ \mathbf{u}_t &= -\boldsymbol{\nabla} p + \nabla^2 \mathbf{u}. \end{aligned}$$

(Note that these are *linear* equations.)

3. A model for earthquakes (Carlson and Langer, 1989) is as follows: N blocks of equal mass are situated at positions $x_j(t)$, $j = 1, \ldots, N$, on a level plane. They are connected to each other and to a plate above by springs (see Fig. 2.2). The plate moves at speed V. The blocks are subject to a friction force $f(\dot{x}_j)$, where f is an odd function, taken as $f(v) = f_0 \operatorname{sgn}(v)/(1 + v/v_0)$.

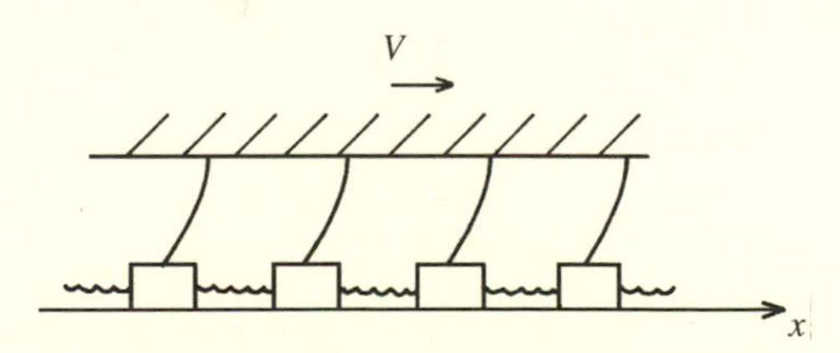

Fig. 2.2. The Carlson–Langer earthquake model.

(i) Show that a model for the motion is

$$m\ddot{x}_j = k_c(x_{j+1} - 2x_j + x_{j-1}) - k_p(x - Vt) - f(\dot{x}_j),$$

where k_c and k_p are the spring constants of the block-block and block-plate springs.
(ii) If the block-block springs are of natural length a, show that, by putting $x_j = Vt + u$, a continuous approximation to the model is

$$mu_{tt} = k_c a^2 u_{xx} - k_p u - f(V + u_t).$$

(iii) By suitably nondimensionalizing the model, show that a dimensionless version is

$$u_{tt} = u_{xx} - u - f(\varepsilon + u_t),$$

where $\varepsilon = V(mk_p)^{1/2}/F_0$, and $f(w) = \operatorname{sgn}(w)/[1+\alpha w]$, $\alpha = f_0/v_0(mk_p)^{1/2}$.
(iv) For values $f_0 = 50$, $v_0 = 1$, $m = 1$, $k_c = 2500$, $k_p = 40$, $a = 1$, and $V = .01$, show that $\varepsilon \approx .001$, $\alpha \approx 9$. Deduce that, for $t \sim 1$, the motion of u is described approximately by

$$u_{tt} = u_{xx} - u - \frac{1}{1 + \alpha u_t}.$$

(v) If $t = \tau/\varepsilon$, show that an approximate model is

$$0 = u_{xx} - u - \operatorname{sgn}(1 + u_\tau).$$

(vi) Show that the steady state $u \approx -1$ is unstable on the fast timescale $t = O(1)$, providing $\alpha > 0$.

4. The Mackey–Glass model of respiratory ventilation (see Glass and Mackey, 1988) is given by

$$\dot{x} = \lambda - \frac{\alpha x V_m x_\tau^n}{\theta^n + x_\tau^n},$$

where $x_\tau = x(t - \tau)$. Values of the parameters are given by

$$\lambda \sim 6 \text{ mm Hg min}^{-1}, \qquad V_m = 80 \text{ liter min}^{-1}, \qquad \tau = 0.25 \text{ min};$$

α, θ, and n are determined from the observed steady-state values of x, $V = V_m x^n/(\theta^n + x^n)$, and $S = dV/dx|_{x^*}$. These are $x^* = 40$ mm Hg, $V^* = 7$ liter min^{-1}, $S^* = 4$ liter min^{-1} (mm Hg)$^{-1}$. Show that if the variables x, V, t are scaled with θ, V_m, τ, then a dimensionless model is

$$\varepsilon\dot{x} = 1 - \beta x v(x_1),$$

where

$$v(x) = \frac{x^n}{1 + x^n}, \qquad \varepsilon = \frac{\theta}{\lambda\tau}, \qquad \beta = \frac{\alpha V_m \theta}{\lambda}.$$

Show that if the dimensionless observed steady value of x is ξ, then

$$\xi = x^*/\theta,$$
$$\frac{\xi^n}{1+\xi^n} = V^*/V_m,$$
$$S^* = \frac{nV^*}{x^*}\left(1 - \frac{V^*}{V_m}\right).$$

Deduce that, roughly,

$$n \approx 25, \qquad \xi^n \approx 0.1,$$

hence

$$\xi \approx 0.9,$$

and thus

$$\theta \approx 44 \text{ mm Hg}.$$

Hence find the corresponding values of ε, β, and α. Is the model sensibly scaled?

5. A model for snow melt runoff is described as follows (Male 1980). The vertical water flux downward (volume per unit area per unit time) is given by *Darcy's law*:

$$u = \frac{k}{\mu}\left[\frac{\partial p_c}{\partial z} + \rho_l g\right],$$

where k is liquid permeability of the snow pack, μ is viscosity, p_c is capillary pressure, ρ_l is density, g is gravity, and z is a coordinate in the vertical downward direction. The permeability is related to effective liquid saturation S (the volume fraction of the pore space occupied by water) by

$$k = k_0 S^3.$$

Conservation of the liquid phase is ensured by

$$\phi\frac{\partial S}{\partial t} + \frac{\partial u}{\partial z} = 0,$$

where ϕ is the porosity (volume fraction of pore space). The capillary pressure is prescribed as a sharply varying function of S: we take

$$p_c = p_0\left(\frac{1}{S} - S\right).$$

(Note that $p_c \to \infty$ as $S \to 0$: some residual water is always present. Also, $p_c \to 0$ as $S \to 1$; in soils, p_c is known as the *effective* pressure, and here it is the pressure transmitted between snow grains.)

The intrinsic permeability k_0 is given by

$$k_0 = 0.077\, d^2 \exp[-7.8\, \rho_s/\rho_l],$$

where ρ_s and ρ_l are snow and water densities, and d is grain size. Take $d = 1$ mm, $\rho_s = 500$ kg m^{-3}, $\rho_l = 10^3$ kg m^{-3}, $p_0 = 1$ kPa, $\phi = 0.4$, $\mu = 1.8 \times 10^{-3}$ Pa s, and $g = 10$ m s^{-2}, and derive a nondimensional model for melting of a one meter thick snow pack at a rate (i.e., u at the top surface $z = 0$) of 10^{-6} m s^{-1}. Determine whether capillary effects are small; describe the nature of the model equation, and find an approximate solution for suitable initial and boundary conditions.

6. By choosing suitable scales, show how one can obtain Eq. (2.6) from Eq. (2.5). Show that periodic traveling waves of the form $\zeta = f(z)$, $z = \xi - (1 + c)\tau$, can be found by using a phase plane analysis, and show in particular that a family of solitary waves ($\zeta \to 0$ as $\xi \to \pm\infty$) exists, and find the relation between their amplitude and speed. (This is the famous wave of translation of John Scott Russell; see, for example, Newell (1985).)

7. Show that the Lotka–Volterra model (1.9) has a nontrivial (i.e., nonzero) steady state that is a center if $\alpha, \beta, \gamma, \delta > 0$. Use the steady state values to nondimensionalize the equations, choosing also a timescale relevant to prey growth in the absence of predators. Show that the dimensionless equations depend on the single parameter $\varepsilon = \gamma/\beta$, and show that the nonzero steady state is a center for all values of ε. By finding a first integral of the equations, show that all the orbits in the positive (x, y) quadrant are closed. If $\varepsilon \ll 1$, describe the form of the orbits, and explain why the predator is likely to die out, both mathematically and ecologically.

3
Asymptotics

3.1 Order notation

Fundamental to the concepts of practical applied analysis are the linked ideas of numerical and asymptotic order of quantities. Practical problems deal with quantities that have physical meaning – velocities, concentrations, densities – and we often have information about the size of such quantities: the temperature variation is about 100°C, the length is about two meters, and so on. The idea of the typical size of a variable is manifested in the use of *order notation*. Specifically, we say that $x \sim 10$ if we mean that x is of the same order as the quantity ten. Note that this does not mean $x \approx 10$ (i.e., approximately equals ten), but rather x is in the region of 10, for example, $5 < x < 15$. This *numerical* use of the idea of order or asymptotic equivalence has a parallel but crucially distinct mathematical usage.

One considers two variables x and y to be *asymptotically equivalent* if, in the limit as some parameter z (which may equal x or y) tends to a quantity z_0 (which may be infinite or zero), x/y tends to a finite, nonzero limit, and we write it thus:

$$x \sim y; \quad \text{or} \quad x = O(y) \quad \text{as } z \to z_0 \tag{3.1}$$

if $\lim_{z\to z_0}(x/y)$ exists and is finite. Often this limit is required to be one. We say 'x is order y,' or colloquially 'x twiddles y.' As examples,

$$\begin{aligned} \sin x &\sim x \quad \text{as } x \to 0, \\ e^{-x} &\sim 1 \quad \text{as } x \to 0, \\ \frac{x}{1+x^2} &\sim \frac{1}{x} \quad \text{as } x \to \infty, \\ \ln\cosh x &\sim x \quad \text{as } x \to \infty. \end{aligned} \tag{3.2}$$

Order notation is explained in any book on asymptotics. Less clearly explained is the relationship between the practical, numerical use of order and its mathematical equivalent.

Consider a simple, algebraic problem such as the solution of the equation

$$x^3 - \frac{1}{10}x - \frac{1}{24} = 0. \tag{3.3}$$

Of course, this poses little numerical difficulty, but often a problem such as this will arise in the parametric form

$$x^3 - \delta_1 x - \delta_2 = 0, \tag{3.4}$$

where δ_1 and δ_2 will be dimensionless combinations, or groups, of design parameters, and although it may be useful to compute the solutions for the *particular* values $\delta_1 = 1/10$, $\delta_2 = 1/24$, it may be of more practical help to find solutions for a *range* of values of δ_1 and δ_2, of which these values are *typical* estimates. In that case, we might wish to make use of the fact that δ_1 and δ_2 are quite small but of roughly the same size, and the formal mathematical way in which this is done is to consider x in the *distinguished limit* where $\delta_1 \to 0$ but δ_1/δ_2 is finite, that is, $\delta_1 = O(\delta_2)$. This is the distinction between the mathematical and the numerical order concept: the numerical idea carries with it an implicit mathematical idea.

Little 'o' notation

As well as the idea of equivalence, there are mathematical and numerical concepts of nonequivalence. We write

$$x = o(y) \quad \text{as } z \to z_0 \tag{3.5}$$

if $x/y \to 0$ as $z \to z_0$ for the mathematical idea, and its numerical equivalent is written $x \ll y$, and spoken 'x is very much less than y.' Examples are

$$\begin{aligned} x &\ll x^2 & &\text{as } x \to \infty, \\ x^2 &\ll x & &\text{as } x \to 0, \\ x^n &\ll e^x & &\text{as } x \to \infty \text{ (for any } n), \\ \ln x &\ll x & &\text{as } x \to \infty. \end{aligned} \tag{3.6}$$

There is a certain subjectivity in deciding when $x \ll y$, but usually an 'order of magnitude' (a factor of ten) is sufficient. That is, $10^2 \gg 5$, $.02 \ll 1$, but we might balk at $5 \gg 1$. It must be emphasized, however, that these choices are to some extent a matter of taste and discretion. For example, formally $10\epsilon \ll 1$ for $\epsilon \ll 1$, because $10\epsilon \sim \epsilon$ (mathematically) as $\epsilon \to 0$. Nevertheless, $10\epsilon = 1$ if $\epsilon = 0.1$. Thus ϵ in this case is not 'small' (i.e., not small enough). A different example is the function $1/\ln\ln(1/\epsilon)$, which is $o(1)$ for $\epsilon \to 0$. Even for very small $\epsilon \ll 1$ (numerically), $1/\ln\ln(1/\epsilon) \sim 1$ (numerically). Equally, $\exp[-\exp(1/\epsilon)] \ll 1$ (numerically) even for $\epsilon = O(1)$ (numerically). For $\epsilon = 10^{-10}$, $1/\ln\ln(1/\epsilon) \approx 0.32$. It is often useful, therefore, to pursue formal approximation procedures that may seem ill-founded, since one does not know in advance how small 'small' has to be in order for a theory to be quantitatively accurate.

3.2 Asymptotic sequences and expansions

A series of terms x_i, dependent on a parameter ϵ, is said to form an *asymptotic sequence* if $x_{n+1} = o(x_n)$ for all n, as $\epsilon \to \epsilon_0$ (usually zero or infinity). For example, the family $\{\epsilon^n\}$ forms an asymptotic sequence as $\epsilon \to 0$, as does the sequence $\{\epsilon/[\log(1/\epsilon)]^n\}$. An asymptotic expansion is then a series generated by an asymptotic sequence $\{x_n\}$, that is, of the form

$$a_0 x_0 + a_1 x_1 + a_2 x_2 + \dots, \tag{3.7}$$

where the a_i are constants. For example, $(1-\epsilon x)^{-1}$ has the asymptotic expansion

$$(1-\epsilon x)^{-1} \sim 1 + \epsilon x + \epsilon^2 x^2 + \dots, \tag{3.8}$$

as $\epsilon \to 0$ (for fixed x).

The above example is in fact a Taylor series, but the important point about an asymptotic expansion is that it need not be convergent, indeed it may be everywhere divergent. An example is the expansion

$$f(x) \sim \sum_1^\infty (-1)^n n!/x^n \quad \text{as } x \to \infty \tag{3.9}$$

where, although we use the $\sum$ notation as shorthand, an infinite convergent sum is not implied. We say $f(x)$ is *asymptotically equivalent* to $\sum_1^\infty a_n\phi_n(x)$ (as $x \to x_0$) if $\{\phi_n\}$ is an asymptotic sequence, and $f \sim \sum_1^N a_n\phi_n$ for all N (and the ratio tends to one). In fact, we make the stronger stipulation that

$$f - \sum_1^N a_n\phi_n = o(\phi_N) \tag{3.10}$$

for each N. Equation (3.9) defines an asymptotic expansion of this type.

The usefulness of asymptotic, as opposed to convergent, expansions is that when we approximate a function $f(x)$ in practice, we will do so with a finite series of N terms. We thus replace the concept of fixing x and letting N tend to infinity with that of fixing N, and considering *asymptotic* limits as $x \to x_0$. Almost invariably, this concept is of more practical utility.

3.3 Convergence versus divergence

As an example, consider the exponential integral

$$E_1(x) = \int_x^\infty e^{-t}\,dt/t \tag{3.11}$$

as $x \to \infty$. Clearly, $E_1(x) \to 0$ as $x \to \infty$. To obtain an asymptotic expansion, we integrate by parts, yielding successively

$$E_1(x) = e^{-x}/x - \int_x^\infty e^{-t}\,dt/t^2,$$

$$E_1(x) = e^{-x}/x - e^{-x}/x^2 + 2\int_x^\infty e^{-t}\,dt/t^3,$$

$$E_1(x) = e^{-x}\left[\frac{1}{x} - \frac{1}{x^2}\dots + (-1)^{n-1}\frac{(n-1)!}{x^n}\right] + (-1)^n n!\int_x^\infty e^{-t}\,dt/t^{n+1}. \tag{3.12}$$

In fact, we have

$$E_1(x) \sim e^{-x} \sum_1^{\infty} (-1)^{n-1} n!/x^n, \tag{3.13}$$

which is clearly an asymptotic series (cf. Eq. (3.9)). The remainder after n terms is

$$R_n = (-1)^n n! \int_x^{\infty} e^{-t}\, dt/t^{n+1}, \tag{3.14}$$

and we prove that Eq. (3.13) is indeed an asymptotic expansion by showing that

$$R_n \ll e^{-x}(n-1)!/x^n \tag{3.15}$$

as $x \to \infty$. In fact,

$$\begin{aligned} |R_n/\{e^{-x}(n-1)!/x^n\}| &= nx^n e^x \int_x^{\infty} e^{-t} dt/t^{n+1} \\ &= n \int_0^{\infty} e^{-xv} \frac{dv}{(1+v)^{n+1}} \end{aligned} \tag{3.16}$$

by writing $t = x(1+v)$, and this is clearly $< n/x$ (and in fact $\sim n/x$). For *fixed* n, the ratio tends to zero as $x \to \infty$. Nevertheless, the series is everywhere divergent. To put it another way, the limits $x \to \infty$, $n \to \infty$ do not commute. The usefulness of an asymptotic expansion lies in the fact that although the infinite series may diverge, the remainder after n terms may decrease to a small value before diverging. For example, in the series (3.13), we have

$$R_n \sim (n!/x^{n+1})e^{-x}. \tag{3.17}$$

Clearly, R_n decreases rapidly when $x \gg 1$ for low values of n, but increases rapidly for very high values of n. Specifically, because $R_n/R_{n-1} \sim n/x$, we find that the remainder terms decrease until $n \sim x$, which gives the optimal number of terms to take in the expansion. In practice, however, one often makes do with a single term. The asymptotic nature of the expansion (3.13) is a manifestation of the fact that $E_1(x)$ has an essential singularity at $x = \infty$, as can be seen by writing $E_1(1/z) = \int_0^z e^{-1/u}\, du/u$.

3.4 An algebraic example

We return to the example (3.4) and assume $\delta_1 \sim \delta_2 \ll 1$, thus

$$x^3 - \epsilon x - c\epsilon = 0, \quad c = O(1), \quad \epsilon \ll 1, \tag{3.18}$$

where $\epsilon = \delta_1$, $c = \delta_2/\delta_1$, and we suppose $c > 0$.

Leading order estimates

The above equation has three roots. If we put $\epsilon = 0$, that is, neglect those terms that we consider to be small, we obtain the degenerate equation $x^3 = 0$, with three roots at zero. This implies that the three exact roots are small. In order to determine how small, we balance terms in Eq. (3.18). That is, x^3 must be the same size as one of the other two terms. Because $x \ll 1$, therefore $\epsilon x \ll \epsilon$, and we have the leading order balance

$$x^3 \sim c\epsilon, \tag{3.19}$$

whence the three roots are, to *leading order* (i.e., in a first approximation),

$$x = \epsilon^{1/3} c^{1/3} e^{2ni\pi/3}, \quad n = 0, 1, 2. \tag{3.20}$$

We can now check *a posteriori* (after the fact) that indeed $\epsilon x \sim \epsilon^{4/3} \ll \epsilon$, so that this estimate is *consistent*. This procedure, balancing terms heuristically, obtaining leading order balances, and confirming that the estimates thus obtained are consistent, is the basic method of obtaining asymptotic approximations. Almost invariably, this constructive method does not permit a proof that the series so obtained actually are asymptotic, although often this can be inferred.

Asymptotic approximation

To obtain higher order approximations, it is convenient to rescale first:

$$x = \epsilon^{1/3} X, \tag{3.21}$$

so that $X = O(1)$ as $\epsilon \to 0$. We then have

$$X^3 - \epsilon^{1/3} X - c = 0, \tag{3.22}$$

and it is more or less obvious that by writing X as a power series expansion in $\epsilon^{1/3}$, we can iteratively obtain asymptotic approximations (which will here in fact be convergent) in the form

$$X \sim X_0 + \epsilon^{1/3} X_1 + \epsilon^{2/3} X_2 + \ldots; \tag{3.23}$$

this is the essence of the method of regular perturbation. We substitute the expansion (3.23) into the equation (3.22), and evaluate the coefficients by equating terms at successive orders. Thus

$$(X_0 + \epsilon^{1/3} X_1 + \epsilon^{2/3} X_2 + \ldots)^3 - \epsilon^{1/3}(X_0 + \epsilon^{1/3} X_1 + \ldots) - c = 0, \tag{3.24}$$

whence, at successive orders $1, \epsilon^{1/3}, \epsilon^{2/3} \ldots$, we obtain

$$\begin{aligned} X_0^3 - c &= 0, \\ 3X_0^2 X_1 - X_0 &= 0, \\ 3X_0^2 X_2 - X_1 &= -3X_0 X_1^2. \end{aligned} \tag{3.25}$$

Notice the common feature in perturbation procedures that the leading order approximation is nonlinear, whereas the higher order terms are determined by linear, inhomogeneous equations. We can easily solve for X_i, and we find that

X is given by

$$X \sim \alpha + \epsilon^{1/3}/3\alpha + O(\epsilon), \tag{3.26}$$

where α is any cube root of c.

Nonuniformity

Although this series expansion can be extended to any order in powers of ϵ, it is evident that it is not uniformly valid as $\alpha \to 0$, as the asymptotic nature of the series breaks down when $\alpha \sim \epsilon^{1/3}/3\alpha$, that is, when $\alpha \sim \epsilon^{1/6}$ ($c \sim \epsilon^{1/2}$). If we are interested in obtaining approximations for smaller values of c, we then extend an investigation to the case when $c \sim \epsilon^{1/2}$,

$$c = \epsilon^{1/2}C, \quad C = O(1), \tag{3.27}$$

and we have to solve

$$X^3 - \epsilon^{1/3}X - \epsilon^{1/2}C = 0. \tag{3.28}$$

Balancing terms as before now yields (as Eq. (3.26) would suggest) $X \sim \alpha \sim \epsilon^{1/6}$, and we put

$$X = \epsilon^{1/6}z \tag{3.29}$$

in Eq. (3.28). Now we find

$$z^3 - z - C = 0, \tag{3.30}$$

and thus we have regained the full equation: No further approximation is possible. This is a *distinguished* limit because it distinguishes a case when, exceptionally, all three terms in the equation balance. For smaller values of c, for which $C \ll 1$, a different balance of terms is appropriate: We have $z^3 - z \approx 0$, so two roots of Eq. (1.30) are $z \approx \pm 1$ and the other is small, whence $z^3 \ll z$, and the leading order estimate is $z \sim -C$.

Summarizing all this, we have the following estimates for x in terms of c and ϵ, when $\epsilon \ll 1$:

$$\begin{aligned} c &= O(1)\colon x \sim \epsilon^{1/3}c^{1/3}e^{2ni\pi/3}, \quad n = 0, 1, 2; \\ c &= O(\epsilon^{1/2})\colon x = \epsilon^{1/2}z, \quad z^3 - z - (c/\epsilon^{1/2}) = 0; \\ c &\ll \epsilon^{1/2}\colon x \sim \pm\epsilon^{1/2}, -c. \end{aligned} \tag{3.31}$$

In fact, the $c = O(1)$ result applies for all $c \gg \epsilon^{1/2}$, because if $x \sim \epsilon^{1/3}c^{1/3}$, then $\epsilon x \ll \epsilon c$ providing $\epsilon^{1/3}c^{1/3} \ll c$, that is, $c \gg \epsilon^{1/2}$. As c decreases, the two complex roots coalesce on the real axis (when $c = O(\epsilon^{1/2})$); this occurs when there is a double root of Eq. (3.30), that is, when $z = -1/\sqrt{3}$ (so $C > 0$), thus $C = 2/3\sqrt{3}$. Also $x = 0$ is never a root of the equation for $c > 0$. Putting all this together, we can establish the behavior of the roots in terms of c when $\epsilon \ll 1$ as shown in Fig. 3.1. We have in fact established the behavior of the roots of Eq. (3.4) for *all* values $0 < \delta_1 \ll 1$, $\delta_2 > 0$.

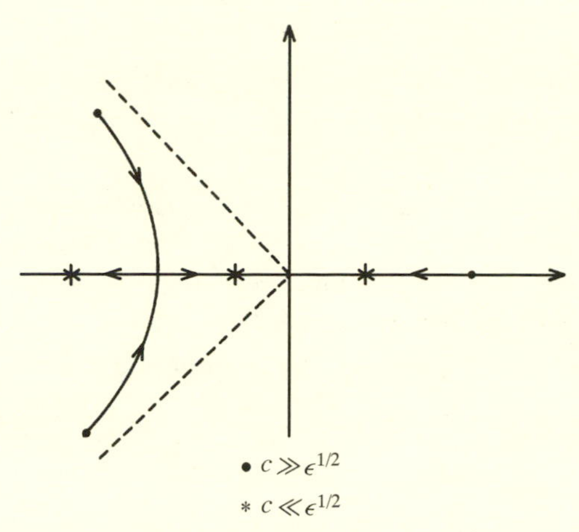

Fig. 3.1. Variation of the roots of Eq. (3.18) with c. For $c \gg \epsilon^{1/2}$, there is only one real positive root and two complex ones with arg $x \sim \pm 2\pi/3$ (circles). As c decreases, the complex roots approach the negative real axis, and for $c \ll \epsilon^{1/2}$, there are three real roots (stars)

3.5 Laplace's method

An interesting way of obtaining asymptotic expansions is via Laplace's method, which can in fact be proven to work (via Watson's lemma). It is used for Laplace type integrals of the form

$$I(x) = \int_a^b e^{x\phi(t)} f(t)\, dt, \tag{3.32}$$

when $x \to \infty$. A particular application is in obtaining Stirling's formula for $n!$ as $n \to \infty$. We introduce the gamma function

$$\Gamma(z) = \int_0^\infty t^{z-1} e^{-t} dt, \tag{3.33}$$

and recall that $\Gamma(n+1) = n!$. Thus

$$\Gamma(z+1) = \int_0^\infty \exp[z \ln t - t]\, dt, \tag{3.34}$$

and as z tends to infinity, the idea is that the exponent undergoes large variation, and thus the dominant part of the integral is determined by the value of the integrand near the maximum of the exponent. Now $z \ln t - t$ is maximum at $t = z$, i.e., it has a *movable* maximum, and for calculational purposes it is convenient to fix this by writing $t = zu$, thus

$$\Gamma(z+1) = z^{z+1} \int_0^\infty \exp[z\{\ln u - u\}]\, du. \tag{3.35}$$

The expression $\ln u - u$ is maximum at $u = 1$, near which (with $u = 1 + v$)

$$\ln u - u = -1 - \frac{v^2}{2} + \frac{v^3}{3} \ldots. \tag{3.36}$$

Thus

$$\Gamma(z+1) = z^{z+1}e^{-z}\int_{-1}^{\infty} \exp\left[z\left\{-\frac{v^2}{2}+\frac{v^3}{3}\dots\right\}\right] dv. \tag{3.37}$$

The idea of the method is that $\exp[z(-v^2/2\dots)]$ is exponentially small for $v \gg O(1/z^{1/2})$, so that the integral can effectively be calculated by truncating the Taylor series of the exponent; moreover, the limits of the integral are irrelevant and can be replaced by $\pm\infty$. This is all justified by Watson's lemma, which additionally shows that higher order terms can be calculated by expanding the exponent as a Taylor series. Thus

$$\Gamma(z+1) \\ \sim z^{z+1}e^{-z}\int_{-\infty}^{\infty} e^{-zv^2/2}\left[1+z\left\{\frac{v^3}{3}-\frac{v^4}{4}\dots\right\}+\frac{z^2}{2!}\left\{\frac{v^3}{3}-\frac{v^4}{4}\dots\right\}^2+\dots\right] dv. \tag{3.38}$$

To make the size of terms explicit, we put $v = (2/z)^{1/2}w$, thus

$$\begin{aligned}\Gamma(z+1) &\sim \sqrt{2}z^{z+(1/2)} \\ &\times e^{-z}\int_{-\infty}^{\infty} e^{-w^2}\left[1+z^{-1/2}\left(\frac{2\sqrt{2}}{3}\right)w^3 - z^{-1}w^4z^{-1}\left(\frac{4}{9}\right)w^6 + O(z^{-3/2})\right] dw \\ &\sim \sqrt{2\pi}z^{z+(1/2)}e^{-z}\left[1+\frac{1}{12}z^{-1}+\dots\right],\end{aligned} \tag{3.39}$$

which is Stirling's formula, using

$$\begin{aligned}\int_{-\infty}^{\infty} e^{-w^2}dw = \sqrt{\pi}, &\qquad \int_{-\infty}^{\infty} w^3e^{-w^2}dw = 0, \\ \int_{-\infty}^{\infty} w^4e^{-w^2}dw = 3\sqrt{\pi}/4, &\qquad \int_{-\infty}^{\infty} w^6e^{-w^2}dw = 15\sqrt{\pi}/8.\end{aligned} \tag{3.40}$$

Two points should be noticed: the approximation is certainly asymptotic, because $z!$ and its approximation diverge from each other as $z \to \infty$. Nevertheless, the ratio tends to 1. Secondly, this approximation is actually accurate even for $z = 1$. In fact, when $z = 1$, Eq. (3.39) gives 0.99898 (as opposed to 1), and when $z = 2$, it gives 1.99896. Such is the power of the asymptotic method.

3.6 Notes and references

Asymptotics There are a number of good books on asymptotic methods and perturbation methods. Classics are the complex variable text by Carrier, Krook, and Pearson (1966) and the book by Bender and Orszag (1978). Hinch's (1991) recent book is short and sweet, and Keener's (1988) book is also recommended, both for this and as a general primer in how to be an elegant applied mathematician.

Asymptotic expansions Asymptotic series expansions were considered by Poincaré (1893) in his study of celestial mechanics. The basic model for the motions of the solar system consists of nine uncoupled central force oscillators (each describing one of the nine planets and the sun) that can be separately solved to give the Keplerian orbits. Weak gravitational interactions between the planets perturb this integrable Hamiltonian system, and in general it is found that perturbation expansions for the perturbed orbits need not converge. Indeed, the perturbed orbits may not even be regular, and there is some numerical evidence that the solar system is mildly chaotic (mostly manifested by the orbit of Pluto). It is presumably the existence of these irregular orbits for arbitrarily small ε that is associated with the nonanalyticity of the solutions and therefore the divergence of the expansions. Despite this, the methods of perturbation theory are extremely powerful, and are of fundamental importance in much of modern applied mathematics. Poincaré's work in celestial mechanics underlies much of modern Hamiltonian mechanics and hence also nonlinear dynamic systems and chaos.

Exercises

1. (a) Show that any power series expansion of the form $y \sim y_0(x) + \varepsilon y_1(x) + \varepsilon^2 y_2(x) + \ldots$ is an asymptotic expansion as $\varepsilon \to 0$, *uniformly* in x, provided $y_i(x)$ is bounded for each i.
(b) A differential equation can be written in the operator form

$$N(u) = L(u) + N_2(u) + N_3(u) + \ldots = 0,$$

where L is a linear differential operator and $N_r(u)$ is of degree r (specifically, $N_r(u) = L_r(u, \ldots u)$, where L_r is linear in each of its r arguments separately). By writing $u = \varepsilon u_1 + \varepsilon^2 u_2 + \ldots$, derive a sequence of equations for u_r in the form

$$L(u_r) = g_r(u_1, u_2, \ldots u_{r-1}),$$

where $g_1 = 0$, and deduce that if these equations have bounded solutions, then an asymptotic expansion for u near $u = 0$ can be obtained.
(c) Show that a differential equation $f(x, y, y') = 0$, where f is analytic in each argument and $f(x, 0, 0) \equiv 0$, can be written in the form of (b).

2. Let $y(t)$ satisfy

$$\frac{dy}{dt} = -y^2, \qquad y(0) = 1/\varepsilon, \qquad \varepsilon \ll 1.$$

Show that a Taylor series expansion for y leads to

$$y = \frac{1}{\varepsilon} - \frac{t}{\varepsilon^2} + \frac{t^2}{\varepsilon^3} \ldots,$$

but that if an asymptotic expansion for y is sought, the leading term is the exact solution. (**Hint:** note that the expansion must be $y = y_{-1}/\varepsilon + y_0 + \varepsilon y_1 + \ldots$ and that this necessitates a rescaling of $t = \varepsilon\tau$.)

3. Suppose $h(x)$ satisfies the first-order differential equation

$$x^3 h' = h - x^2,$$

with $h(0) = 0$. Show that an asymptotic expansion for small x is a power series in x that is everywhere divergent (for $x \neq 0$). Explain why by solving the equation exactly.

4. *An example of Lighthill* (see Van Dyke, 1975)

Show that the solution of the equation

$$(x + \varepsilon y)y' + y = 1, \qquad y(1) = 2,$$

has an asymptotic expansion in powers of ε,

$$y \sim \frac{1+x}{x} - \varepsilon \frac{(1-x)(1+3x)}{2x^3} + \frac{\varepsilon^2(1+x)(1-x)(1+3x)}{2x^5} \cdots .$$

Deduce that the expansion is not uniformly asymptotic near $x = 0$.

5. Find an asymptotic expansion for the solution $x(t)$ of

$$\ddot{x} + \varepsilon \dot{x} + x = 0, \quad \varepsilon \ll 1,$$

with $x(0) = 1, \dot{x}(0) = 0$. Show that the solution is asymptotic as $\varepsilon \to 0$ for $0 < t < T$, where T is fixed (but arbitrary). How big *in practice* can T be? By examining the exact solution, identify the cause of the nonuniformity.

6. (i) Let $f(x)$ be continuously differentiable and 2π-periodic, with Fourier series $\sum_{-\infty}^{\infty} c_k e^{ikx}$. Show that

$$c_k = \frac{1}{2\pi i k} \int_0^{2\pi} f'(x) e^{-ikx}\, dx,$$

and deduce that $c_k \leq O(|k|^{-1})$ as $k \to \infty$.

(ii) If $f \in C^r$ (is r times continuously differentiable), show that

$$c_k = \frac{1}{2\pi (ik)^r} \int_0^{2\pi} f^{(r)}(x) e^{-ikx}\, dx,$$

and thus $c_k \leq O(|k|^{-r})$ as $k \to \infty$.

(iii) If f is analytic on the real line (has a convergent Taylor series), show that f extends analytically to a 2π-periodic function in $|\mathrm{Im}\, z| < \alpha$ for some α (**hint:** consider its Fourier series). Hence show that $c_k = O(e^{-\beta|k|})$ for all $\beta < \alpha$, as $|k| \to \infty$.

(iv) How can a function be $O(e^{-\beta k})$ as $k \to \infty$ for all $0 < \beta < \alpha$, but not be $O(e^{-\alpha k})$? Give an example.

4

Perturbation methods

We have already seen, in Chapter 3, some simple ideas of perturbation theory. In Chapter 2, we have come across the idea of heuristic discussion of the structure of solutions based on the order of magnitude of different terms. In this chapter, we will briefly introduce the formal methods of singular perturbation theory, specifically as used in boundary layer theory. These provide the analytic platform upon which much of the later discussion in this book is based.

4.1 Elementary boundary layer theory

Consider the linear two-point boundary value problem

$$\begin{aligned} \epsilon y'' + a(x)y' + b(x)y &= 0, \\ y(0) = A, \quad y(1) &= B, \end{aligned} \tag{4.1}$$

where $\epsilon \ll 1$ and we initially assume $a(x) \neq 0 \; \forall \, x \in [0, 1]$, let us say $a > 0$. We assume a, b, A, B are of $O(1)$.

Leading order approximations

If we put $\epsilon = 0$ in Eq. (4.1) (on the basis that it is small), we find

$$a(x)y' + b(x)y = 0, \tag{4.2}$$

with solution

$$y = C \exp\left[-\int_0^x \frac{b(t)\,dt}{a(t)}\right], \tag{4.3}$$

C being a constant. It is obvious that, except for special choices of A and B, we can only choose C to satisfy one boundary condition. This is associated with the loss of the highest derivative in Eq. (4.1) and is why this is called a *singular* perturbation problem: The approximate solution is not uniformly valid.

Let us choose C to satisfy the boundary condition at $x = 1$, thus

$$y = B \exp\left[\int_x^1 \{b(t)/a(t)\}\,dt\right]. \tag{4.4}$$

Now, as $x \to 0$, $y \to A'$, where

$$A' = B \exp\left[\int_0^1 \{b(t)/a(t)\}\, dt\right]. \tag{4.5}$$

If $A' \neq A$, the approximation breaks down, and our neglect of $\epsilon y''$ must be wrong. Because this requires that $|y''| \gg 1$, and because y is only required to jump by $O(1)$ to satisfy the boundary condition, we then naturally expect that the rapid jump in y occurs in a thin region, termed a *boundary layer*, and moreover that this is located where the problem is, namely, at $x = 0$. As always in asymptotic approximation methods, one proceeds with the simplest assumptions possible. Only if they are found to be inconsistent does one resort to more complicated analyses.

For a boundary layer located at $x = 0$, we bring back the highest derivative term by rescaling x there as

$$x = \delta X, \tag{4.6}$$

where $\delta(\epsilon) \ll 1$ is chosen to balance the highest derivative with the other principal terms in the equation. Substituting into Eq. (4.1), we have

$$(\epsilon/\delta^2)y'' + a(\delta X)\delta^{-1}y' + b(\delta X)y = 0, \tag{4.7}$$

where now $y' = dy/dX$, and to balance terms, we choose $\delta = \epsilon$. By Taylor expanding a, we then have the leading order approximation

$$y'' + a_0 y' = 0, \tag{4.8}$$

where $a_0 = a(0) > 0$, and the solution is

$$y = D + Ee^{-a_0 X}. \tag{4.9}$$

To satisfy the condition at $x = 0$ (i.e., $X = 0$), we choose

$$D + E = A, \tag{4.10}$$

thus

$$y = D + (A - D)e^{-a_0 X}. \tag{4.11}$$

Matching

We now have two approximations (4.4) and (4.11) to the solution (we hope) valid in $0 < x = O(1)$ and $0 < X = O(1)$, respectively. Eqs. (4.4) can be written as

$$y = A' \exp\left[-\int_0^x \{b(t)/a(t)\}\, dt\right]. \tag{4.12}$$

It remains to choose the constant D in Eq. (4.11). This is accomplished by matching the two outer ($x = O(1)$) and inner ($X = O(1)$) expansions in an intermediate region where $x \ll 1$ but $X \gg 1$. We should hope that each expansion is valid there, and moreover, each is the same. We then have the inner expansion

$$y \sim D, \tag{4.13}$$

and the outer expansion

$$y \sim A' + \ldots, \tag{4.14}$$

and so we choose

$$D = A', \tag{4.15}$$

which completes the leading order approximation to the solution.

Composite approximation

It is possible to write a uniform approximation to the solution. We do this by summing the inner and outer expansions and subtracting the common part, valid in the intermediate region. Thus a uniform approximation is

$$y = (A - A')e^{-a_0 X} + A' \exp\left[-\int_0^x \{b(t)/a(t)\}\,dt\right]. \tag{4.16}$$

4.2 Matched asymptotic expansions

The above gives a recipe for computing leading order approximations. To extend the procedure in a systematic manner, we introduce inner and outer expansions:

$$\begin{aligned} y &\sim y_0(x) + \epsilon y_1(x) + \ldots, \quad x = O(1); \\ y &\sim Y_0(X) + \epsilon Y_1(X) + \ldots, \quad X = O(1). \end{aligned} \tag{4.17}$$

Outer expansion

Expanding and equating powers of ϵ, we have successively

$$\begin{aligned} a(x)y_0' + b(x)y_0 &= 0, \\ a(x)y_1' + b(x)y_1 &= -y_0'', \end{aligned} \tag{4.18}$$

and so on. Thus, choosing y to satisfy $y(1) = B$, we have

$$\begin{aligned} y_0 &= A' \exp\left[-\int_0^x \{b(t)/a(t)\}\,dt\right], \\ y_1 &= -\exp\left[-\int_0^x \{b(t)/a(t)\}\,dt\right] \int_0^x \frac{y_0''(t)}{a(t)} \exp\left\{\int_0^t \{b(u)/a(u)\}\,du\right\} dt, \end{aligned} \tag{4.19}$$

and so on.

Inner expansion

With $x = \epsilon X$, the equation is

$$y'' + \left[a_0 + \epsilon a_0' X + \ldots\right] y' + \epsilon[b_0 + \ldots] y = 0, \tag{4.20}$$

and by expanding and equating powers of ϵ, we have

$$\begin{aligned} Y_0'' + a_0 Y_0' &= 0, \\ Y_1'' + a_0 Y_1' &= -b_0 Y_0 - a_0' X Y_0', \end{aligned} \tag{4.21}$$

whence (with $y = A$ at $x = 0$)

$$
\begin{aligned}
Y_0 &= D + (A - D)e^{-a_0 X}, \\
Y_1 &= \left(\frac{F}{a_0} + \frac{b_0 D}{a_0^2}\right)(1 - e^{-a_0 X}) - \frac{b_0 D}{a_0} X \\
&\quad + (A - D)\left\{\frac{(b_0 - a_0')}{a_0} X - \frac{1}{2} a_0' X^2\right\} e^{-a_0 X},
\end{aligned}
\tag{4.22}
$$

etc.

Matching principle

We now identify an intermediate matching region by writing

$$
x = (\epsilon/\eta) x_\eta, \qquad X = x_\eta/\eta, \tag{4.23}
$$

where $\epsilon \ll \eta \ll 1$ and $x_\eta = O(1)$. The matching procedure is to write each expansion in terms of the intermediate variable and choose any free constants (here D and F) in order to render the expansions asymptotically equivalent.

Thus, the inner expansion is

$$
y \sim A' \exp\left[-\frac{b_0}{a_0} x + O(x^2)\right] - \epsilon \exp[O(x)]\left[\frac{x y_0''(0)}{a_0} + O(x^2)\right] \ldots, \tag{4.24}
$$

whence

$$
y \sim A'\left[1 - \frac{b_0}{a_0}(\epsilon/\eta) x_\eta\right] + O(\epsilon^2), \tag{4.25}
$$

where $O(\epsilon^2)$ may include powers of η. The outer expansion is

$$
y \sim D + \epsilon\left\{\frac{F}{a_0} + \frac{b_0 D}{a_0^2}\right\} - \frac{\epsilon}{\eta}\frac{b_0 D}{a_0} x_\eta + O(\epsilon^2, TST); \tag{4.26}
$$

here *TST* denotes *transcendentally small terms* proportional to $\exp(-a_0 x_\eta/\eta)$, which are smaller than any power of ϵ provided $\eta = O(\epsilon^\alpha)$ for some $\alpha \in (0, 1)$.

To match up to terms of $O(\epsilon)$, we choose

$$
D = A', \qquad F = -b_0 A'/a_0^3. \tag{4.27}
$$

Notice that the term in x matches automatically.

4.3 Interior layers

The alert reader will realize that the assumption $a(x) > 0$ was tacitly invoked in the matching procedure, because the exponential term $\exp[-a_0 X]$ only decays if $a_0 > 0$. This, in fact, is the reason that the boundary layer is at $x = 0$ rather than $x = 1$. If $a < 0$ everywhere, then a boundary layer will exist at $x = 1$, but not at $x = 0$.

What if $a(0) < 0$ but $a(1) > 0$? In this case, there can be no boundary layer at *either* end. We are forced to suppose

$$
\begin{aligned}
y &\sim A \exp\left[-\int_0^x \{b(t)/a(t)\}\,dt\right] \quad \text{near } x = 0,\\
y &\sim B \exp\left[\int_x^1 \{b(t)/a(t)\}\,dt\right] \quad \text{near } x = 1.
\end{aligned}
\tag{4.28}
$$

Suppose a has a simple zero at $x = x^* \in (0, 1)$, thus $a \gtrless 0$ for $x \gtrless x^*$. It is natural to associate x^* with a region where transition between the two solutions in Eq. (3.28) occurs. Moreover, if $b = b^* + \ldots$, $a = a'^*(x - x^*) + \ldots$ near $x = x^*$, then

$$
\begin{aligned}
y &\sim c_-(x^* - x)^\beta, \quad x \to x^*-,\\
y &\sim c_+(x - x^*)^\beta, \quad x \to x^*+,
\end{aligned}
\tag{4.29}
$$

where $\beta = b^*/a'^*$. $a'^* > 0$, but b^* may have either sign, so β may be positive or negative. In either event, it is clear that the solutions (4.28) break down when $x \to x^*$.

We therefore seek a *transition layer* near x^*, where $x = x^* + \delta X$ and $\delta \ll 1$. Moreover, in view of Eq. (4.29), we put

$$y = \delta^\beta Y, \tag{4.30}$$

then

$$(\epsilon/\delta^2)Y'' + [\delta a'^* X + \ldots]\delta^{-1} Y' + [b^* + \ldots]Y = 0, \tag{4.31}$$

and a distinguished limit in which the highest derivative is brought back can be selected by choosing

$$\delta = \epsilon^{1/2}. \tag{4.32}$$

To leading order, we then have

$$Y'' + a'^* X Y' + b^* Y = 0, \tag{4.33}$$

and at leading order, the matching conditions are that

$$
\begin{aligned}
Y &\sim c_-(-X)^\beta, \quad X \to -\infty,\\
Y &\sim c_+ X^\beta, \quad X \to +\infty,
\end{aligned}
\tag{4.34}
$$

where $\beta = b^*/a'^*$. Solutions of Eq. (4.33) are related to parabolic cylinder functions, which are tabulated. Thus, although closed-form solutions do not exist, Eqs. (4.33) and (4.34) determine a transition layer structure between the two outer solutions in Eq. (4.28).

The case $a' < 0$

In this case, it turns out that no transition layer structure is possible at x^*. Therefore, there is one outer solution, and in fact this must be $y = 0$. Because boundary layers are possible at both ends, they will exist, so that y jumps to A at $x = 0$ and to B at $x = 1$.

4.4 A nonlinear example

We now consider the nonlinear boundary value problem

$$\epsilon y'' + yy' - y = 0,$$
$$y(0) = A, \quad y(1) = B, \quad \epsilon \ll 1. \tag{4.35}$$

The basic techniques are as before, but the nonlinearity provides for additional flexibility.

At leading order, an outer solution satisfies

$$yy' - y = 0, \tag{4.36}$$

with *two* solutions

$$y = 0 \quad \text{or} \quad y = x + c, \tag{4.37}$$

of which only the latter can satisfy (one) boundary condition. We would then naturally expect a boundary layer at the other. If there is a boundary (or transition) layer at x_d, then we put

$$x = x_d + \epsilon X, \tag{4.38}$$

so that y satisfies ($y' = dy/dX$)

$$y'' + yy' = \epsilon y, \tag{4.39}$$

and at leading order

$$y'' + yy' = 0, \tag{4.40}$$

whence

$$y' + y^2/2 = \beta^2/2, \tag{4.41}$$

where we require $\beta^2 > 0$ in order that this solution match to an outer solution. A further integration yields

$$y = \beta \tanh\left[\frac{\beta}{2}(X + k)\right]$$

or

$$y = \beta \coth\left[\frac{\beta}{2}(X + k)\right], \tag{4.42}$$

which are illustrated in Fig. 4.1 (we can take $\beta > 0$ in Eq. (4.42)). These transition structures can be used to match inner and outer expansions, depending on the values of A and B.

For example, the outer solution $y = x + B - 1$ satisfies the $x = 1$ boundary condition and tends to $B - 1$ at $x = 0$. A coth boundary layer at $x = 0$ *descending* to $\beta = B - 1$ then completes the solution, providing $A > B - 1 > 0$ and k is chosen appropriately. Similarly, there will be a tanh boundary layer at $x = 0$ *ascending* to $\beta = B - 1$, providing

$$-(B - 1) < A < B - 1, \quad B > 1. \tag{4.43}$$

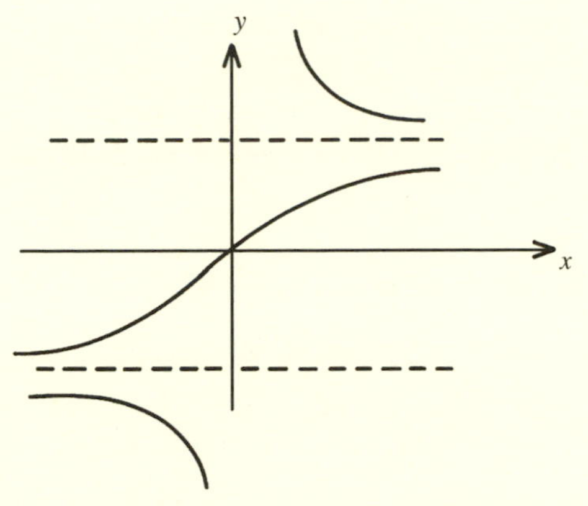

Fig. 4.1. The solutions (given by Eq. (4.42)) of Eq. (4.41)

Symmetrical kinds of solution with boundary layers attached to $x = 1$ exist with equivalent conditions on A and B.

In certain circumstances, we require an interior tanh transition layer, when no boundary layer can exist at either end. In this case, we have

$$\begin{aligned} y &= x + A, \quad x < x_d, \\ y &= x + B - 1, \quad x > x_d, \end{aligned} \tag{4.44}$$

and in order for the tanh layer to match $-\beta$ to $+\beta$ at x_d, we require

$$x_d + A = -\beta, \qquad x_d + B - 1 = +\beta, \tag{4.45}$$

thus

$$x_d = (1 - A - B)/2, \tag{4.46}$$

and in order that $\beta > 0$ and $x_d \in (0, 1)$, we require

$$\begin{aligned} -1 < A + B < 1, \\ B - A > 1. \end{aligned} \tag{4.47}$$

Other parts of parameter space (e.g. $A > 0$, $B < 0$) are inaccessible to solutions of this type, and another boundary layer structure is required, in which $x - x_0 \sim \sqrt{\epsilon}$, $y \sim \sqrt{\epsilon}$. These so-called 'corner layers' can be used to connect the linear outer solutions $y = x + c$ with the zero outer solution $y = 0$. Further discussion of the details is beyond the scope of the present text.

4.5 Nonlinear oscillations

The *Van der Pol oscillator* is represented by the nonlinear second-order equation

$$\ddot{x} + \varepsilon(x^2 - 1)\dot{x} + x = 0. \tag{4.48}$$

The origin is linearly stable for $\varepsilon < 0$, but unstable for $\varepsilon > 0$, and a stable periodic solution exists. One talks of equations such as (4.48) as 'nonlinear oscillators,' or more generally, the perturbed (Hamiltonian) equation $\ddot{x} + V'(x) = \varepsilon g(x, \dot{x}, t)$; $V(x)$ is the potential.

Straightforward perturbation methods give results that are not uniformly asymptotic for all t. Indeed, the 'energy' $E = (x^2 + \dot{x}^2)/2$ of Eq. (4.48) satisfies

$$\dot{E} = -\varepsilon(x^2 - 1)\dot{x}^2, \tag{4.49}$$

and because the leading order solution of Eq. (4.48) is just $x \approx a \cos t$, we have

$$\dot{E} \approx \varepsilon a^2[a^2 \cos^2 t \sin^2 t - \sin^2 t], \tag{4.50}$$

and the change ΔE over a period 2π is

$$\Delta E \approx -2\pi\epsilon a^2 \left[\frac{1}{8}a^2 - \frac{1}{2}\right]. \tag{4.51}$$

Because $E = a^2/2$, this suggests that a changes on the slow timescale $\tau = \varepsilon t$, and in fact we can approximate the evolution of $a(\tau)$ by a continuous approximation to Eq. (4.51), thus

$$\frac{da}{d\tau} \approx -\frac{a}{8}(a^2 - 4), \tag{4.52}$$

and we see that a stable limit cycle of amplitude $A = 2$ exists. This is essentially the method of *averaging*.

Probably the most common method used to analyze equations of this type is the method of *multiple scales*, so-called because of the explicit dependence of the solution on the separate timescales t and εt. In fact, because $a(\varepsilon t) \cos t = a_0 \cos t + \varepsilon a_0' t \cos t \ldots$, we see that a straightforward expansion in powers of ε leads to *secular terms* $t \cos t$ that render the series nonasymptotic when $t \sim 1/\varepsilon$. The method of multiple scales allows for the systematic removal of these terms.

To apply the method to Eq. (4.48), we assume an expansion

$$x \sim x^{(0)}(t, \tau) + \varepsilon x^{(1)}(t, \tau) + \ldots, \tag{4.53}$$

where each term is a function of both the fast time t and the slow time $\tau = \varepsilon t$. Then $d/dt = \partial/\partial t + \varepsilon \partial/\partial \tau$ by the chain rule, and the equations for $x^{(0)}$, $x^{(1)}$, etc., are

$$\begin{aligned} x_{tt}^{(0)} + x^{(0)} &= 0, \\ x_{tt}^{(1)} + x^{(1)} &= -2x_{t\tau}^{(0)} - \left(x^{(0)2} - 1\right)x_t^{(0)}, \end{aligned} \tag{4.54}$$

whose solutions are $x^{(0)} = Ae^{it} + (cc)$, where (cc) is the complex conjugate; thus

$$x_{tt}^{(1)} + x^{(1)} = -2iA'e^{it} - iA^3e^{3it} - i|A|^2Ae^{it} + iAe^{it} + (cc), \tag{4.55}$$

and secular terms $te^{\pm it}$ will occur unless the coefficient of e^{it} (and thus also e^{-it}) is zero, so we choose

$$\frac{dA}{d\tau} = \frac{1}{2}A - \frac{1}{2}|A|^2 A. \tag{4.56}$$

Take A real, without loss of generality, and put $A = a/2$ (so $x^{(0)} = a \cos t$); then $da/d\tau = a(4 - a^2)/8$, just as in Eq. (4.52).

4.6 Partial differential equations

Boundary layer theory applies to partial differential equations as much as to ordinary differential equations, and the principles are the same. As an example, we consider the convective diffusion equation

$$\mathbf{u}.\nabla T = \epsilon \nabla^2 T \tag{4.57}$$

in a closed two-dimensional domain D, where the velocity field $\mathbf{u}$ is irrotational, that is,

$$\mathbf{u} = (\psi_y, -\psi_x), \tag{4.58}$$

where $\psi(x, y)$ is the stream function. We assume that the flow field perfuses the domain, as shown in Fig. 4.2, where the streamlines (ψ = constant) of the flow are shown. On the boundary, we prescribe

$$T = g(s), \tag{4.59}$$

where s is a coordinate directed round the boundary of D.

At leading order, Eq. (4.57) is the hyperbolic equation

$$\psi_y T_x - \psi_x T_y = 0, \tag{4.60}$$

whose solution is

$$T = T_0(\psi), \tag{4.61}$$

and the curves ψ = constant are the characteristics; they are called *subcharacteristics* of the full equation, because it is in fact elliptic.

Because the streamlines cut the boundary S of D in two segments, S_- and S_+, where the streamlines are directed from S_- toward S_+ (that is, $\mathbf{u}.\mathbf{n} < 0$ on S_-, $\mathbf{u}.\mathbf{n} > 0$ on S_+, where $\mathbf{n}$ is the outward normal to S), there are two possible choices for $T_0(\psi)$. It will turn out that the correct choice is to assign T_0 via the values of T on S_-; that is, if $T = g_-(s)$ on s_-, and $\psi = \psi_-(s)$ there, then we define $T_0(\psi)$ via

$$T_0[\psi_-(s)] \equiv g_-(s). \tag{4.62}$$

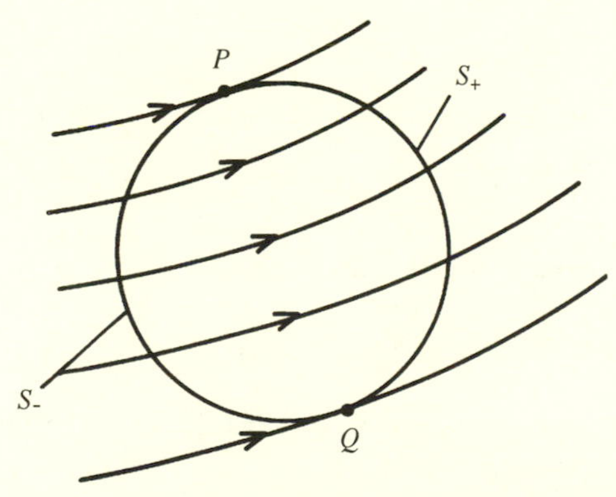

Fig. 4.2. Streamlines of flow in a circular domain. They originate on S_- and terminate on S_+. At P and Q they are tangent to the boundary.

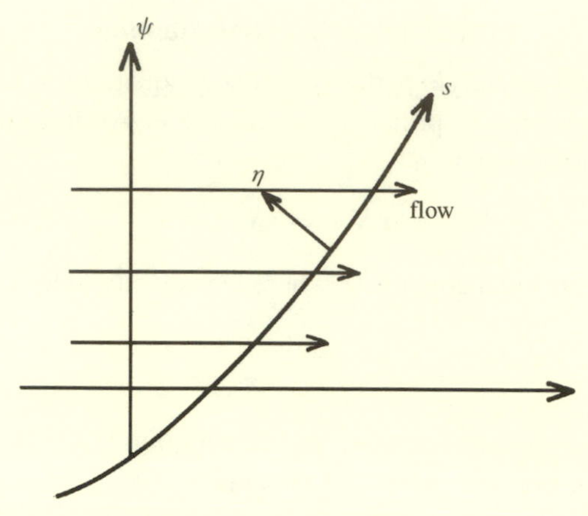

Fig. 4.3. Geometry near the boundary S_+, in terms of the stream function coordinate ψ and the normal coordinate η.

Evidently, we cannot satisfy the boundary condition on S_+, and we therefore seek a boundary layer structure near S_+. We choose ψ and η as independent coordinates, where η is the coordinate normal to S_+,

$$\begin{aligned}\partial_x &= \eta_x \partial_\eta + \psi_x \partial_\psi,\\ \partial_y &= \eta_y \partial_\eta + \psi_y \partial_\psi,\end{aligned} \tag{4.63}$$

and we find

$$-JT_\eta = \epsilon[(\eta_x \partial_\eta + \psi_x \partial_\psi)^2 + (\eta_y \partial_\eta + \psi_y \partial_\psi)^2]T, \tag{4.64}$$

where J is the Jacobian

$$J = \psi_x \eta_y - \psi_y \eta_x, \tag{4.65}$$

and $J \neq 0$. Notice that with the directions indicated in Fig. 4.3, we have $\partial\psi/\partial s > 0$ and $J > 0$.

In the boundary layer, we put

$$\eta = \epsilon N, \tag{4.66}$$

so that to leading order, Eq. (4.64) is (noting $|\nabla\eta|$, $J = O(1)$)

$$-JT_N \sim |\nabla\eta|^2 T_{NN}, \tag{4.67}$$

and the solution that matches to the outer solution $T_0(\psi)$ is just

$$T = T_0(\psi) + \{T_+(\psi) - T_0(\psi)\}\exp[-JN/|\nabla\eta|^2], \tag{4.68}$$

where $T = T_+(\psi)$ on S_+. Notice that it is crucial for exponential decay that $J > 0$; that is, the flow is directed into S_+.

This is analogous to the requirement that in ordinary differential equations, the boundary layers be located where exponential decay is possible.

4.7 Notes and references

There are a variety of good and, by now, classic textbooks on the use of perturbation methods in applied mathematics. Chief among these is the text by Kevorkian and Cole (1982). Another excellent book is that by Bender and Orszag (1978), which also has many numerical illustrations; in the context of fluid mechanics, Van Dyke (1975) is excellent. Hinch's (1991) book is short but to the point. Nayfeh's (1973) book has many examples, a lot of them concerned with nonlinear oscillations.

Exercises

1. Find a regular perturbation expansion for the scaled equation (2.17), that is,

$$\ddot{\theta} + \dot{\theta} + \varepsilon^2 \beta \sin \gamma\theta = 0$$
$$\theta(0) = 1, \quad \dot{\theta}(0) = \varepsilon,$$

where $\varepsilon \ll 1$, $\beta, \gamma \sim O(1)$. Hence show that there is a transient response on a time $t = O(1)$ but that secular terms appear at $O(\varepsilon^2)$. Derive the appropriate equation for $t \sim 1/\varepsilon^2$, and show that it is of boundary layer type. Show that the leading order outer equation for $t \sim 1/\varepsilon^2$ is

$$\dot{\theta} + \beta \sin \gamma\theta = 0, \quad \theta(0) = 1,$$

where now $\dot{\theta} = d\theta/d\tau$, $t = \tau/\varepsilon^2$. If $\gamma \ll 1$, $\beta \sim 1/\gamma$, what is the approximate solution?

2. (i) Find a leading order approximation to the solution of

$$\varepsilon y'' + a(x)y' + b(x)y = 0,$$
$$y(0) = A, \quad y(1) = B,$$

when $a < 0 \; \forall \, x \in [0, 1]$.

(ii) Find a leading order approximation to the solution of the equation

$$\varepsilon y'' + (1 - 2x)y' + x^2 y = 0,$$
$$y(0) = -1, \quad y(1) = 1.$$

3. *(Bender and Orszag)* For what real values of α does the equation

$$\varepsilon y'' + y' - x^\alpha y = 0,$$
$$y(0) = 1, \quad y(1) = 1,$$

have a solution with a boundary layer near $x = 0$?

4. *Exponential asymptotics*

(i) Find the exact solution(s) to the equation

$$\varepsilon^2 y'' + \lambda e^y = 0,$$

with the boundary conditions

$$y(-1) = y(1) = 0,$$

and show that the maximum temperature (at $x = 0$) is $y = 2\ln\xi$, where ξ satisfies

$$\xi = \cosh\left[\frac{1}{\varepsilon}\left(\frac{\lambda}{2}\right)^{1/2}\xi\right].$$

Deduce that no solution exists if $\lambda > \varepsilon^2\lambda_c$, where $\lambda_c \approx 0.878$. Why is a boundary layer type solution inappropriate?
(ii) Define $\lambda = \exp(-p/\varepsilon)$, where p is $O(1)$. Show that for sufficiently small ε, there are two solutions for y, and show that one of them can be written as a regular perturbation expansion in powers of $\delta = \varepsilon^{-2}e^{-p/\varepsilon} \ll 1$.
(iii) By using the exact solution, or otherwise, show that an approximation to the other solution can be determined by finding inner and outer expansions near and far from $x = 0$. (First rescale $y \sim 1/\varepsilon$, then write (two) outer expansions to satisfy the boundary conditions at $x = \pm 1$. Show that a transition layer to join the two is possible near $x = 0$ and that $y(0) \approx p/\varepsilon$.) Deduce the shape of the graph of $y(0)$ versus λ.
(iv) A more physically realistic model (for combustion; see Chapter 12) is

$$\varepsilon^2 y'' + \lambda\exp[y/(1+\varepsilon y)] = 0.$$

Following the methodology of (iii), with $\lambda = \exp(-p/\varepsilon)$, show that on the 'warm' branch $y(0) \approx p/(1-p)\varepsilon$, and deduce that the warm branch only exists for $\lambda \geq \exp(-1/\varepsilon)$.

5. *Resonance*

The equation

$$\ddot{x} + \beta\dot{x} + x + \delta x^3 = \Gamma\cos\omega t$$

represents a damped, forced, nonlinear oscillator.

(i) Solve for x when $\beta = \delta = 0$, and plot the *response a* from the forcing (i.e., the amplitude of the particular solution arising from the Γ term) versus ω: this exhibits the phenomenon of resonance at $\omega = 1$.
(ii) For ω near 1, $\beta = 0$ and $\delta \ll 1$, show that there is a *distinguished limit* (see Section 2.1) in which $\omega = 1 + \varepsilon\mu$, $\mu = O(1)$, $\varepsilon \ll 1$, $x \sim \Gamma/\varepsilon$, and $\varepsilon = (\delta\Gamma^2)^{1/3}$. By expanding the resultant equation in powers of ε, using multiple scales t and $\tau = \varepsilon t$, show that the complex amplitude $A(\tau)$ of the leading order approximation, $x \sim Ae^{it} + (cc)$, satisfies

$$\frac{dA}{d\tau} = \frac{3i}{2}|A|^2 A + \frac{1}{4}ie^{i\mu\tau}.$$

Deduce that the steady response curve of $|A|$ versus μ is a deformed version of the resonant response, as shown in Fig. 4.4(a).
(iii) With parameters as above but also with $\beta = b\varepsilon$, $b = O(1)$, show that a term $-\frac{1}{2}bA$ is added to the amplitude equation above, and deduce that the response curve is as shown in Fig. 4.4(b).

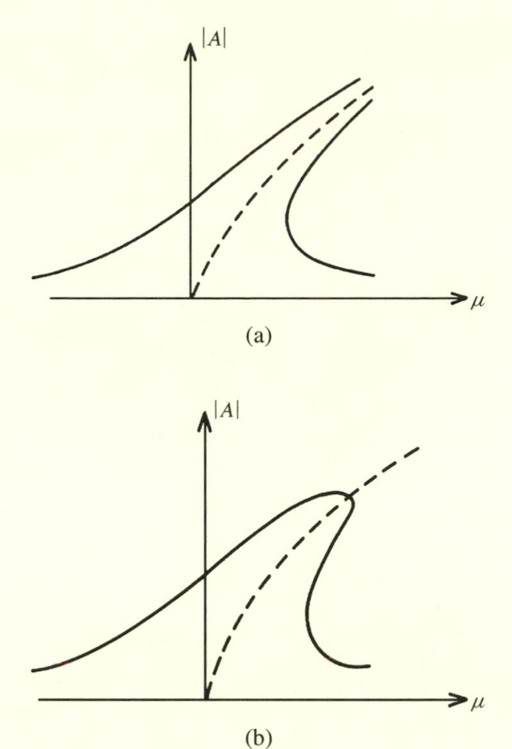

Fig. 4.4. Response diagram of a nonlinear oscillator with (a) no damping and (b) with damping

6. *Relaxation oscillations*

By rescaling x and t when $\Lambda \gg 1$, show that the Van der Pol oscillator

$$\ddot{x} + \Lambda(x^2 - 1)\dot{x} + x = 0$$

can be written as

$$\delta\ddot{x} + (x^2 - 1)\dot{x} + x = 0,$$

where $\delta = 1/\Lambda^2 \ll 1$. The Liénard phase plane is determined by writing this as

$$\frac{d}{dt}\left[\delta\dot{x} + \frac{1}{3}x^3 - x\right] = -x,$$

and then

$$\delta\dot{x} = y - \left(\frac{1}{3}x^3 - x\right),$$
$$\dot{y} = -x.$$

Justify the following arguments: if $\delta \ll 1$, then x moves rapidly so that $y = \frac{1}{3}x^3 - x = f(x)$. Only the parts of $f(x)$ where $f'(x) > 0$ are stable. Thus y moves slowly along the positive slope parts of $f(x)$ toward $x = 0$, until $x = \pm 1$, when a rapid transition to the other branch takes place. (Thus the system 'relaxes' rapidly to the other slow branch.) Show that in the resulting periodic motion, the period is approximately (in the original time units) $P \approx (3 - 2\ln 2)\Lambda$.

7. A fluid flows with two-dimensional velocity field $\mathbf{u}$ past a bluff body S (for example, a cylinder). Dimensionless equations governing the temperature are

$$\mathbf{u}.\boldsymbol{\nabla}T = \frac{1}{Pe}\nabla^2 T,$$

with

$$T \to 0 \quad \text{at} \infty, \qquad \frac{\partial T}{\partial n} = -\alpha(1 - T) \quad \text{on } S,$$

and the prescribed velocity $\mathbf{u}$ satisfies $\boldsymbol{\nabla}.\mathbf{u} = 0$ and is zero on S. Show that, if $Pe \gg 1$, a boundary layer of thickness $Pe^{-1/3}$ exists near S and that if $\alpha Pe^{-1/3} \gtrsim 1$, then $T = O(1)$, but if $\alpha Pe^{-1/3} \ll 1$, then $T \sim \alpha Pe^{-1/3}$. Show that the scaled equation for the temperature θ in the boundary layer is

$$\psi_n \theta_s - \psi_s \theta_n = \theta_{nn},$$

where $\mathbf{u} = (\psi_y, -\psi_x)$ (and we have $\psi = \partial\psi/\partial n = 0$ on S); and, by introducing the *Von Mises* transformation $(s, n) \to (s, \psi)$, show that θ satisfies a diffusion equation

$$\theta_\xi = (\psi^{1/2}\theta_\psi)_\psi,$$

where we take $\psi = \frac{1}{2}n^2\tau(s)$ and $\xi = \int_0^s (2\tau)^{1/2}\, ds$. Show that a similarity solution can be found if $\alpha Pe^{-1/3} \gtrsim 1$. Can a similarity solution exist in the case $\alpha Pe^{-1/3} \ll 1$?

Part three

Classical models

5

Heat transfer

5.1 The diffusion equation

A conservation law

Consider a solid occupying a volume V, with temperature T being a function of space $\mathbf{x}$ and time t. The quantity of heat (energy) per unit mass is c_pT, where c_p is the specific heat (at constant pressure), and so the heat per unit volume is $\rho c_p T$, where ρ is the density. Insofar as energy is conserved, we pose a conservation law

$$\frac{d}{dt}\int_V \rho c_p T\, dV = -\int_S \mathbf{J}.\mathbf{n}\, dS \tag{5.1}$$

applicable to arbitrary closed volumes V with boundary S. This equation expresses the fact that the rate of change of heat in V is exactly given by the rate of heat transfer through the surface S, described by the *heat flux vector* $\mathbf{J}$. On the assumption that $\mathbf{J}$ is continuously differentiable, the divergence theorem implies that

$$\int_V \frac{\partial}{\partial t}(\rho c_p T)\, dV = -\int_V \operatorname{div} \mathbf{J}\, dV, \tag{5.2}$$

and if the integrands are both continuous, then the arbitrariness of V means that they must be equal at every point. Hence we have the point form of the conservation law,

$$\frac{\partial}{\partial t}(\rho c_p T) = -\boldsymbol{\nabla}.\mathbf{J}. \tag{5.3}$$

The reduction of the integral conservation law to a differential equation is a standard procedure. It is important to remember that Eq. (5.3) only applies if T and $\mathbf{J}$ are continuously differentiable; where this is not the case, Eq. (5.1) still applies.

Fourier's law

It is also common that conservation laws introduce extra variables, whose form must be prescribed by *constitutive laws*, which are usually empirical or experimentally based. Prescription of the heat flux vector $\mathbf{J}$ is most simply based on Newton's law of cooling, which says that the rate of heat loss in a one-dimensional bar is proportional to the temperature gradient, with a negative sign. This law is naturally generalized to

Fourier's law in three dimensions, on the assumption that the medium is isotropic and that heat flux in one direction does not affect the flux in any other direction. There follows *Fourier's law*:

$$\mathbf{J} = -k\boldsymbol{\nabla}T, \tag{5.4}$$

where the quantity k is known as the *thermal conductivity*. It may be variable, most often being taken as a function of T; but in most applications, it is taken as constant, as are ρ and c_p.

The heat equation

Putting the above together, we have the heat equation

$$\rho c_p \frac{\partial T}{\partial t} = k\nabla^2 T, \quad \text{or} \quad T_t = \kappa \nabla^2 T, \tag{5.5}$$

where $\kappa = k/\rho c_p$ is the *thermal diffusivity*. Also known as the diffusion equation, Eq. (5.5) arises in other contexts, most notably diffusion of solute in a liquid. In this case, the flux is determined by *Fick's law*: $\mathbf{J}_c = -D\boldsymbol{\nabla}c$, where c is concentration, leading via $c_t = -\boldsymbol{\nabla}.\mathbf{J}_c$ to $c_t = \boldsymbol{\nabla}.[D\boldsymbol{\nabla}c]$.

Nonlinear diffusion

If $k = k(T)$, then the heat equation

$$\rho c_p \frac{\partial T}{\partial t} = \boldsymbol{\nabla}.(k(T)\boldsymbol{\nabla}T) \tag{5.6}$$

is nonlinear. A more dramatic example occurs in the flow of gas in a porous medium. Here the conservation equation (of mass) is

$$\rho_t + \boldsymbol{\nabla}.\mathbf{q} = 0, \tag{5.7}$$

where $\mathbf{q}$ is the mass flux. In analogy to Fourier's and Fick's laws, this is given (for slow flow) by *Darcy's law*:

$$\mathbf{q} = -\rho \frac{k}{\mu} \boldsymbol{\nabla} p, \tag{5.8}$$

where k is known as the permeability and μ is the gas viscosity. Here p is the pressure that drives the flow (down pressure gradients). Finally, a simple constitutive relation is the perfect gas law

$$p = \rho R T, \tag{5.9}$$

so that for an isothermal flow and constant properties, we have

$$\rho_t = \boldsymbol{\nabla}.\left[\left\{\frac{kRT}{\mu}\right\}\rho\boldsymbol{\nabla}\rho\right]. \tag{5.10}$$

Here the diffusion coefficient goes to zero if ρ goes to zero, a situation known as *degeneracy*. It has some interesting consequences, as we shall see later.

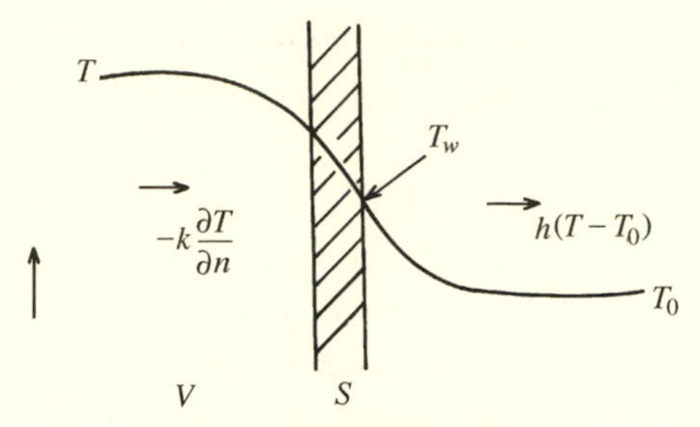

Fig. 5.1. Physical origin of a mixed thermal boundary condition at a wall

Boundary conditions

The diffusion equation (5.5) is parabolic. In a closed domain, it requires an initial condition and a boundary condition applied on the surface of the domain. There are two main types of boundary condition. *Dirichlet* conditions apply to a *conducting* surface, where there is no thermal resistance between the surface and its surroundings. In this case, we require

$$T = f(\mathbf{x}) \quad \text{on } S. \tag{5.11}$$

Neumann boundary conditions arise for an insulating boundary when the heat flux is prescribed, thus

$$-k\frac{\partial T}{\partial n} = q(\mathbf{x}) \quad \text{on } S. \tag{5.12}$$

More generally, a mixed condition of the typical form

$$-k\frac{\partial T}{\partial n} = h(T - T_0) \quad \text{on } S \tag{5.13}$$

can arise, as is shown in Fig. 5.1.

Suppose a thin wall lies between the domain V and the surroundings, which are at temperature T_0. If the wall temperature is T_w, then a plausible assumption is that the heat loss to the surroundings is $h(T_w - T_0)$, where h is the *heat transfer coefficient*. This must equal the heat flux outward, or $-k\partial T/\partial n$ (with $\mathbf{n}$ being the unit *outward* normal); hence we derive Eq. (5.13), with $T_w = T$.

Scaling

The diffusion equation in a domain of linear dimension l has a natural associated timescale $t \sim l^2/\kappa$. In practical situations, this gives a measure of the timescale over which T will change appreciably. For example, $\kappa \sim 10^{-2}$ cm^2 s^{-1} for water, so that the conductive timescale for a cup of tea of dimension 5 cm (e.g., the cup radius) to cool down is $t \sim d^2/\kappa \sim 2500$ s, that is, 40 minutes. We leave as an exercise the consideration of whether this (for tea) is actually correct!

The process of nondimensionalization for the heat equation allows its reduction to the parameter free form

$$T_t = \nabla^2 T \tag{5.14}$$

by choosing the length scale $\mathbf{x} \sim l$, and timescale $t \sim l^2/\kappa$.

Fundamental solution

Consider now the heat equation in one dimension

$$T_t = T_{xx} \tag{5.15}$$

on the infinite domain $-\infty < x < \infty$, with $T \to 0$ as $x \to \pm\infty$. Note that $\int_{-\infty}^{\infty} T\,dx$ is constant, so that the area under the graph of T as a function of x is the same for all time. We shall see that in fact diffusion has a *spreading* effect, so that an initially concentrated peak of T spreads (or *diffuses*) away from its center.

A particular (important) example is the release of a unit quantity of heat at position $x = \xi$ at time τ; we represent this by prescribing T at $t = \tau$ in terms of a *delta function*, where the delta function $\delta(x)$ is 'defined' by $\delta(x) = 0, x \neq 0, \int_{-\infty}^{\infty} \delta(x)\,dx = 1$. It is singular in a special way at $x = 0$ and can be defined as the limit of a sequence of functions, for example, $f_n(x) = 0$, $|x| > 1/2n$; $f_n = n$, $|x| < 1/2n$. It is an example of a *generalized function*. By inspection, the solution corresponding to the prescription of T as a delta function at $t = \tau$, $x = \xi$, called the *fundamental solution*, is

$$G(x, t; \xi, \tau) = \frac{1}{2\sqrt{\pi(t-\tau)}} \exp\left[\frac{-(x-\xi)^2}{4(t-\tau)}\right], \quad t > \tau, \tag{5.16}$$

with $G = 0$ for $t < \tau$. For $t > \tau$, we see that G spreads out in the way we have described.

However, the fundamental solution can be used to build more general solutions, owing to the *linearity* of the heat equation. If $T = T_0(\xi)$ at $t = 0$, then we can view T as resulting from the superposition of the solutions corresponding to release of delta functions $T_0(\xi)\,d\xi\,\delta(x-\xi)$ at $t = 0$, whence the solution is

$$T(x, t) = \int_{-\infty}^{\infty} G(x, t; \xi, 0)\, T_0(\xi)\, d\xi. \tag{5.17}$$

In more general situations, the fundamental solution can be used to establish existence of solutions by reformulation as an integral equation, or as above by explicit construction of the solution. There is never any problem of existence for $t > 0$ for the diffusion equation with appropriate data.

Smoothing property

If $T_0(\xi)$ is localized about zero (for example, if it is of compact support, though this is not necessary), then at large times, we have Eq. (5.17) as

$$\begin{aligned} T &= \frac{1}{2\sqrt{\pi t}} \int_{-\infty}^{\infty} T_0(\xi) \exp\left[\frac{-(x-\xi)^2}{4t}\right] d\xi \\ &\sim \frac{1}{2\sqrt{\pi t}} \left\{ \int_{-\infty}^{\infty} T_0(\xi)\, d\xi \right\} \exp(-x^2/4t), \end{aligned} \tag{5.18}$$

so that the solution eventually relaxes to the fundamental solution. This indicates the smoothing property of diffusion. A different perspective follows from taking a Fourier transform, $\hat{T} = \int_{-\infty}^{\infty} e^{ikx} T(x, t)\, dx$, whence $\hat{T}_t = -k^2 \hat{T}$. We see that the high wavenumber components (corresponding to sharp gradients) are rapidly damped.

Uniqueness

Solutions to the heat equation exist for $t > 0$, and they are unique too. One nice way of showing this is to use an energy method; thus if T_1, T_2 are two solutions, their difference $T = T_1 - T_2$ satisfies

$$\frac{1}{2}\frac{d}{dt}\int_V T^2\,dV = \int_V T\nabla^2 T\,dV = \int_S T\frac{\partial T}{\partial n}\,dS - \int_V |\boldsymbol{\nabla} T|^2\,dV. \tag{5.19}$$

For physically sensible boundary conditions of the types (5.11)–(5.13), we have $T = 0$, $\partial T/\partial n = 0$, or $\partial T/\partial n = -\alpha T, \alpha > 0$, respectively, on S. In any event, $\int_V T^2\,dV$ is strictly decreasing unless $T = T_1 - T_2 = 0$, whence we obtain uniqueness; since for the same initial conditions, $T_1 - T_2 = 0$ at $t = 0$.

Finite domains

The basic method of solution in finite domains is (or can be) the Fourier series. Sturm–Liouville theory tells us that the Helmholtz equation

$$\nabla^2 u + \lambda u = 0 \tag{5.20}$$

in a closed domain V with (for example) $u = 0$ on its boundary S has a denumerable sequence of increasing eigenvalues $0 < \lambda_1 < \lambda_2 \ldots$ with a corresponding complete set of eigenfunctions $\phi_1, \phi_2 \ldots$. In this case, T satisfying the heat equation (and homogeneous boundary conditions) can be represented as

$$T = \sum_1^\infty a_n e^{-\lambda_n t}\phi_n(\mathbf{x}). \tag{5.21}$$

For an inhomogeneous boundary condition $T = T_0(\mathbf{x})$, we first of all subtract off the steady solution of Laplace's equation with the inhomogeneous boundary condition.

In one dimension, we obtain Fourier series; for example, take $T_t = T_{xx}$ in $[0, 1]$ with $T = 0$ at $t = 0$, $T = 0$ at $x = 0$, $T = 1$ at $x = 1$. Then

$$T = x + \sum_1^\infty a_n e^{-n^2\pi^2 t}\sin n\pi x \tag{5.22}$$

with $\sum_1^\infty a_n \sin n\pi x = -x$ giving a_n as the Fourier sine coefficients of $-x$.

Blowup

Diffusion becomes interesting when other processes occur. If, for example, a chemical reaction releases heat internally at a rate $f(T)$ per unit volume (in appropriate units), then we derive

$$T_t = \nabla^2 T + f(T), \tag{5.23}$$

a so-called nonlinear diffusion equation (if f is not linear). Although solutions will exist and be unique for reasonable f, they need not exist for all time. By excluding spatial dependence, we can see that T may become singular at a finite time if

$f'' > 0$, for example, for $f = T^2$, and this phenomenon is known as blowup. The most celebrated example is when $f = \lambda e^T$, an example which occurs in combustion theory.

Similarity solutions

On an infinite domain, where there is no preferred length scale, we see from Eq. (5.18) that the solution becomes self-similar, depending in shape only on the ratio $x/\sqrt{t}$. This is the hallmark of diffusive processes, where the spatial spread of a disturbance is confined to a region $x \sim \sqrt{t}$. The variable $x/\sqrt{t}$ is a *similarity variable*, and it is sometimes advisable to seek similarity solutions whose spatial dependence is controlled by a similarity variable. This will be the case when there is no intrinsic length scale to the problem.

Degeneracy

Now recall the nonlinear diffusion equation for gas seepage in porous media, given by Eq. (5.10), and in appropriate nondimensional units,

$$\rho_t = \boldsymbol{\nabla}.[\rho \boldsymbol{\nabla} \rho]. \tag{5.24}$$

Let us consider the solution corresponding to a release of a unit quantity of gas at a point at time $t = 0$. Suppose this is in three dimensions, thus

$$\rho_t = \frac{1}{r^2}\frac{\partial}{\partial r}\left[r^2 \rho \frac{\partial \rho}{\partial r}\right] \tag{5.25}$$

and also

$$4\pi \int_0^\infty r^2 \rho \, dr = 1, \tag{5.26}$$

with $\rho = 0$ for $t = 0$, $r > 0$. We can expect a similarity solution to be appropriate, of the form

$$\rho = t^{-\beta} f(\eta), \quad \eta = r/t^\alpha, \tag{5.27}$$

where, in order to allow Eqs. (5.25) and (5.26) to be time independent, we require the exponents α and β to satisfy

$$\begin{aligned} -\beta - 1 &= -2\alpha - 2\beta, \\ 3\alpha - \beta &= 0, \end{aligned} \tag{5.28}$$

that is,

$$\alpha = 1/5, \qquad \beta = 3/5 \tag{5.29}$$

(note the difference to the linear diffusive case, which would have $\alpha = 1/2$). Then f satisfies

$$\begin{aligned} &\frac{d}{d\eta}\left[\eta^2 f \frac{df}{d\eta}\right] + \frac{1}{5}[\eta^3 f' + 3\eta^2 f] = 0, \\ &\int_0^\infty \eta^2 f \, d\eta = 1/4\pi, \quad f(\infty) = 0. \end{aligned} \tag{5.30}$$

A first integral is

$$\eta^2 f f' + \frac{1}{5}\eta^3 f = 0, \tag{5.31}$$

and the solution has the typical degenerate structure

$$f = \left(\eta_0^2 - \eta^2\right)/10 \; (\eta < \eta_0), \quad \text{or} \quad f = 0 \; (\eta > \eta_0), \tag{5.32}$$

where η_0 is determined from $\int_0^\infty \eta^2 f \, d\eta = 1/4\pi$. The degeneracy of the diffusion coefficient allows f to be piecewise smooth, and we find $\eta_0 = (75/4\pi)^{1/5} \approx 1.43$. The solution for ρ is thus

$$\rho = t^{-3/5}\left[\eta_0^2 - \frac{r^2}{t^{2/5}}\right], \tag{5.33}$$

and the fingerprint of diffusion here is that $r \sim t^{1/5}$. The hallmark of degenerate nonlinear diffusion is the existence of a front that propagates with finite speed.

Stefan problems

Quiescent water in a lake at a temperature T_l is suddenly subjected to a surface temperature $T_0 < 0$, where zero is the freezing temperature. We can expect the resulting temperature to depend only on vertical distance z as well as time; moreover, where $T < 0$, the lake freezes. Thus for $t > 0$, we can expect a frozen region $0 < z < s(t)$ and an unfrozen region $s < z < l$, say, with the *free boundary* $z = s(t)$ (which is to be determined) being a function of time. A model to describe this situation is of heat diffusion in both ice and water, as follows:

$$\begin{aligned}
T_t &= \kappa_w T_{zz} \quad \text{in } 0 < z < s(t),\\
T &= -|T_0| \quad \text{on } z = 0,\\
T &= 0 \quad \text{on } z = s,\\
T_t &= \kappa_i T_{zz} \quad \text{in } s(t) < z < l,\\
\frac{\partial T}{\partial z} &= 0 \quad \text{at } z = l,
\end{aligned} \tag{5.34}$$

if we suppose the lake bottom is insulating. Initial conditions are

$$s = 0, \qquad T = T_l \quad \text{at } t = 0. \tag{5.35}$$

Suppose that s is prescribed. Then we have exactly the right number of boundary and initial conditions to solve for T, at least in principle. But because s is unknown, an extra condition is necessary to determine s. This is known as the *Stefan condition.* The net heat flux accumulating at the ice-water interface is $\left[-k\frac{\partial T}{\partial z}\right]_+^-$, where $[]_+^-$ represents the jump across $z = s$ ($s-$ value minus $s+$ value); as water freezes, latent heat (L, per unit mass) is removed. It follows that the energy balance condition is

$$\rho L\dot{s} = -\left[-k\frac{\partial T}{\partial z}\right]_+^- = \left[-k\frac{\partial T}{\partial z}\right]_-^+, \tag{5.36}$$

and it must be emphasised that, in writing Eq. (5.36), it is already assumed that the densities of ice and water are equal (otherwise there must be a net flow of water away from $z = s$, because it expands on freezing – or the lake surface rises, as will happen in practice).

For simplicity, suppose that ice and water properties are equal. We nondimensionalize by scaling as follows:

$$T \sim |T_0|, \qquad z \sim l, \qquad t \sim l^2/\kappa; \tag{5.37}$$

the dimensionless model is then

$$\begin{aligned} T_t = T_{zz} \quad &\text{in } 0 < z < s, \quad \text{and} \quad s < z < 1, \\ T = -1 \quad &\text{on } z = 0, \\ T = 0 \quad &\text{on } z = s, \\ \frac{\partial T}{\partial z} = 0 \quad &\text{on } z = 1, \\ T = \theta_l, \qquad s = 0 \quad &\text{at } t = 0, \\ St\,\dot{s} = -\left[\frac{\partial T}{\partial z}\right]_-^+ \quad &\text{at } z = s, \end{aligned} \tag{5.38}$$

where the dimensionless parameters are

$$\begin{aligned} \theta_l &= T_l/|T_0|, \\ St &= L/c_p|T_0|, \end{aligned} \tag{5.39}$$

the latter of which is known as the Stefan number. This kind of a problem is a *Stefan problem*, where the diffusion equation is solved in a domain bounded by a free boundary, which is determined by an extra boundary condition determined by a latent heat condition. Without going into details, analytic solutions are possible for $t \ll 1$, when also $s \ll 1$, and a similarity solution is possible, in which $s \sim \sqrt{t}$.

If $St \gg 1$, then s moves slowly, so that the temperature fields are quasi-static, that is,

$$\begin{aligned} T &\approx -1 + (z/s), \quad z < s, \\ T &\approx 0, \quad z > s, \end{aligned} \tag{5.40}$$

whence

$$\dot{s} \approx St^{-1}s^{-1}, \qquad s \sim (2t/St)^{1/2}. \tag{5.41}$$

If $St \ll 1$, then $\partial T/\partial z$ is essentially continuous, so that we simply solve the diffusion equation in $0 < z < 1$; thus

$$T = -1 + \sum_0^\infty a_n e^{-(2n+1)^2\pi^2 t/4} \sin[(2n+1)\pi z/2], \tag{5.42}$$

with

$$a_n = 2(1+\theta_l)\int_0^1 \sin[(2n+1)\pi z/2]\,dz = \frac{4(1+\theta_l)}{(2n+1)\pi}, \tag{5.43}$$

and s is determined by putting $T = 0$ in Eq. (5.42). The lake freezes in a finite time, given implicitly by

$$\frac{\pi}{4(1+\theta_l)} = \sum_0^\infty \frac{(-1)^n}{(2n+1)} e^{-(2n+1)^2\pi^2 t/4}, \tag{5.44}$$

or approximately (using the first term)

$$t \approx \frac{4}{\pi^2} \ln\left[\frac{4(1+\theta_l)}{\pi}\right], \tag{5.45}$$

which is reasonably accurate for $\theta_l \gtrsim 1$.

5.2 Notes and references

Basic texts Any book on partial differential equations or methods of mathematical physics will discuss the heat equation. For example, the book by Carrier and Pearson (1976) is an easy introduction. The classic text is that of Carslaw and Jaeger (1959), and a similar but more recent book is that of Crank (1975). Both of these deal with practical methods of solving the diffusion equation. Applications are discussed in Tayler's (1986) book. A very thorough (but hard) book of functional analytic type is that by Friedman (1964).

Distributions A generalized function is defined (rather in the way real numbers are defined) as an equivalence class of sequences of functions $\{f_n(x)\}$ having the property that $\lim_{n\to\infty} \int_{-\infty}^{\infty} f_n(x)F(x)\,dx$ exists for all 'good' functions $F(x)$ (basically C^∞ smooth functions that decay exponentially at $\pm\infty$) and is the same for all such sequences. For example, the Dirac delta function $\delta(x)$ is defined through its 'integral' $\int_{-\infty}^{\infty} \delta(x)F(x)\,dx = F(0)$ and corresponds to function sequences equivalent to $\{(n/\pi)^{1/2}e^{-nx^2}\}$. By use of the integral, one can extend operations to generalized functions; for example, $f'(x)$ is defined by $\int f'(x)F(x)\,dx = -\int f(x)F'(x)\,dx$, and Fourier transforms are defined by the transforms of the sequences. One of the important uses of generalized functions is that Fourier transforms can be used without worry. Classical analysis (see, for example, Titchmarsh (1937)) shows that one needs all sorts of special conditions on a function $f(x)$ in order that its Fourier transform be invertible. As an instance, the Fourier transform $\hat{f}$ exists if $f \in L^1$ (is Lebesgue integrable), but in general $\hat{f}$ is not in L^1, so the inverse may not exist. Generalized functions save the day (as well as providing a rigorous background for the use of such models as '$u_t = u_{xx}$ on $\mathbf{R}$, $u = \delta(x)$ at $t = 0$'). The little book by Lighthill (1958) is a succinct summary.

Blowup Further discussion of blowup in exothermic chemical reactions is given in Chapter 12. See also Exercise 5. Blowup can also occur in equations of Ginzburg–Landau type:

$$A_t = \alpha A + \beta\,|A|^2\,A + \gamma A_{xx}, \tag{5.46}$$

where α, β, γ may be complex. Equations of this type arise in the study of nonlinear stability of parallel shear flows (Stuart, 1960; Stewartson and Stuart, 1971; Hocking, Stewartson, and Stuart, 1972; Hocking and Stewartson, 1971, 1972) where blowup occurs as $A \to \infty$ at one value of x at finite t: This is known as a *burst*. Similar kinds of equations occur in other weakly nonlinear systems (Gibbon and McGuinness, 1981; Bretherton and Spiegel 1983), and one such, the *nonlinear Schrödinger (NLS) equation* occurs ubiquitously in weakly nonlinear, dispersive systems (Newell, 1985); for it, we may take $\alpha = 0$, $\beta = -i$. In one dimension, solutions are bounded and in fact relax to a collection of *solitons*, which are stable waves of permanent form. In two or more dimensions, however (with $\partial^2/\partial x^2 \to \nabla^2$), *focusing* occurs: the amplitude A blows up in finite time (Le Mesurier *et al.*, 1988).

Stefan problems Free or moving boundary problems have a long history, and Stefan gave his name to such problems through his early work on freezing (Stefan, 1891). In the last twenty years, there has been an explosion of interest in free boundary problems, much of it stemming from applications to industrial problems. Much of the literature is in various conference proceedings, the first of which is that of a conference in Oxford in 1974 (Ockendon and Hodgkins, 1974), and there are now a large number of such volumes, including a series published as Pitman Research Notes (see volumes 59, 78, 79, 106, 120, 121, 185, 186, 280, 281, and 282). Crank (1984) gives a good summary of both analytical and numerical methods.

Exercises

1. (i) Burgers' equation is

$$u_t + uu_x = \varepsilon u_{xx}.$$

Show that if $\varepsilon = 0$, shocks can develop in finite time. Show that if $\varepsilon \ll 1$, a shock structure can be found where $x = x_s(t) + \varepsilon X$, $u = U(X)$, with $U \to U_\pm$ as $X \to \pm\infty$. (Here $x_s(t)$ is the 'shock' position, and $U_\pm = u|_{x_s\pm}$.) Does a shock structure exist if $\varepsilon < 0$?

(ii) A thin drop controlled by surface tension has thickness h determined by

$$h_t + hh_x = -\varepsilon h_{xxxx}, \quad \varepsilon \ll 1.$$

By considering the Fourier transform of the linear terms, show that the fourth derivative term is smoothing, and show that it is plausible that a local shock structure exists. (**Hint:** derive the differential equation in the shock by writing $x = x_s(t) + \delta(\varepsilon)X$ for some suitable δ, and then consider the problem of solving the resulting equation for $h = H(X)$ by shooting from $H = h_-$ as $X \to -\infty$ to $H = h_+$ as $X \to +\infty$ (how many conditions need to be specified for H near $h_\pm$?).)

(iii) A fluid-filled crack propagating in an elastic medium has width h satisfying

$$h_t = (h^3 P_x)_x,$$

where

$$P = -\frac{1}{\pi} \int_{-\infty}^{\infty} \frac{\partial h}{\partial s} \frac{ds}{s-x} = -H(h_x),$$

where $H(g)$ is the Hilbert transform of g. Show that the Fourier transform of $H(g)$ is $-i\,\mathrm{sgn}(k)\hat{g}$, where $\hat{g}(k)$ is the Fourier transform $\int_{-\infty}^{\infty} g(x)e^{ikx}dx$ of g, and deduce that $\widehat{(P_{xx})} = -k^2|k|\hat{h}$. Hence deduce that the singular integral smooths out high wave number perturbations.

(a) Show that similarity solutions of the form $h = t^{-\beta} f(\eta)$, $\eta = x/t^{\alpha}$, exist and find a relation between α and β. If $h = \delta(x)$ at $t = 0$ (so that $\int_{-\infty}^{\infty} h\,dx$ is conserved), show that $\alpha = \beta = 1/6$, and derive the equation

$$\eta = \frac{6f^2}{\pi} \frac{d}{d\eta} \int_{-\infty}^{\infty} \frac{f'(\zeta)\,d\zeta}{\zeta - \eta}.$$

What might be a good method to solve this equation numerically?

(b) Derive a linearized equation for perturbations about the steady solution $h = 1$ by writing $h = 1 + \rho$, $\rho \ll 1$ and neglecting quadratic terms. Hence derive the general solution

$$\rho = \frac{1}{2\pi} \int_{-\infty}^{\infty} \hat{\rho}_0(k) e^{-ikx - k^2|k|t}\,dk,$$

and deduce that the uniform crack is stable.

(See Emerman, Turcotte, and Spence, 1986; Spence, Sharp, and Turcotte, 1987; and Lister 1990 for a discussion of this problem in a geophysical context.)

2. Water flow in saturated soil is governed by *Darcy's law*:

$$\mathbf{q} = -\frac{k}{\mu}[\boldsymbol{\nabla} p - \rho_l g \mathbf{k}],$$

where $\mathbf{q}$ is water flux, k is permeability, μ is viscosity, p is pressure, ρ_l is water density, g is gravity, and $\mathbf{k}$ is a unit vector in the vertical direction (downward). Conservation of mass is described by

$$\phi_t + \boldsymbol{\nabla}.\mathbf{q} = 0,$$

where ϕ is the porosity; and the capillary relation gives $p_e = p_e(\phi)$, where p_e is the effective pressure, $p_e = P - p$, P is the overburden pressure (due to weight of soil, etc.), and usually $dp_e/d\phi < 0$.

Assuming $k = k(\phi)$, $p_e = p_e(\phi)$, $P = P_0 + \rho_s g z$, and other properties are constant, derive the equation for one-dimensional flow in the (downward) z direction:

$$\phi_t - V\phi_z = (D\phi_z)_z,$$

where

$$V = \frac{(\rho_s - \rho_l) g k'(\phi)}{\mu}, \qquad D = -\frac{k p_e'}{\mu}.$$

Taking $p_e = p_0(\phi_0 - \phi)/\phi_0$, $k = k_0(\phi/\phi_0)^m$, $m > 0$, derive a dimensionless model suitable to describe infiltration over a length scale l, and show that the advective term can be neglected if the Péclet number

$$Pe = \frac{\phi_0 \Delta\rho g l}{p_0} \ll 1,$$

where $\Delta\rho = \rho_s - \rho_l$. If $p_0 = 1$ bar (10^5 N m^{-2}), $\rho_s = 3$ gm cm^{-3}, $\rho_l = 1$ gm cm^{-3}, and $\phi_0 = 0.4$, for what values of l is this satisfied?
If $Pe \ll 1$, show that the resulting degenerate diffusion equation

$$\phi_t = (\phi^m \phi_z)_z$$

has similarity solutions $\phi = t^{-\beta} f(\eta)$, $\eta = z/t^{\alpha}$, where $m\beta + 2\alpha = 1$. Deduce that if $\phi = \delta(z)$ at $t = 0$, so that $\int_{-\infty}^{\infty} \phi \, dz$ is constant and $\phi \to 0$ as $z \to \pm\infty$, then $\alpha = \beta = 1/(m+2)$; and hence find $f(\eta)$ if $m > 0$. (**Hint:** f is not smooth). If $\int_{-\infty}^{\infty} \phi \, dz = 1$, show that $f = 0$ for $\eta > \eta_0$, where

$$\eta_0 = \left[\frac{(2\gamma)^{(\gamma-1)/2}}{2 I_\gamma} \right]^{1/\gamma},$$

where $I_\gamma = \int_0^{\pi/2} \cos^\gamma \theta \, d\theta$, $\gamma = (m+2)/m$. What happens when $m \gg 1$? Is this consistent with the equation? Can you think of an asymptotic approach to get around the problem (if you did not have the exact solution)? What if $m \ll 1$ (the so-called mesa problem)?

3. Prove that if

$$\begin{aligned} \psi_t &= \psi_{xx}, & \phi_t &= \phi_{xx}, & x, t > 0, \\ \psi &= 0, & \phi &= 0 & \text{on } t = 0, \\ \psi &= 1, & \phi &= f(t) & \text{on } x = 0, \end{aligned}$$

then

$$\phi = \int_0^t f(t-s) \frac{\partial \psi}{\partial s}(x, s) \, ds.$$

Find ψ and deduce an expression for ϕ.

4. The temperature in a freezing lake (cf. Eq. (5.38)) satisfies

$$\begin{gathered} T_t = T_{zz} \quad \text{in } 0 < z < s, \\ T = -1 \quad \text{on } z = 0, \qquad T = 0 \quad \text{in } z \geq s, \\ \dot{s} = \left. \frac{\partial T}{\partial z} \right|_{s-}. \end{gathered}$$

find a similarity solution for T in which $s = 2(\alpha t)^{1/2}$, and show that α satisfies the transcendental equation

$$\alpha \int_0^1 \frac{e^{\alpha u} \, du}{(1-u)^{1/2}} = 1.$$

Deduce that a unique positive value of α can be found to satisfy this, and compute it numerically.

5. Show that the solution of $\theta_t = e^\theta$ is $\theta = -\ln(t_0 - t)$. Derive a similarity type solution of the equation

$$\theta_t = \theta_{xx} + e^\theta$$

by writing θ in the form

$$\theta = -\ln(t_0 - t) + g(\eta, \tau),$$

$\eta = (x - x_0)/2(t_0 - t)^{1/2}$, $\tau = -\ln(t_0 - t)$; and show that steady solutions $g_0(\eta)$ exist, where $g_0 \sim -2\ln|\eta| + A + \dots$ as $\eta \to \pm\infty$. By linearizing about g_0, consider whether it is likely that g tends to g_0 as $\tau \to \infty$. In fact, it is known (Bebernes, Bressan, and Eberly, 1987) that $g \to 0$ as $\tau \to \infty$. Show that for $g \ll 1$, the general solution of the linearized equation for g is

$$g = \sum_0^\infty a_n H_n(\eta) \exp\left[\left(1 - \frac{n}{2}\right)\tau\right],$$

where H_n is the nth Hermite polynomial, providing $\int_{-\infty}^{\infty} g^2 e^{-\eta^2} d\eta < \infty$. Show that if $g \to 0$ as $\tau \to \infty$, then $a_0 = a_1 = a_2 = 0$.
More generally, show that if

$$g = \sum_0^\infty a_n(\tau) H_n(\eta),$$

then

$$\sum \dot{a}_n H_n = -\sum a_n \left(\frac{n}{2} - 1\right) H_n + \psi(g),$$

where $\psi = e^g - 1 - g = \frac{1}{2}g^2 + \dots$. Assuming $a_2 \neq 0$ (but $a_0 = a_1 = 0$), show that $a_2 \sim 1/\tau$ as $\tau \to \infty$.
A different form of expansion (Dold, 1985) uses the variable $z = \eta/\tau^{1/2}$ together with τ. Derive an equation for $g(z, \tau)$ and show that a formal asymptotic expansion for g as

$$g \sim g_0(z) + \psi_1(\tau) g_1(z) + \dots$$

can be found, and find g_0. Show that for fixed η, $g_0 \sim 1/\tau$ as $\tau \to \infty$, but that the asymptotic expansion becomes disordered in this limit.

6. *Counter-current flow*
In the analysis of salt fingers (see Howard and Veronis, 1987), one has to solve the mixed diffusion equation

$$x S_z = S_{xx}$$

in $0 < z < 1$, $-\infty < x < \infty$, with conditions

$$S = \frac{1}{2} \quad \text{at } z = 1,\ x < 0,$$

$$S = -\frac{1}{2} \quad \text{at } z = 0,\ x > 0,$$

$$S \to \mp\frac{1}{2} \quad \text{as } x \to \pm\infty;$$

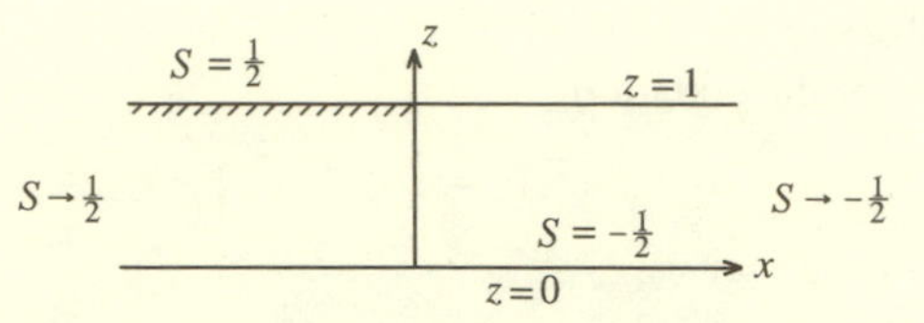

Fig. 5.2. Domain and bounday conditions for the counter-current flow problem in Exercise 6

also S, S_x are continuous (see Fig. 5.2). Why might one expect this problem to be well-posed?

Let $S = f(z)$, $S_x = g(z)$ on $x = 0$. Show that if $f = 1$, then a similarity solution exists for $x > 0$ given by

$$S = \int_{x/z^{1/3}}^{\infty} \exp[-\zeta^3/9]\, d\zeta.$$

By using Laplace transforms or otherwise, deduce that for general $f(z)$,

$$S = \frac{3^{1/3}}{\Gamma(1/3)} \int_{x/z^{1/3}}^{\infty} \exp[-\zeta^3/9] f\left(z - \frac{x^3}{\zeta^3}\right) d\zeta,$$

and find a similar expression for S in terms of $g(z)$. Hence show that

$$f(z) = \frac{-\Gamma(1/3)}{3^{1/3}\Gamma(2/3)} I^{1/3}\{g\},$$

where $I^n\{g\}$ is the Riemann–Liouville fractional integral

$$I^n\{g\} = \int_0^z (z-\zeta)^{n-1} g(\zeta)\, d\zeta.$$

By writing a comparable expression derived from the solution in $x < 0$, show that the function $G(z) = -g(z)/\{3^{1/3}\Gamma(2/3)\}$ satisfies the singular integral equation

$$\int_0^1 |z-\zeta|^{-2/3} G(\zeta)\, d\zeta = 1.$$

Solve the integral equation. (See, for example, Carrier, Krook, and Pearson, 1966).

7. Consider a saturated liquid at (freezing) temperature T_f of infinite extent outside a sphere of radius R_w. At $t = 0$, the surface at R_w is suddenly lowered to a subfreezing temperature T_0. Write down a suitable model, and show that it can be nondimensionalized as

$$\theta_t = \frac{1}{r}\frac{\partial^2}{\partial r^2}(r\theta),$$

$$\theta = 0 \quad \text{on } r = 1,$$

$$\theta = 1, \qquad \frac{\partial \theta}{\partial r} = St\dot{R} \quad \text{on } r = R(t),$$

where the Stefan number is $St = L/c_p(T_f - T_0)$.

If $St \gg 1$, define $\varepsilon = 1/St$, $t = (St)\tau$, and show that the freezing front $R(\tau)$ is described by

$$\tau = \frac{2R^3 - 3R^2 + 1}{6} + \frac{1}{6}\varepsilon(R-1)^2 - \frac{1}{45}\varepsilon^2\frac{(R-1)^2}{R} + O(\varepsilon^3).$$

Write down the equivalent problem for freezing saturated liquid in the interior of a sphere from the outside, and deduce that the corresponding expansion for $\tau(R)$ is nonuniform as $R \to 0$. (See Lunardini and Aziz (1993) for further discussion and other examples where perturbation methods are useful in freezing problems.)

6

Viscous flow

6.1 The Navier–Stokes equation

In describing the motion of a fluid (either liquid or gas, and in fact, even in some circumstances, solid) it is assumed that the locally averaged velocity vector field $\mathbf{u}$ is a twice continuously differentiable function of $\mathbf{x}$ and t. This is the continuum hypothesis. Equations describing the motion of a fluid are those of conservation of mass and momentum, together with an equation of state. In addition, there is an energy equation for the temperature T, but in most cases it uncouples and is of no concern.

Conservation of mass follows from the integral conservation law

$$\frac{d}{dt}\int_V \rho\, dV = -\int_S \rho \mathbf{u}.\mathbf{n}\, dS, \tag{6.1}$$

whence we derive the point form

$$\rho_t + \nabla.(\rho \mathbf{u}) = 0, \tag{6.2}$$

where ρ is the density and $\mathbf{u}$ is the velocity. The term on the right of Eq. (6.1) is the mass flux out of S.

Conservation of momentum follows from Newton's second law, in the form

$$\frac{d}{dt}\int_V \rho u_i\, dV = -\int_S \rho u_i \mathbf{u}.\mathbf{n}\, dS + \int_S \sigma_{ni}\, dS, \tag{6.3}$$

for each component $i = 1, 2, 3$. Here, the first term on the right is the loss through S of momentum flux. The second term represents the forces acting on the volume, conceptualized as being applied at the surface. The nature of these forces depends both on the direction considered (hence the suffix i) and the orientation of the surface (hence the suffix n representing the normal). In fact, $\boldsymbol{\sigma}$ is a (second-order) tensor (otherwise the equations become coordinate frame dependent), and we can write $\sigma_{ni} = \sigma_{ij} n_j$ (summed over j) $= \boldsymbol{\sigma}_i.\mathbf{n}$ (where $\boldsymbol{\sigma}_i = (\sigma_{i1}, \sigma_{i2}, \sigma_{i3})$). Applying the divergence theorem to Eq. (6.3), we obtain

$$\frac{\partial}{\partial t}(\rho u_i) + \nabla.[\rho \mathbf{u} u_i] = \nabla.\boldsymbol{\sigma}_i, \tag{6.4}$$

or in vector form, using Eq. (6.2) and writing the tensor $\boldsymbol{\sigma} = (\boldsymbol{\sigma}_1, \boldsymbol{\sigma}_2, \boldsymbol{\sigma}_3)^T$,

$$\rho[\mathbf{u}_t + (\mathbf{u}.\nabla)\mathbf{u}] = \nabla.\boldsymbol{\sigma}. \tag{6.5}$$

Particular forms of this equation follow from the choice of $\boldsymbol{\sigma}$, which is called the *stress tensor*. Denote the *Kronecker delta* by δ_{ij}, with $\delta_{ij} = 0$ if $i \neq j$, $\delta_{ij} = 1$ if $i = j$. Then for an inviscid fluid, the only internal forces are due to *pressure* p, and

$$\sigma_{ij} = -p\delta_{ij} \tag{6.6}$$

(because then $\sigma_{ni} = \sigma_{ij}n_j = -p\delta_{ij}n_j = -pn_i$, an inwardly acting normal pressure on S in Eq. (6.3)). Then $\boldsymbol{\nabla}.\boldsymbol{\sigma} = -\boldsymbol{\nabla} p$, and we have the *Euler equation*

$$\mathbf{u}_t + (\mathbf{u}.\boldsymbol{\nabla})\mathbf{u} = -\frac{1}{\rho}\boldsymbol{\nabla} p. \tag{6.7}$$

A *viscous fluid* has an additional *deviatoric* stress tensor τ_{ij}, so we write

$$\sigma_{ij} = -p\delta_{ij} + \tau_{ij}, \tag{6.8}$$

and $\boldsymbol{\nabla}.\boldsymbol{\sigma} = -\boldsymbol{\nabla} p + \boldsymbol{\nabla}.\boldsymbol{\tau}$. The deviatoric stress tensor represents the effects of internal friction. Naively, the *shear stress* τ_{12} in a two-dimensional shearing motion $u_1 = u_1(x_2)$ depends on the *shear strain rate* $\partial u_1/\partial x_2$, and we define a (dynamic) viscosity μ by writing

$$\tau_{12} = \mu\frac{\partial u_1}{\partial x_2} \tag{6.9}$$

in this case. More generally, one relates the deviatoric stress tensor τ_{ij} to the *strain rate tensor*

$$\dot{\varepsilon}_{ij} = \frac{1}{2}\left(\frac{\partial u_i}{\partial x_j} + \frac{\partial u_j}{\partial x_i}\right), \tag{6.10}$$

and one generalisation of Eq. (6.9) (in fact, the most general linear relationship) is

$$\tau_{ij} = 2\mu\dot{\varepsilon}_{ij} + \lambda\dot{\varepsilon}_{kk}\delta_{ij}, \tag{6.11}$$

summed over k.

The *equation of state* relates ρ to p and T. Mostly, compressibility effects (varying ρ) are important for gases, where, however, viscosity is often ignored. For liquids, it is often the case that ρ varies only weakly with p and T, and a common assumption is that of incompressibility, and more strongly, that $\rho = $ constant. Strictly, we define an incompressible fluid to be one for which $d\rho/dt = 0$ (see below), or equivalently,

$$\boldsymbol{\nabla}.\mathbf{u} = 0. \tag{6.12}$$

Examining Eq. (6.10), the trace of the strain rate tensor, $\dot{\varepsilon}_{kk}$, is just $\dot{\varepsilon}_{kk} = \partial u_k/\partial x_k = \boldsymbol{\nabla}.\mathbf{u}$. Thus $\tau_{ij} = 2\mu\dot{\varepsilon}_{ij}$, and

$$(\boldsymbol{\nabla}.\boldsymbol{\tau})_i = \frac{\partial \tau_{ij}}{\partial x_j} = \frac{\partial^2 u_i}{\partial x_j^2} + \frac{\partial^2 u_j}{\partial x_i \partial x_j} = \nabla^2 u_i + \frac{\partial}{\partial x_i}\boldsymbol{\nabla}.\mathbf{u}; \tag{6.13}$$

because $\boldsymbol{\nabla}.\mathbf{u} = 0$, we finally obtain the *Navier–Stokes equations* (for an incompressible fluid):

$$\boldsymbol{\nabla}.\mathbf{u} = 0, \tag{6.14a}$$

$$\mathbf{u}_t + (\mathbf{u}.\boldsymbol{\nabla})\mathbf{u} = -\frac{1}{\rho}\boldsymbol{\nabla} p + \nu\nabla^2\mathbf{u}, \tag{6.14b}$$

where $\nu = \mu/\rho$ is the *kinematic viscosity*.

The material derivative

The left-hand side of the Navier–Stokes equation (6.14b) above is the fluid acceleration. The operator $\partial_t + \mathbf{u}.\boldsymbol{\nabla}$ is known as the *material*, or *Lagrangian* time derivative, and one writes

$$\frac{d}{dt} = \frac{\partial}{\partial t} + \mathbf{u}.\boldsymbol{\nabla} \tag{6.15}$$

for the material derivative. It is a time derivative following a fluid element, as can be seen as follows. Let $\boldsymbol{\xi}$ be a spatial coordinate that describes the fluid at a fixed initial instant, $t = 0$, and let $\mathbf{x}$ be a spatial coordinate that gives the location at time t of the fluid element that was at $\boldsymbol{\xi}$ at $t = 0$. Thus $\mathbf{x} = \mathbf{x}(\boldsymbol{\xi}, t)$. The usual (Eulerian) time derivative is then

$$\frac{\partial}{\partial t} = \left.\frac{\partial}{\partial t}\right|_{\mathbf{x}\text{ fixed}}, \tag{6.16}$$

whereas the Lagrangian derivative is

$$\frac{d}{dt} = \left.\frac{\partial}{\partial t}\right|_{\boldsymbol{\xi}\text{ fixed}}. \tag{6.17}$$

Using the chain rule,

$$\frac{d}{dt} = \left.\frac{\partial}{\partial t}\right|_{\boldsymbol{\xi}\text{ fixed}} = \left.\frac{\partial}{\partial t}\right|_{\mathbf{x}\text{ fixed}} + \left.\frac{\partial x_i}{\partial t}\right|_{\boldsymbol{\xi}\text{ fixed}} \frac{\partial}{\partial x_i}, \tag{6.18}$$

which reproduces Eq. (6.15), since $\partial \mathbf{x}/\partial t|_{\boldsymbol{\xi}} = \mathbf{u}$.

As an illustration of the use of the material derivative, let $J = (\partial x_i/\partial \xi_j)$ be the Jacobian of the transformation from $\boldsymbol{\xi}$ to $\mathbf{x}$. Thus J is a measure of the volume of a fluid element. We find

$$\frac{dJ}{dt} = J \operatorname{div} \mathbf{u}, \tag{6.19}$$

so that for an incompressible fluid, the volume is constant. More generally, Eq. (6.2) can be written

$$\frac{d\rho}{dt} + \rho \operatorname{div} \mathbf{u} = 0, \tag{6.20}$$

whence ρJ is constant for a fluid element (as is obvious).

Boundary conditions

At a stationary solid boundary, a viscous fluid will satisfy both a no-flow-through condition and a no-slip condition, thus

$$\mathbf{u} = \mathbf{0}. \tag{6.21}$$

At a free boundary, for example, when the fluid is bounded by a deformable medium, one would prescribe continuity of velocity and of stress. In practice (e.g. at a water–air interface), this often means that one allows slip and zero shear stress, that is,

$$\tau_{nt} = \tau_{ij} n_i t_j = 0, \tag{6.22}$$

where $\mathbf{n}$, $\mathbf{t}$ are normal and tangent vectors to the interface.

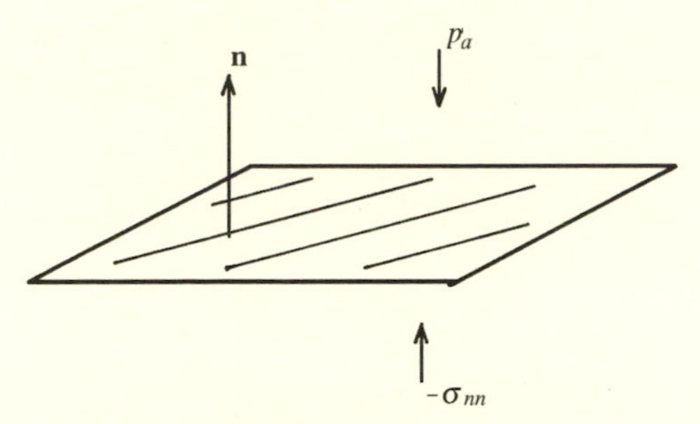

Fig. 6.1. Normal stresses at an interface

Similarly, the normal stress is continuous, and at an air-water interface, one takes the air to have zero strength, so that its pressure is constant: then

$$\sigma_{nn} = \sigma_{ij} n_i n_j = -p_a, \tag{6.23}$$

where p_a is the atmospheric pressure.

Particularly for small drops of fluid, *surface tension* is important. The surface tension γ arises as an energy associated with interfacial surfaces: to increase the surface area, energy must be supplied, that is, work must be done, and this is manifested by the surface energy as a force per unit length, which acts along the interface. To be specific, consider an interface $z = \eta(x, y, t)$ between air and water, as shown in Fig. 6.1. The normal stress exerted *on* the water at the interface is σ_{nn} in the normal (here upward) direction, thus the upward force exerted *by* the fluid is $-\sigma_{nn}$. If each point on $z = \eta$ is moved infinitesimally a distance δn in the normal direction to $z = \eta$ at that point, then the work done on a surface element of area δS is $(-\sigma_{nn} - p_a)\delta n \delta S$, and this equals the change in surface energy $\gamma \delta[\delta S]$. Therefore the appropriate boundary condition is

$$-\sigma_{nn} = p_a + 2\gamma\kappa, \tag{6.24}$$

where $\kappa = (\delta S)^{-1} \partial(\delta S)/2\partial n$ is the mean curvature measured from *below* (Segal, 1986). It can be defined by (Do Carmo, 1976; Finn, 1986)

$$2\kappa = \frac{\left(1 + \eta_x^2\right)\eta_{yy} - 2\eta_x \eta_y \eta_{xy} + \left(1 + \eta_y^2\right)\eta_{xx}}{\left(1 + \eta_x^2 + \eta_y^2\right)^{3/2}}, \tag{6.25}$$

where, however, this definition gives curvature measured from the *top*. More succinctly, twice the curvature of a surface $\phi(x, y, z) =$ constant is simply $2\kappa = \boldsymbol{\nabla}.\mathbf{n}$, where $\mathbf{n} = \boldsymbol{\nabla}\phi/|\boldsymbol{\nabla}\phi|$. Specifically, if $\mathbf{n}$ points from fluid 1 to fluid 2, then this formula defines the curvature relative to fluid 1.

Kinematic boundary condition

At a free surface, such as at a water–air interface, the surface can move, and its determination is therefore part of the problem. Just as for a Stefan problem, an extra condition is therefore necessary to specify the problem. This is called a *kinematic condition*, and reflects the fact that in a smooth (continuously differentiable) transformation from $\boldsymbol{\xi}$ to $\mathbf{x}$, fluid elements at the boundary must remain in the boundary for all time. Mathematically, this is formulated for a free surface $f(\mathbf{x}, t) = 0$ as

$$\frac{df}{dt} = 0, \tag{6.26}$$

where d/dt is the material derivative. Equivalently,

$$\frac{\partial f}{\partial t} + \mathbf{u}.\nabla f = 0, \tag{6.27}$$

or if the free surface is $z = \eta(x, y, t)$, so $f = \eta - z$, then

$$w = \eta_t + u\eta_x + v\eta_y, \tag{6.28}$$

where $\mathbf{u} = (u, v, w)$.

Vorticity

The vorticity $\boldsymbol{\omega}$ is defined as the curl of the velocity field,

$$\boldsymbol{\omega} = \text{curl}\,\mathbf{u}. \tag{6.29}$$

It is a measure of the rotationality of the flow. For example, a *vortex* is described by a two-dimensional velocity field (in polar coordinates) $u_r = 0$, $u_\theta = 1/r$. Here in fact the vorticity is concentrated at the origin as a delta function. Notice for this flow that

$$\oint_C \mathbf{u}.d\mathbf{r} = 2\pi \tag{6.30}$$

for any closed curve C surrounding the origin.

By taking the curl of the Navier–Stokes equation, we can find that, for an incompressible fluid,

$$\frac{d\boldsymbol{\omega}}{dt} = (\boldsymbol{\omega}.\nabla)\mathbf{u} + \nu\nabla^2\boldsymbol{\omega}, \tag{6.31}$$

an equation that displays several important properties. Viscosity enables vorticity to *diffuse*. In a fluid flow, the vorticity diffuses from solid boundaries, because, particularly for fluids with small viscosity, it is there that the shear is concentrated – and strong shear implies strong vorticity. Vorticity can be shed from fixed boundaries, at trailing edges (for example, of aerofoils), or through boundary layer separation (which we do not discuss here). For low-viscosity fluids, vorticity in the fluid interior moves with the fluid in the following sense. For (relatively) inviscid fluids, vorticity is drawn off solid walls as *vortex lines*, or in two dimensions, vortex sheets. Vortex lines are idealizations for inviscid flow of concentrated filaments of vorticity (like tornadoes) with local cross-sectional velocity fields corresponding to Eq. (6.30). We denote the *circulation* through a loop of fluid as

$$\Gamma = \oint_C \mathbf{u}.d\mathbf{r}; \tag{6.32}$$

from the Euler equation, it follows that Γ is conserved following the fluid (i.e., $d\Gamma/dt = 0$, with C being a *material* loop). This is Kelvin's theorem. Moreover, vortex lines move with the fluid: this is Helmholtz's theorem and really follows from Kelvin's theorem. In two dimensions, $\boldsymbol{\omega}.\boldsymbol{\nabla} \equiv 0$, and so (if $\nu = 0$) vorticity itself is conserved. Hence if $\boldsymbol{\omega} = \mathbf{0}$ initially, it remains zero. Such flow fields are called irrotational. In reality, however, even small viscosity enables vorticity to be generated at boundaries.

The Reynolds number

If a typical flow velocity is U, and a typical flow geometry is of dimension l, then if we nondimensionalize the variables as follows:

$$\mathbf{u} \sim U, \qquad \mathbf{x} \sim l, \qquad p - p_\infty \sim \rho U^2, \qquad t \sim l/U \tag{6.33}$$

(write $\mathbf{u} = U\mathbf{u}^*$, etc., substitute in, drop asterisks), we obtain

$$\begin{aligned} \boldsymbol{\nabla}.\mathbf{u} &= 0, \\ \mathbf{u}_t + (\mathbf{u}.\boldsymbol{\nabla}) &= -\boldsymbol{\nabla} p + \frac{1}{Re}\nabla^2\mathbf{u}, \end{aligned} \tag{6.34}$$

where the dimensionless parameter

$$Re = Ul/\nu, \tag{6.35}$$

called the *Reynolds number*, is the single parameter that determines the flow solution. Here p_∞ is an ambient pressure, for example, at infinity.

If $Re \ll 1$, one refers to the motion as *slow flow*, or *Stokes flow*, and the momentum equation can be approximated as

$$\boldsymbol{\nabla} p = \nabla^2\mathbf{u}, \tag{6.36}$$

with p rescaled with $1/Re$. This is a regular perturbation, at least in finite domains.

If $Re \gg 1$, then the approximation of Eq. (6.34) is the Euler equation, but this is a singular approximation, as it reduces the order of the equation. In fact, one cannot in general satisfy the no-slip boundary condition with the Euler equation; and it is necessary to introduce thin viscous boundary layers of thickness $O(Re^{-1/2})$ in order to bring the viscous term back. These boundary layers exist near solid boundaries and enable the no-slip condition to be satisfied. They generate vorticity as we have mentioned, which is injected into the inviscid flow outside by separation at corners or points of zero 'skin friction' (wall shear stress).

Instability and turbulence

A plane parallel flow, or shear flow, is one where $\mathbf{u} = (U(z), 0, 0)$. Examples are plane Poiseuille flow and pipe Poiseuille flow, and an approximately one-dimensional flow is the Blasius boundary-layer flow over a flat plate. Over a hundred years ago, Reynolds observed in pipe flow experiments that as Re is increased, the flow becomes irregular: Dye streaks injected into the flow become sinuous, a manifestation of the turbulence of the fluid flow. Similar results are found in plane Poiseuille flow. Initial efforts to understand the onset of turbulence focused on the stability of the flow. If we consider the plane flow problem and put $\mathbf{u} = (U(z), 0, 0) + \mathbf{u}'$, then $\mathbf{u}'$ satisfies the linearized equations

$$\nabla.\mathbf{u}' = 0,$$
$$\left(\frac{\partial}{\partial t} + U\frac{\partial}{\partial x}\right)\mathbf{u}' + w'U'(z)\mathbf{i} = -\nabla p' + \frac{1}{Re}\nabla^2\mathbf{u}', \tag{6.37}$$

providing $\mathbf{u}'$ is small (here p' is the pressure perturbation). The Reynolds number for both pipe and plane Poiseuille flow is determined in terms of maximum velocity and half-channel width (or pipe radius). Squire's theorem tells us that two-dimensional disturbances are the most unstable, and for these, we have a stream function ψ such that $u' = \psi_z$, $w' = -\psi_x$ and curl $\mathbf{u}' = -\nabla^2\psi\mathbf{k}$. Taking the curl of Eq. (6.37), we then have (because $\nabla^2 =$ curl curl $-$ grad div)

$$\left(\frac{\partial}{\partial t} + U\frac{\partial}{\partial x}\right)\nabla^2\psi - U''\psi_x = \frac{1}{Re}\nabla^4\psi, \tag{6.38}$$

and for normal modes of the form $\psi = \phi(z)\exp[i\alpha(x - ct)]$ (here α is the *wave number* and c is the *wave speed*), ϕ satisfies the Orr–Sommerfeld equation

$$(U - c)(D^2 - \alpha^2)\phi - U''\phi = (i\alpha Re)^{-1}(D^2 - \alpha^2)^2\phi, \tag{6.39}$$

where $D \equiv d/dz$, and we require the no-slip conditions $\phi = D\phi = 0$ on the walls.

Equation (6.39) defines an eigenvalue problem: Given α, we determine $c(\alpha, Re)$, and the curve(s) $\mathrm{Im}\, c(\alpha, Re) = 0$ define stability boundaries in the α, Re plane that separate regions of stability ($\mathrm{Im}\, c_i < 0$) from instability ($\mathrm{Im}\, c_i > 0$). For plane Poiseuille flow, $U = 1 - z^2$ in $-1 \le z \le 1$, the stability curve is as shown in Fig. 6.2. Instability sets in at the nose of the curve, where $\alpha = \alpha_c = 1.02$ and $Re = Re_c = 5772$. In practice, however, turbulent 'spots' first appear at Reynolds numbers as low as $Re = 1000$; and for pipe Poiseuille flow $U = 1 - r^2$, $0 \le r \le 1$ (r being the cylindrical radius), turbulence sets in at $Re \approx 2000$ (if disturbances are sufficiently large), whereas the flow is always linearly stable. Linear stability theory, subtle though it is, fails to explain the mechanism of the onset of turbulence.

In fact, this is not so surprising, because the regular 'cats-eye' pattern of the linear disturbances, and their two-dimensionality, bears little resemblance to the violent

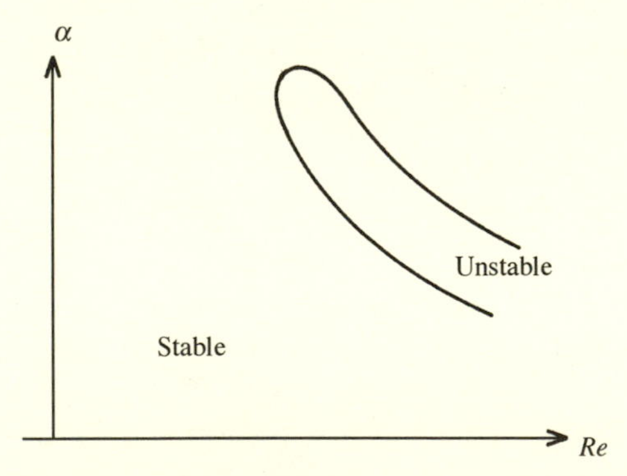

Fig. 6.2. A schematic illustration of the instability curve for plane Poiseuille flow. The axes are Reynolds number Re and wave number α

three-dimensional disturbances that are in fact seen, and it is now thought that the onset of turbulence is associated with chaotic behavior of the Navier–Stokes equations. In other situations, however, instability may occur in the form of regular traveling waves – for example, in Taylor–Couette flow.

Lubrication theory

A particular situation where inertia (acceleration) terms are negligible, even at high *Re*, is when the flow is shallow, for example in lubrication bearings or shallow rivulet flow. In a two-dimensional situation, where the transverse (y) length scale (relative to the longitudinal (x) one) is small, we rescale the equations by writing

$$y \sim \varepsilon, \qquad v \sim \varepsilon, \qquad p \sim 1/\varepsilon^2 Re, \tag{6.40}$$

where $\varepsilon \ll 1$ is the aspect ratio; the rescaled equations (write $v = \varepsilon \tilde{v}$, etc., substitute in, drop overtildes) are then (with $\mathbf{u} = (u, v)$)

$$u_x + v_y = 0, \tag{6.41a}$$

$$\varepsilon^2 Re(du/dt) = -p_x + u_{yy} + \varepsilon^2 u_{xx}, \tag{6.41b}$$

$$\varepsilon^4 Re(dv/dt) = -p_y + \varepsilon^2[v_{yy} + \varepsilon^2 v_{xx}], \tag{6.41c}$$

whence we deduce $p \approx p(x)$; and if $\varepsilon^2 Re \ll 1$, then

$$u_{yy} \approx p'. \tag{6.42}$$

Integration of the equation leads to *Reynolds' equation*. For example, in a journal bearing, we would specify

$$\begin{aligned} u = v = 0 \quad &\text{on } y = 0, \\ u = 1, \qquad v = h' \quad &\text{on } y = h \end{aligned} \tag{6.43}$$

(the last being the no-flow-through condition). Integration of the continuity equation (6.41a) gives

$$\int_0^h u\,dy = Q, \tag{6.44}$$

where Q is constant, while three integrations of Eq. (6.42) give Reynolds' equation

$$Q = \frac{1}{2}h - \frac{1}{12}p'h^3, \tag{6.45}$$

which determines p', given Q and h.

Droplet dynamics

Consider a thin, two-dimensional droplet $y = h(x, t)$ whose shape is controlled by surface tension. From Eqs. (6.24) and (6.25) with $p_a = 0$, we then have

$$p = \frac{-\lambda h_{xx}}{\left[1 + \varepsilon^2 h_x^2\right]^{3/2}}, \tag{6.46}$$

where

$$\lambda = \frac{\gamma \varepsilon^3}{\mu U}, \tag{6.47}$$

and $d = \varepsilon l$ is the depth scale. Boundary conditions for (u, v) are now

$$\begin{aligned} u = v = 0 \quad & \text{at } y = 0, \\ u_y = 0 \quad & \text{at } y = h \end{aligned} \tag{6.48}$$

(the latter being the leading order zero shear stress condition), and the kinematic boundary condition

$$v = h_t + u h_x \text{ on } y = h. \tag{6.49}$$

Thus

$$\begin{aligned} u_y &= -p_x(h - y), \\ u &= -p_x\left(hy - \frac{1}{2}y^2\right), \\ \int_0^h u\,dy &= -\frac{1}{3}p_x h^3; \end{aligned} \tag{6.50}$$

mass conservation integrates to yield

$$h_t + \frac{\partial}{\partial x}\int_0^h u\,dy = 0, \tag{6.51}$$

whence, putting $\varepsilon = 0$ in Eq. (6.46) so that $p \approx -\lambda h_{xx}$, we have

$$h_t + \frac{1}{3}\lambda\frac{\partial}{\partial x}[h^3 h_{xxx}] = 0, \tag{6.52}$$

a fourth-order equation. Small perturbations to a constant solution h = constant of the form $e^{\sigma t + ikx}$ satisfy $\sigma \propto -k^4$, indicating stability. The equation is degenerate, having a coefficient h^3 of the highest derivative that goes to zero when $h = 0$ at the edges. Consequently, we cannot prescribe the usual kinds of boundary conditions.

We prescribe

$$\begin{aligned} \int_{x_-}^{x_+} h\,dx &= 1 \text{ (prescribed mass)}, \\ h &= 0 \quad \text{at } x = x_+, \\ h &= 0 \quad \text{at } x = x_-, \\ h^3 h_{xxx} &\to 0 \quad \text{at } x = x_\pm \text{ (zero flux at edges)}, \end{aligned} \tag{6.53}$$

and that is it. With a fourth-order equation and two unknown boundaries, we expect six conditions, but there are only five, and in fact only four are independent. Nevertheless, these are sufficient to solve the problem, or so we hope.

Ice sheets

Ice sheets flow (slowly) as a viscous medium (an example of solids deforming like fluids) under their own weight, owing to gravity. Gravity is a body force, ρg per unit volume, so that Eq. (6.41) is modified at leading order to

$$0 = -p_y - 1, \tag{6.54}$$

if the depth scale is chosen appropriately: Specifically, we choose

$$d = \left(\frac{\mu U l}{\rho g}\right)^{1/3}; \tag{6.55}$$

for ice sheets, we have $\mu \sim 10$ bar year $\sim 10^6$ N m^{-2} y, $U \sim 100$ m y^{-1}, $l \sim 10^3$ km, $\rho \sim 10^3$ kg m^{-3}, $g \sim 10$ m s^{-2}, whence $d \sim 2$ km. Then $d/l = \varepsilon \sim 10^{-3} \ll 1$, as supposed (and observed).

From Eq. (6.54)

$$p = h - y, \tag{6.56}$$

and we have the same boundary conditions as for the droplet, except that the ice surface is supplied by an accumulation due to snowfall. Suppose this is constant; then we can take it (dimensionlessly) as equal to one, so that the appropriate kinematic condition is

$$\frac{d}{dt}(h - y) = 1. \tag{6.57}$$

There follows

$$\frac{\partial h}{\partial t} + \frac{\partial}{\partial x}\int_0^h u\,dy = 1, \tag{6.58}$$

and finally

$$h_t = \frac{\partial}{\partial x}\left[\frac{1}{3}h^3 h_x\right] + 1, \tag{6.59}$$

a degenerate nonlinear diffusion equation. We prescribe

$$h = 0 \quad \text{at } x = x_{\pm}, \tag{6.60}$$

which might represent continental margins. In a steady state, we choose the origin so that $x_+ = -x_-$: h is then symmetric; thus

$$h^3 h_x = -3x, \tag{6.61}$$

and so

$$h = \left[6\left(x_+^2 - x^2\right)\right]^{1/4}, \tag{6.62}$$

and this steady profile is stable to perturbations.

6.2 Notes and references

There are many books on fluid mechanics. Batchelor's (1967) book is a classic, though very wordy at first sight. Paterson (1983) is a nice introduction, as is Acheson (1990) and Tritton (1988). Pedlosky (1979) is an excellent compendium of work on rotating, shallow flows applied to ocean and atmospheric motions. Turner (1973) deals with buoyancy effects associated with stratified flows and convection. Other classics are Landau and Lifschitz (1959), with the perspective of physicists, Yih (1979), an exhaustive introduction, and Schlichting (1979), a voluminous engineering approach to boundary layer theory. Stability theory is treated by Drazin and Reid (1981), much of it concerning parallel shear flows. An older book is that of Chandrasekhar (1961), more concerned with convection, and the presence of rotation and magnetic fields, while Townsend (1956) describes turbulence. Ghil and Childress (1987) is a nice selection of some geophysical fluid dynamics topics. Ockendon and Ockendon (1995) is a short gloss with some novel applications and insights. Lastly, one should mention Van Dyke's *An Album of Fluid Motion* (1982), a stunning sequence of photographs of fluids in motion.

Tensors One does not need to know much about tensors to study fluids, and most comprehensive texts include a discussion. Jeffreys and Jeffreys (1953) is a useful alternative source. Second-order tensors are labeled by two subscripts, as for instance σ_{ij}, as opposed to the single component of a vector (which is a first-order tensor) and they obey the transformation rule

$$\sigma_{ij} = l_{ii'} l_{jj'} \sigma_{i'j'}, \tag{6.63}$$

where $l_{ij'}$ are the direction cosines $\mathbf{e}_i.\mathbf{e}_{j'}$ between two sets of orthonormal coordinates $\{\mathbf{e}_i\}$ and $\{\mathbf{e}_{i'}\}$. In this expression the *summation convention* is used, whereby summation over repeated suffixes is understood. It is necessary that Eq. (6.63) be satisfied in order that σ_{ij} be a coordinate independent quantity. The analogous relation for a vector $\mathbf{r}$ is that $r_i = l_{ij'} r_{j'}$, because we must have $\mathbf{r} = r_i \mathbf{e}_i = l_{ij'} r_{j'} \mathbf{e}_i = (l_{j'i} \mathbf{e}_i) r_{j'} = r_{j'} \mathbf{e}_{j'}$; that is, $\mathbf{r}$ is independent of the coordinate system. One can in fact deduce in a plausible manner that Eq. (6.11) is the most general isotropic linear tensor constitutive relation between τ_{ij} and $\dot{\varepsilon}_{ij}$.

Turbulence Turbulence is *the* unsolved problem of fluid mechanics. Considering the case of plane Poiseuille flow, it is found that the instability at $Re = 5772$ is associated with a *subcritical bifurcation* (see Drazin 1992), and unstable two-dimensional traveling wave patterns exist down to $Re \approx 2900$. Below this value, ghostly traveling wave perturbations exist but are viscously stable; that is, they decay to zero on a long timescale $t \sim Re$. They are, however, *inviscidly* unstable to three-dimensional disturbances on a fast timescale $t \sim O(1)$ (Orszag and Kells, 1980). It seems that this phenomenology can explain the onset of transition at $Re \sim 1000$, but it sheds no light on what the transition is *to*. Presumably, the equations admit chaotic solutions, but there is no real understanding as to what these might be.

Droplet dynamics and the no-slip condition The study of the contact angle between a droplet of fluid and a solid substrate has led to interesting theoretical problems

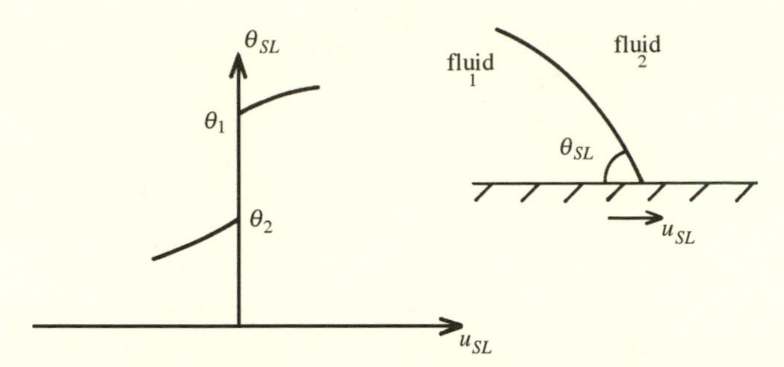

Fig. 6.3. Contact angle hysteresis as the contact line velocity u_{SL} varies

in recent years. Observationally, fluids appear to have a finite contact angle at a contact line between solid, fluid one, and fluid two. This contact angle appears to exhibit hysteresis, that is (see Fig. 6.3) $\theta_{SL} = \theta_1$ if $u_{SL} > 0$, $\theta_{SL} = \theta_2 < \theta_1$ if $u_{SL} < 0$, and if the contact line velocity $u_{SL} = 0$, then a range of values $\theta_2 < \theta < \theta_1$ is possible. However, there is a problem. Dussan V. and Davis (1974) showed that there must be a velocity discontinuity in the displaced fluid and hence an infinite force. To overcome this, several authors have proposed a local slip condition at the wall, which, however, does not seem to affect the dynamics to any great extent. Nevertheless, the issue of the correct prescription of contact line conditions remains a puzzle.

The no-slip condition itself, which one tends to apply automatically as a boundary condition for viscous fluids at a wall, was in fact an issue of contention for a long time. Goldstein (1938) gives a lucid account of the history of the controversy on this topic. Apart from the no-slip condition, Navier advanced the slip condition

$$\beta u = \mu \frac{\partial u}{\partial n}, \tag{6.64}$$

where μ/β has dimensions of length. Maxwell suggested (in work on gases) that μ/β would be of the order of the molecular mean free path. If this were even crudely accurate also for liquids, then effectively, β would be very large, so that the no slip condition should normally apply – except in circumstances, such as at a contact line, where this leads to a singularity.

Exercises

1. An interface between two fluids is given by $z = \eta(x, y)$ and is bounded by a prescribed closed curve C' (think of a soap film on a loop of wire). If C is the projection of C' on to the (x, y) plane and S is the interior of C, show that the surface energy of the film is

$$E = \gamma \iint_S [1 + |\nabla\eta|^2]^{1/2}\, dx\, dy,$$

where γ is the surface energy per unit area. By using the calculus of variations (e.g. Courant and Hilbert, 1953), show that E is minimized (subject to the film being attached to C') if

$$\nabla.\left[\frac{\nabla\eta}{[1+|\nabla\eta|^2]^{1/2}}\right]=0, \quad \eta=\eta|_{C'} \text{ for } (x,y)\in C,$$

and show that (by comparison with Eq. (6.25)) this implies that the mean curvature is everywhere zero. Do you expect a solution to exist in all cases?

In reality, fluid interfaces are curved (for example, the meniscus in a capillary tube). This can be associated with the existence of surface energy associated with solid–fluid interfaces, which causes a nonzero contact angle θ to exist (see Fig. 6.3). Show that if the fluid boundary (the contact line) C lies in the (x, y) plane and there is a contact angle θ at C, then

$$\frac{-\mathbf{n}.\nabla\eta}{[1+|\nabla\eta|^2]^{1/2}}=\sin\theta \quad \text{at } C,$$

where $\mathbf{n}$ is the outward normal (in the (x, y) plane) to C.

A drop of liquid rests on a plane, bounded by a contact line. Show that the surface energy associated with the drop is

$$E=\gamma\iint_S[1+|\nabla\eta|^2]^{1/2}\,dxdy+(\gamma_{SL}-\gamma_{SV})\int_S dxdy$$

(plus a constant), where γ_{SL} and γ_{SV} are the surface energies of the solid–liquid and solid–vapor interfaces, respectively. Show also that if η is not constrained to be zero at the contact line C, then the first variation of a functional $\int_S I\,dS$ is

$$\delta\int_S I\,dS=\int_S \delta I\,dS+\int_C I\delta n\,ds,$$

where $\delta n=-\delta\eta/(\partial\eta/\partial n)$. Use the method of Lagrange multipliers to show that E is minimized subject to the constraint of constant fluid volume, $V = \iint_S \eta\,dxdy$, provided twice the mean curvature,

$$2\kappa=\nabla.\left[\frac{\nabla\eta}{[1+|\nabla\eta|^2]^{1/2}}\right],$$

is constant, *and*

$$\gamma\cos\theta=\gamma_{SV}-\gamma_{SL},$$

where θ is the contact angle. Show that the latter relation is interpretable as a horizontal force balance if $\gamma, \gamma_{SV}, \gamma_{SL}$ are interpreted as surface tensions in the fluid interfaces.

2. *Shape of a meniscus*

Hydrostatic pressure in a liquid is given by $p = p_a - \rho g z$, where p_a is atmospheric pressure and $z = 0$ is the liquid surface far from a wall. If the liquid surface is given by $z = \eta(x, y)$, show that η satisfies

$$\nabla.\left[\frac{\nabla\eta}{[1+|\nabla\eta|^2]^{1/2}}\right]=\eta,$$

where lengths are made dimensionless with $d = (\gamma/\rho g)^{1/2}$. If $\mathbf{n}'$ is a unit normal to a contact line at a wall, parallel to the wall and pointing away from the fluid, and $\hat{\mathbf{n}}$ is a unit normal to the contact line that is normal to the liquid air interface and directed into the air, show that $\hat{\mathbf{n}}.\mathbf{n}' = \sin\theta$, where θ is the contact angle. Assuming the contact line is horizontal, show that at a vertical wall

$$(1 + |\nabla\eta|^2)^{1/2} = \operatorname{cosec}\theta,$$

and thus $\partial\eta/\partial n = \cot\theta$, where $\mathbf{n}$ is a unit normal to the wall, pointing away from the liquid. Solve this problem for the case of a plane wall at $x = 0$ with $\eta \to 0$ as $x \to \infty$, and hence deduce that the capillary rise of the fluid is

$$2(\gamma/\rho g)^{1/2} \sin\left[\frac{\pi}{4} - \frac{\theta}{2}\right].$$

3. Show that for a two-dimensional incompressible flow, there is a stream function ψ such that $\mathbf{u} = (\psi_y, -\psi_x, 0)$. Show that the vorticity is $\boldsymbol{\omega} = (0, 0, -\nabla^2\psi)$. Deduce that for an inviscid fluid, the vorticity following the fluid is conserved. A uniform stream ($\mathbf{u}$ = constant at infinity) flows past a bluff body. Show that the vorticity is zero everywhere (the flow is irrotational) and deduce that there is a velocity potential ϕ such that $\mathbf{u} = \nabla\phi$. For a two-dimensional flow, show that $w = \phi + i\psi$ is an analytic function of $z = x + iy$ (this is the complex velocity potential), and describe the flows associated with $w = (m/2\pi)\log z$ (a source of strength m) and $w = (-i\kappa/2\pi)\log z$ (a vortex of circulation κ). Show that a vortex induces no motion at its center. Show that the complex velocity potential associated with vortices of circulation κ at the points $z_1, z_2, \ldots, z_n$ is

$$w = -\frac{i\kappa}{2\pi}\ln\left[\prod_1^n (z - z_i)\right],$$

and deduce that the motions of the vortices are determined by the equations

$$\dot{\bar{z}}_i = -\frac{i\kappa}{2\pi}\sum_{j\neq i}\frac{1}{(z_i - z_j)}.$$

By using $z_i, \bar{z}_i$ as generalized coordinates, show that the motion is Hamiltonian with

$$H = \frac{i\kappa}{\pi}\sum_{i=1}^{n}\sum_{j<i}\ln|z_i - z_j|.$$

4. Show that if the vorticity of an unbounded, incompressible inviscid fluid with $\mathbf{u} \to \mathbf{0}$ at ∞ is $\boldsymbol{\omega}$, then

$$\mathbf{u} = -\frac{1}{4\pi}\int \frac{(\mathbf{r} - \mathbf{r}') \times \boldsymbol{\omega}(\mathbf{r}')}{|\mathbf{r} - \mathbf{r}'|^3}\, dV(\mathbf{r}'),$$

where the integral is over $\mathbf{R}^3$. Now suppose that vorticity is concentrated in a thin vortex tube γ of strength Γ, so that $\boldsymbol{\omega}\, dS = \Gamma\mathbf{t}$, where $\mathbf{t} = d\mathbf{r}/ds$ is the unit tangent vector to the vortex tube, and dS is its cross-sectional area. Then the velocity field is (because $dV = dS\,ds$, and $\mathbf{t}\,ds = d\mathbf{r}'$)

$$\mathbf{u} = -\frac{\Gamma}{4\pi}\int_\gamma \frac{(\mathbf{r}-\mathbf{r}')\times d\mathbf{r}'}{|\mathbf{r}-\mathbf{r}'|^3}.$$

To find the induced motion of the vortex itself by its own curvature, first subtract the local rotational velocity by finding the value of $\mathbf{u}$ near γ, as follows. Select a point O on γ, and let $\mathbf{t}$, $\mathbf{n}$, $\mathbf{b}$ be the Serret–Frenet basis vectors there (thus $d\mathbf{t}/ds = \kappa\mathbf{n}$, $d\mathbf{n}/ds = \tau\mathbf{b} - \kappa\mathbf{t}$, and $d\mathbf{b}/ds = -\tau\mathbf{n}$, where s is arc length on γ, κ is the curvature, and τ is the torsion). Then with $s = 0$ at O, we have

$$\mathbf{r}' \approx s\mathbf{t} + \frac{1}{2}\kappa s^2\mathbf{n} + \ldots;$$

thus $d\mathbf{r}' \approx (\mathbf{t} + \kappa s\mathbf{n})ds$, and if $\mathbf{r} = (0, x_2, x_3)$ ($x_1 = 0$ by choice of O), then

$$\frac{(\mathbf{r}-\mathbf{r}')\times d\mathbf{r}'}{|\mathbf{r}-\mathbf{r}'|^3} \approx \frac{\left[-x_3\kappa s\mathbf{t} + x_3\mathbf{n} - \left(x_2 + \frac{1}{2}\kappa s^2\right)\mathbf{b}\right]ds}{\left[x_2^2 + x_3^2 + s^2(1-\kappa x_2) + \frac{1}{4}\kappa^2 s^4\right]^{3/2}}.$$

Show that the velocity field as $\rho = (x_2^2 + x_3^2)^{1/2} \to 0$ can thus be approximated by (if $x_2 + ix_3 = \rho e^{i\theta}$)

$$\mathbf{u} \approx \frac{\Gamma}{4\pi}\int_{|\mathbf{r}|>L} \frac{\mathbf{r}'\times d\mathbf{r}'}{|\mathbf{r}'|^3}$$
$$+ \int_{-L/\rho}^{L/\rho} \frac{\Gamma}{4\pi(1+\xi^2)^{3/2}}\left[\frac{1}{\rho}(\mathbf{b}\cos\theta - \mathbf{n}\sin\theta) + \frac{1}{2}\kappa\xi^2\mathbf{b}\right]d\xi,$$

for any value of L. Deduce that for fixed (small) L,

$$\mathbf{u} \approx \frac{\Gamma}{2\pi\rho}(\mathbf{b}\cos\theta - \mathbf{n}\sin\theta) + \frac{\Gamma\kappa}{4\pi}\mathbf{b}\log(L/\rho) + \ldots.$$

The first term corresponds to the local circulation of a line vortex. Deduce that the second term determines the self-induced vortex motion in the form

$$\frac{\partial\mathbf{r}}{\partial t} \approx C\kappa\mathbf{b},$$

where $C = \Gamma\log(1/\rho)/4\pi$ and we take ρ to be the local radius of the vortex tube. Show that this partial differential equation can be written in the form

$$\mathbf{r}_t = C\mathbf{r}_s \times \mathbf{r}_{ss}.$$

Hasimoto (1972) has shown that if $\psi = \kappa\exp[i\int_0^s \tau\,ds]$, then ψ satisfies the nonlinear Schrödinger equation, and thus there are soliton (solitary wave) solutions of the above equation. Corresponding kinks have been observed on tornadoes (see Aref and Flinchem, 1984).

5. If a coordinate frame rotates with angular velocity $\boldsymbol{\Omega}$, then time derivatives of vectors $\mathbf{a}$ are related by

$$\left(\frac{d\mathbf{a}}{dt}\right)_F = \left(\frac{d\mathbf{a}}{dt}\right)_R + \boldsymbol{\Omega} \times \mathbf{a},$$

where F, R mean relative to the fixed and rotating frames, respectively. Show that the Euler equations can be written relative to a rotating frame in the form

$$\begin{aligned} \boldsymbol{\nabla}.\mathbf{u} &= 0, \\ \frac{d\mathbf{u}}{dt} + 2\boldsymbol{\Omega} \times \mathbf{u} &= -\frac{1}{\rho}\boldsymbol{\nabla} P, \end{aligned}$$

where $P = p - \frac{1}{2}\rho|\boldsymbol{\Omega} \times \mathbf{r}|^2$ is the reduced pressure and $\boldsymbol{\Omega}$ is constant. Estimate the size of $|\boldsymbol{\Omega} \times \mathbf{r}|^2$ from the estimates $\boldsymbol{\Omega} \sim 1\ \text{day}^{-1}$, $r \sim 7000$ km, $\rho \sim 1\,\text{kg}\,\text{m}^{-3}$, and $p \sim 1$ bar $(= 10^5$ Pa), and deduce that $\frac{1}{2}\rho|\boldsymbol{\Omega} \times \mathbf{r}|^2/p \ll 1$. The *Rossby number* is defined as $\varepsilon = U/2\Omega L$, where U is a typical velocity scale and L is a relevant length scale. Take values $L = 1000$ km, $U \sim 20$ m s^{-1}, and $\Omega \sim 10^{-4}$ s^{-1} to show that the Rossby number for such *mesoscale* motions is relatively small. Deduce that in this case (called geostrophic flow), isobars (lines of constant pressure) are also streamlines (lines parallel to the velocity field): fluid flows *perpendicularly* to pressure gradients.

6. Take local Cartesian coordinates (x, y, z) at latitude θ (> 0) on the Earth, with z vertically upward, y pointing north, and x pointing eastward. Show that $\boldsymbol{\Omega} = (0, \Omega\cos\theta, \Omega\sin\theta)$, and deduce that the Euler equations for an incompressible fluid flow with velocity $\mathbf{u} = (u, v, w)$ are

$$\begin{aligned} \boldsymbol{\nabla}.\mathbf{u} &= 0, \\ \frac{du}{dt} - fv + 2w\Omega\cos\theta &= -\frac{1}{\rho}p_x, \\ \frac{dv}{dt} + fu &= -\frac{1}{\rho}p_y, \\ \frac{dw}{dt} - 2u\Omega\cos\theta &= -\frac{1}{\rho}p_z - g, \end{aligned}$$

where $f = 2\Omega\sin\theta$. The Earth's atmosphere is shallow, so that it is appropriate to scale $x, y \sim L$, $z \sim H \ll L$ (e.g., $L \sim 10^3$ km, $H \sim 10$ km). Show that if U is a suitable horizontal velocity scale, then δU, $\delta = H/L$ is a suitable vertical velocity scale. Deduce that an approximate dimensional model is given by the following rotating shallow-water equations:

$$\begin{aligned} \boldsymbol{\nabla}.\mathbf{u} &= 0, \\ \frac{du}{dt} - fv &= -\frac{1}{\rho}p_x, \\ \frac{dv}{dt} + fu &= -\frac{1}{\rho}p_y, \\ p_z &= -\rho g. \end{aligned}$$

If the atmosphere is considered to be an incompressible layer of fluid of height h with $p = 0$ at $z = h$, show that $p = \rho g(h - z)$; and by integrating from $z = 0$ (where $w = 0$) to $z = h(x, y, t)$ (on which $w = h_t + uh_x + vh_y$) show that, if $\partial u/\partial z \approx \partial v/\partial z \approx 0$, then

$$u_t + uu_x + vu_y - fv = -gh_x,$$
$$v_t + uv_x + vv_y + fu = -gh_y,$$
$$h_t + (uh)_x + (vh)_y = 0.$$

On the Earth, the polar air is colder, hence denser, and we can model this by reference to a basic state $h = h_0[1 - \alpha y/L]$. Show that such a basic state is consistent with the above model, together with a zonal wind $(u, v) = (U, 0)$, where $U = \alpha g h_0/fL$ (assume f is constant). Nondimensionalize these equations, and show that the resulting set can be written in terms of two parameters, α and $\varepsilon = u/fL$, the Rossby number. By linearizing about the basic state, show that disturbances proportional to $\exp[i(kx + ly + \omega t)]$ exist, find the dispersion relation between ω and k, and show that $\omega(k)$ has three possible values. If α is small, show that $l = l_R + il_I$ can be chosen complex with $l_I = -\alpha/2$ so that ω is real, and that in this case, two roots are $\omega = O(1/\varepsilon)$ and the other is given by

$$\omega + k \approx \frac{k}{1 + \frac{\varepsilon}{\alpha}\left(k^2 + l_R^2\right)}.$$

The first pair of frequencies correspond to *Poincaré waves*, while the third frequency is that of *Rossby waves*. Relative to the Earth, their velocity is $-\omega/k$, and thus these waves travel east (in the northern hemisphere), but drift westward relative to the zonal flow. These are the isobaric waves that wander across the weather charts each night on the television.

7. *The Blasius boundary layer*

Fluid flows past a flat plate $y = 0, x > 0$ at high Reynolds number, such that the two-dimensional velocity (u, v) satisfies $u = v = 0$ on $y = 0, x > 0$, $u \to 1, v \to 0, p \to 0$ as $y \to \infty$. Show that the outer (inviscid) flow away from $y = 0$ satisfying Eq. (6.34) with $1/Re = 0$ is $(u, v) = (1, 0)$, $p = 0$. Show that a boundary layer exists near $y = 0$, where $v \sim \delta$, $y \sim \delta$, where $\delta = Re^{-1/2}$, and show that the corresponding equations for u and (rescaled) V are

$$u_x + V_Y = 0,$$
$$uu_x + Vu_Y = u_{YY}.$$

By introducing a stream function $u = \psi_Y$, $V = -\psi_x$, deduce that

$$\psi_Y \psi_{xY} - \psi_x \psi_{YY} = \psi_{YYY}.$$

Show that a similarity solution of the form $\psi = (2x)^{1/2} f(\eta)$, $\eta = Y/(2x)^{1/2}$ exists, where f satisfies

$$f''' + ff'' = 0,$$
$$f(0) = f'(0) = 0, \quad f'(\infty) = 1.$$

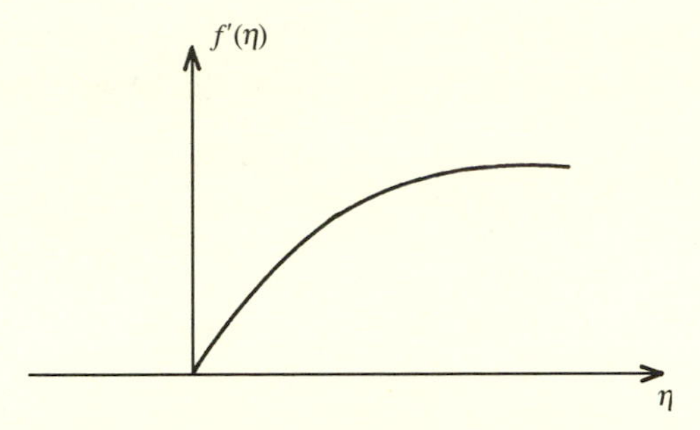

Fig. 6.4. Blasius boundary layer profile

The function $f(\eta)$ must be found numerically; and, in common with many similarity solutions, there is a trick to doing this by rescaling. Solve

$$F'''(\xi) + F(\xi)F''(\xi) = 0,$$
$$F(0) = F'(0) = 0, \qquad F''(0) = 1$$

(this can be done easily as an initial value problem, providing (as is the case) F' cannot blow up at ∞). Put $f(\eta) = bF(a\eta)$; thus

$$f''' + ff'' = ba^3F''' + b^2a^2FF'' = 0$$

if $b = a$, and then

$$f(0) = 0, \qquad f'(0) = 0, \qquad f'(\infty) = a^2F'(\infty).$$

Thus we compute F and then define $f(\eta)$ as above with $a = \{F'(\infty)\}^{-1/2}$. As computed, f' is as shown in Fig. 6.4, resembling an error function.

8. *Bulk viscosity*
The pressure p in a fluid is *defined* as $-\sigma_{ii}/3$, where σ_{ij} is the stress tensor. It is an experimental observation that, for a fluid in equilibrium, the pressure depends on density and temperature, and we may call this observed equilibrium pressure p_e. There is, however, no guarantee that $p = p_e$ in general. The simplest constitutive relation (Batchelor, 1967) relates p to p_e by a bulk viscosity κ, specifically

$$p - p_e = -\kappa \operatorname{div} \mathbf{u}.$$

Consider, for example, the motion of an incompressible fluid of viscosity μ containing bubbles of a compressible gas. If we consider one such bubble of radius a, with far-field pressure p_l (when $r \gg a$, the origin being at the center of the bubble), and a (uniform) gas pressure p_g inside the bubble, show that for a purely radial velocity $u(r)$,

$$\frac{\partial}{\partial r}(r^2u) = 0,$$
$$\rho[u_t + uu_r] = -p_r + \mu\left\{u_{rr} + \frac{2}{r}u_r - \frac{2}{r^2}u\right\}.$$

(The form of the Navier–Stokes equations in spherical polar coordinates can be found in Batchelor (1967), for example.) Deduce that

$$u = a^2\dot{a}/r^2,$$
$$p_l - p = \rho\left[\frac{1}{2}u^2 - (\dot{a^2\dot{a}})/r\right],$$

and that the normal stress $\sigma_{rr} = -p + 2\mu\partial u/\partial r$ is given by

$$p_l + \sigma_{rr} = \rho\left[\frac{1}{2}u^2 - (\dot{a^2\dot{a}})/r\right] - 4\mu a^2\dot{a}/r^3.$$

At the bubble wall, we require the normal stress $\sigma_{rr} = -p_g$. Deduce that

$$p_l - p_g = -\rho\left[\frac{3}{2}\dot{a}^2 + a\ddot{a}\right] - 4\mu\dot{a}/a.$$

If a fluid contains n bubbles of small diameter a per unit volume (which move with the fluid), so that the *void fraction* $\alpha = 4\pi na^3/3$, show that div $\mathbf{u} = d\alpha/dt$ (following the fluid), and hence deduce that

$$\text{div}\,\mathbf{u} = 3\alpha\dot{a}/a.$$

If L, U are macroscopic length and velocity scales for the fluid, and $a \sim \varepsilon L$, $\varepsilon \ll 1$, $t \sim L/U$, $p \sim \rho U^2$, show that a scaled version of the equation for $p_l - p_g$ is

$$p_l - p_g = -\varepsilon^2\left[\frac{3}{2}\dot{a}^2 + n\ddot{a}\right] - \frac{4}{Re}\frac{\dot{a}}{a}.$$

Deduce that if $\varepsilon \ll 1$ (and specifically if $\varepsilon \ll Re^{-1/2}$) and if p_l is identified with the dynamic pressure p and p_g with the equilibrium pressure p_e, then the appropriate bulk viscosity is given by

$$\kappa = \frac{4\mu}{3\alpha}.$$

9. *Bubble collapse*

Consider the collapse of a vacuous bubble of radius a and pressure p_g in an incompressible fluid of viscosity μ and with far-field pressure p_l. From Exercise 8, the radius of the bubble is determined by

$$\Delta p = -\rho\left[\frac{3}{2}\dot{r}^2 + r\ddot{r}\right] - 4\mu\dot{r}/r,$$

where ρ is the density of the fluid. By scaling r and t appropriately, show that a dimensionless form of this equation can be written as

$$-1 = \frac{3}{2}\dot{r}^2 + r\ddot{r} + \varepsilon\dot{r}/r,$$

where $\varepsilon = 4\mu/a\{\rho\Delta p\}^{1/2}$. Deduce that $\varepsilon \ll 1$, if $Re \gg 4$, where $Re = Ua/\nu$ and $U = (\Delta p/\rho)^{1/2}$. Estimate the bubble size for which this inequality is valid if $\Delta p = 1$ bar ($= 10^5$ Pa), $\rho = 10^3$ kg m^{-3}. Show that for such bubbles, collapse occurs at a finite time, given dimensionlessly by

$$t_c = \sqrt{\frac{3}{2}} \int_0^1 \left(\frac{r^3}{1 - r^3} \right)^{1/2} dr.$$

Is this affected by $\varepsilon \neq 0$? What if the bubble is not vacuous?

10. An ice sheet of thickness h satisfies the equation

$$h_t = \frac{\partial}{\partial x} \left[\frac{1}{3} h^3 h_x \right] + a(x),$$

where at the margins, the accumulation rate is negative (see also Chapter 18). Show that, if the margin $x_m(t)$ is nonstationary, then (if $h > 0$ for $x < x_m$)

$$h \sim c(x_m - x),$$
$$\dot{x}_m \sim -|a(x_m)|/c,$$

if $\dot{x}_m < 0$ (retreat), whereas

$$h \sim c(x_m - x)^{1/3},$$
$$\dot{x}_m \sim c^3/9,$$

if $\dot{x}_m > 0$. Thus the slope is singular in advance but finite in retreat. This distinction causes the degenerate diffusion equation above to have waiting-time behavior, because following a retreat, the margin slope must rebuild itself before another advance is possible. (See Lacey, Ockendon, and Tayler, (1982) and Lacey (1983) for further discussion.)

11. The temperature of a rapidly circulating fluid in a closed box is described by the equation

$$\mathbf{u}.\boldsymbol{\nabla} T = \varepsilon \nabla^2 T,$$

where the two-dimensional incompressible flow field is given by

$$\mathbf{u} = (\psi_y, -\psi_x),$$

where ψ is the stream function (and the streamlines are closed). If $\varepsilon \ll 1$, show that $T =$ constant ($= T_0$, say) to all orders in an asymptotic expansion in powers of ε (this is an example of the Prandtl–Batchelor theorem). If T is prescribed on the boundary, show that one can expect boundary layers of width $\varepsilon^{1/2}$. Can you think of a reason why the temperature should decay exponentially away from the boundary? Show that in this case, the exponentially small connection to T_0 in the interior is given by

$$T = T_0 + \exp[-\phi/\varepsilon^{1/2}],$$

where ϕ satisfies

$$\mathbf{u}.\boldsymbol{\nabla}\phi \approx |\boldsymbol{\nabla}\phi|^2.$$

Discuss the solution of this equation by the method of characteristics.

7

Solid mechanics

7.1 Stress and strain

An elastic body is one that, when deformed, undergoes a finite recoverable *strain*. If a material point at $\mathbf{x}$ is deformed to a position $\mathbf{x}'$, then the displacement is defined to be

$$\mathbf{u} = \mathbf{x}' - \mathbf{x}. \tag{7.1}$$

It is clearly reasonable to expect the internal forces of a deformed elastic body to depend (only) on the differential displacement, whereby the line element $d\mathbf{x}$ is deformed to $d\mathbf{x}'$. From first principles, and using the summation convention, we find that the distances $ds = |d\mathbf{x}|$ and $ds' = |d\mathbf{x}'|$ are related by

$$ds'^2 = ds^2 + 2\varepsilon_{ij}\, dx_i\, dx_j, \tag{7.2}$$

where

$$\varepsilon_{ij} = \frac{1}{2}\left(\frac{\partial u_i}{\partial x_j} + \frac{\partial u_j}{\partial x_i} + \frac{\partial u_l}{\partial x_i}\frac{\partial u_l}{\partial x_j}\right) \tag{7.3}$$

is the (Lagrangian) *strain tensor*. For small displacements, we have the approximation

$$\varepsilon_{ij} = \frac{1}{2}\left(\frac{\partial u_i}{\partial x_j} + \frac{\partial u_j}{\partial x_i}\right), \tag{7.4}$$

which we shall adopt henceforth.

In a linearly elastic body, the strains are related to the *stress tensor* σ_{ij}, where σ_{ij} is the jth component of the force per unit area exerted on a surface element with a normal in the i direction. (That the forces can be so represented follows from Newton's third law, which asserts that if the internal force acting on a material element is $\mathbf{F}$, then

$$\int_V F_i\, dV = 0 \tag{7.5}$$

for all volumes, whence there exists a tensor σ_{ij} satisfying $F_i = \partial\sigma_{ij}/\partial x_j$, which is the stress tensor (see Landau and Lifschitz (1986) for further discussion of this).)

7.2 Linear elasticity

In large elastic deformations, one makes a distinction between the Cauchy stress tensor, which refers to the current configuration of the body, and the first and second Piola–Kirchhoff stress tensors, which refer to the original configuration. In proposing a constitutive relationship between stress and strain that is coordinate frame indifferent, it is necessary to relate the second Piola–Kirchhoff stress tensor to the Lagrangian strain tensor. However, when the strains are small (as we suppose here), the necessity for distinguishing between these descriptions disappears.

For an isotropic material, we relate the stress σ_{ij} and the strain ε_{ij} by the linear constitutive law

$$\sigma_{ij} = 2\mu\varepsilon_{ij} + \lambda\varepsilon_{kk}\delta_{ij}, \tag{7.6}$$

where λ and μ are known as Lamé coefficients. Contraction of Eq. (7.6) gives

$$\sigma_{ll} = (2\mu + 3\lambda)\varepsilon_{ll}. \tag{7.7}$$

It is not difficult to show that div $\mathbf{u} = \varepsilon_{ll}$ is the relative change of volume $\delta V/V$ of a material element, whereas $-\sigma_{ll}/3 = p$ is the isotropic pressure; thus

$$p = -K \operatorname{div} \mathbf{u}, \tag{7.8}$$

where

$$K = \lambda + \frac{2}{3}\mu \tag{7.9}$$

is called the bulk modulus and relates volume changes to pressure in isotropic compression (or expansion).

In terms of K and μ, the constitutive relation can be written as

$$\sigma_{ij} = 2\mu\left(\varepsilon_{ij} - \frac{1}{3}\delta_{ij}\varepsilon_{kk}\right) + K\varepsilon_{kk}\delta_{ij}; \tag{7.10}$$

thus K is associated with bulk compression only, whereas μ is called the shear modulus and is associated with *deviatoric* strains (involving no volume change). We can invert this relation to obtain

$$\begin{aligned} \varepsilon_{11} &= \frac{1}{E}[\sigma_{11} - \nu(\sigma_{22} + \sigma_{33})], \\ \varepsilon_{12} &= \frac{(1+\nu)}{E}\sigma_{12}, \end{aligned} \tag{7.11}$$

with corresponding expressions for ε_{22}, ε_{33}, and ε_{13} and ε_{23}. Here E is *Young's modulus* and ν is *Poisson's ratio*, defined by

$$\begin{aligned} E &= \frac{\mu(3\lambda + 2\mu)}{\lambda + \mu} = \frac{9K\mu}{3K + \mu}, \\ \nu &= \frac{\lambda}{2(\lambda + \mu)} = \frac{3K - 2\mu}{2(3K + \mu)}. \end{aligned} \tag{7.12}$$

We require $K, \mu > 0$, and usually one also has $\lambda > 0$, in which case $0 < \nu < 1/2$, and the limit $\nu \to 1/2$ corresponds to $K \to \infty$, that is, an incompressible medium. In terms of E and ν, Eq. (7.10) can be written

$$\begin{aligned}\sigma_{11} &= \frac{E}{(1+\nu)(1-2\nu)}[(1-\nu)\varepsilon_{11} + \nu(\varepsilon_{22}+\varepsilon_{33})],\\ \sigma_{12} &= \frac{E}{(1+\nu)}\varepsilon_{12},\end{aligned} \tag{7.13}$$

and other relations for $\sigma_{22}, \sigma_{33}, \sigma_{13}$, and σ_{12} by rotation.

Equation of motion

For small displacements, the acceleration is just $\partial^2\mathbf{u}/\partial t^2$; thus conservation of momentum for a material element leads to the equation

$$\rho\frac{\partial^2 u_i}{\partial t^2} = \rho F_i + \frac{\partial \sigma_{ij}}{\partial x_j}, \tag{7.14}$$

where F_i is a body force, most usually due to gravity. Substitution of the constitutive law (7.10) leads to the vector form

$$\rho\frac{\partial^2 \mathbf{u}}{\partial t^2} = \rho\mathbf{F} + (\lambda + 2\mu)\text{grad div}\,\mathbf{u} - \mu\,\text{curl curl}\,\mathbf{u}, \tag{7.15}$$

or equivalently

$$\rho\frac{\partial^2 \mathbf{u}}{\partial t^2} = \rho\mathbf{F} + \mu\nabla^2\mathbf{u} + (\lambda + \mu)\text{grad div}\,\mathbf{u}, \tag{7.16}$$

or, in terms of E and ν,

$$\rho\frac{\partial^2 \mathbf{u}}{\partial t^2} = \rho\mathbf{F} + \frac{E}{2(1+\nu)}\nabla^2\mathbf{u} + \frac{E}{2(1+\nu)(1-2\nu)}\text{grad div}\,\mathbf{u}. \tag{7.17}$$

In equilibrium, we simply ignore the time derivatives.

Principal axes

Because the stress and strain tensors are symmetric, they can be diagonalized. The corresponding eigenvectors are orthogonal and are called *principal axes*. In terms of these principal axes, the corresponding strains $\varepsilon_1, \varepsilon_2, \varepsilon_3$ and stresses $\sigma_1, \sigma_2, \sigma_3$ satisfy

$$\sigma_i = 2\mu\varepsilon_i + \lambda\Delta, \tag{7.18}$$

where we write $\Delta = \text{div}\,\mathbf{u} = \varepsilon_1 + \varepsilon_2 + \varepsilon_3$ for the dilatation. The inversion is

$$\begin{aligned}\varepsilon_1 &= (\sigma_1 - \nu\sigma_2 - \nu\sigma_3)/E,\\ \varepsilon_2 &= (\sigma_2 - \nu\sigma_3 - \nu\sigma_1)/E,\\ \varepsilon_3 &= (\sigma_3 - \nu\sigma_1 - \nu\sigma_2)/E.\end{aligned} \tag{7.19}$$

Principal axes are useful in calculating simple stress and strain fields, particularly in plane stress or strain conditions, when σ_3 or ε_3, respectively, is zero.

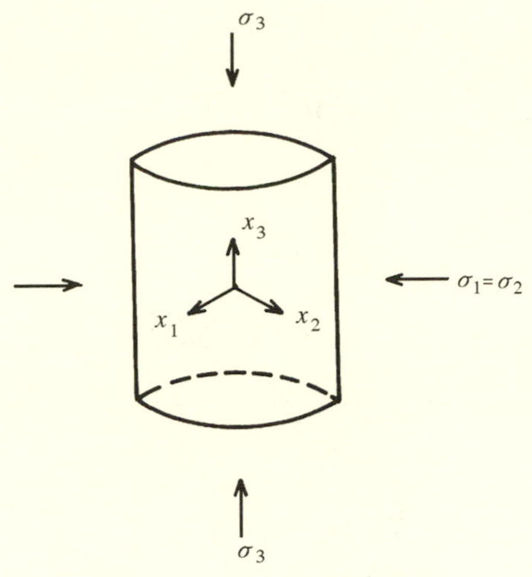

Fig. 7.1. Triaxial compression

As an example, consider the so-called triaxial compression tests in which a cylindrical rock is stressed along its (x_3) axis by a compressive stress σ_3, as shown in Fig. 7.1.

By rotational symmetry, the set of axes x_i is a principal axis set. The confining jacket forces $u_1 = u_2 = 0$, and hence $\varepsilon_1 = \varepsilon_2 = 0$. Thus (because $\sigma_1 = \sigma_2$)

$$\begin{aligned} 0 &= [(1-\nu)\sigma_1 - \nu\sigma_3]/E, \\ \varepsilon_3 &= (\sigma_3 - 2\nu\sigma_1)/E, \end{aligned} \tag{7.20}$$

and the equilibrium equations are

$$\frac{\partial \sigma_1}{\partial x_1} = \frac{\partial \sigma_3}{\partial x_3} = 0, \tag{7.21}$$

in the absence of gravity. Thus σ_1 and σ_3 satisfy

$$\begin{aligned} \sigma_1 &= \frac{\nu\sigma_3}{1-\nu}, \\ \varepsilon_3 &= \frac{(1+\nu)(1-2\nu)E}{(1-\nu)}\sigma_3. \end{aligned} \tag{7.22}$$

Thus, for given σ_3, measurements of σ_1 and ε_3 will give values for E and ν. Note that for a *compressive* stress, $\sigma_3 < 0$.

The beam equation

The Euler–Bernoulli beam equation describes oscillations of an elastic rod. We assume plane strain in (x, z) coordinates; thus $\varepsilon_{22} = \varepsilon_{12} = \varepsilon_{23} = \sigma_{12} = \sigma_{23} = 0$. Writing

$$\sigma_1 = \sigma_{11}, \qquad \sigma_3 = \sigma_{33}, \qquad \tau = \sigma_{13}, \qquad \varepsilon_1 = \varepsilon_{11}, \qquad \varepsilon_3 = \varepsilon_{33}, \tag{7.23}$$

we have

$$\sigma_{22} = \frac{E}{(1+\nu)(1-2\nu)}[\nu(\varepsilon_1 + \varepsilon_3)], \tag{7.24}$$

whence

$$\tau = \frac{E}{(1+\nu)}\varepsilon_{13}, \tag{7.25a}$$

$$\sigma_1 = \frac{E}{(1+\nu)(1-2\nu)}[(1-\nu)\varepsilon_1 + \nu\varepsilon_3], \tag{7.25b}$$

$$\sigma_3 = \frac{E}{(1+\nu)(1-2\nu)}[\nu\varepsilon_1 + (1-\nu)\varepsilon_3], \tag{7.25c}$$

$$\varepsilon_1 = u_x, \tag{7.25d}$$

$$\varepsilon_3 = w_z, \tag{7.25e}$$

$$\varepsilon_{13} = \frac{1}{2}(u_z + w_x), \tag{7.25f}$$

$$\frac{\partial \sigma_1}{\partial x} + \frac{\partial \tau}{\partial z} = \rho u_{tt}, \tag{7.25g}$$

$$\frac{\partial \tau}{\partial x} + \frac{\partial \sigma_3}{\partial z} + f = \rho w_{tt}, \tag{7.25h}$$

where f is a transverse load. Derivations of the beam equation are in fact usually done from first principles, on the basis that the beam is thin. Here we sketch the flavor of a derivation from the above equations, though without detailed justification (for which see Exercise 1). The approximations are ultimately based on the assumption that the beam thickness $h \ll l$, its length.

We suppose ε_3 is small in Eq. (7.25e), so that $w \approx w(x,t)$; we assume ε_{13} is small in (7.25f), whence

$$u \sim -zw_x, \tag{7.26}$$

where we take the z origin locally to be at the *neutral surface* in the beam where $u = 0$. Next, σ_3 is small in Eq. (7.25c), so

$$\varepsilon_3 \approx -\frac{\nu\varepsilon_1}{1-\nu}, \tag{7.27}$$

and thus Eq. (7.25b) is

$$\sigma_1 \approx \frac{E}{(1-\nu^2)}\varepsilon_1 = \frac{E}{(1-\nu^2)}u_x = -\frac{zEw_{xx}}{(1-\nu^2)}. \tag{7.28}$$

We define the *bending moment* M as

$$M = \int_{-h/2}^{h/2} z\sigma_1\, dz = -Dw_{xx}, \tag{7.29}$$

where the *flexural rigidity* is defined as

$$D = \frac{Eh^3}{12(1-\nu^2)}. \tag{7.30}$$

Two equations for M and the shear force

$$T = \int_{-h/2}^{h/2} \tau \, dz \tag{7.31}$$

now follow from integrating Eqs. (7.25g) and (7.25h): with $h \approx$ constant, we get

$$\frac{\partial M}{\partial x} - T \approx 0, \tag{7.32}$$

by neglecting $\int \rho z u_{tt} \, dz$, and

$$\frac{\partial T}{\partial x} + F = \rho h w_{tt}, \tag{7.33}$$

where $F = \int f dz$ is the force per unit length. Thus, finally,

$$\rho h w_{tt} + D w_{xxxx} = F, \tag{7.34}$$

which is the Euler–Bernoulli beam equation. It requires two boundary conditions at each end. For example, for a springboard embedded at $x = 0$, we have

$$w = w_x = 0 \quad \text{at } x = 0, \tag{7.35}$$

and at the free end $x = l$, $M = T = 0$ because no stresses are applied ($\sigma_{11} = \sigma_{13} = 0$); thus

$$w_{xx} = w_{xxx} = 0 \quad \text{at } x = l. \tag{7.36}$$

For a heavy springboard with $F = -\rho g h$, we can then deduce the steady-state configuration

$$w = -\frac{\rho g h x^2}{D}\left[\frac{l^2}{4} - \frac{lx}{6} + \frac{x^2}{24}\right], \tag{7.37}$$

and the downward deflection at $x = l$ is

$$-\Delta w = \frac{\rho g h l^4}{8D}. \tag{7.38}$$

Excitation of the springboard leads to oscillations $\exp(\pm ikx + i\omega t)$, $\exp(\pm kx + i\omega t)$ with frequency ω given by the *dispersion relation*

$$\omega = k^2 \left(\frac{D}{\rho h}\right)^{1/2}, \tag{7.39}$$

where k satisfies

$$1 + \cosh kl \cos kl = 0. \tag{7.40}$$

Elastic waves

We reconsider the elastic wave equation in the form (7.17):

$$\rho\ddot{\mathbf{u}} = \frac{E}{2(1+\nu)}\nabla^2\mathbf{u} + \frac{E}{2(1+\nu)(1-2\nu)}\,\text{grad div}\,\mathbf{u}. \tag{7.41}$$

Consider a disturbance propagating in the x direction; thus $\mathbf{u} = \mathbf{u}(x,t)$, and we write the longitudinal (primary) scalar component as $u_p(x,t)$ and the transverse (secondary) vector component as $\mathbf{u}_s(x,t)$. Thus $\mathbf{u} = (u_p, \mathbf{u}_s)$. Thus grad div $\mathbf{u} = \mathbf{i}\partial^2 u_p/\partial x^2$, so that we have

$$\begin{aligned}\frac{\partial^2 u_p}{\partial t^2} &= c_p^2\frac{\partial^2 u_p}{\partial x^2},\\ \frac{\partial^2 \mathbf{u}_s}{\partial t^2} &= c_s^2\frac{\partial^2 \mathbf{u}_s}{\partial x^2},\end{aligned} \tag{7.42}$$

where

$$\begin{aligned}c_p &= \left[\frac{E(1-\nu)}{\rho(1+\nu)(1-2\nu)}\right]^{1/2},\\ c_s &= \left[\frac{E}{2\rho(1+\nu)}\right]^{1/2}.\end{aligned} \tag{7.43}$$

There are thus two types of wave with wave speeds c_p and c_s, and (for $\nu > 0$) $c_p > \sqrt{2}c_s$. They are called primary and secondary waves, or just P and S waves, and can be thought of as *pressure* and *shear* waves. More generally, $\mathbf{u} = \mathbf{u}_p + \mathbf{u}_s$, where div $\mathbf{u}_s = 0$, curl $\mathbf{u}_p = 0$; thus

$$\mathbf{u}_p = \boldsymbol{\nabla}\phi, \qquad \mathbf{u}_s = \text{curl}\,\mathbf{A}. \tag{7.44}$$

Such a representation always exists.

At an interface between media labeled one and two, elastic waves are reflected and refracted. For example, an S wave approaching a plane wall at an angle (of incidence) θ to the normal in medium one is reflected and refracted as both P and S waves. The angles of reflection or refraction θ' (to the normal) are given by Snell's law:

$$\frac{\sin\theta}{\sin\theta'} = \frac{c_{S1}}{c_{ij}}, \tag{7.45}$$

where c_{ij} refers to the wave speed ($i = P$ or S) in medium j ($j = 1, 2$).

Rayleigh waves

In a semi-infinite medium $z < 0$, surface waves of the form $\exp[\kappa z + i(kx - \omega t)]$ exist if

$$\kappa = [k^2 - (\omega^2/c^2)]^{1/2}, \tag{7.46}$$

c being the wave speed for the relevant component of $\mathbf{u}_p$ or $\mathbf{u}_s$. At the free surface $z = 0$, we require that zero stress conditions apply. This leads to a restriction on the components of the displacement vector and a dispersion relation that can be written as

$$\omega = kc_s\xi, \tag{7.47}$$

where ξ is a function of c_s/c_p and hence ν. The propagation speed is $\omega/k = c_s\xi$ (and the waves are not dispersive). The value of ξ varies from 0.874 to 0.955 as ν increases from zero to one half.

Cracks

Cracks in solids are usually modeled by ruptures in the fabric of the continuum, across which the displacement and (less likely) stress may be discontinuous. There are three distinct ways in which a crack can be stressed, and these are referred to as modes I, II, and III; the modes refer to in-plane extensional stress, in-plane shear, and antiplane shear, respectively. More simply, they are associated with opening, sliding, and tearing. The first two are essentially two-dimensional (plane strain) problems, and there is a particularly nice analysis of these that uses complex variable theory. Denote $z = x + iy$, and put the complex displacement $D = u + iv$ ((u, v) being the x and y components of $\mathbf{u}$). Then the steady-state equation (7.16) can be written

$$2(\lambda+\mu)\frac{\partial}{\partial z}\left(\frac{\partial D}{\partial z} + \overline{\frac{\partial D}{\partial z}}\right) + 4\mu\frac{\partial^2 D}{\partial z\partial \bar{z}} = 0, \tag{7.48}$$

where we ignore $\mathbf{F}$ for convenience (though this is not necessary – see Exercise 3). Integrating with respect to $\bar{z}$, we have

$$2(\lambda+\mu)\left(\frac{\partial D}{\partial z} + \overline{\frac{\partial D}{\partial z}}\right) + 4\mu\frac{\partial D}{\partial z} = \theta'(z), \tag{7.49}$$

where θ is holomorphic. Taking real and imaginary parts, we deduce

$$8\mu(\lambda+\mu)\frac{\partial D}{\partial z} = (\lambda+3\mu)\theta'(z) - (\lambda+\mu)\overline{\theta'(z)}. \tag{7.50}$$

Integrating this and putting $\theta = 4(\lambda+2\mu)\Omega(z)/(\lambda+\mu)$, we find

$$2\mu D = \kappa\Omega - z\overline{\Omega'(z)} - \overline{\omega(z)}, \tag{7.51}$$

where Ω and ω are holomorphic functions and $\kappa = (\lambda+3\mu)/(\lambda+\mu)$. The functions Ω and ω are determined by the boundary conditions. From the constitutive relationship (7.10), we also derive the normal stresses σ_1, σ_2 and the shear stress τ in the form

$$\begin{aligned} \sigma_1 + \sigma_2 &= 2\Omega' + 2\overline{\Omega'}, \\ \sigma_1 - \sigma_2 + 2i\tau &= -2z\overline{\Omega''} - 2\overline{\omega'}. \end{aligned} \tag{7.52}$$

Suppose a crack exists on the y axis in $-l < x < l$. We allow for possible discontinuities in Ω and ω by defining these functions differently in $y > 0$ and $y < 0$. We denote their definitions in $y > 0$ and $y < 0$ by subscripts $+$ and $-$, respectively.

Now σ_2 and τ must be continuous across $y = 0$ in $|x| > l$ (why?). From Eq. (7.52) it follows that on $z = x \in \mathbf{R}$, $|x| > l$,

$$\Omega'_+(x) - \overline{\Omega'_-}(x) - x\overline{\Omega''_-}(x) - \overline{\omega'_-}(z) = \Omega'_-(x) - \overline{\Omega'_+}(x) - x\overline{\Omega''_+}(x) - \overline{\omega'_+}(x). \tag{7.53}$$

The Schwarz reflection principle says that $\overline{F(\bar{z})}$ is analytic if $F(z)$ is; it follows that

$$\Omega'(z) - \overline{\Omega'(\bar{z})} - z\overline{\Omega''(\bar{z})} - \overline{\omega'(\bar{z})} = \theta'(z) \tag{7.54}$$

for some analytic function $\theta(z)$. We then find

$$\sigma_2 - i\tau = \Omega'(z) + \Omega'(\bar{z}) - (\bar{z} - z)\overline{\Omega''} - \theta'(\bar{z}), \tag{7.55a}$$

$$2\mu(u + iv) = \kappa\Omega(z) - \Omega(\bar{z}) + (\bar{z} - z)\overline{\Omega'} + \theta(\bar{z}). \tag{7.55b}$$

Let us suppose that stresses are zero at infinity, so that $\Omega, \theta \to 0$ as $z \to \infty$. We prescribe a crack pressure p and assume the crack width h is small ($h \ll l$), so that $\sigma_2 - i\tau = p$ on $y = 0$, $|x| < l$. (This corresponds to a mode I crack.) We then evaluate Eq. (7.55a) on $y = 0\pm$, from which we obtain two Hilbert problems, $\theta'_+ - \theta'_- = 0$, whence $\theta = 0$, and

$$\Omega'_+ + \Omega'_- = -p. \tag{7.56}$$

Alternatively, we use Eq. (7.55b); the crack width is $h = [v]_-^+$, and the crack slip is $S = [u]_-^+$. Then we derive

$$2\mu(S + ih) = (1 + \kappa)(\Omega_+ - \Omega_-). \tag{7.57}$$

We can now solve Eq. (7.57) for Ω to find an expression for p or solve Eq. (7.56) for Ω to find an expression for $S + ih$. If we define Poisson's ratio $\nu = \lambda/\{2(\lambda + \mu)\}$, then we find $S = 0$ and

$$p = -\frac{\mu}{2(1 - \nu)\pi} \fint_{-l}^{l} \frac{\partial h}{\partial s} \frac{ds}{s - x}. \tag{7.58}$$

The inversion of this (from solving Eq. (7.57)) is

$$\frac{\mu}{2(1 - \nu)} \frac{\partial h}{\partial x} = \frac{1}{\pi(l^2 - x^2)^{1/2}} \fint_{-l}^{l} \frac{p(t)(l^2 - t^2)^{1/2}}{t - x} \, dt. \tag{7.59}$$

From these results, one can show that the stresses are singular at each crack tip, as $\Omega \sim (z^2 - l^2)^{1/2}$ there.

7.3 Plasticity

The infinite stress at a crack tip would not be expected to occur in practice: nature tends to avoid singularities. Usually, the occurrence of such a singularity in a model indicates either the effect of an approximation, or the neglect of some physical phenomenon. In practice, when the stresses are very large, the response of the material ceases to be elastic, and this is modeled by assuming that the material fails. There are two types of failure: brittle failure causes internal rupturing and causes cracks to form; ductile failure leads to irrecoverable deformation, termed plastic deformation. The usual assumption is then that some function of the stresses cannot be exceeded. There are two main *yield criteria* used: that of Tresca,

$$\sigma_1 - \sigma_3 = Y; \tag{7.60}$$

and the possibly more realistic one of Von Mises,

$$\sigma_1'^2 + \sigma_2'^2 + \sigma_3'^2 = 2Y^2. \tag{7.61}$$

In these expressions, Y is the yield stress, $\sigma_1 \geq \sigma_2 \geq \sigma_3$ are the principal stress components, and the stress deviators are $\sigma_i' = \sigma_i + p$, where p is the pressure, $p = -\frac{1}{3}(\sigma_1 + \sigma_2 + \sigma_3)$. The extra condition implied by Eqs. (7.60) or (7.61) requires an added flexibility in the constitutive relation. In an elastic-plastic material, the increment of elastic strain due to an increment of stress is (inverting Eq. (7.10))

$$d\varepsilon_{ij}^e = \frac{d\sigma_{ij}'}{2\mu} + \frac{(1-2\nu)}{3E}\, d\sigma_{kk}\,\delta_{ij}. \tag{7.62}$$

The total incremental strain is then assumed to be given by

$$d\varepsilon_{ij} = d\varepsilon_{ij}^e + d\varepsilon_{ij}^p, \tag{7.63}$$

where $d\varepsilon_{ij}^p$ is the plastic incremental strain. In an ideal plastic material, the Prandtl–Reuss equations are then obtained from the assumption that the plastic strain is induced parallel to the applied deviatoric stress; thus $d\varepsilon_{ij}^p = \sigma_{ij}' d\lambda$, where λ is like a Lagrange multiplier. This assumes that the contraction $d\varepsilon_{ii}^p = 0$; that is, the plastic volume change is zero. The Prandtl–Reuss equations are

$$d\varepsilon_{ij} = \sigma_{ij}' d\lambda + \frac{d\sigma_{ij}'}{2\mu} + \frac{(1-2\nu)}{3E}\, d\sigma_{kk}\,\delta_{ij}, \tag{7.64}$$

together with Eq. (7.61), which is $\sigma_{ij}'\sigma_{ij}' = 2Y^2$. Notice that the physical significance of λ is deliberately unspecific. Mostly, one examines the behavior of an elastic solid with a yield stress. The plastic deformation is then irrecoverable, but there is no permanent creep. On the other hand, one could have a material that is *visco-plastic*, in which case ductile failure will lead to permanent plastic creep.

As an example, consider the torsion of a cylindrical bar of cross-section S, as indicated in Fig. 7.2. We take the z axis along the axis of the cylinder. We expect the only nonzero stresses to be σ_{13} and σ_{23}, whence

$$\frac{\partial \sigma_{13}}{\partial x} + \frac{\partial \sigma_{23}}{\partial y} = 0, \tag{7.65}$$

and there is a stress function ϕ with $\sigma_{13} = \partial\phi/\partial y$, $\sigma_{23} = -\partial\phi/\partial x$. The other components of the force balance equation imply $\phi = \phi(x, y)$. In a plastic region, $\sigma_{13}^2 + \sigma_{23}^2 = Y^2$; thus

$$|\nabla\phi| = Y, \tag{7.66}$$

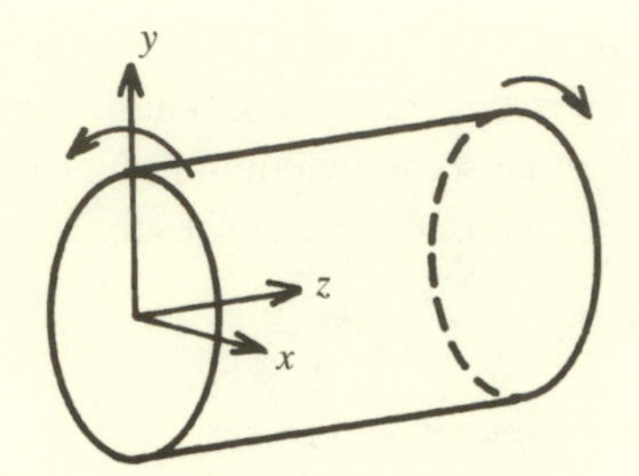

Fig. 7.2. Torsion of a cylindrical bar

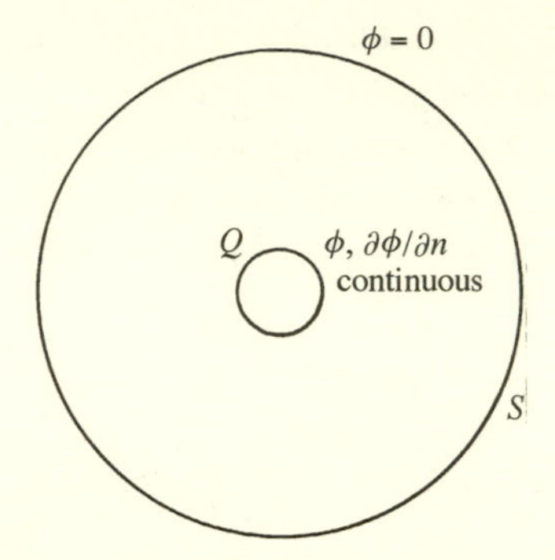

Fig. 7.3. Elastic-plastic domains in the torsion of a cylinder. Within Q the deformation is elastic, whereas between Q and S it is plastic

which is called the eikonal equation and occurs in other physical contexts, notably that of geometrical optics (see also Eq.(18.41)). There must be an elastic region at the center of the bar (otherwise the stresses are not continuous), and in this we solve the normal elastic equations. One can show that, if θ is the twist per unit length, then

$$u = -\theta yz, \qquad v = \theta xz, \qquad w = w(x, y), \tag{7.67}$$

whence

$$\begin{aligned} \sigma_{13} &= 2\mu\varepsilon_{13} = \mu(w_x - \theta y), \\ \sigma_{23} &= 2\mu\varepsilon_{23} = \mu(w_y + \theta x), \end{aligned} \tag{7.68}$$

and thus ϕ satisfies

$$\nabla^2\phi = -2\mu\theta. \tag{7.69}$$

Because $(\sigma_{13}, \sigma_{23}).\boldsymbol{\nabla}\phi = 0$, the stress is directed along lines of constant ϕ. We prescribe zero normal stress at the boundary; thus we can take $\phi = 0$ on S. There is also an unknown boundary Q (see Fig. 7.3) where the elastic-plastic transition takes place, at which the stress must be continuous, that is, $|\boldsymbol{\nabla}\phi| = Y$ and ϕ is continuous.

Simplification occurs for the situation where $\mu, E \to \infty$ in Eq. (7.64), also known as the rigid-plastic model, and also if the problem is two-dimensional (plane plastic strain). Then $\sigma_3 = \frac{1}{2}(\sigma_1 + \sigma_2)$, and

$$\frac{\partial \sigma_1}{\partial x} + \frac{\partial \tau}{\partial y} = 0, \qquad \frac{\partial \tau}{\partial x} + \frac{\partial \sigma_2}{\partial y} = 0, \tag{7.70}$$

where τ is the shear stress σ_{12}, $\sigma_1 = \sigma_{11}$, and $\sigma_2 = \sigma_{22}$. We write σ for the deviatoric part of σ_1; thus

$$\sigma_1 = -p + \sigma, \qquad \sigma_2 = -p - \sigma, \tag{7.71}$$

and we have

$$\begin{aligned} -p_x + \sigma_x + \tau_y &= 0, \\ -p_y - \sigma_y + \tau_x &= 0, \\ \tau^2 + \sigma^2 &= k^2, \end{aligned} \tag{7.72}$$

where the last equation is the yield criterion. Define ϕ by

$$\tau = k \cos 2\phi, \qquad \sigma = -k \sin 2\phi. \tag{7.73}$$

After some manipulation we find

$$\left[\frac{\partial}{\partial x} + \alpha_{\pm}\frac{\partial}{\partial y}\right][p \pm 2k\phi] = 0, \tag{7.74}$$

where $\alpha_{\pm} = (\pm 1 - \cos 2\phi)/\sin 2\phi$; thus the equations are hyperbolic and $p \pm 2k\phi =$ constant on the (characteristic) lines $dy/dx = \alpha_{\pm}, = \tan\phi$ (+), and $-\cot\phi$ (−), respectively. In general, the solution of Eq. (7.74) must be found numerically. The characteristics are called *slip-lines*, and it is possible for the tangential compressive stress to be discontinuous across them (and also the velocity).

Soils

It is a principle of soil mechanics (Terzaghi's principle) that the state of a soil is described by the *effective stresses*

$$\sigma'_{ij} = \sigma_{ij} + p\delta_{ij}. \tag{7.75}$$

Here p is the pressure of the pore fluid phase, which will normally be water. The idea is that the shear stress exerted on the soil is effected by intergranular tangential friction and is unaffected by pore pressure. However, the applied normal stress σ_n at a surface is partly supported by the fluid pressure p, so that the normal stress that the soil particles actually experience is $\sigma_n - p$. (We denote $\sigma_n = -\sigma_{nn}$ here, in keeping with the fluid mechanical usage, whereby $-\sigma_{nn}$ is a (compressive) pressure. It is often the case in solid mechanics that the signs are reversed.)

Thus the effective stresses σ'_{ij} are those transmitted through the soil grains, and it is these that effect the deformation. Terzaghi's principle is plausible for granular structures where the grains have point contacts. More generally, one might expect that the effective stresses would be given by

$$\sigma'_{ij} = \sigma_{ij} + (1 - a)p\,\delta_{ij}, \tag{7.76}$$

where a is the specific grain surface contact area per grain surface area. Although an expression such as Eq. (7.76) is likely to be appropriate, the precise definition of a will in fact depend on the material.

Poroelasticity

Biot's theory of deformation of an elastic porous medium generalizes linear elasticity theory to include the effect of the pore pressure. His equations are

$$\sigma_{ij} = 2\mu\varepsilon_{ij} + \left[\left\{\frac{2\mu\nu}{1 - 2\nu}\right\}\varepsilon_{kk} - \alpha p\right]\delta_{ij}, \tag{7.77}$$

together with a constitutive relation for the variation in water content, or porosity:

$$\phi - \phi_0 = \alpha\varepsilon_{ii} + p/Q. \tag{7.78}$$

Here Q and α are material constants and ϕ_0 is the reference state porosity. We see that Eq. (7.76) is equivalent to an ordinary elastic medium, providing the effective stresses are defined by

$$\sigma'_{ij} = \sigma_{ij} + \alpha p\,\delta_{ij}; \tag{7.79}$$

thus α plays the role of $(1 - a)$ in the Terzaghi description. We define the effective pressure

$$p_e = P - \alpha p = -\frac{1}{3}\sigma_{ii} - \alpha p, \tag{7.80}$$

and then

$$p_e = -\frac{2\mu(1 - \nu)}{3(1 - 2\nu)}\varepsilon_{kk} \tag{7.81}$$

(and $\varepsilon_{kk} = \nabla.\mathbf{u}$ is the dilatation), and

$$\phi - \phi_0 = -\frac{3\alpha(1 - 2\nu)}{\mu(1 - \nu)}p_e + \frac{p}{Q}. \tag{7.82}$$

For a saturated clay, Biot suggests $Q = \infty$ and $\alpha = 1$. Then $\delta\phi = -\delta p_e/K_c$, where $K_c\ (= \mu(1 - \nu)/\{3(1 - 2\nu)\})$ is a compressive modulus. Ignoring inertia and gravity, the elasticity equations are (with $\alpha = 1$)

$$0 = -\frac{\partial p}{\partial x_i} + \frac{\partial \sigma'_{ij}}{\partial x_j}, \tag{7.83}$$

so that $-\boldsymbol{\nabla} p$ is a body force in the stress equations.

Consolidation

Consider a transient situation in which a saturated soil is squeezed by application of a weight (this is called *consolidation*). Then conservation of water mass requires

$$\phi_t + \nabla.\mathbf{v} = 0 \tag{7.84}$$

where $\mathbf{v}$ is the water flux given by *Darcy's law* (see also Chapter 13)

$$\mathbf{v} = -\frac{k}{\eta}\nabla p; \tag{7.85}$$

k is called the permeability of the soil, and η is the viscosity of water. From these two equations, we have

$$\phi_t = \nabla.[(k/\eta)\nabla p], \tag{7.86}$$

whereas Eqs. (7.83) and (7.15) give

$$\nabla p = (\lambda + 2\mu)\text{grad div }\mathbf{u} - \mu \text{ curl curl }\mathbf{u}. \tag{7.87}$$

If k/η is constant, then we have

$$\phi_t = \frac{(\lambda + 2\mu)k}{\eta}\nabla^2\varepsilon_{kk}, \tag{7.88}$$

and from Eq. (7.78) (with $Q = \infty$, $\alpha = 1$)

$$\phi_t = c\nabla^2\phi, \tag{7.89}$$

where the *consolidation coefficient* is defined by

$$c = \frac{k(\lambda + 2\mu)}{\eta} = \frac{2\mu(1-\nu)k}{(1-2\nu)\eta}; \tag{7.90}$$

we see that the material relaxes diffusively to equilibrium.

7.4 Viscoelasticity

Many common materials such as paints, polymers, plastics, and more exotic ones such as silicic magma, saturated soils, and the Earth's lithosphere (the part from the surface to a depth of about 100 km), behave as viscoelastic fluids. On a short timescale, or for small stresses, they behave like elastic solids, but for larger stresses, or on longer timescales, they behave rather as viscous fluids. Conceptually, it is easy to understand how such behavior can occur. Polymers have long-chain molecules, which get tangled up like hair when it is twirled. Subjected to small stresses, the molecules remain locked and behave elastically. Over longer times, however, they slither past each other and the polymer behaves as a fluid. Viscoelastic fluids have a range of unlikely behaviors. In conditions of shear flow, they develop transverse normal stresses that cause such phenomena as shear thinning or thickening, rod-climbing, and extrudate swell.

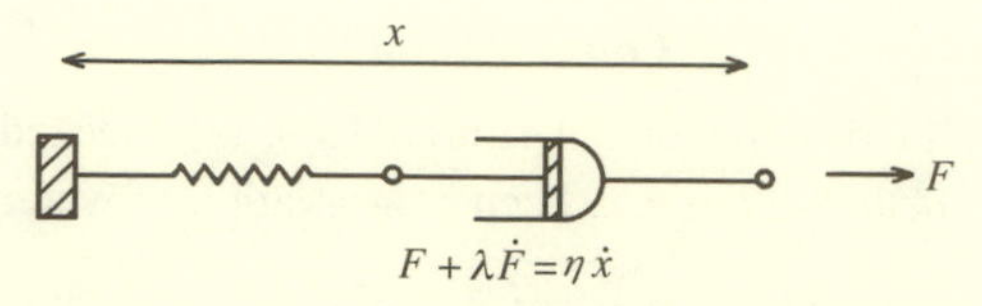

Fig. 7.4. Spring-and-dashpot model for the viscoelastic Maxwell model

Linear viscoelastic fluids: Maxwell and Jeffreys models

The simplest representation of the combined rheological behavior is for conditions of small disturbances, when the strains remain small and a linear constitutive law can be expected to apply. The Maxwell model is simply

$$\tau_{ij} + \lambda \frac{\partial \tau_{ij}}{\partial t} = \eta \dot{\varepsilon}_{ij}, \tag{7.91}$$

where $\dot{\varepsilon}_{ij}$ is the strain-rate tensor. We identify η with the viscosity, and $\lambda = \eta/\mu$, where μ is the shear modulus (we consider here only incompressible materials). Evidently, Eq. (7.91) represents the strain rate as the sum of that produced viscously and elastically, and one can understand this behavior by the 'spring-and-dashpot' arrangement shown in Fig. 7.4.

Different arrangements of springs and dashpots can be used, in series and/or in parallel, to generate more complicated models. Note that Eq. (7.91) can be written in the 'memory-integral' form

$$\tau_{ij} = \int_{-\infty}^{t} G(t-s)\dot{\varepsilon}_{ij}(s)\, ds, \tag{7.92}$$

where $G(\tau) = (\eta/\lambda)\exp[-\tau/\lambda]$, and in fact Eq. (7.92) (for arbitrary G) represents the most general linear viscoelastic model. Another common model is the Jeffreys model:

$$\tau_{ij} + \lambda_1 \frac{\partial \tau_{ij}}{\partial t} = \eta \left[\dot{\varepsilon}_{ij} + \lambda_2 \frac{\partial \dot{\varepsilon}_{ij}}{\partial t} \right], \tag{7.93}$$

for which $G(\tau) = (\eta/\lambda_1^2)[(\lambda_1 - \lambda_2)\exp(-\tau/\lambda_1) + 2\lambda_1\lambda_2\delta(\tau)]$, where $\delta(\tau)$ is the Dirac delta function. In the limit $\lambda_1 \sim \eta \to \infty$, the Jeffreys model reduces to that of the 'Kelvin solid' – not much used, though.

The solution of the equations of motion in the generalized case is facilitated by the observation that Eq. (7.92) is a Laplace convolution. Therefore a Laplace transform of the equations reduces the solution to that of linear elasticity in the Laplace transform space.

Viscoelastic flow

In conditions of permanent creep, the strains are not small, and we can expect that the proper generalization of models such as those of Maxwell or Jeffreys will convert the partial time derivatives to material derivatives. There is a problem here, however. If we simply replace $\partial/\partial t$ in Eq. (7.93), for example, by the material derivative $d/dt = \partial/\partial t + \mathbf{u}.\nabla$ (where $\mathbf{u}$ is the velocity, not the strain), then the corresponding rheological equations of state are not *objective*, in the sense that they are not invariant when arbitrary (rigid-body) rotations of the material are imposed. Put another way, we do not expect the rheological behavior of a fluid to depend on whether it sits on a (slowly) revolving turntable or not. We thus modify the time derivatives in the rheological equations of state to satisfy this principle of objectivity.

Corotating and co-deformational derivatives

There is no unique way to generalize d/dt to guarantee objectivity. In the corotating models, we define the (Jaumann) derivatives of tensors τ_{ij} by

$$\frac{\mathcal{D}\tau_{ij}}{\mathcal{D}t} = \frac{d\tau_{ij}}{dt} + \frac{1}{2}(\omega_{ik}\tau_{kj} - \tau_{ik}\omega_{kj}) \tag{7.94}$$

(summed over k), where the vorticity tensor is defined by

$$\omega_{ij} = \frac{\partial u_j}{\partial x_i} - \frac{\partial u_i}{\partial x_j} \tag{7.95}$$

(and $\mathbf{u}$ is the velocity vector). Alternatively, the co-deformational models define two possible derivatives:

$$\begin{aligned} \frac{\mathcal{D}\tau_{ij}}{\mathcal{D}t_+} &= \frac{d\tau_{ij}}{dt} + \tau_{ik}\frac{\partial u_k}{\partial x_j} + \frac{\partial u_k}{\partial x_i}\tau_{kj}, \\ \frac{\mathcal{D}\tau_{ij}}{\mathcal{D}t_-} &= \frac{d\tau_{ij}}{dt} - \left\{\tau_{ik}\frac{\partial u_j}{\partial x_k} + \frac{\partial u_i}{\partial x_k}\tau_{kj}\right\}. \end{aligned} \tag{7.96}$$

Notice that

$$\frac{\mathcal{D}\tau_{ij}}{\mathcal{D}t} = \frac{1}{2}\left[\frac{\mathcal{D}\tau_{ij}}{\mathcal{D}t_+} + \frac{\mathcal{D}\tau_{ij}}{\mathcal{D}t_-}\right], \tag{7.97}$$

and also that

$$\frac{\mathcal{D}\tau_{ii}}{\mathcal{D}t_\pm} = \frac{d\tau_{ii}}{dt} \pm \tau_{ij}\dot{\varepsilon}_{ij}; \tag{7.98}$$

because $\tau_{ii} = 0$ by definition for incompressible fluids (where τ_{ij} is the deviatoric stress tensor), we see that the co-deformational derivatives do not preserve this property.

The various generalized models have different names: thus we have the corotational Jeffreys model:

$$\tau + \lambda_1\frac{\mathcal{D}\tau}{\mathcal{D}t} = \eta_0\left[\dot{\varepsilon} + \lambda_2\frac{\mathcal{D}\dot{\varepsilon}}{\mathcal{D}t}\right]. \tag{7.99}$$

If $\lambda_2 = 0$, then we have the corotational Maxwell model, also known as the ZFD model. In the co-deformational case, the models

$$\tau + \lambda_1\frac{\mathcal{D}\tau}{\mathcal{D}t_\pm} = \eta_0\left[\dot{\varepsilon} + \lambda_2\frac{\mathcal{D}\dot{\varepsilon}}{\mathcal{D}t_\pm}\right] \tag{7.100}$$

are known as Oldroyd's models A (+) and B (−), and if $\lambda_2 = 0$, we have the White–Metzner model (with $\mathcal{D}/\mathcal{D}t_-$).

Shear flow

To illustrate the effect of such rheological descriptions, consider a simple shearing flow $\mathbf{u} = (u(z), 0, 0)$ in a corotational Jeffreys model. The nonzero components of ω_{ij} and $\dot{\varepsilon}_{ij}$ are

$$\dot{\varepsilon}_{13} = \dot{\varepsilon}_{31} = u', \qquad \omega_{13} = -\omega_{31} = -u', \tag{7.101}$$

and the Jaumann derivatives of τ_{ij} are (with $\tau_{11} + \tau_{22} + \tau_{33} = 0$, $\tau_{ij} = \tau_{ji}$)

$$\begin{aligned} \frac{\mathcal{D}\tau_{11}}{\mathcal{D}t} &= -\tau_{13}u', \qquad \frac{\mathcal{D}\tau_{12}}{\mathcal{D}t} = -\frac{1}{2}\tau_{23}u', \\ \frac{\mathcal{D}\tau_{13}}{\mathcal{D}t} &= \frac{1}{2}(\tau_{11} - \tau_{33})u', \qquad \frac{\mathcal{D}\tau_{22}}{\mathcal{D}t} = 0 \ (\Rightarrow \tau_{22} = 0), \\ \frac{\mathcal{D}\tau_{23}}{\mathcal{D}t} &= \frac{1}{2}\tau_{12}u'. \end{aligned} \tag{7.102}$$

The equations of slow flow are

$$\begin{aligned} 0 &= -p_x + \frac{\partial \tau_{13}}{\partial z}, \\ 0 &= \frac{\partial \tau_{23}}{\partial z}, \\ 0 &= -p_z + \frac{\partial \tau_{33}}{\partial z}, \end{aligned} \tag{7.103}$$

and the constitutive relations are

$$\begin{aligned} \tau_{11} - \lambda_1\tau_{13}u' &= -\eta_0\lambda_2 u'^2, \\ \tau_{12} - \frac{1}{2}\lambda_1\tau_{23}u' &= 0, \\ \tau_{13} + \frac{1}{2}\lambda_1 u'(\tau_{11} - \tau_{33}) &= \eta_0 u', \\ \tau_{23} - \frac{1}{2}\lambda_1 u'\tau_{12} &= 0, \end{aligned} \tag{7.104}$$

thus we can take $\tau_{12} = \tau_{22} = \tau_{23} = 0$, and we find

$$\begin{aligned} \tau_{13} &= \frac{\eta_0 u'[1 + \lambda_1\lambda_2 u'^2]}{1 + \lambda_1^2 u'^2}, \\ \tau_{11} - \tau_{33} &= \frac{\eta_0(\lambda_1 - \lambda_2)u'^2}{1 + \lambda_1^2 u'^2}. \end{aligned} \tag{7.105}$$

Thus this model predicts a nonlinear dependence of viscosity on strain rate and generation of normal stresses, which are necessary to explain phenomena such as rod-climbing or extrudate swell. Note, however, that for a range of strain rates, the stress will decrease with u' if λ_2 is small enough. In particular, the corotational Maxwell (ZFD) model has $\partial\tau_{13}/\partial u' < 0$ for all $u' > 1/\lambda_1$ (see Exercise 10).

7.5 Notes and references

There are perhaps fewer books on solid mechanics than on fluids. Landau and Lifschitz (1986) is a succinct introduction to linear elasticity, while Timoshenko and Goodier (1970) is directed toward engineers. Atkin and Fox (1980) is short but to the point and includes introductory material on large elastic deformations. Gould (1994) is similar but less ambitious and directed toward engineers. Lai, Rubin, and Krampl (1993) is rather wordy but has a lot of material both on large elastic deformations, as well as viscoelasticity. There are some nice geophysical applications of elasticity theory in the book by Turcotte and Schubert (1982). Advanced books on the rheological behaviour of porous media are those by Chen and Mizuno (1990) and Coussy (1995).

Cracks Complex variable methods in elasticity are treated by England (1971). In the text we have discussed only static cracks, but of course cracks *propagate*. The theory of crack propagation divides into two parts: dynamic and subcritical. Motivating this theory is the concept of fracture energy: the stress field caused by the presence of a crack implies stored elastic energy in the solid. There is also surface energy associated with the cleavage surface of the crack. A crack will propagate if the release of the stored elastic energy is sufficient to extend the crack via the production of new fracture surface energy.

The Griffith energy criterion enunciates this, and gives the equilibrium length of a crack in plane strain conditions as

$$l = \frac{2\gamma\mu}{\pi(1-\nu)\sigma^2}, \tag{7.106}$$

where γ is the fracture surface energy, and σ is the applied tensile stress. An alternative viewpoint can be used to describe the propagation speed of a nonequilibrium crack. The stress at crack tips has a singularity of the form $\sigma_{ij} \sim K/r^{1/2}$, where r is distance from the crack tip and K is known as the stress intensity factor. K depends on the nature of the applied stress, but in general $K \sim \sigma l^{1/2}$. Furthermore, the rate of energy release by the crack is just $G = K^2(1-\nu)/2\mu$. A generalization of Griffith's criterion is then that a crack propagates if $G > G_c$, or equivalently, $K > K_c$, where K_c is called the critical stress intensity factor.

We now distinguish between *subcritical* crack propagation, when $K < K_c$ but the extra energy supply can be provided by other means, for example, crack fluid buoyancy, or diffusion and reaction of chemical species (when the process is called stress corrosion); and *dynamic* fracture, when $K > K_c$, and in general, acceleration terms are important: The stress fields are transient, and elastic waves propagate from the crack tip. In this case, $K = K(v)$, more specifically $K = \sigma l^{1/2}\phi[v/c_p, v/c_s]$, where c_p and c_s are the elastic wave speeds. For $K < K_c$, propagation speeds are low and a quasi-static approach is suitable. Fracture mechanics is described by Lawn and Wilshaw (1975) and Freund (1990), see also the book edited by Atkinson (1987), and the review articles by Rudnicki (1980) and Anderson and Grew (1977); these last three have applications in rock mechanics.

Plasticity The classic text on plasticity is the book by Hill (1950). A more recent book is by Khan and Huang (1995). Plasticity occurs in crack tip stress fields, where the singularity is thought to be removed by application of a yield stress. Soils, or granular flows in general, also deform beyond a yield stress. This leads to what is called critical state soil mechanics (see also Chapter 13), whereby the critical yield stress lies on a surface in (τ, p_e, ϕ) space, where τ is shear stress, p_e is effective pressure (or effective compressive stress), and ϕ is porosity. The intersection of this surface with the plane $\tau = 0$ is the normal consolidation line, on which $p_e = p_e(\phi)$, which is locally given in Biot's (1941) poroelasticity theory by Eq. (7.82) above (with $Q = \infty$). Critical state soil mechanics is discussed by Atkinson and Bransby (1978) and Schofield and Wroth (1968). The nature of Terzaghi's effective pressure is discussed by Skempton (1960) and by Bear and Bachmat (1990).

Viscoelasticity The viscoelastic character of some polymeric fluids is often characterized by describing the fluids as a *Bingham fluid*; that is, a fluid for which there is no strain rate for $\tau < \tau_c$ and then $\dot{\varepsilon} > 0$ for $\tau > \tau_c$. One speaks of τ_c as a yield stress, but it needs to be distinguished from a plastic yield stress, because stresses $\tau > \tau_c$ are now allowed. An extreme case of this rheology is a rigid-plastic medium, for which $\dot{\varepsilon} = 0$ for $\tau < \tau_c$, and $\dot{\varepsilon} \geq 0$ for $\tau = \tau_c$ (but always $\tau \leq \tau_c$). Early studies of ice sheets (Nye, 1951) used this rheology. The medium behaves as a solid or a fluid, depending on whether a strain or a strain rate is imposed. Viscoelastic rheologies, and in particular the nonlinear corotational and co-deformational theories are discussed by Bird, Armstrong, and Hassager (1977).

Exercises

1. Derive the Euler–Bernoulli equation by first nondimensionalizing the eight equations in (7.25) with scales chosen by the balances indicated by the text discussion. That is, anticipate that $\sigma_3, \varepsilon_3, \varepsilon_{13}, \rho u_{tt}$ are small in Eqs. (7.25c, e, f, g) respectively, and show that the resultant scales are

$$\tau \sim fl, \qquad \sigma_3 \sim fh, \qquad t \sim (\rho l^4/Eh^2)^{1/2}, \qquad \sigma_1 \sim fl^2/h,$$

$$u \sim fl^3/Eh, \qquad \varepsilon_1 \sim \varepsilon_3 \sim fl^2/Eh, \qquad \varepsilon_{13} \sim fl/E, \qquad w \sim fl^4/Eh^2.$$

Use these (and $x \sim l$) to scale the problem, and show that the indicated terms are indeed small providing $\varepsilon = h^2/l^2 \ll 1$. Show that the boundary conditions of no stress on the top surface of the bar can be linearized, providing a second parameter $\delta = fl^4/Eh^3 \ll 1$; and in this case, deduce the beam equation.

2. Demonstrate graphically that the transcendental equation (7.40) has an infinite number of positive roots $k_n = \xi_n/l$, $n = 1, 2, \ldots,$ where $\xi_n = -\frac{\pi}{2} + n\pi + (-1)^{n-1}\theta_n$ and $\theta_n \in (0, \frac{\pi}{2})$ with $\theta_n \to 0$ as $n \to \infty$. Show that $\theta_n \approx 2e^{-n\pi+(\pi/2)}$ when $n \gg 1$, and deduce that a crude approximation for ξ_1 is

$$\xi_1 \approx \frac{\pi}{2} + 2e^{-\pi/2} \approx 1.99.$$

Show that if a solution of $F(\xi) = G(\xi)$ is sought, then the map $F(\xi_{n+1}) = G(\xi_n)$ will converge to a nearby fixed solution, providing $|G'(\xi)| < |F'(\xi)|$.

Hence show that $\xi_1 \approx 1.8751$ (**hint:** $F = \cos$, $G = -\text{sech}$ will work). Show also that the estimate for ξ_2 is accurate to three decimal places. Why is that?

3. Show that if $\mathbf{F} = (F_1, F_2)$ is conservative (i.e., derived from a (real) potential) then $F_1 + iF_2$ can be written as $\partial V/\partial \bar{z}$ for some real V. If also $V = \partial X/\partial z$, show that the stresses and displacements in plane strain can be written as

$$2\mu(u + iv) = \kappa\Omega - z\overline{\Omega'} - \bar{\omega} - \frac{\mu}{2(\lambda + 2\mu)}X,$$

$$\sigma_1 + \sigma_2 = 2\Omega' + 2\overline{\Omega'} - \left(\frac{\lambda + \mu}{\lambda + 2\mu}\right)V,$$

$$\sigma_1 - \sigma_2 + 2i\tau = -2z\overline{\Omega''} - 2\overline{\omega'} - \frac{\mu}{\lambda + 2\mu}\frac{\partial X}{\partial \bar{z}}.$$

If $\mathbf{F} = (g, 0)$ (x points downward), find X and V; and if $\frac{1}{2}(\sigma_1+\sigma_2) \to -\rho g x$, $\sigma_1 - \sigma_2 + 2i\tau \to 0$ as $z \to \infty$, show that for a crack L on $y = 0, \theta'$ is holomorphic and

$$\theta' = \frac{\mu}{2(\lambda + \mu)}\rho g z.$$

Derive two Hilbert problems for Ω, and deduce that the crack pressure p satisfies

$$p - \rho g x = -\frac{\mu}{2(1 - \nu)\pi} ⨍_L \frac{\partial h}{\partial s}\frac{ds}{s - x}.$$

Note that ρ is the *solid* density.

4. For a crack in $(-l, l)$ the *stress intensity factor* is defined as K, where

$$\frac{1}{2}(\sigma_1 + \sigma_2) \sim \frac{K}{[2(x - l)]^{1/2}} \quad \text{as } x \to l+$$

(with a similar expression near $x = -l$). Show that if the stress intensity factor is K, then $h \sim \alpha(l - x)^{1/2}$ as $x \to l-$, where

$$\alpha = \frac{2K(1 - \nu)\sqrt{2}}{\mu}.$$

5. Consider an infinite crack on $y = 0$ of width h and pressure p. Ignoring gravity and assuming zero stress at infinity, show that the Hilbert problems (7.56) and (7.57) can be solved to obtain

$$p = -\frac{\mu}{2(1 - \nu)\pi} ⨍_{-\infty}^{\infty} \frac{\partial h}{\partial s}\frac{ds}{s - x},$$

and find the inverse relation. (**Hint:** note that this is just a Hilbert transform.)

6. Show in detail that if σ_{13} and σ_{23} are the only nonzero stresses and are independent of z, then Eq. (7.67) applies. Show also that the torque required to produce a twist with stress function ϕ is

$$T = 2\iint_S \phi \, dx \, dy.$$

Find the solution for the twist of a bar of circular cross section, radius a, and show that plastic failure occurs if the twist per unit length θ satisfies $\theta > Y\mu a$, where Y is the Von Mises yield stress. Show that in this case, the twist per unit length is

$$\theta = \frac{Y}{\mu}\left[\frac{\pi}{6(T_c - T)}\right]^{1/2},$$

where T is the applied torque and $T_c = \frac{2}{3}\pi Y a^3$. What do you think happens if $T > T_c$?

7. A granular material (e.g., coal in a hopper, salt in an hourglass) flows downward under gravity toward an exit orifice. For a two-dimensional flow, show that suitable dimensionless equations (neglecting inertia terms) are

$$\frac{\partial \sigma_1}{\partial x} + \frac{\partial \tau}{\partial y} = 0, \qquad \frac{\partial \tau}{\partial x} + \frac{\partial \sigma_2}{\partial y} = 1.$$

The flow law of the material is governed by Coulomb friction; that is, for any line in the material, if $-\sigma_{nn}$ is the normal stress exerted on the plane and σ_{nt} is the shear stress, then $|\sigma_{nt}| \leq -\sigma_{nn} \tan\phi$, where ϕ is the angle of friction, and if the material is flowing,

$$\max_{\mathbf{n}} |\sigma_{nt}| = -\sigma_{nn} \tan\phi$$

and the corresponding line is the slip line for the flow. Show that if a plane is inclined at an angle θ to the horizontal, then $\mathbf{n} = (-\sin\theta, \cos\theta)$, $\mathbf{t} = (\cos\theta, \sin\theta)$ are the normal and tangential vectors, and deduce that (if $\sigma_1 = -p + \sigma$, $\sigma_2 = -p - \sigma$) then

$$-\sigma_{nn} = p + \sigma\cos 2\theta + \tau \sin 2\theta,$$
$$\sigma_{nt} = \tau \cos 2\theta - \sigma \sin 2\theta.$$

By writing $\sigma = k\cos 2\psi$, $\tau = k \sin 2\psi$, show that $|\sigma_{nn}|$ and σ_{nt} lie on a circle in the $(\sigma_{nt}, |\sigma_{nn}|)$ plane, termed the *Mohr circle*, of radius $(\sigma^2 + \tau^2)^{1/2}$ and center p, and deduce that the Coulomb condition requires

$$\tau = p \sin\phi \sin 2\psi,$$
$$\sigma = p \sin\phi \cos 2\psi,$$

and hence that the stress equations can be reduced to two equations for p and ψ. Show that the equations are hyperbolic, and find their characteristics. What are appropriate boundary conditions for flow in a wedge-shaped hopper with an outlet at the bottom? (See also Tayler (1986).)

8. Show that if $\alpha \neq 1$ and $Q < \infty$ in Biot's poroelasticity theory, then the consolidation equation (7.89) can still be obtained, but with the coefficient of consolidation equal to

$$c = \frac{(\lambda + 2\mu)k}{\eta}\left[\alpha^2 + \frac{\lambda + 2\mu}{Q}\right]^{-1}.$$

9. Show that if the spring in Fig. 7.4 is a linear (Hookean) spring and the dashpot has linear (viscous) resistance proportional to its piston speed, then the series arrangement in the figure gives Maxwell's model. Show also that a dashpot in series with an element consisting of a spring and dashpot in parallel represents the Jeffreys model, Eq. (7.93). Write down the model corresponding to N Maxwell elements in parallel (called the generalized Maxwell model).
10. Show that the corotational Jeffreys model for a unidirectional shear flow $(u(z), 0, 0)$ yields $\partial\tau_{13}/\partial u' < 0$ over a range of values of u' if $\lambda_2 < \lambda_1/9$. Why does this seem unphysical? Show also that if the flow is transient, then it is unstable if $|u'| > 2\lambda_1$. (Assume $u = u(z, t)$ and that τ_{ij} depends on t only.)

8

Electromagnetism

8.1 Fundamentals

Electrostatics

In analogy to gravitation, stationary point charges of strength q_i at position $\mathbf{r}_i$ satisfy *Coulomb's law*: the force exerted on q_i by q_j is

$$\mathbf{f}_i = \frac{q_i q_j (\mathbf{r}_i - \mathbf{r}_j)}{4\pi\varepsilon_0 |\mathbf{r}_i - \mathbf{r}_j|^3}, \tag{8.1}$$

where ε_0 is known as the *permittivity*. The electric field due to a distribution of charge is then the force per unit charge exerted at a point $\mathbf{r}$ and is denoted $\mathbf{E}(\mathbf{r})$. If the charge density is ρ, then

$$\mathbf{E} = \frac{1}{4\pi\varepsilon_0} \int_V \frac{\rho(\mathbf{r}')(\mathbf{r} - \mathbf{r}')\, dV}{|\mathbf{r} - \mathbf{r}'|^3}, \tag{8.2}$$

whence there follow

$$\boldsymbol{\nabla} \times \mathbf{E} = 0, \tag{8.3a}$$

$$\boldsymbol{\nabla}.\mathbf{E} = \rho/\varepsilon_0, \tag{8.3b}$$

so that there is an electrostatic potential ϕ given by $\mathbf{E} = -\boldsymbol{\nabla}\phi$, and ϕ satisfies Poisson's equation:

$$\nabla^2\phi = -\rho/\varepsilon_0, \tag{8.4}$$

and in fact

$$\phi = \frac{1}{4\pi\varepsilon_0} \int_V \frac{\rho\, dV}{|\mathbf{r} - \mathbf{r}'|}. \tag{8.5}$$

Poisson's equation can be solved in the usual way.

In conductors (e.g. metals), charges are mobile and move (rapidly) to dispel the electric field. Hence ϕ is constant, and one prescribes ϕ on the surface to solve $\nabla^2\phi = 0$ in the space between conductors, for example, between the electrodes of a battery.

Magnetism

A moving charge constitutes a *current*. A magnetic field causes a force to act on a current-carrying wire. We define the magnetic field strength $\mathbf{B}$ by

$$d\mathbf{F} = I d\mathbf{l} \times \mathbf{B}, \tag{8.6}$$

where $d\mathbf{F}$ is the force due to a current I in a wire $d\mathbf{l}$ located at $\mathbf{r}$. If there is a current density $\mathbf{J}$, which may be defined as a charge flux through the conservation of charge equation

$$\rho_t + \nabla.\mathbf{J} = 0, \tag{8.7}$$

then the force per unit volume is

$$\mathbf{f} = \mathbf{J} \times \mathbf{B}. \tag{8.8}$$

Equally, current in a wire causes a magnetic field to exist, given by

$$d\mathbf{B} = \frac{\mu_0 I \, d\mathbf{l} \times (\mathbf{r} - \mathbf{r}')}{4\pi |\mathbf{r} - \mathbf{r}'|^3}, \tag{8.9}$$

where the current element is at $\mathbf{r}'$. Volumetrically,

$$\mathbf{B} = \int_V \frac{\mu_0 \mathbf{J} \times (\mathbf{r} - \mathbf{r}') \, dV}{4\pi |\mathbf{r} - \mathbf{r}'|^3}. \tag{8.10}$$

The constant μ_0 is known as the permeability. A combination of these results gives the *Biot–Savart law*, which gives the force on a current element i at $\mathbf{r}_i$ due to a current element j at $\mathbf{r}_j$:

$$\mathbf{f}_i = \frac{\mu_0 I_i I_j \, d\mathbf{l}_i \times [d\mathbf{l}_j \times (\mathbf{r}_i - \mathbf{r}_j)]}{4\pi |\mathbf{r}_i - \mathbf{r}_j|^3}. \tag{8.11}$$

From Eq. (8.10), we find

$$\nabla.\mathbf{B} = 0 \tag{8.12}$$

($\mathbf{B}$ is *solenoidal*), and

$$\nabla \times \mathbf{B} = \mu_0 \mathbf{J}. \tag{8.13}$$

Eq. (8.12) implies that $\mathbf{B}$ is the curl of a vector, and in fact

$$\mathbf{B} = \text{curl}\left\{ \frac{\mu_0}{4\pi} \int_V \frac{\mathbf{J} \, dV}{|\mathbf{r} - \mathbf{r}'|} \right\}. \tag{8.14}$$

The expression in curly brackets is called the vector potential.

In the case that both electric and magnetic fields are present, then the force per unit volume is the *Lorentz force*, given by

$$\mathbf{F} = \rho \mathbf{E} + \mathbf{J} \times \mathbf{B}, \tag{8.15}$$

or for a unit charge moving at velocity $\mathbf{v}$,

$$\mathbf{F} = \mathbf{E} + \mathbf{v} \times \mathbf{B}. \tag{8.16}$$

Electromagnetic induction

When a conductor moves in a magnetic field, the Lorentz force induces motion in the mobile charges, which induces an emf. (electromotive force). The total emf. induced in a closed loop L is

$$V = \oint_L \mathbf{v} \times \mathbf{B}.d\mathbf{l} = \int_L d\mathbf{l} \times \mathbf{v}.\mathbf{B}. \tag{8.17}$$

Now $\mathbf{v} \times d\mathbf{l}$ is the rate at which $d\mathbf{l}$ sweeps out area. Therefore

$$V = -\dot{\Psi}, \tag{8.18}$$

where

$$\Psi = \int_S \mathbf{B}.d\mathbf{S} \tag{8.19}$$

is the *magnetic flux* and may be conceptualized as the 'density' of B lines linking the loop.

Faraday's law arises from the (natural) experimental result that finds that it is only the relative motion that is important. Thus the induced emf. in the wire is

$$\oint_L \mathbf{E}.d\mathbf{l} = -\int_S \dot{\mathbf{B}}.d\mathbf{S} \tag{8.20}$$

where now the loop L is stationary. Applying Stokes' theorem, we have Faraday's law:

$$\boldsymbol{\nabla} \times \mathbf{E} = -\frac{\partial \mathbf{B}}{\partial t}, \tag{8.21}$$

which generalizes Eq. (8.3a) to time-dependent situations.

8.2 Maxwell's equations

If we compare the conservation of charge equation (8.7), which applies for unsteady currents as well as steady ones, with Eq. (8.13), we see that there is an inconsistency, as (8.13) implies $\rho_t = 0$. Thus Eq. (8.13) can only apply in a steady state. Adopting Eq. (8.3), we have

$$\nabla.[\mathbf{J} + \varepsilon_0 \mathbf{E}_t] = 0, \tag{8.22}$$

and it follows that the time-dependent generalization of Eq. (8.13) is

$$\boldsymbol{\nabla} \times \mathbf{B} = \mu_0 \mathbf{J} + \mu_0 \varepsilon_0 \mathbf{E}_t. \tag{8.23}$$

The additional term $\varepsilon_0 \mathbf{E}_t$ was called by Maxwell the *displacement current*. Of course, Eq. (8.22) only requires that the right-hand side of Eq. (8.23) be the curl of *some* vector field. The assumption of Maxwell in writing Eq. (8.23) is that the displacement current produces a magnetic field in the same way as the current.

Collecting the various equations together, we have Maxwell's equations:

$$\begin{aligned} \boldsymbol{\nabla}.\mathbf{B} &= 0, \\ \boldsymbol{\nabla} \times \mathbf{E} &= -\partial \mathbf{B}/\partial t, \\ \varepsilon_0 \boldsymbol{\nabla}.\mathbf{E} &= \rho, \\ \boldsymbol{\nabla} \times \mathbf{B} - \mu_0 \varepsilon_0\, \partial \mathbf{E}/\partial t &= \mu_0 \mathbf{J}. \end{aligned} \tag{8.24}$$

In a medium (e.g. a vacuum) with no charge or current, we find

$$\frac{\partial^2 \mathbf{B}}{\partial t^2} = c^2 \nabla^2 \mathbf{B}, \qquad \frac{\partial^2 \mathbf{E}}{\partial t^2} = c^2 \nabla^2 \mathbf{E}, \tag{8.25}$$

where

$$c = (\mu_0 \varepsilon_0)^{-1/2} \tag{8.26}$$

is the speed of light. Thus **E** and **B** propagate as linear electromagnetic waves, and, in particular, there is no energy loss.

Ohm's law

The relation $V = IR$ is familiar from elementary physics. Experimentally, current is proportional to the electric field strength, and one writes

$$\mathbf{J} = \sigma \mathbf{E}, \tag{8.27}$$

where σ is called the conductivity. This law is analogous to Fourier's law in heat conduction, as **J** is a charge flux and must be constituted.

In electrodynamics, we have

$$\rho_t = -\boldsymbol{\nabla}.\mathbf{J} = -(\sigma/\varepsilon_0)\rho, \tag{8.28}$$

and the displacement current relaxes on a time ε_0/σ. If σ is large, this time is correspondingly short, and the medium is a conductor. For metals, this *dielectric* relaxation time is $\sim 10^{-17}$ s. Beyond such short times, the interior electric field is zero, and charge is located in an (atomically) thin skin near the surface. For insulators, σ is small; and for plastics, $\varepsilon_0/\sigma \sim 10^6$ s.

It is clear from Eq. (8.28) that the current is a dissipative term, and consequently, we can expect an energy loss associated with it. If N charged particles per unit volume of charge q have velocity **v**, then $\mathbf{J} = Nq\mathbf{v}$, and the Lorentz force per unit volume is $\mathbf{F} = Nq(\mathbf{E} + \mathbf{v} \times \mathbf{B})$. It follows that the particles do work at a rate

$$\mathbf{F}.\mathbf{v} = \mathbf{J}.\mathbf{E}, \tag{8.29}$$

and this is the rate at which energy is supplied by the emf., for example, by a battery. We come back to this in the following.

Dielectrics and magnets

Electric and magnetic fields in matter modify Maxwell's equations. The effect of an external electric field is modified by an insulating medium, as some part of it is used in redistributing bound charges of the atoms themselves. This phenomenon is called *polarization*. The corresponding phenomenon in a magnetic medium (e.g. iron) is called *magnetization*. We modify Maxwell's equations by introducing two vector fields, the polarization **P** and the magnetization **M**, and we then define the *displacement field*

$$\mathbf{D} = \varepsilon_0 \mathbf{E} + \mathbf{P}, \tag{8.30}$$

and the *magnetic field*

$$\mathbf{H} = \frac{1}{\mu_0}\mathbf{B} - \mathbf{M}, \tag{8.31}$$

so that Maxwell's equations become

$$\begin{aligned} \nabla.\mathbf{B} &= 0, \\ \nabla \times \mathbf{E} &= -\partial \mathbf{B}/\partial t, \\ \nabla.\mathbf{D} &= \rho, \\ \nabla \times \mathbf{H} - \partial \mathbf{D}/\partial t &= \mathbf{J}. \end{aligned} \tag{8.32}$$

Evidently, **M** and **P**, or equivalently **H** and **D**, must be related to **B** and **H** by constitutive relations, and for simple linear media we define

$$\mathbf{D} = \varepsilon_0 \varepsilon \mathbf{E}, \qquad \mathbf{B} = \mu_0 \mu \mathbf{H}, \tag{8.33}$$

where ε and μ are known as the dielectric constant and the relative permeability, respectively. There are many media in which these simple linear relationships do not hold, for example, piezoelectrics and permanent steel magnets.

Boundary conditions

At an interface between two media, the use of 'pillbox integration' gives the appropriate boundary conditions: these are

$$\begin{aligned} [\mathbf{D}.\mathbf{n}]_-^+ &= \sigma_c, \\ [\mathbf{B}.\mathbf{n}]_-^+ &= 0, \\ [\mathbf{E}.\mathbf{t}]_-^+ &= 0, \\ [\mathbf{H} \times \mathbf{n}]_-^+ &= \mathbf{J}_{sc}, \end{aligned} \tag{8.34}$$

where **n** is a unit normal vector from $-$ to $+$, **t** is any tangent vector, σ_c is the surface charge (zero unless the lower medium is a metal), and $\mathbf{J}_{sc}$ is the surface current, normally zero, unless one medium is a superconductor ($\sigma = \infty$).

Electromagnetic energy

In a vacuum, by analogy with (for example) elastic energy in a string, we define the electromagnetic energy as the sum of the electric energy $\frac{1}{2}\varepsilon_0|\mathbf{E}|^2$ and the magnetic energy $|\mathbf{B}|^2/2\mu_0$. The electric energy follows from first principles and the definition of **E**, as does the magnetic energy (with the definition of **B**). The appropriate generalization for dielectric or magnetic media is, at least for linear media satisfying Eq. (8.33) with constant μ and ε, that the electromagnetic energy is

$$U = \frac{1}{2}\int_V [\mathbf{E}.\mathbf{D} + \mathbf{H}.\mathbf{B}]\, dV. \tag{8.35}$$

It follows from Maxwell's equations that (using Eq. (8.33))

$$\dot{U} = -\int_V \mathbf{J}.\mathbf{E}\, dV - \int_S \mathbf{N}.d\mathbf{S}, \tag{8.36}$$

where **N** is the Poynting vector

$$\mathbf{N} = \mathbf{E} \times \mathbf{H} \tag{8.37}$$

and gives the energy flux. We see that electromagnetic energy is lost at a rate **J**.**E** (as we deduced previously). Eq. (8.36) remains valid in general (even if Eq. (8.33) is invalid), although Eq. (8.35) is not then the correct definition of U.

Electromagnetic waves

For linear, homogeneous, isotropic media for which Eq. (8.33) holds and ε and μ are constant, we derive as before Eq. (8.25), with the wave speed c now being given by

$$c = c_0/(\mu\varepsilon)^{1/2}, \tag{8.38}$$

where c_0 is the speed of light in vacuo. We define the *refractive index* of a medium

$$n = c_0/c = (\mu\varepsilon)^{1/2}. \tag{8.39}$$

Reflection and refraction of plane waves can be studied just as for elastic waves. Snell's law of refraction is then

$$n_1 \sin\theta_i = n_2 \sin\theta_t, \tag{8.40}$$

where an incident wave at angle θ_i to the normal in medium one is transmitted at an angle θ_t in medium two.

In conductors we put $\mathbf{J} = \sigma\mathbf{E}$ in Eq. (8.32), so that

$$\nabla \times \mathbf{B} = \mu\mu_0\sigma\mathbf{E} + \mu\varepsilon\mu_0\varepsilon_0\, \partial\mathbf{E}/\partial t, \tag{8.41}$$

and hence

$$\mu\varepsilon\mu_0\varepsilon_0\frac{\partial^2\mathbf{E}}{\partial t^2} + \mu\mu_0\sigma\frac{\partial\mathbf{E}}{\partial t} = \nabla^2\mathbf{E} - \text{grad div}\,\mathbf{E}. \tag{8.42}$$

If we seek solutions $\mathbf{E} = (E_1, E_2, E_3)\exp(\alpha z + i\omega t)$, which vary in the direction normal to a metal in $z > 0$, then

$$-\omega^2 \mu\varepsilon\mu_0\varepsilon_0(E_1, E_2, E_3) + i\omega\mu\mu_0\sigma(E_1, E_2, E_3) \\ = \alpha^2(E_1, E_2, E_3) - (0, 0, \alpha^2 E_3), \tag{8.43}$$

whence

$$[\alpha^2 + \omega^2\mu\varepsilon\mu_0\varepsilon_0 - i\omega\mu\mu_0\sigma]E_i = 0, \quad i = 1, 2, \tag{8.44}$$

and

$$[\omega^2\mu\varepsilon\mu_0\varepsilon_0 - i\omega\mu\mu_0\sigma]E_3 = 0. \tag{8.45}$$

Therefore $E_3 = 0$, and

$$\alpha = \pm[i\omega\mu\mu_0\sigma - \omega^2\mu\varepsilon\mu_0\varepsilon_0]^{1/2}. \tag{8.46}$$

In a conductor, σ is so large that we can neglect the second derivative in Eq. (8.42) and equivalently

$$\alpha \approx -(1 + i)/\delta, \tag{8.47}$$

where

$$\delta = \left(\frac{2}{\omega\mu\mu_0\sigma}\right)^{1/2} \tag{8.48}$$

is the penetration distance. At 50 Hz, this is about 1 cm in copper and less than 1 mm in iron.

Circuits

In electric circuits, Maxwell's equations take integrated forms, and the relevant variables are V, potential or voltage; I, current; Q, charge; and we have material properties R, resistance; C, capacitance; and L, inductance.

The capacitance of a pair of conductors is given by

$$C = Q/V, \tag{8.49}$$

where V is the potential difference and Q is the (positive) charge on one of the conductors. The resistance of a circuit (between two points) is given by Ohm's law

$$R = V/I, \tag{8.50}$$

where V is the potential difference between two points. Self-inductance (or simply inductance) is related to electromagnetic induction (Faraday's law) as follows. The emf. generated by a moving circuit or by a changing magnetic field is given by Eq. (8.18), $V = -\dot{\Psi}$, where Ψ is the magnetic flux linking the circuit. In the absence of magnetic materials, the $\mathbf{B}$ field generated by a circuit is proportional to the current (because of Eq. (8.14)). We therefore define the *self-inductance* of a circuit by

$$L = \Psi/I. \tag{8.51}$$

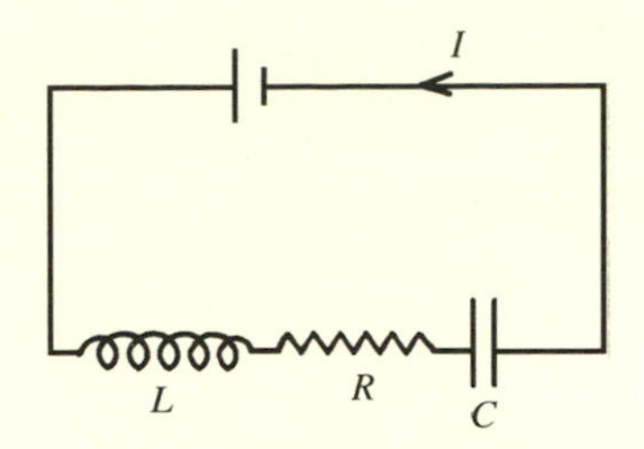

Fig. 8.1. A simple electrical circuit

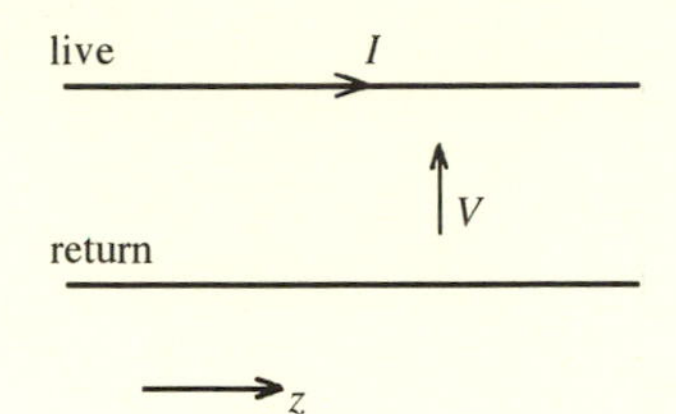

Fig. 8.2. A two-wire transmission cable

Equivalently, we define *mutual inductance* M of two circuits, 1 and 2, by

$$\Psi_1 = MI_2, \qquad \Psi_2 = MI_1. \tag{8.52}$$

These elements enable us to write differential equations for linear circuits, such as in Fig. 8.1. For a circuit driven by an emf. E (from a battery) containing *in series* an inductor, a resistor, and a capacitor, the relevant equation is just

$$L\ddot{Q} + R\dot{Q} + Q/C = E. \tag{8.53}$$

For more complicated networks, we apply *Kirchhoff's laws*:

$$\begin{aligned} \Sigma I &= 0 \quad \text{at a junction,} \\ \Sigma V &= 0 \quad \text{around a loop.} \end{aligned} \tag{8.54}$$

A continuously distributed example is a transmission line, as illustrated in Fig. 8.2. A two-wire cable has an inductance L, resistance R, and capacitance C per unit length. If the potential difference between the live and return wires is V, and if z is distance along the wire, then

$$\begin{aligned} \frac{\partial V}{\partial z} &= -L\frac{\partial I}{\partial t} - RI, \\ \frac{\partial I}{\partial z} &= -C\frac{\partial V}{\partial t}, \end{aligned} \tag{8.55}$$

whence

$$V_{zz} = LCV_{tt} + RCV_t, \tag{8.56}$$

and propagating waves of speed $\sim (LC)^{-1/2}$ are attenuated in a time $\sim (L/R)$.

Joule heating

The energy dissipation rate **J.E** is converted into heat in conductors; thus the heat equation in a solid conductor is modified to

$$\rho c_p \frac{\partial T}{\partial t} = k\nabla^2 T + \sigma |\mathbf{E}|^2, \tag{8.57}$$

and for most applications the timescale associated with heat conduction is much longer than that associated with electromagnetic waves, whence the $\partial \mathbf{D}/\partial t$ term can be dropped from Eq. (8.32). In some materials, Joule heating is significant, and the conductivity σ depends strongly on temperature, thus the problem is thermoelectrically coupled.

In a circuit, the corresponding integrated heating rate is $I^2 R$, which gives the familiar wattage, or power rating, of electrical appliances.

Magnetohydrodynamics

In an incompressible, electrically conducting fluid such as liquid iron, the momentum equation is now supplemented by the Lorentz force, so that

$$\rho[\mathbf{u}_t + (\mathbf{u}.\boldsymbol{\nabla})\mathbf{u}] = -\boldsymbol{\nabla} p + \mathbf{J} \times \mathbf{B} + \mu \nabla^2 \mathbf{u}, \tag{8.58}$$

where (for Galilean invariance)

$$\mathbf{J} = \sigma[\mathbf{E} + \mathbf{u} \times \mathbf{B}]. \tag{8.59}$$

The 'pre-Maxwell' equations neglect the rapid timescale associated with displacement currents; thus

$$\begin{aligned} \boldsymbol{\nabla} \times \mathbf{B} &= \mu \mathbf{J}, \\ \boldsymbol{\nabla} \times \mathbf{E} &= -\partial \mathbf{B}/\partial t, \\ \boldsymbol{\nabla}.\mathbf{B} &= 0, \\ \boldsymbol{\nabla}.\mathbf{E} &= \rho/\varepsilon, \end{aligned} \tag{8.60}$$

supplement Eq. (8.59). Here we have written (as is sometimes the custom) $\mathbf{B} = \mu \mathbf{H}$, $\mathbf{D} = \varepsilon \mathbf{E}$. Elimination of **E** leads to

$$\frac{\partial \mathbf{B}}{\partial t} + (\mathbf{u}.\boldsymbol{\nabla})\mathbf{B} = (\mathbf{B}.\boldsymbol{\nabla})\mathbf{u} + \eta \nabla^2 \mathbf{B}, \tag{8.61}$$

an equation analogous to the vorticity equation, with which it thus shares many properties. The magnetic diffusivity is defined by

$$\eta = (\mu\sigma)^{-1}. \tag{8.62}$$

The first term on the right enables magnetic field generation by a convecting fluid, and it is this mechanism that is responsible for generating the Earth's magnetic field by virtue of motions in the Earth's liquid iron (alloy) outer core.

8.3 Notes and references

There are plenty of undergraduate-level textbooks on electromagnetism. A nice, short introduction is the book by Robinson (1973). Longer, but elementary, introductions are by Solymar (1984) and Duffin (1980), and a more advanced treatment is that by Duffin (1968). For applications in magnetohydrodynamics, see Cowling (1957) or Roberts (1967).

Units Electricity and magnetism developed separately over a long time and were not finally unified until Maxwell's theory. In consequence, two separate systems of units, electromagnetic units (emu.) and electrostatic units (esu.) were developed. These are now of historical interest, having been superseded by the SI (Système International) system, which, apart from meter, kilogram, and second as units of length, mass, and time, defines the unit of current to be the ampere (A), or *amp*, in terms of which the unit of charge is the coulomb (C), and $1\ \mathrm{A} = 1\ \mathrm{C\ s^{-1}}$. The unit of electric potential is the volt (V), which can be defined through its relation to power: 1 volt-ampere $= 1\ \mathrm{V\ A} = 1$ watt (W) $= 1\ \mathrm{J\ s^{-1}}$. Thus $1\ \mathrm{V} = 1\ \mathrm{J\ A^{-1}\ s^{-1}}$.

Other derived units are thus for charge density ρ ($\mathrm{C\ m^{-3}}$), electric potential $\mathbf{E}$ ($\mathrm{V\ m^{-1}}$), and permittivity ε_0 ($\mathrm{C\ V^{-1}\ m^{-1} = F\ m^{-1}}$), where $\mathrm{F = C\ V^{-1}}$ is the *farad*, the unit of capacitance. There is also the magnetic field $\mathbf{B}$ ($\mathrm{V\ s\ m^{-2} = T}$), where T is the *tesla*: more familiar is the cgs. unit, the *Gauss*, and $1\ \mathrm{T} = 10^4\ \mathrm{G}$. The current density is $\mathbf{J}$ ($\mathrm{A\ m^{-2}}$), and the permeability is μ_0 ($\mathrm{V\ s\ A^{-1}\ m^{-1} = H\ m^{-1}}$), where $1\ \mathrm{H} = 1\ \mathrm{V\ s\ A^{-1}} = 1\ \Omega\ \mathrm{s}$ is the henry (the unit of inductance), and $1\ \Omega = 1\ \mathrm{V\ A^{-1}}$ is the unit of resistance (called the *ohm*). The permeability of free space is defined to be $\mu_0 = 4\pi \times 10^{-7}$. We also define the conductivity σ ($\mathrm{S\ m^{-1}} = \Omega^{-1}\ \mathrm{m^{-1}}$), where the *Siemens* $\mathrm{S} = \Omega^{-1}$ ($\mathrm{ohm^{-1}}$).

Magnetic materials For linear media, $\mathbf{B} = \mu_0\mu\mathbf{H}$, and the relative permeability of free space is $\mu = 1$. If $\mu - 1$ is small, then a medium is *diamagnetic* if $\mu < 1$ and *paramagnetic* if $\mu > 1$. Materials for which $\mu - 1$ is large and history dependent are *ferromagnetic*. Here B is a nonlinear function of H, as indicated in Fig. 8.3.

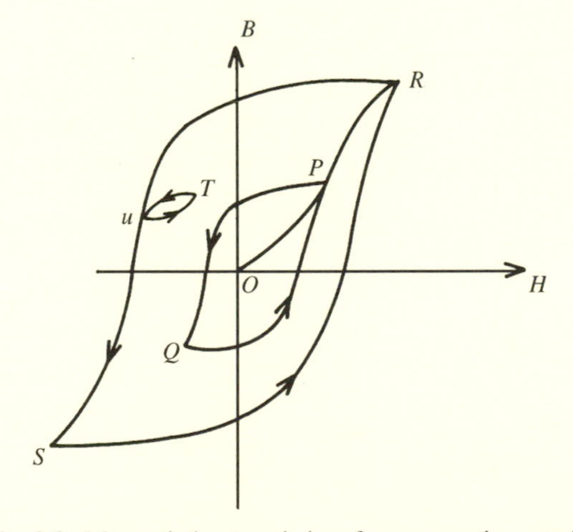

Fig. 8.3. Magnetic hysteresis in a ferromagnetic material

Hysteresis occurs as indicated as H is increased from O to P, down to Q, up to R, and down to S. The outer loop RS is the limit of this behavior, and the area of the curve $\oint H\,dB$ is the work done to go around the loop. Small loops TU are traversed if small reversals in H are made (compare the consolidation curve Fig. 13.3 in Chapter 13). In particular, we see that the magnetic B-field is nonzero when the applied field is zero: this is the phenomenon of permanent magnetization. Ferromagnetic materials are metallic and become paramagnetic above a critical temperature called the *Curie point*. It was the 'freezing in' of the Earth's magnetic field as rocks cooled below the Curie point in the Earth's mantle that enabled palaeomagnetic data to be used to confirm the hypothesis of sea-floor spreading in the early 1960s. Materials with ferromagnetic behavior, but that are nonmetallic are called *ferrimagnetic*; or if $\mu - 1$ is small and positive, *antiferromagnetic*.

Exercises

1. Show that if $\mathbf{E}$ is given by Eq. (8.2) and ϕ by Eq. (8.5), then $\nabla\phi = -\mathbf{E}$. Show also that if $\mathbf{B}$ is given by Eq. (8.14), then $\nabla \times \mathbf{B} = \mu_0\mathbf{J}$.
2. From the definitions of $\mathbf{E}$ and $\mathbf{B}$ in Eqs. (8.2) and (8.10), show that the electric energy is $\frac{1}{2}\varepsilon_0|\mathbf{E}|^2$, and the magnetic energy is $|\mathbf{B}|^2/2\mu_0$. Obtain also the generalization (8.35) when μ and ε are not equal to one.
3. Show that in a slowly varying medium, that is, one where $\varepsilon \neq 1$ and $\mu \neq 1$ are functions that vary 'slowly' in space, the electromagnetic wave equation can be derived in the approximate form

$$\frac{\partial^2\mathbf{D}}{\partial t^2} = \frac{c_0^2}{n^2}\nabla^2\mathbf{D},$$

where $n(\mathbf{x})$ is the *refractive index*. Hence show that the amplitude ψ of one-dimensional oscillators $D = \psi e^{i\omega t}$ satisfies *Helmholtz's equation*:

$$\nabla^2\psi = -\left(\omega^2 n^2/c_0^2\right)\psi.$$

Show that if n is constant, electromagnetic waves $e^{i\mathbf{k}.\mathbf{x}}$ of wavenumber $\mathbf{k}$ exist with $|\mathbf{k}| = \omega n/c_0$. Use the length scale c_0/ω to nondimensionalize the problem, and deduce that n is indeed slowly varying (for transmission of waves of this wavelength) if $c_0|\nabla n|/\omega n \ll 1$. If n varies by $O(1)$ on a length scale L, rescale $\mathbf{x}$ with L, and show that the dimensionless equation is

$$\nabla^2\psi = -\lambda^2 n^2(\mathbf{x})\psi,$$

where $\lambda = \omega L/c_0 \gg 1$. Show that an approximate solution can be generated by writing

$$\psi = \exp[i\lambda S_0 + S_1 + S_2/\lambda + \ldots],$$

and show that at leading order, S_0 satisfies

$$|\nabla S_0| = n(\mathbf{x}).$$

The approximation method is called the *WKB method*, and this is the *eikonal equation*, which can be solved using the method of characteristics (see also the discussion following Eq. (18.41) in Chapter 18).

4. In metals, charges reside on the surface (for $t \gg \varepsilon_0/\sigma$). A capacitor consists of two metal plates a distance λ apart, separated by an air gap in which ε_0/σ is large. Show that if a potential difference V is applied between the plates, then the charges $\pm Q$ induced on the plates are equal and opposite, and calculate the capacitance $C = Q/V$.

   More generally, suppose that two finite conductors with surfaces S_1 and S_2 are maintained at potentials $\phi_1 = V$ and $\phi_2 = 0$, respectively. Deduce that the induced surface charges are $q_1 = -q_2$, and show that the capacitance q_1/V is defined by

$$C = -\varepsilon_0 \int_{S_1} \frac{\partial \phi}{\partial n} \, dS,$$

   where ϕ satisfies $\nabla^2\phi = 0$, $\phi = 1$ on S_1, $\phi = 0$ on S_2, $\phi \to 0$ as $\mathbf{r} \to \infty$.

5. A thin conducting wire of length l and cross-sectional area A has a voltage V applied between ends. The current at any point is defined by

$$I = \int_A \mathbf{J}.\mathbf{t} \, dS,$$

   where $\mathbf{t}$ is a tangent vector along the wire. Show that I depends only on time if there is no leakage of charge from the wire. Derive Ohm's law in the form $V = I/\sigma A$, where σ is the conductivity.

6. An inductance is represented (see Fig. 8.1) as a series of loops. Similarly, a mutual inductance is represented as a pair of coils, as shown in Fig. 8.4. Let the voltage drop be V_1 in a loop of the curve γ_1 due to a current I_2 in the other loop γ_2. Taking the ends of the loops as nearly coincident, we have

$$V_1 = -\oint_{\gamma_1} \mathbf{E}.d\mathbf{r}.$$

   Use the vector potential to show that

$$V_1 = M\dot{I}_2,$$

   where the *mutual inductance* M is defined as

$$M = \frac{\mu_0}{4\pi} \oint_{\gamma_1} \oint_{\gamma_2} \frac{d\mathbf{r}_1.d\mathbf{r}_2}{|\mathbf{r}_1 - \mathbf{r}_2|}.$$

   What happens if this formula is used to calculate the (self-) inductance L of a loop with $\gamma_1 = \gamma_2$? What could be wrong?

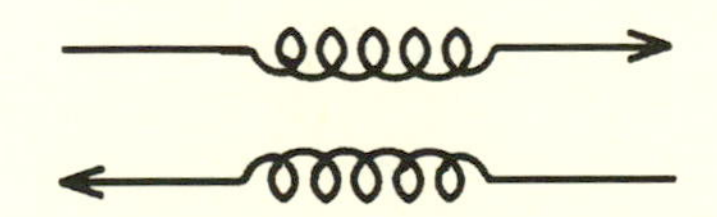

Fig. 8.4. Mutual inductance of a pair of coils

7. Show that the MHD equations for a Boussinesq, electrically conducting fluid in a rotating frame (such as the Earth's outer core) with angular velocity $\boldsymbol{\Omega}$, can be written in the form

$$\begin{aligned}
\boldsymbol{\nabla}.\mathbf{u} &= 0,\\
\frac{d\mathbf{u}}{dt} + 2\boldsymbol{\Omega}\times\mathbf{u} &= -\frac{1}{\rho_0}\boldsymbol{\nabla}\left[p + \frac{B^2}{2\mu}\right] + \frac{1}{\mu\rho_0}(\mathbf{B}.\boldsymbol{\nabla})\mathbf{B}\\
&\quad - \frac{\Delta\rho g}{\rho_0}\mathbf{k} + \nu\nabla^2\mathbf{u},\\
\frac{d\mathbf{B}}{dt} &= (\mathbf{B}.\boldsymbol{\nabla})\mathbf{u} + \eta\nabla^2\mathbf{B},
\end{aligned}$$

where $\rho = \rho_0 + \Delta\rho$ is the density, and the Boussinesq approximation neglects $\Delta\rho$ except in the momentum equation, based on the assumption $\Delta\rho \ll \rho_0$ (where ρ_0 is a constant reference density); $\mathbf{k}$ is a unit vector upwards; p is the *reduced* pressure; μ is magnetic permeability; ν is kinematic viscosity; and $\eta = (\mu\sigma)^{-1}$ is the magnetic diffusivity.

Scale the equations using a length scale l, velocity scale $[u]$, magnetic field scale $[B]$, timescale $l/[u]$, and by choosing $[B]$ and $[u]$ appropriately. (The westward drift of the Earth's field has a speed $\sim 3\times 10^{-4}$ m s^{-1}, *suggestive* of $[u] \sim 10^{-4}$ m s^{-1}. The linear nature of the $\mathbf{B}$ equation requires $(\mathbf{B}.\boldsymbol{\nabla})\mathbf{B}$ to balance the largest term in the momentum equation, so that $\mathbf{u}$ is coupled to $\mathbf{B}$. Therefore choose $[u] = [B]^2/2\rho_0\mu l\Omega$, and note that this implies $[u] \sim \frac{1}{6}\times 10^{-4}$ m s^{-1} if $[B] \sim 10^{-2}$ T (100 Gauss), where also $\rho_0 = 10^4$ kg m^{-3}, $\mu = 4\pi\times 10^{-7}$ kg m C^{-2}, $l = 3.5\times 10^6$ m, and $\Omega = 7\times 10^{-5}$ s^{-1}. This compares with an observed surface magnetic field $B \sim .3\times 10^{-4}$ T, but it is thought that the Earth's mantle acts as an electrical insulator and dampens the field significantly. Take other values

$$\Delta\rho/\rho_0 = 1/4, \qquad g = 5 \text{ m s}^{-2},$$
$$\sigma = 3\times 10^5 \text{ m}^{-3}\text{ kg}^{-1}\text{ s C}^2, \qquad \nu = 10^{-6}\text{ m}^2\text{ s}^{-1},$$

and deduce that suitable scaled equations are

$$\begin{aligned}
\boldsymbol{\nabla}.\mathbf{u} &= 0,\\
\varepsilon\frac{d\mathbf{u}}{dt} + \boldsymbol{\Omega}\times\mathbf{u} &= -\boldsymbol{\nabla}p + (\mathbf{B}.\boldsymbol{\nabla})\mathbf{B} - R\mathbf{k} + E\nabla^2\mathbf{u},\\
\frac{d\mathbf{B}}{dt} &= (\mathbf{B}.\boldsymbol{\nabla})\mathbf{u} + \frac{1}{R_m}\nabla^2\mathbf{B},
\end{aligned}$$

where

$$\varepsilon = \frac{[u]}{2l\Omega}, \qquad E = \frac{\nu}{2\Omega l^2}, \qquad R = \frac{(\Delta\rho/\rho_0)g}{2\Omega[u]}, \qquad R_m = \frac{[u]l}{\eta}$$

are the Rossby, Ekman, Rayleigh, and magnetic Reynolds numbers. Find estimates for them, and show that

$$\varepsilon \ll 1, \qquad R \gg 1, \qquad E \ll 1, \qquad R_m \gg 1.$$

Is this a sensible scaling?

Part four

Continuum models

9

Enzyme kinetics

Chemical reactions in biological systems are often catalyzed by proteins called *enzymes*. They are highly specific and react only with particular *substrates* and are very effective at very low concentrations. The simplest enzyme reaction mechanism is that due to Michaelis and Menten (1913), and is represented schematically as follows:

$$S + E \underset{k_{-1}}{\overset{k_1}{\rightleftharpoons}} C \overset{k_2}{\rightarrow} E + P. \tag{9.1}$$

Here S and E represent substrate and enzyme, respectively; they combine to form a complex C, which then breaks down into a product P and the original enzyme. There is no net consumption of enzyme, so it acts as a catalyst in the reaction.

The first reaction is reversible, and the quantities k_1 and k_{-1} are *rate constants* that quantify the rate at which the reaction proceeds. Similarly, k_2 measures the rate at which product is (irreversibly) created. In order to quantify the rate of change of chemical concentrations, we invoke the law of mass action, which states that in a well-stirred medium, the rate at which two chemical constituents combine to form a third is proportional to the product of their concentrations. (This is a simple probabilistic argument; the rate depends on how often molecules of each species collide, which (if the medium is sufficiently dilute) is simply proportional to the concentration of each.)

Let s, e, c, p denote the concentrations of S, E, C, P (in moles per liter, or molar (written M)). Then the law of mass action states that the rates of change of these variables are given by the equations

$$\frac{ds}{dt} = -k_1 se + k_{-1}c, \tag{9.2a}$$

$$\frac{de}{dt} = -k_1 se + (k_{-1} + k_2)c, \tag{9.2b}$$

$$\frac{dc}{dt} = k_1 se - (k_{-1} + k_2)c, \tag{9.2c}$$

$$\frac{dp}{dt} = k_2 c, \tag{9.2d}$$

and if the reaction is initiated at time $t = 0$ in a medium with concentrations $e = e_0$,

$s = s_0$, then the initial conditions are

$$e = e_0, \qquad p = 0, \qquad c = 0, \qquad s = s_0, \quad \text{at } t = 0. \tag{9.3}$$

The terms in Eq. (9.2) arise as follows: for instance, the rate of formation of C from S and E (in the left-hand reaction) is $k_1 se$, hence the corresponding source term in Eq. (9.2c) and the sink term in (9.2a). Similar considerations apply to other terms.

9.1 Pseudo-steady state hypothesis

The system of equations (9.2) is simple but nonlinear, and some simplification needs to be made to solve it. (Of course, it can be solved numerically, but this is rather beside the point.) The pseudo-steady state hypothesis is that the enzyme reacts so fast with the substrate that it can be taken as being in equilibrium, that is to say, $de/dt \approx 0$. In addition, we can see that conservation of enzyme takes the form

$$e + c = e_0, \tag{9.4}$$

and that the equation for p uncouples from the rest. We only need consider the equations for c and s, and the pseudo-steady state hypothesis suggests that we solve

$$\begin{aligned} k_1 s(e_0 - c) - (k_{-1} + k_2)c &\approx 0, \\ \frac{ds}{dt} &= -k_1 s(e_0 - c) + k_{-1}c, \end{aligned} \tag{9.5}$$

whence

$$\frac{ds}{dt} \approx -k_2 c \approx \frac{-k_1 k_2 e_0 s}{k_1 s + (k_{-1} + k_2)}. \tag{9.6}$$

We define the Michaelis constant K_m as

$$K_m = (k_{-1} + k_2)/k_1, \tag{9.7}$$

so that Eq. (9.6) is

$$\frac{ds}{dt} = -\frac{k_2 e_0 s}{s + K_m}, \tag{9.8}$$

which can be solved implicitly for s.

In practice, one wants to determine the rate constants k_1, k_2, and k_{-2}. This cannot be done uniquely with the present theory, but k_2 and K_m can be determined as follows. We take a sequence of different initial values of s_0, and measure the corresponding variation of s with t, $s = s(t; s_0)$ (this may also be done by measuring $p(t)$, since we see from Eq. (9.2d) and (9.6) that $p + s \approx s_0$). The *rate of reaction* v is given by

$$v = \frac{dp}{dt} \approx -\frac{ds}{dt} = \frac{k_2 e_0 s}{s + K_m}. \tag{9.9}$$

If s (or p) is measured at regular time intervals and the resulting curve extrapolated back toward $t = 0$, as indicated in Fig. 9.1, we obtain for each experiment a measurement of the initial rate v_0. Eq. (9.9) implies that

$$\frac{1}{v_0} = \frac{1}{k_2 e_0} + \frac{K_m}{k_2 e_0}\left(\frac{1}{s_0}\right), \tag{9.10}$$

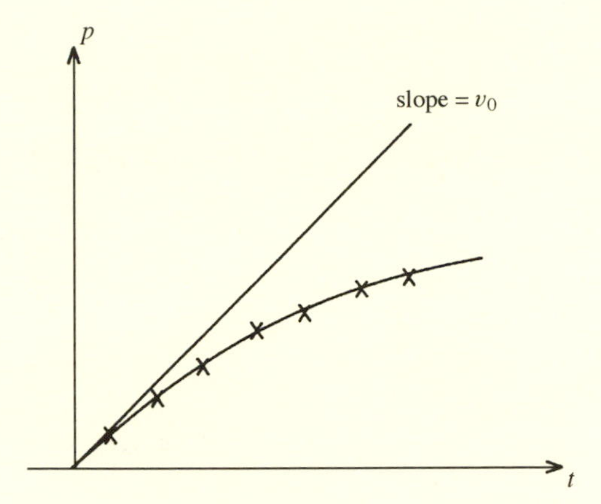

Fig. 9.1. Estimation of the initial reaction rate v_0

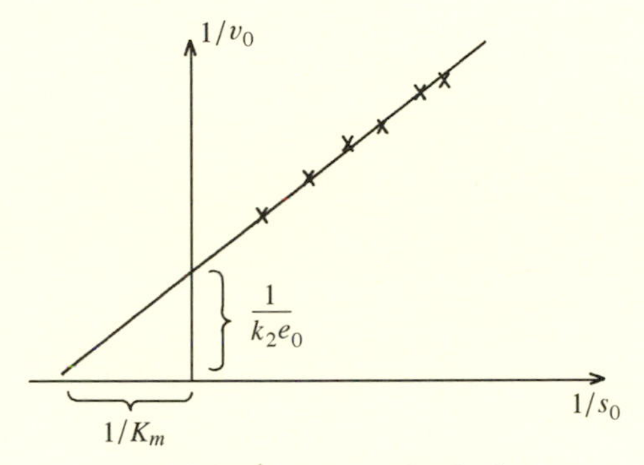

Fig. 9.2. Lineweaver–Burk plot

and hence (knowing e_0) k_2 and K_m can be determined from a sequence of experiments by drawing the *Lineweaver–Burk* plot, as shown in Fig. 9.2.

This procedure is very successful, but the astute reader will notice some untoward features. The system (9.2) (using Eq. (9.4) and uncoupling p) is a second-order system with two initial conditions, $s = s_0$ and $c = 0$, but the simplification to the first-order system (9.6) can only satisfy one of them. In deriving Eq. (9.10), we have taken this to be $s = s_0$, but we see that Eq. (9.6) then implies

$$c = \frac{k_2 e_0 s_0}{s_0 + K_m} \tag{9.11}$$

at $t = 0$, which is evidently false. Why should we not choose $c = 0$ at $t = 0$ in determining Eq. (9.10)? How can we be sure that Eq. (9.10) is right?

9.2 Nondimensionalization

In order to provide an answer to these questions, we must rationalize the approximation. The approximation is that dc/dt is 'small', but small compared to what? To answer this, it is *essential* to nondimensionalize the equations, and we aim to do this in such a way that the variables are $O(1)$. Thus we put

$$s = s_0 s^*, \qquad c = e_0 c^*, \qquad t = t_0 t^*, \tag{9.12}$$

where the timescale t_0 has yet to be chosen. Thus we anticipate $s^*, c^* = O(1)$, and that s^* decays significantly on a time $t^* = O(1)$; in fact, the *modus operandi* for nondimensionalizing equations is the *a priori* hypothesis that if $s \sim s_0$, $t \sim t_0$, then $ds/dt \sim s_0/t_0$. The validity of this assumption is then borne out (or not) in the ensuing calculation.

Substituting these variables in, we obtain the dimensionless equations

$$\begin{aligned} \frac{ds^*}{dt^*} &= -(k_1 e_0 t_0) s^*(1 - c^*) + (k_{-1} e_0 t_0 / s_0) c^*, \\ (e_0/c_0)\frac{dc^*}{dt^*} &= (k_1 e_0 t_0) s^*(1 - c^*) - \{(k_{-1} + k_2) t_0 e_0 / s_0\} c^*, \end{aligned} \tag{9.13}$$

for the dimensionless variables c^* and s^*. We wish to choose t_0 so that $ds^*/dt^* = O(1)$. We cannot guarantee how to do this until the numerical values of the constants are used, but one obvious choice is

$$t_0 = 1/k_1 e_0. \tag{9.14}$$

With this choice, the equations become

$$\begin{aligned} \frac{ds}{dt} &= -s(1 - c) + (\kappa - \lambda)c, \\ \varepsilon\frac{dc}{dt} &= s - (s + \kappa)c, \end{aligned} \tag{9.15}$$

where we now omit the unnecessary asterisks. The parameters ε, κ, λ are dimensionless, and are defined as

$$\varepsilon = e_0/s_0, \qquad \kappa = (k_{-1} + k_2)/k_1 s_0, \qquad \lambda = k_2/k_1 s_0; \tag{9.16}$$

typically $e_0 \ll s_0$, so that $\varepsilon \ll 1$ (e.g., 10^{-6}). Moreover, if s_0 is large enough, we can guarantee that $\kappa, \lambda \lesssim O(1)$. In that case, the problem is correctly scaled: no large parameters appear. If κ or $\lambda \gg 1$, the problem is not correctly scaled, and a judicious rescaling is advisable (see the exercises).

To be precise, let us now suppose that $\kappa, \lambda = O(1)$, $\varepsilon \ll 1$. Then the term $\varepsilon dc/dt$ is small (if $dc/dt = O(1)$) and can be neglected – this is precisely the pseudo-steady state hypothesis. However, it is a singular approximation, as we lose the ability to satisfy both initial conditions, and we can expect a 'boundary layer' near $t = 0$ where dc/dt is large and $\varepsilon dc/dt$ cannot be neglected. The formal procedure for establishing this result is the method of matched asymptotic expansions.

9.3 Singular perturbation theory

We solve Eq. (9.15) approximately, as follows. For time t of $O(1)$, we hope that s, c and their derivatives will be $O(1)$. Then we seek an (asymptotic) expansion in powers of ε

$$\begin{aligned} s &\sim s_0 + \varepsilon s_1 + \dots, \\ c &\sim c_0 + \varepsilon c_1 + \dots, \end{aligned} \tag{9.17}$$

where $\sim$ signifies asymptotic equivalence; we substitute these expansions into the equations, and by equating powers of ε, we obtain a sequence of equations for the successive terms. Thus at leading order (terms of $O(1)$), we have the pseudo-steady state hypothesis

$$c_0 = \frac{s_0}{s_0 + \kappa}, \tag{9.18}$$

whence

$$\frac{ds_0}{dt} = \frac{-\lambda s_0}{s_0 + \kappa}, \tag{9.19}$$

which has solution

$$s_0 + \kappa \ln s_0 = A - \lambda t; \tag{9.20}$$

A is a constant that would be equal to one if s_0 satisfies the condition $s_0 = 1$ at $t = 0$. However, as an approximation to the full system, we cannot satisfy both initial conditions (because $c_0 \neq 0$ at $t = 0$), and therefore the formal approximation must break down near $t = 0$. We anticipate a *boundary layer* where the neglected $\varepsilon dc/dt$ term is important. Within this boundary layer, c must change by $O(1)$ to a value $c = 0$ at $t = 0$. This suggests a *rescaling* of time by writing

$$t = \varepsilon \tau. \tag{9.21}$$

The equations are then

$$\begin{aligned} \frac{ds}{d\tau} &= \varepsilon[-s(1-c) + (\kappa - \lambda)c], \\ \frac{dc}{d\tau} &= s - (s + \kappa)c, \end{aligned} \tag{9.22}$$

and within this *inner layer* we seek a series expansion as before;

$$\begin{aligned} s &= s^0 + \varepsilon s^1 + \dots, \\ c &= c^0 + \varepsilon c^1 + \dots, \end{aligned} \tag{9.23}$$

whence

$$\begin{aligned} \frac{ds^0}{d\tau} &= 0, \\ \frac{dc^0}{d\tau} &= s^0 - (s^0 + \kappa)c^0, \end{aligned} \tag{9.24}$$

and so on. Because both initial conditions can be satisfied, we do so, and thus

$$\begin{aligned} s^0 &= 1, \\ c^0 &= \frac{1}{(1+\kappa)}[1 - e^{-(1+\kappa)\tau}] \end{aligned} \tag{9.25}$$

constitutes the boundary layer solution. The *outer* solution, Eq. (9.18) and (9.20), can now be completed by *matching* the two solutions in an intermediate layer. The idea is that each expansion should represent the solution in a region 'between' $t = O(\varepsilon)$ and $t = O(1)$. That is, consider $\varepsilon \ll t \ll 1$, and expand both inner and outer solutions as asymptotic series. For the variable s we have

$$\text{(outer) } s \sim s_0 \sim s_{00} + O(t), \tag{9.26}$$

where

$$s_{00} + \kappa \ln s_{00} = A, \tag{9.27}$$

whereas the inner solution is just

$$s \sim s^0 = 1. \tag{9.28}$$

Matching requires that we choose

$$s_{00} = 1, \tag{9.29}$$

and thus

$$A = 1. \tag{9.30}$$

Carrying out the same procedure for c, we find that the inner and outer expansions match automatically.

9.4 Enzyme-substrate-inhibitor system

The simple Michaelis–Menten reaction studied above is a classic example of how applied mathematics is done. The physical system leads to a model, which is then analyzed, and the resulting predictions (here, the Lineweaver–Burk plots) tell us how to design experiments to substantiate the results. Another telling lesson is that the biochemists hit on the right way to analyze the problem, even though they did it 'wrong': *never* underestimate the scientist's intuitive understanding. On the other hand, there are (numerous) occasions when mathematics can penetrate more deeply than intuition; it is part of the applied mathematician's job to persuade the applied scientist that this is so.

Suppose we now consider a situation when two substrates compete for a single enzyme. In an experiment, one would measure the rate of change of one of the substrates; the other substrate then inhibits this reaction, and is called the inhibitor: Clearly, the roles can be reversed.

We write the two competing reaction schemes as

$$\begin{aligned} S + E &\underset{k_{-1}}{\overset{k_1}{\rightleftharpoons}} C_S \overset{k_2}{\rightarrow} P_S + E, \\ I + E &\underset{k_{-3}}{\overset{k_3}{\rightleftharpoons}} C_I \overset{k_4}{\rightarrow} P_I + E, \end{aligned} \tag{9.31}$$

where S and I are the substrate and inhibitor, respectively. Applying the law of mass action we find

$$\begin{aligned}
\frac{ds}{dt} &= -k_1 se + k_{-1} c_s, \\
\frac{de}{dt} &= -k_1 se + (k_{-1} + k_2) c_s - k_3 ie + (k_{-3} + k_4) c_i, \\
\frac{dc_s}{dt} &= k_1 se - (k_{-1} + k_2) c_s, \\
\frac{di}{dt} &= -k_3 ie + k_{-3} c_i, \\
\frac{dc_i}{dt} &= k_3 ie - (k_{-3} + k_4) c_i, \\
\frac{dp_s}{dt} &= k_2 c_s, \\
\frac{dp_i}{dt} &= k_4 c_i,
\end{aligned} \tag{9.32}$$

where s, e, c_s, i, c_i, p_s, and p_i are the concentrations of S, E, C_S, I, C_I, P_S, and P_I, respectively. We see that p_s and p_i uncouple from the other variables, and conservation of enzyme takes the form

$$e + c_s + c_i = e_0, \tag{9.33}$$

where we suppose

$$e = e_0, \qquad s = s_0, \qquad i = i_0, \qquad c_i = c_s = p_i = p_s = 0 \quad \text{at } t = 0. \tag{9.34}$$

Let us now follow through the Michaelis–Menten procedure as before, firstly nondimensionalizing the system. We choose scales

$$s \sim s_0; \qquad i \sim i_0; \qquad c_s, c_i \sim e_0; \qquad t \sim 1/k_1 e_0 \tag{9.35}$$

much as before. This means that we write $s = s_0 s^*, t = t^*/k_1 e_0$, etc., and write dimensionless equations as before. We again omit the asterisks, and we also write $i = i_0 z^*$ to avoid the inconvenient use of i. The *dimensionless* equations are thus

$$\begin{aligned}
\frac{ds}{dt} &= -s + (s + K_s - L_s) c_s + s c_i, \\
\varepsilon \frac{dc_s}{dt} &= s - (s + K_s) c_s - s c_i, \\
\frac{dz}{dt} &= \gamma[-z + z c_s + (z + K_i - L_i) c_i], \\
\varepsilon \frac{dc_i}{dt} &= \beta\gamma[z - z c_s - (z + K_i) c_i],
\end{aligned} \tag{9.36}$$

with corresponding *dimensionless* parameters defined by

$$\begin{aligned} \varepsilon = e_0/s_0, \qquad \beta = i_0/s_0, \qquad \gamma = k_3/k_1, \\ K_s = \frac{k_{-1}+k_2}{k_1 s_0}, \qquad K_i = \frac{k_{-3}+k_4}{k_3 i_0}, \\ L_s = \frac{k_2}{k_1 s_0}, \qquad L_i = \frac{k_4}{k_3 i_0}. \end{aligned} \tag{9.37}$$

In order to analyze the system, we suppose that $\varepsilon \ll 1$ and that $\beta, \gamma, K_s, K_i, L_s, L_i = O(1)$. Although this is unlikely always to be strictly true, it is attainable by judicious choice of e_0, s_0, and i_0. In particular, $\varepsilon \sim 10^{-6}$ is a typical experimental value.

We can now follow the same procedure as before. Formally, one writes $s \sim s_0 + \varepsilon s_1 + \ldots$, etc., but as we are only concerned with leading order approximations, it suffices to truncate the procedure. Because this is often how one works in practice, we illustrate the ideas here.

For the outer solution, when $t \sim 1$, we have (neglecting $O(\varepsilon)$ in Eq. (9.36))

$$\begin{aligned} 0 &\approx s - (s+K_s)c_s - sc_i, \\ 0 &\approx z - zc_s - (z+K_i)c_i, \end{aligned} \tag{9.38}$$

whence

$$\begin{aligned} c_s &\sim \frac{K_i s}{K_s z + K_i s + K_s K_i}, \\ c_i &\sim \frac{K_s z}{K_s z + K_i s + K_s K_i}, \end{aligned} \tag{9.39}$$

and thus s and z satisfy approximately

$$\begin{aligned} \frac{ds}{dt} &\sim \frac{-L_s K_i s}{K_s z + K_i s + K_s K_i}, \\ \frac{dz}{dt} &\sim \frac{-\gamma K_s L_i z}{K_s z + K_i s + K_s K_i}. \end{aligned} \tag{9.40}$$

As before, we expect s and z to satisfy their respective initial conditions, because there is no boundary layer behavior associated with the equations (no ε multiplying the highest derivatives). Thus we satisfy Eq. (9.40), together with $z = s = 1$ at $t = 0$; in particular,

$$z \sim s^{\delta}, \tag{9.41}$$

where

$$\delta = \frac{\gamma K_s L_i}{K_i L_s} = \frac{(k_{-1}-k_2)k_3 k_4}{(k_{-3}+k_4)k_1 k_2}. \tag{9.42}$$

Now the *dimensionless* reaction rate for s is given by $v = |ds/dt|$; thus

$$\frac{1}{v} \sim \frac{1}{L_s} + \left(\frac{K_s}{L_s}\right)\frac{1}{s} + \left(\frac{K_s}{L_s K_i}\right)\frac{z}{s}. \tag{9.43}$$

In dimensional terms (denoted by a suffix D), we have $v = v_D/k_1e_0s_0$, and thus

$$\frac{1}{v_D} = \frac{1}{k_1e_0}\left[\frac{k_1}{k_2} + \frac{(k_{-1}+k_2)}{k_2}\{1+z/K_i\}\frac{1}{s_0s}\right], \tag{9.44}$$

where z and s are still dimensionless. In order to use this result, we would carry out a series of experiments using different initial values of s_0 (and i_0). By extrapolation to $t = 0$, we can then measure the predicted initial velocity, which (because $z = s = 1$ at $t = 0$) is given by

$$\frac{1}{v_D} = \frac{1}{k_2e_0} + \frac{(k_{-1}+k_2)}{k_1k_2e_0}\{1+1/K_i\}\frac{1}{s_0} \tag{9.45}$$

and yields the usual Lineweaver–Burk straight line.

Some systems appear anomalous; one example is the enzyme asparaginase, which reacts with asparagine to form aspartase or with glutamine to form glutamate. When asparagine is the inhibitor, Lineweaver–Burk plots do not yield straight lines, which might suggest that the reaction scheme is more complicated than Eq. (9.31). The solution is rather different, however. It turns out that for the asparagine/glutamine system, with asparagine as inhibitor, $\delta \approx 160$. Thus z changes very rapidly with s, and this affects the measurements in the following way. If one is measuring the change of s (or p_s) with t, then naturally one will obtain a series of measurements, such as shown in Fig. 9.1. Of necessity, the initial (dimensionless) velocity is estimated from $v \approx |s(\Delta t)/\Delta t|$, where Δt is small and s is close to one. Then $s \approx 1 - s'\Delta t \approx \exp[-s'\Delta t]$, where $s' \approx L_s/(1+K_s)$, so that

$$z \approx \exp[-\delta s'\Delta t]. \tag{9.46}$$

Therefore, *practical measurements* will give a Lineweaver–Burk plot in the form

$$\frac{1}{v_D} = \frac{1}{k_2e_0} + \frac{(k_{-1}+k_2)}{k_2}\left[1 + \frac{1}{K_i}\exp\{-\delta s'\Delta t\}\right]\frac{1}{s_0}. \tag{9.47}$$

Because s' depends on s_0 through K_s and L_s, it is clear that when $s'\Delta t = \Delta s \sim 1/\delta$, the plots will not be straight lines when $\delta \gg 1$ if, for example, measurements are taken at constant time increments Δt. To obtain the desired result, Eq. (9.45) requires $\Delta s \ll 1/\delta$, which is impractical if $\delta = 160$. On the other hand, if $\Delta s \gg 1/\delta$, then Eq. (9.47) collapses to a straight line whose slope is independent of i_0. However, notice that for small values of $1/s_0$ (cf. Fig. 9.2), both K_s and L_s become small, so that Eq. (9.40) implies $s' \approx -L_s$ for small t. Thus the rate of change of s decreases, and in practice one would have smaller values of Δs. Only with judicious use of the data could Eq. (9.47) provide useful results. When $\delta \ll 1$, the problem does not arise, because then Eq. (9.45) does indeed apply.

It may be thought that the pseudo-steady state hypothesis should become invalid when $\delta \gg 1$, because we did make the explicit assumption that γ, K_s, K_i, L_s, L_i were all $O(1)$ and hence also $\delta = O(1)$. Nevertheless, the analysis is still accurate. This is a good example of the distinction between parameters that are *asymptotically* $O(1)$, as opposed to *numerically* $O(1)$. In a formal asymptotic procedure, one studies solutions in terms of a small parameter (here ε) as $\varepsilon \to 0$. The formal series expansions one obtains are not generally convergent, but become asymptotically equivalent to the actual solution (we hope) as $\varepsilon \to 0$. This formal structure provides the framework for making specific predictions, for specific values of ε. The problem thus generally arises as to whether a specific *numerical* value of a parameter is sufficiently small to render an asymptotic approximation numerically valid. Typically, one does not know. Fluid flow is considered to have low Reynolds number *Re* if $Re < 1$, or even higher, and expansions in powers of *Re* are useful. The onset of Rayleigh–Bénard convection between stress-free boundaries occurs at a critical value of the Rayleigh number *Ra* of 657.511 ...; although numerically large, there is no asymptotic limit associated with this value, which is in fact given by $Ra = Ra_c = 27\pi^4/4$. It just so happens that the numerically $O(1)$ numbers 27, π, and 4 combine to form a numerically large value of *Ra*, but from the asymptotic point of view, *Ra* must be as large as $O(10^6)$ before asymptotic approximations based on $Ra \to \infty$ become quantitatively meaningful.

In the present context, inspection of Eq. (9.36) suggests that the formal asymptotic procedure based on $\varepsilon \to 0$ and β, γ, K_s, etc. $= O(1)$ will be numerically valid even for values of K_s, K_i, etc., of $O(10)$ or $O(10^2)$, for example, provided ε is sufficiently small, and if $\varepsilon \sim 10^{-6}$, this will be the case. In other words, the important *numerical* principle is that a neglected term should be relatively very small. This is *indicated* by the smallness of ε (here), but there is some flexibility in how one interprets the size of the parameters. Such flexibility can be of great use in obtaining quantitative results and qualitative insight and forms the underlying theme of this book.

Alternatively, one can argue as follows. Suppose that $\varepsilon \ll 1$, $\delta \gg 1$ but our analysis is as above, based on $\varepsilon \ll 1$, $\delta \sim 1$. Because, in general, assumption of $O(1)$ values of certain parameters should include the extreme cases (i.e., $\delta \gg 1$, $\delta \ll 1$), we can expect the analysis to be valid, *provided* the slow/fast behavior associated with $\delta \gg 1$ or $\delta \ll 1$ does not interfere with the asymptotic structure of the analysis. Suppose, for example, that δ is large because γ is large, whereas the other parameters are $O(1)$. It is evident from Eq. (9.36) that the outer solutions of Eq. (9.40) change on timescales of $O(1)$ (for s) and $O(1/\gamma)$ (for z), whereas the boundary layer scales for c_s and c_i are $O(\varepsilon)$ and $O(\varepsilon/\gamma)$, respectively. Thus the assumption that the pseudo-steady state has been reached requires $t \gg \varepsilon$ ($\gg \varepsilon/\gamma$), whereas the outer solution describes changes on timescales $O(1)$ (for s) and $O(1/\gamma)$ (for z). So long as $1/\gamma \gg \varepsilon$ (and more generally $1/\delta \gg \varepsilon$), the previous analysis is therefore valid. It is of course possible to carry out asymptotic analyses based on multiple parameter expansions. This is generally messy and is not usually done. Typically, one either gives an asymptotic analysis based on a single limiting parameter, or, where this leads to inconsistent results, one chooses a *distinguished limit*, which effectively reduces the number of independent parameters to one. In the present example, the distinguished limit is when $\varepsilon \sim 1/\gamma$, because then z and c_s both evolve on the same timescale ($\sim\varepsilon$). The formal procedure then replaces the independent parameters ε and γ by ε and $\bar{\gamma}$, where $\bar{\gamma} = \varepsilon\gamma$ and one considers the limiting behavior $\varepsilon \to 0$, $\bar{\gamma} = O(1)$. Provided ε and $1/\gamma$ are comparable *numerically*, this gives sensible asymptotic results.

9.5 Notes and references

There are a number of good books on mathematical biology. A voluminous monograph is that by Murray (1989), which includes useful chapters on enzyme kinetics (as well as the Belousov reaction and the spruce budworm model). Other books of note are those by Segel (1984) and Rubinow (1975).

Units Chemical concentrations are measured in a variety of ways. Usually, one uses either the mass or volume fraction of a substance (for gases, one may also use the partial pressure) or the mass or volume per unit volume. Because the chemical reactions themselves are described *stoichiometrically*, that is, in terms of numbers of molecules, it is convenient to have a concentration measure that is really the number of molecules per unit volume. This is the *mole*, which consists of 6.023×10^{23} molecules, the number (Avogadro's number) of molecules in 32 grams of natural O_2. Evidently, a mole of O_2 weighs 32 grams, and equally, a mole of any substance has a fixed weight, called its (gram) molecular weight. Sometimes the mole is called a gram-mole; 10^3 moles is then a kg-mole. The common unit of chemical concentration is then mole/liter, and a solution of concentration one mole per litre (of solution) is called a molar solution. The concentration thus referred to is the *molarity*, and the unit is denoted by M. Alternatively, the *molality* is the concentration in moles per kilogram of solute and is denoted by m.

Michaelis–Menten theory Michaelis and Menten (1913) described the pseudo-steady state hypothesis, and Briggs and Haldane (1925) gave the first mathematical treatment. The asymptotic treatment was described by Heineken, Tsuchiya, and Aris (1967) and Ó Mathúna (1971). More recently, other asymptotic limits have been described by Frenzen and Maini (1988) and Segel and Slemrod (1989); see Exercise 5 below.

Exercises

1. Consider in turn the following reaction sequences:

$$\text{(i)}\quad S + E \underset{k_{-1}}{\overset{k_1}{\rightleftharpoons}} C_1 \underset{k_{-2}}{\overset{k_2}{\rightleftharpoons}} C_2 \to E + P$$

$$\text{(ii)}\quad \begin{aligned} S + E &\rightleftharpoons C_1 \to E + P \\ S + C_1 &\rightleftharpoons C_2 \to C_1 + P. \end{aligned}$$

Derive the relevant rate equations, and by making suitable assumptions, deduce the equivalent pseudo-steady state results. What do Lineweaver–Burk plots yield?

2. A continuous flow reactor of volume V_R introduces enzyme and substrate at concentrations e_0 and s_0 at a rate V (volume per unit time) and removes the stirred mixture at the same rate. Model the reaction and show that a steady state can exist. Is it stable?

3. A well-known and often-cited reaction mechanism is the formation of HBr from H_2 and Br_2. At about 572°F (300°C) this reaction is known to take place in a series of steps:

$$\left.\begin{array}{ll} (a) & Br_2 \to 2Br \\ (b) & Br + H_2 \to HBr + H \\ (c) & H + Br_2 \to HBr + Br \\ (d) & H + HBr \to H_2 + Br \\ (e) & 2Br \to Br_2 \end{array}\right\} \quad \text{overall, } H_2 + Br_2 \to 2HBr.$$

Applying the law of mass action to each elementary step and noting that, experimentally, steps (a) and (b) have been found to be rapid and essentially at equilibrium and that atomic hydrogen is present only in a small amount, the following rate equation can be developed:

$$r_{HBr} = \frac{K_1 C_{H_2} C_{Br_2}^{1/2}}{1 + K_2 C_{HBr}/C_{Br_2}},$$

where K_1, K_2 are combinations of rate constants. Investigate the derivation of this expression. Note: if B is the concentration of Br and Y is that of Br_2, and if k_1 and k_{-1} are the rates of steps (a) and (e), respectively, then Br is produced at a rate $2k_1 Y$ in (a) and lost at a rate $2k_{-1}B^2$ in (e), whereas Br_2 is lost at a rate $k_1 Y$ in (a), and gained at a rate $k_{-1}B^2$ in (e). Explain this.

4. Analyze the substrate-inhibitor system (9.36) when $\varepsilon \sim 1/\gamma \ll 1$ and other parameters are $O(1)$.

5. Sometimes experiments are done with $e_0 \sim s_0$, that is, $\varepsilon \sim 1$. Show how a Michaelis–Menten type theory can be derived when $\varepsilon \sim 1$, $\lambda \sim 1$, and $\kappa \gg 1$. Can a Lineweaver–Burk plot be useful in this circumstance? What if also $\lambda \gg 1$?

   Show that if $\varepsilon \sim 1$, $\lambda \gg 1$, and $\kappa \sim 1$, then s reaches zero in a finite time $t = t_0 \approx \pi/2(\varepsilon\lambda)^{1/2}$, and show the maximum value of c is approximately $c(t_0) \approx 2t_0/\pi$. What happens for $t > t_0$? Note that this limit is physically meaningless, because in fact $\kappa > \lambda$, from Eq. (9.16). (See also the papers by Frenzen and Maini (1988) and Segel and Slemrod (1989).)

6. Suppose the reaction $C \to E + P$ in Eq. (9.1) is reversible, with backward rate constant k_{-2}. Do the corresponding Michaelis–Menten analysis, and show that as $t \to \infty$, the concentrations of S and P satisfy

$$\frac{P}{S} = \frac{k_1 k_2}{k_{-1} k_{-2}},$$

which is known as Haldane's relation.

10

The Belousov–Zhabotinskii reaction

The reaction discovered by Belousov in 1951 is a chemical reaction, which, in a well-stirred medium, exhibits regular and striking oscillations with a period on the order of a minute. When the reaction takes place in a petri dish (i.e., in a thin layer) the oscillation takes the form of traveling periodic waves, which form elegant spiral patterns. From a chemical point of view, the significance of the oscillation lies in the fact that most chemical reactions proceed rapidly and monotonically to equilibrium. The existence of an oscillation (which can be dramatically observed with appropriate dyes) lasting for up to an hour is in striking contrast to this and led to an explosive growth in the study of biological and chemical oscillators. More recently, the Belousov–Zhabotinskii (BZ) reaction has been a focus for part of the general subject known as 'nonlinear dynamics.'

From a biological point of view, both the oscillations and the petri dish waves are useful analogs to biochemical oscillators and pulsatory phenomena, such as the aggregation of the slime mould *Dictyostelium discoideum*, and the observation of chemical waves led to a whole theory of reaction-diffusion driven waves and instabilities, with many potential applications in biology and ecology.

The Belousov reaction consists of the oxidation of malonic acid [$CH_2(COOH)_2$] by bromate ions (BrO_3^-) in a medium of sulphuric acid (so that there are plenty of hydrogen ions available). The reaction needs a catalyst; Belousov used cerium, and found that he could observe the oscillation by the fact that the oxidized cerium ion Ce^{4+} gives a pale yellow color, whereas in the reduced state Ce^{3+} is clear. The solution thus changes color regularly. A more dramatic variant of the same basic reaction uses iron as a catalyst. In the presence of the dye phenanthroline, the color change is from dark red to a bright Oxford blue. In what follows, we describe the basic reaction of the original Belousov reaction, as described by Field, Körös, and Noyes (1972) and subsequently simplified by Field and Noyes (1974); this model is the so-called *Oregonator*.

Figure 10.1 shows measured values of cerium and bromide ion (Br^-) concentrations during a typical experiment. There is a transient phase lasting about ten minutes, and then the concentrations begin to oscillate (almost) periodically, with a period on the order of a minute. The oscillations can go on for an hour or more before the reaction finally reaches equilibrium.

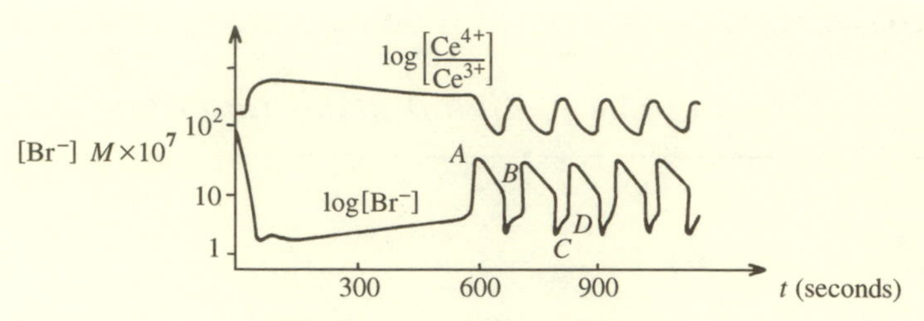

Fig. 10.1. Belousov reaction

10.1 Reaction mechanism

The overall reaction may be thought of as consisting of two competing processes, I and II. During process I, Br^- is consumed (AB in Fig. 10.1), while during process II, Br^- is produced (CD). Evidently, there is a fast switching between these two processes. In more detail, during process I, bromate (BrO_3^-) and bromide (Br^-) combine to produce bromine (Br_2), while malonic acid ($CH_2(COOH)_2$) is brominated by Br_2; the cerium is not involved at this stage. The main reactions can be represented as follows:

Process I

$$Br^- + BrO_3^- + 2H^+ \overset{k_3}{\to} HBrO_2 + HOBr, \tag{10.1a}$$

$$Br^- + HBrO_2 + H^+ \overset{k_2}{\to} 2HOBr, \tag{10.1b}$$

$$Br^- + HOBr + H^+ \overset{k_1}{\to} Br_2 + H_2O, \tag{10.1c}$$

$$Br_2 + CH_2(COOH)_2 \to Br(H(COOH)_2 + Br^- + H^+. \tag{10.1d}$$

The important reactions here are (thought to be) the first two, in which bromine is consumed.

In the second process, cerium (Ce^{3+}) is oxidized to Ce^{4+} by reaction with a bromate radical ($BrO_2^{\bullet}$). Thus we have the following:

Process II

$$\begin{aligned} BrO_3^- + HBrO_2 + H^+ &\overset{k_5}{\to} 2BrO_2^{\bullet} + H_2O, \\ Ce^{3+} + BrO_2^{\bullet} + H^+ &\to Ce^{4+} + HBrO_2; \end{aligned} \tag{10.2}$$

the overall reaction is thus

$$2Ce^{3+} + BrO_3^- + HBrO_2 + 3H^+ \overset{k_5}{\to} 2Ce^{4+} + 2HBrO_2 + H_2O, \tag{10.3}$$

which we suppose is rate limited by the first reaction of Eq. (10.2) (i.e., it is slower than the second). In addition, we have

$$2HBrO_2 \overset{k_4}{\to} HOBr + BrO_3^- + H^+. \tag{10.4}$$

There are many other reactions going on (Field et al. (1972) list twenty-two), but those above are thought to be the most important, in particular, they are *rate controlling*. In the above, Ce^{4+} is produced from Ce^{3+}. The reduction of Ce^{4+} to Ce^{3+} is effected by the following composite reactions (for example):

$$4Ce^{4+} + BrCH(COOH)_2 + 2H_2O \to 4Ce^{3+} + HCOOH + 2CO_2 + 5H^+ + Br^-, \tag{10.5a}$$

$$6Ce^{4+} + CH_2(COOH)_2 + 2H_2O \to 6Ce^{3+} + HCOOH + 2CO_2 + 6H^+. \tag{10.5b}$$

In what follows, we shall identify key reactants in the oscillation. We can establish these by discussing the mechanism whereby this reaction scheme could oscillate. As already stated, process I consumes Br^- and produces Br_2, HOBr (hypobromous acid), and bromomalonic acid ($BrCH(COOH)_2$). Thus, while $[Br^-]$ (the concentration of Br^-) is high and $[HBrO_2]$ is low, process I dominates (because process II requires $HBrO_2$ to operate). Eventually, however, $[Br^-]$ decreases sufficiently, and $[HBrO_2]$ increases sufficiently, that process II takes over. Now process II contains an *autocatalytic* step in reaction (10.3), that is, $HBrO_2 \to 2HBrO_2$; hence $HBrO_2$ is produced rapidly and explosively, and as this happens, $[Br^-]$ decreases rapidly (via Eq. (10.1b)). This is the phase BC in Fig. 10.1; in addition, the increasing level of $HBrO_2$ concentration enables the oxidation of cerium, $Ce^{3+} \to Ce^{4+}$, to take place. Finally, process II draws to a close because the increasing levels of Ce^{4+} enable bromide production to resume via Eq. (10.5a), and once there is enough bromide to allow (10.1b) to become significant, the $HBrO_2$ production shuts off and process I takes over again.

This is of course a schematic description, but it provides a plausible mechanism whereby an oscillation could occur. It tacitly assumes that other species are present in sufficient quantities that their production or consumption does not affect the reaction rates: this is, for example, the case with sulphuric acid (i.e., hydrogen ions H^+) and is supposed true for bromomalonic acid $BrCH(COOH)_2$, although it is produced in process I and absorbed in process II (and might thus be rate limiting). On the basis of our discussion of the mechanism of the reaction, there are three principal reactants: Br^-, $HBrO_2$, and Ce^{4+}. (Because cerium is a catalyst, it is conserved, i.e., $[Ce^{3+}] + [Ce^{4+}] = \text{constant}$.) If we assume (possibly unrealistically) that $[Ce^{4+}] \ll [Ce^{3+}]$, then only the variation of $[Ce^{4+}]$ will have significance. Field and Noyes also include the bromate ion BrO_3^- and the hypobromous acid HOBr, and we follow their example here, although these latter species will be supposed irrelevant. We write the reactions (10.1a,b), (10.3), (10.4), and (10.5) in the form

$$BrO_3^- + Br^- \; [+2H^+] \xrightarrow{k_3} HBrO_2 + HOBr,$$

$$HBrO_2 + Br^- \; [+H^+] \xrightarrow{k_2} 2HOBr,$$

$$[2Ce^{3+}] + BrO_3^- + HBrO_2 \; [+3H^+] \xrightarrow{k_5} 2Ce^{4+} + 2HBrO_2 \; [+H_2O],$$

$$2HBrO_2 \xrightarrow{k_4} BrO_3^- + HOBr \; [+H^+],$$

$$Ce^{4+}\left[+ f\text{BrMA} + \frac{1}{6}(1-4f)\,\text{MA} + \frac{1}{3}(1+2f)H_2O\right] \to \left[Ce^{3+} + \frac{1}{6}(1+2f)HCOOH + \frac{1}{3}(1+2f)CO_2 + (1+f)H^+\right] + fBr^-, \tag{10.6}$$

where MA $= CH_2(COOH)_2$, BrMA $= BrCH(COOH)_2$, and the stoichiometric coefficient f reflects the (unknown) relative importance of the two reactions in Eq. (10.5), in the proportions $4f$ to $1-4f$. Thus $f=\frac{1}{4}$ if Eq. (10.5a) dominates, and $0<f<\frac{1}{4}$ if Eq. (10.5b) is important. As we shall see below (following Eq. (10.26)), this is a drawback of the model; Noyes and Jwo (1975) suggested the 'correct' value should be $f=0.5$. This can of course be justified on the basis of other reactions involving bromide production by the oxidation of cerium or through some intermediary effect of the malonic acid concentration.

The reactants in square brackets, if present in sufficiently large quantities, may be ignored, and the basic reaction scheme is then the following simplified version, where we write the concentrations as

$$\begin{gathered} A=[BrO_3^-], \qquad P=[HOBr], \qquad X=[HBrO_2], \\ Y=[Br^-], \qquad Z=[Ce^{4+}]: \end{gathered} \tag{10.7}$$

$$\begin{aligned} A+Y &\xrightarrow{k_3} X+P \quad \text{(production of } HBrO_2 \text{ from } Br^-\text{)}, \\ X+Y &\xrightarrow{k_2} 2P \quad \text{(production of HOBr)}, \\ A+X &\xrightarrow{k_5} 2X+2Z \quad \text{(oxidation of } Ce^{3+}\text{, autocatalysis of } HBrO_2\text{)}, \\ 2X &\xrightarrow{k_4} A+P \quad \text{(reduction of } HBrO_2\text{, production of HOBr)}, \\ Z &\rightarrow fY \quad \text{(reduction of } Ce^{4+}\text{, regeneration of } Br^-\text{)}. \end{aligned} \tag{10.8}$$

Adding suitable multiples of the reaction equations, we find that the overall reaction is given by

$$A \rightarrow \frac{(1+6f)}{(1+2f)}P. \tag{10.9}$$

The reaction proceeds so long as A (bromate) is present, and the eventual product is P (hypobromous acid). Thus P uncouples from the reaction scheme, and we assume that A changes sufficiently slowly that its concentration can be taken as approximately constant. Applying the law of mass action to the reaction scheme (10.8) gives the following three equations for X, Y, and Z:

$$\frac{dX}{dt} = K_1AY - K_2XY + K_3AX - 2K_4X^2, \tag{10.10a}$$

$$\frac{dY}{dt} = -K_1AY - K_2XY + fK_5Z, \tag{10.10b}$$

$$\frac{dZ}{dt} = 2K_3AX - K_5Z. \tag{10.10c}$$

The rate constants K_1–K_5 are related to those given in Eq. (10.6) by the relations

$$\begin{aligned} K_1 &= k_3[H^+]^2 \sim 2\ M^{-1}\ s^{-1}, \\ K_2 &= k_2[H^+] \sim 3\times 10^6\ M^{-1}\ s^{-1}, \\ K_3 &= k_5[H^+] \sim 40\ M^{-1}\ s^{-1}, \\ K_4 &= k_4 \sim 3\times 10^3\ M^{-1}\ s^{-1}, \end{aligned}$$

and we also take the rate of reduction of Ce^{4+}, K_5, to be

$$K_5 \approx 0.6[\mathrm{BrMA}]\ \mathrm{M}^{-1}\ \mathrm{s}^{-1} \approx 1.8 \times 10^{-2}\ \mathrm{s}^{-1}, \tag{10.11}$$

using experimentally relevant values of $[H^+] \approx 1$ M, [BrMA] ≈ 0.03 M, where M denotes a molar solution (1 mole/liter). The above values are those determined by Tyson (1994). We take A as constant, with $A \approx 6 \times 10^{-2}$ M.

10.2 Relaxation oscillation analysis

The first step is to nondimensionalize the equations. We do this by choosing scales for X, Y, and Z so as to balance important terms on the right-hand sides of (10.10). How this is done is really a matter of trial and error, and in problems that have relaxational behavior (as is often the case), there is no single 'correct' method of scaling. To illustrate this, we will suggest one plausible scaling of the equations and then show in the analysis how other scales naturally occur.

If we balance terms in (10.10c), we have $Z \sim K_3AX/K_5$. Because $f \sim O(1)$, the first three terms in (10.10a) are of comparable order to those in (10.10b). An obvious possible choice of scales is to choose $K_1AY \sim K_2XY \sim K_3AX$, so that all terms in (10.10b) balance and also the first three in (10.10a). This scaling will be consistent if we then find $K_4X^2 \lesssim K_3AX$; otherwise a different scaling needs to be chosen. Each reaction will generally occur on a different timescale, and we simply choose the timescale of one of them. Here we pick the second equation's timescale, so that $t \sim 1/K_1A$. Thus we now write

$$\begin{aligned} X &\sim K_1A/K_2 \sim 4 \times 10^{-8}\ \mathrm{M}, \\ Y &\sim K_3A/K_2 \sim 0.8 \times 10^{-6}\ \mathrm{M}, \\ Z &\sim 2K_1K_3A^2/K_2K_5 \sim 1.1 \times 10^{-5}\ \mathrm{M}, \\ t &\sim 1/K_1A \sim 8.3\ \mathrm{s}; \end{aligned} \tag{10.12}$$

here we deliberately abbreviate the scaling procedure. The choice of the scales in (10.12) is shorthand for the following: write $X = (K_1A/K_2)x$, etc., and substitute in to derive the dimensionless equations (where also we write $t = (1/K_1A)t^*$, and then drop the asterisk for convenience):

$$\varepsilon\dot{x} = x + y - xy - qx^2, \tag{10.13a}$$

$$\dot{y} = 2fz - y - xy, \tag{10.13b}$$

$$p\dot{z} = x - z, \tag{10.13c}$$

where

$$\begin{aligned} \varepsilon &= K_1/K_3 \sim 0.05, \\ p &= K_1A/K_5 \sim 6.7, \\ q &= 2K_1K_4/K_2K_3 \sim 10^{-4}. \end{aligned} \tag{10.14}$$

We see that $q \ll 1$, which corroborates the choice of scaling adopted: for at least some of the time, we can expect $x, y, z = O(1)$. The parameters ε and p are small and large, respectively; thus the reactions apparently happen on different timescales: Eq. (10.13a) rapidly attains quasi-equilibrium, then Eq. (10.13b) less rapidly attains equilibrium, and Eq. (10.13c) is the (slowest) rate-controlling step.

One can use classical analytic techniques to discuss solutions of Eq. (10.13), for example, the existence of different steady states and their stability (see Murray, 1989). For nonlinear behavior, however, one requires either direct numerical computation or the use of singular perturbation theory. Given the existence of ε, p, q, all either small or large, it makes common sense to use this information to construct approximate solutions. In a formal sense, this requires a multiple scales expansion procedure in three independent parameters. In practice, a leading order description can be obtained without such effort, and this is presented below.

For $x, y, z \sim O(1)$, the x reaction is fast and proceeds (on a timescale $O(\varepsilon)$) to a quasi-equilibrium given approximately (because $q \ll 1$) by

$$x \approx y/(y-1), \tag{10.15}$$

providing $y > 1$. Recalling $y \propto [\mathrm{Br}^-]$, this is process I (where y is 'large'). On the intermediate timescale $t \sim 1$, y proceeds to a quasi-equilibrium given by

$$y \approx \frac{2fz}{1+x}, \tag{10.16}$$

that is,

$$y\left[\frac{2y-1}{y-1}\right] \approx 2fz, \tag{10.17}$$

providing $y > 1$. Finally, on the long timescale $O(p)$, z evolves according to

$$p\dot{z} = X(z) - z, \tag{10.18}$$

where X is given parametrically by Eqs. (10.15) and (10.17), that is,

$$X = y/(y-1), \qquad y[(2y-1)/(y-1)] = 2fz. \tag{10.19}$$

In order to understand how z behaves, we proceed graphically by first establishing what $X(z)$ looks like. X as a function of y is shown in Fig. 10.2(i), and $2fz = Z = 2y + 1 + 1/(y-1)$ as a function of y is shown in Fig. 10.2(ii). From these we infer that X as a function of Z $(= 2fz)$ is as shown in Fig. 10.3. We define $t = p\tau$, and write Eq. (10.18) in the form

$$\frac{dZ}{d\tau} = X(Z) - Z/2f. \tag{10.20}$$

We require $Z > 3 + 2\sqrt{2}$ in order that $y > 1$ so that the approximation is consistent. There are two clear cases, depicted in Figs. 10.4(i) and 10.4(ii), corresponding to the cases $f < \frac{1}{2}$ and $f > \frac{1}{2}$. If $f < \frac{1}{2}$, then $X < Z/2f$ for all $Z > 3 + 2\sqrt{2}$, and thus Z decreases inexorably until it reaches $3 + 2\sqrt{2}$, at which point quasi-equilibrium in the fast equations can no longer be maintained, and the assumption that x, y, z remain $O(1)$ must become invalid and the neglected term qx^2 in (10.13a) must become important. On the other hand, if $f > \frac{1}{2}$, an equilibrium exists

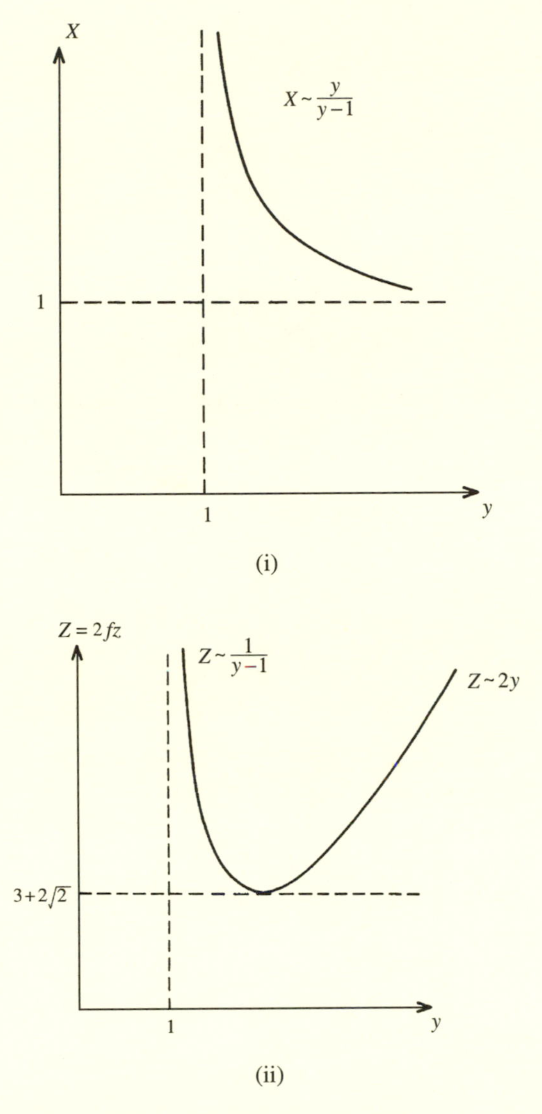

Fig. 10.2. (i) X versus y for process I; (ii) $Z = 2fz$ versus y for process I

at which $X = Z$. By consideration of the *stability* of the quasi-steady solution of (10.13b) (see the exercises), one can show that a stable quasi-equilibrium can only exist on the lower branch in Fig. 10.3. Thus if $1/2f > (\sqrt{2}+1)/(3+2\sqrt{2})$, that is, if $f < (1+\sqrt{2})/2 \approx 1.207$, although there exists an (approximate) equilibrium of Eq. (10.20), it is unstable by virtue of (10.13b). For $f > 1.207$, there exists a stable equilibrium of Eq. (10.20) on the lower branch, that is stable to $O(1)$ disturbances, so long as $\varepsilon, q \ll 1$, $p \gg 1$.

If $f < 1.207$, any initial values $x, y, z = O(1)$ lead to rapid quasi-equilibrium on the lower branch of the $X(Z)$ curve and then Z decreases till $Z = 3 + 2\sqrt{2}$, at which point x (being the fastest variable) must change rapidly, and the term qx^2 must become important. This suggests a rescaling of the equations to make this explicit.

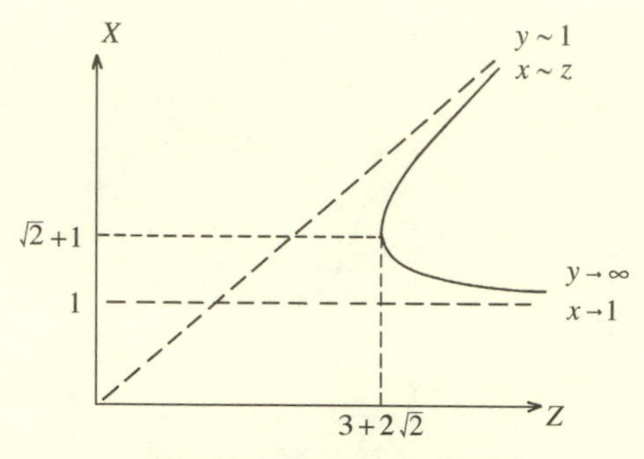

Fig. 10.3. $X(Z)$; process I

We may expect that x goes rapidly toward the other equilibrium of Eq. (10.13a), which (although $y \sim O(1)$) is $x \sim 1/q$; then the equations for y and z suggest that y remains $O(1)$, but $z \sim O(1/q)$. Therefore we rescale Eq. (10.13) as

$$x = \xi/q, \qquad z = \zeta/q; \tag{10.21}$$

we obtain

$$\begin{aligned} \varepsilon\dot{\xi} &= \xi[1 - y - \xi] + qy, \\ q\dot{y} &= 2f\zeta - \xi y - qy, \\ p\dot{\zeta} &= \xi - \zeta. \end{aligned} \tag{10.22}$$

Note that for $\xi, y, \zeta \sim O(1)$, y now reacts on the fastest timescale $t \sim O(q)$. Neglecting $O(q)$, y tends to the quasi-equilibrium value

$$y \sim 2f\zeta/\xi; \tag{10.23}$$

following this, ξ rapidly tends (on a timescale $t \sim O(\varepsilon)$) to a quasi-equilibrium given approximately by

$$\xi \sim 1 - y, \tag{10.24}$$

provided $y > 1$. Thus $\xi \approx X(Z)$, where here

$$\begin{aligned} Z &= 2f\zeta, \\ X(1 - X) &= Z, \end{aligned} \tag{10.25}$$

as illustrated in Fig. 10.5.

Again there are two branches $X(Z)$, and only the upper branch is (quasi-) stable. Thus as y decreases through 1, we suggest that x will rapidly jump to the positive branch of $X(Z)/q$. As before, with $t = p\tau$, Z then satisfies

$$\frac{dZ}{d\tau} = X(Z) - Z/2f. \tag{10.26}$$

If $f < \frac{1}{4}$, then a fixed point of Eq. (10.26) exists on the upper branch and is stable. However if $\frac{1}{4} < f < \frac{1}{2}$, the fixed point is on the lower, unstable branch; and if $f > \frac{1}{2}$, no fixed point exists. In any event, if $f > \frac{1}{4}$, then solutions of Eq. (10.26) will migrate along the upper branch until $Z = \frac{1}{4}$, when they will undergo a rapid transition to process I again.

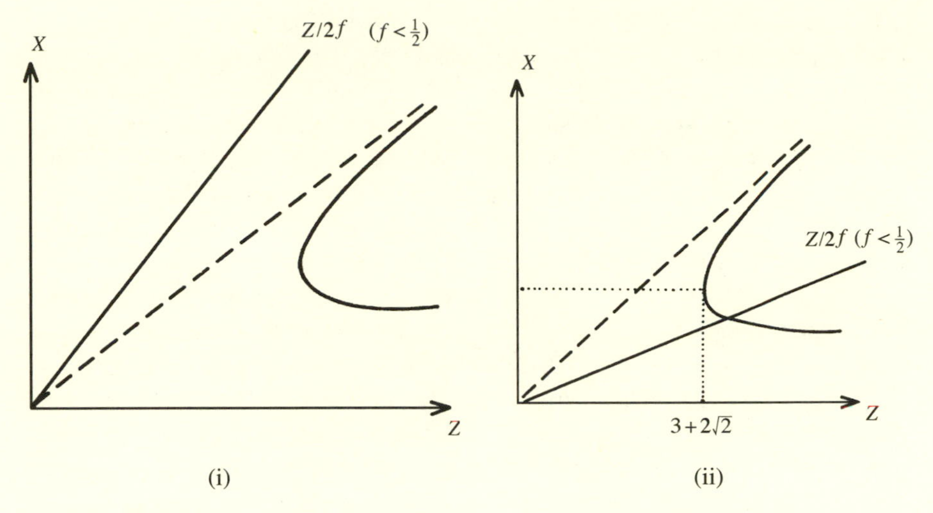

Fig. 10.4. (i) Process I equilibria: $f < \frac{1}{2}$; (ii) Process I equilibria: $f > \frac{1}{2}$

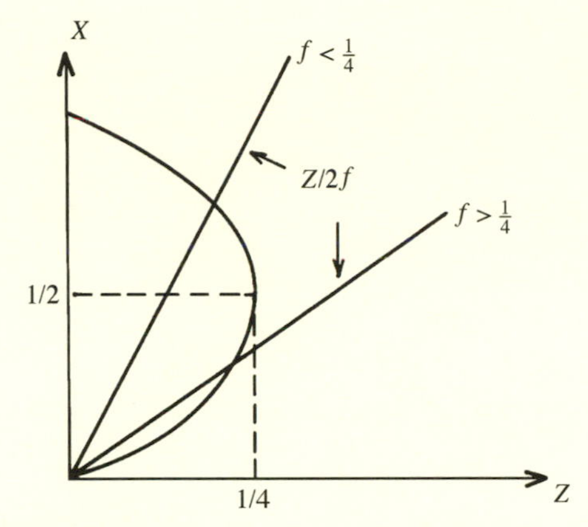

Fig. 10.5. $X(1 - X) = Z$ (see Eqs. (10.25) and (10.26))

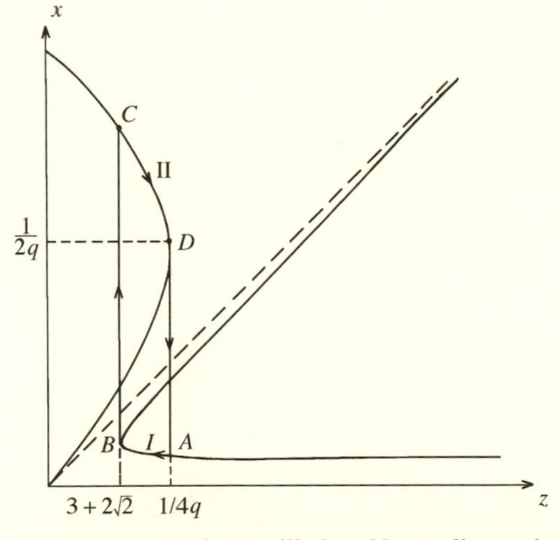

Fig. 10.6. Relaxation oscillation. Not at all to scale

In summary (see Fig. 10.6), we expect a stable process I reaction if $f > 1.207$, a stable process II reaction if $f < 0.25$, and a stable periodic relaxation oscillation between processes I and II if $0.25 < f < 1.207$, providing q is small enough, roughly $q < 1/[4(3+2\sqrt{2})] \approx 0.04$ (and if also ε is small and p is large). We can also see that the characteristics of the oscillation illustrated in Fig. 10.6 are consistent with those in Fig. 10.1. In particular, in process I there is a slow decline of z (i.e., Ce^{4+}) (this is AB in Fig. 10.1). At B, x (and also y, i.e., $[Br^-]$) has a sudden change to the upper branch, whereas z (i.e., Ce^{4+}) stays the same. In process II, z (Ce^{4+}) declines slowly (CD in Fig. 10.1) until D, where y ($\approx Z/X$) jumps sharply, whereas z is essentially unchanged. Furthermore, the timescale for transition from process I to process II is $O(1)$ (cf. Eq. (10.13)), whereas that from process II to process I is $O(\varepsilon)$ (cf. Eq. (10.22)). When ferroin is used as the catalyst, process II corresponds to the reduced (red) state, whereas process I corresponds to the oxidized (blue) state. One observes during the experiment a gradual reddening of the solution, whereas the transition from red to blue is virtually instantaneous. This corresponds to the transition from D to A on the timescale $O(\varepsilon)$.

To quantify the oscillation, we find that process I is approximately described by

$$\frac{dZ}{d\tau} = X(Z) - Z \approx -Z \tag{10.27}$$

(for $Z \gg 1$), with the phase AB corresponding to Z decreasing from $\frac{1}{4}$ to about 0. With $\tau = 0$ at A, then

$$Z \approx \frac{1}{4}e^{-\tau}, \tag{10.28}$$

and at B, $Z = (3+2\sqrt{2})q$; thus

$$\tau \approx \tau_B \approx \ln[1/\{4(3+2\sqrt{2})q\}]. \tag{10.29}$$

At B, x jumps to C, and from C to D, Z approximately satisfies

$$\frac{dZ}{d\tau} = X(Z) - Z, \qquad Z = X(1-X), \tag{10.30}$$

with $Z \approx 0$ ($X = 1$) at τ_B and $Z \approx \frac{1}{4}$ ($X = \frac{1}{2}$) at $\tau = \tau_D$. Thus

$$\frac{dX}{d\tau} = \frac{X^2}{1-2X}, \tag{10.31}$$

whence

$$1 - \frac{1}{X} - 2\ln|X| = \tau - \tau_B; \tag{10.32}$$

thus

$$\tau_D - \tau_B = -1 + 2\ln 2. \tag{10.33}$$

We therefore find that the period of oscillation is, to leading order,

$$P \sim \ln\left[\frac{1}{(3+2\sqrt{2})q}\right] - 1, \tag{10.34}$$

of which a time $\ln[O(1/q)]$ is spent in process I (the oxidized (blue) state), whereas $O(1)$ is spent in process II (the reduced (red) state). This corresponds qualitatively to observations, as described above. However, the value of q (10^{-4}) gives a period $P \approx 6.5$ corresponding to six minutes, rather longer than observed. If a higher value $q = 0.005$ is assumed, then $P \approx 2.5$, corresponding to 140 seconds, close to that observed.

10.3 Notes and references

The Belousov equation is discussed, for example, in the book by Murray (1989). An early book on the reaction is by Tyson (1976) and a more recent one is that edited by Field and Burger (1985). The original Field–Körös–Noyes (FKN) model (Field, Körös, and Noyes, 1972; Field and Noyes, 1974) does portray the basic mechanism, but it has quantitative shortcomings. In particular, appropriate values of the rate constants to be used are uncertain. Later work (Field and Försterling, 1986; Tyson, 1985) suggests the values used here. Other investigators have analyzed higher-order models. Some of these are mentioned by Argoul, Arnéodo, and Richetti (1989), whose seven variable model forms the basis of the fifth question in the exercises; see also Edelson, Field, and Noyes (1975) and Györgi and Field (1991).

Although the batch-run reaction can oscillate periodically, the dynamics in a continuously stirred tank reactor (CSTR) can be chaotic, and this has been experimentally observed (see Argoul et al., 1989). BZ models that can oscillate chaotically have been analyzed by Györgi and Field (1991, 1992). In a petri dish, spatial diffusion causes the oscillation to occur as traveling waves, which propagate slowly in distinctive blue and red striped patterns, often in the form of *spiral waves*. The discovery of such spatial structure has had an enormous impact in the study of pattern formation and morphogenesis in biological systems. See Murray (1989) for a discussion of these topics.

A more detailed asymptotic analysis of the relaxation FKN oscillator is given by Stanshine and Howard (1975) and extended to traveling waves by Stanshine (1976).

Recipes It is easy to demonstrate the BZ reaction in the classroom. Here is a recipe given in Tyson's (1976) book. First, the ingredients:

150 ml	1 M H_2SO_4	1 M
0.175 g	$Ce(NO_3)_6(NH_4)_2$	0.002 M
4.292 g	$CH_2(COOH)_2$	0.28 M
1.415 g	$NaBrO_3$	0.063 M

The first column gives the quantity, the second the substance, and the third the initial molar concentrations. To make a one-molar solution of sulphuric acid (H_2SO_4), dissolve (approximately!) 7.57 ml of concentrated H_2SO_4 in distilled water to a total of 150 ml. Add acid to water! It is an exothermic reaction (and will warm the beaker), and if you do it the other way round, it is likely to explode.

In a beaker with a magnetic stirrer, dissolve malonic acid $CH_2(COOH)_2$ (a powder) and cerium ammonium nitrate ($Ce(NO_3)_4(NH_4)_2$ – also a powder) in the sulphuric acid. The solution will first be yellow, and then, after a few minutes, turn clear. When it is clear, add the sodium bromate (powder); potassium bromate should do as well. The solution will oscillate between yellow (oxidized, Ce^{4+}) and clear (reduced, Ce^{3+}) states. For a more dramatic color change, between red and blue, add a few milliliters of 0.025 M ferroin (1,10 phenanthroline ferrous sulphate).

Next, we give Winfree's (1972) recipe for producing wave patterns in a petri dish. To 67 ml of water add 2 ml of sulphuric acid and 5 g of sodium bromate. To 6 ml of this solution, add 0.5 ml of sodium bromide solution (1 g/10 ml). Add 1 ml of malonic acid solution (1 g/10 ml) and wait for the brownish bromine color to vanish. Add 1 ml of .025 M ferroin (and a drop of 1 g/1000 ml Triton X-100 surfactant to facilitate spreading: not really necessary, however). Mix, pour into a covered petri dish. The patterns are easily displayed by putting the dish on an overhead projector, but it does need a perspex cover to prevent convection of the mixture. Bubbles that are produced are CO_2.

Belousov's recipe uses citric acid rather than malonic acid. The ingredients are (dry, except for H_2SO_4) as follows:

2 g	citric acid,
0.16 g	cerium sulphate,
0.2 g	sodium bromate,
2 ml	sulphuric acid.

Then add to water to a total of 10 ml.

Finally, we give a recipe used by Lou Howard, formerly of M.I.T., and now at Florida State University. There are five stock solutions (use distilled water to avoid chloride ions):

1. 0.32 M potassium bromate solution:
   26.7 g $KBrO_3$ in 500 ml solution.
2. 1.4 M malonic acid:
   72.8 g $CH_2(COOH)_2$ in 500 ml solution.
3. 0.01 M cerous sulphate:
   3.56 g $Ce_2(SO_4)_3.8H_2O$ in 500 ml solution.
4. 4.2 M sulphuric acid solution:
   212 ml 97% H_2SO_4 added to 788 ml H_2O.
5. 2M sodium bromide:
   102.9 g KBr in 500 ml solution.

From there we make

6. Half and half malonic/bromomalonic acids:
   mix 10 ml # 4, 40 ml # 1, 43 ml # 2; add 10 ml # 5 in a glass-stoppered flask and put in stopper; wait till color clears.

Howard's recipes are then

for oscillations: 10 ml each of # 1,2, and 3, and 20 ml of # 4. (For greater visibility add 1–2 ml ferroin.)

for waves: mix 7 ml # 1, 3 ml # 3, 6 ml ferroin, 16 ml # 6, 14 ml # 4, and pour into a petri dish.

Other oscillatory reactions Periodic chemical reactions are not plentiful. They can occur in oxidation reactions in an acidic medium. Three groups of such reactions are characterized by the oxidant: alkaline iodate, bromate, and chlorite. For further information, see the review by Vidal and Pacault (1982) and the book edited by Field and Burger (1985).

Exercises

1. Write down the equation for $A = [BrO_3^-]$ in the Field–Noyes five-reaction model of the BZ reaction. Using a suitable nondimensionalization, examine whether it is consistent to neglect dA/dt in the Field–Noyes model.
2. Show that the Oregonator model of the BZ reaction has a nontrivial steady state, and examine its linear stability. (The Routh–Hurwitz criterion says that the roots of $\lambda^3 + A\lambda^2 + B\lambda + C = 0$ all have negative real part iff $A > 0$, $C > 0$, $AB - C > 0$.) (You may assume $p \gg 1$, $q \ll 1$, $\varepsilon \ll 1$.)
3. The BZ reaction is set up in a thin layer of *unstirred* fluid in a petri dish. Justify modeling this situation by the reaction-diffusion equation

$$\mathbf{c}_t = \mathbf{f}(\mathbf{c}) + D\nabla^2\mathbf{c},$$

where $\mathbf{c}$ is the vector of reactant concentrations and $\mathbf{f}$ is given by the Field–Noyes model. If the dimensionless concentrations of $HBrO_2$, Br^-, and Ce^{4+} are denoted by u, v, and w, respectively, derive the model

$$\begin{aligned} \varepsilon u_t &= u + v - uv - qu^2 + \varepsilon\nabla^2 u, \\ v_t &= 2fw - v - uv + \nabla^2 v, \\ pw_t &= u - w + p\nabla^2 w, \end{aligned}$$

for motions with a length scale $l \sim (D/K_1 A)^{1/2}$. If $D \sim 10^{-5}$ cm^2 s^{-1}, show that $l \sim 10^{-1}$ mm.

Discuss the following description of a propagating wavefront for this system: we anticipate rapid regions of change that join regions where process I and process II apply; that is, in process I we expect $u, v, w \sim 1$, whereas in process II, we expect $u, w \sim 1/q$. We ignore variation of w at the wave front and put $u = \tilde{u}/q$, $w = \tilde{w}/q$; then, approximately,

$$\begin{aligned} \varepsilon\tilde{u}_t &= \tilde{u} - \tilde{u}v - \tilde{u}^2 + \varepsilon\nabla^2\tilde{u}, \\ qv_t &= 2f\tilde{w} - \tilde{u}v + q\nabla^2 v, \\ p\tilde{w}_t &= \tilde{u} - \tilde{w} + p\nabla^2\tilde{w}. \end{aligned}$$

Rescale t via $t = \varepsilon\tilde{t}$ and space via $\mathbf{x} = \varepsilon^{1/2}\tilde{\mathbf{x}}$, drop tildes, to obtain (with $\tilde{w}$ constant)

$$\begin{aligned} u_t &= u(1 - u - v) + \nabla^2 u, \\ (q/\varepsilon)v_t &= 2f\tilde{w} - uv + (q/\varepsilon)\nabla^2 v. \end{aligned}$$

If $q \ll \varepsilon$, we may put $v \approx 2f\tilde{w}/u$, so that u satisfies (with $\lambda = 2f\tilde{w}$)

$$u_t = u(1 - u) - \lambda + \nabla^2 u;$$

discuss what should be plausible boundary conditions for this equation. Examine this equation in one space dimension for possible traveling wave solutions of the form $u = u(x - ct)$. (See, for example, Murray, 1977.) Show that the dimensional wave speed is then $c(K_3 AD)^{1/2}$ (experiments confirm this, with $c = 1.7$; see Tyson (1985)).

4. Write down a model set of equations for the BZ reaction that include the concentration of bromo-malonic acid BrMA as a variable. Discuss the relevance of this variable to the system in terms of the relevant rate constants in Eqs. (10.1) and (10.5). (Field et al. (1972) give rates $d[\text{BrMA}]/dt = 1.3 \times 10^{-2}[\text{H}_+][\text{MA}]\ \text{M}^{-1}\ \text{s}^{-1}$ for Eq. (10.1d) and $-d[\text{BrMA}]/dt = k'_{10}[\text{Ce}^{4+}][\text{BrMA}]/\{K''_{10} + [\text{BrMA}]\}$, $k'_{10} = 1.7 \times 10^{-2}\ \text{s}^{-1}$, $K''_{10} = 0.2$ M, for Eq. (10.5a).)

5. The model considered by Argoul et al. (1991) is the following

$$
\begin{aligned}
&R_1 : \text{BrO}_3^- + \text{Br}^- + 2\text{H}^+ &&\xrightarrow{k_1} \text{HOBr} + \text{HBrO}_2;\\
&R_2 : \text{HBrO}_2 + \text{Br}^- + \text{H}^+ &&\xrightarrow{k_2} 2\text{HOBr};\\
&R_3 : \text{HOBr} + \text{Br}^- + \text{H}^+ &&\xrightarrow{k_3} \text{Br}_2 + \text{H}_2\text{O};\\
&R_4 : \text{BrO}_3^- + \text{HBrO}_2 + \text{H}^+ &&\underset{k_{-4}}{\overset{k_4}{\rightleftharpoons}} 2\text{BrO}_2^\bullet + \text{H}_2\text{O};\\
&R_5 : 2\text{HBrO}_2 &&\xrightarrow{k_5} \text{HOBr} + \text{BrO}_3^- + \text{H}^+;\\
&R_6 : \text{BrO}_2^\bullet + \text{Ce}^{3+} + \text{H}^+ &&\xrightarrow{k_6} \text{Ce}^{4+} + \text{HBrO}_2;\\
&R_7 : \text{HOBr} + \text{MA} &&\xrightarrow{k_7} \text{BrMA} + \text{H}_2\text{O};\\
&R_8 : \text{BrMA} + \text{Ce}^{4+} &&\xrightarrow{k_8} \text{Br}^- + \text{R}^\bullet + \text{Ce}^{3+} + \text{H}^+;\\
&R_9 : \text{R}^\bullet + \text{Ce}^{4+} &&\xrightarrow{k_9} \text{Ce}^{3+} + \text{P}.
\end{aligned}
$$

Write down the equations governing the seven variables

$$
X_1 = [\text{Br}^-], \quad X_2 = [\text{HBrO}_2], \quad X_3 = [\text{HOBr}], \quad X_4 = [\text{BrO}_2^\bullet],
$$
$$
X_5 = [\text{Ce}^{4+}], \quad X_6 = [\text{BrMA}], \quad X_7 = [\text{R}^\bullet],
$$

where $R^\bullet$ is an oxidized derivative of malonic acid. Analyze the model and compare with the Oregonator. (Argoul et al. use the values

$$
\begin{aligned}
k_1 &= 2.1\,\text{M}^{-3}\,\text{s}^{-1},\\
k_2 &= 2 \times 10^9\,\text{M}^{-2}\,\text{s}^{-1},\\
k_3 &= 8 \times 10^9\,\text{M}^{-2}\,\text{s}^{-1},\\
k_4 &= 10^4\,\text{M}^{-2}\,\text{s}^{-1},\\
k_{-4} &= 2 \times 10^7\,\text{M}^{-2}\,\text{s}^{-1},\\
k_5 &= 4 \times 10^7\,\text{M}^{-1}\,\text{s}^{-1},\\
k_6 &= 10^3\,\text{M}^{-1}\,\text{s}^{-1},\\
k_7 &= 10^6\,\text{M}^{-1}\,\text{s}^{-1},\\
k_8 &= 5 \times 10^3\,\text{M}^{-1}\,\text{s}^{-1},\\
k_9 &= 10^8\,\text{M}^{-1}\,\text{s}^{-1},\\
[\text{BrO}_3^-] &= 5 \times 10^{-2}\,\text{M},\ [\text{MA}] = 10^{-3}\,\text{M},\\
[\text{Ce}^{3+}] &= 1.05 \times 10^{-4}\,\text{M},\ [\text{H}^+] = 1.5\,\text{M}.)
\end{aligned}
$$

6. The BZ experiment is carried out in a CSTR, so that the reaction mixture is drained at a constant rate and replaced by bromate at the same rate. Derive the corresponding Oregonator model, and analyze the dynamics as a function of the residence time.

7. To the FKN (Oregonator) model, add the reactions

$$R1 : H^+ + HOBr + Br^- \rightleftharpoons Br_2 + H_2O,$$
$$R8 : Br_2 + CH_2(COOH)_2 \rightarrow BrCH(COOH)_2 + Br^- + H^+,$$

with rate constants for $R1$:

$$k_{R1}\text{ (forward) } = 9 \times 10^9\ M^{-2}\ s^{-1}, \qquad k_{-R1}\text{ (reverse) } = 110\ s^{-1}.$$

Assume $[H^+] \sim 1$ M, $[BrMA] \sim 10^{-3}$ M, and $[MA] \sim 0.03$ M, are essentially constant. Field et al. (1972) quote the rate of reaction R8 as

$$r_{R8} = k_8^0[H^+][MA]\ M\ s^{-1}, \qquad k_8^0 = 1.3 \times 10^{-2}\ M^{-1}\ s^{-1}.$$

This is based on rapid equilibration of $R1$. Assuming $R1$ is in equilibrium, show that the rate constant k_{R8} for $R8$ is related to the rate of reaction r_{R8} $(= d[BrMA]/dt)$ by

$$k_{R8} = \frac{k_8^0}{[HOBr][Br^-]}(k_{R1}/k_{-R1}).$$

Use the FKN model (dimensionless version) to estimate typical values for [HOBr] and $[Br^-]$, and use these to estimate k_{R8}. Now derive an augmented FKN model by appending an equation for $U = [Br_2]$ to those for $X = [HBrO_2]$, $Y = [Br^-]$, and $Z = [Ce^{4+}]$. (*Don't* assume equilibrium for $R1$, *a priori*.) Nondimensionalize and scale (using $[H^+]$, etc., as before).

Examine the following questions: Is it reasonable to have $R1$ in equilibrium? What dynamic effect does the extra equation have on the oscillation? For example, look at the linear stability of the steady state and the singular approximation to the limit cycle. Feel free to look at variations in k_{R8}, [MA], etc., in order to facilitate approximations (and hence insight).

8. The Lotka system of chemical reactions is governed by the scheme

$$A + X \xrightarrow{k_1} 2X,$$
$$X + Y \xrightarrow{k_2} 2Y,$$
$$Y \xrightarrow{k_3} P,$$

where k_1, k_2, k_3 denote rate constants and $2X$ means $X + X$, etc.

Write down a set of equations for the concentrations $[A]$, $[X]$, $[Y]$, and $[P]$ of the chemical species. If, during the reaction, $[A]$ is approximately constant, show that $[X]$ and $[Y]$ are determined by two coupled differential equations, and show that a nontrivial steady state $[X] = X_0$, $[Y] = Y_0$ exists for these.

Use the steady states to nondimensionalize the equations, and show that dynamic behavior of $[X]$ and $[Y]$ is determined by a single parameter $\delta = k_3/k_1[A]$.

Show that small perturbations to the steady state are oscillatory, and show that the *dimensional* period of oscillation is $2\pi/(k_1k_3[A])^{1/2}$.

If k_3 is known, how could you use this result to find k_1 experimentally? What is the restriction on the value of k_1 that allows $[A]$ to be considered as constant? (You may assume $\delta = O(1)$.)

9. A reaction scheme is given by

$$\begin{aligned} U + Y &\xrightarrow{k_1} X + P, \\ X + Y &\xrightarrow{k_2} 3X, \\ U + X &\xrightarrow{k_3} Z, \\ Z &\xrightarrow{k_4} fY. \end{aligned}$$

Write down a mathematical model based on the law of mass action.

By scaling the model, show that the equations can be written in the form

$$\begin{aligned} \dot{u} &= -u(y + x), \\ \delta_1\dot{x} &= uy + 2xy - ux, \\ \delta_2\dot{y} &= -uy - xy + fz, \\ \delta_3\dot{z} &= ux - z. \end{aligned}$$

and define the parameters δ_1, δ_2, and δ_3. (**Hint:** choose $U = U_0u$, where U_0 is the initial concentration of U.)

Use the values

$$\begin{aligned} k_1 &= 10^6\ \mathrm{M}^{-1}\ \mathrm{s}^{-1}, \\ k_2 &= 10^7\ \mathrm{M}^{-1}\ \mathrm{s}^{-1}, \\ k_3 &= 10^5\ \mathrm{M}^{-1}\ \mathrm{s}^{-1}, \\ k_4 &= 10^3\ \mathrm{s}^{-1}, \\ U_0 &= 10^{-4}\ \mathrm{M}, \end{aligned}$$

to estimate values of δ_1, δ_2, and δ_3, and hence show that

$$\dot{u} \approx -\frac{-2f(1-f)u^2}{(2f-1)}, \quad \text{with } u(0) = 1,$$

provided that $\frac{1}{2} < f < 1$.

10. Tyson (1985) uses a different scaling for the FKN model, where he balances

$$K_2XY \sim K_3AX \sim 2K_4X^2 \sim \frac{1}{2}K_5Z$$

and chooses $t \sim K_5^{-1}$. Derive the corresponding dimensionless model and analyze it. Is there any advantage to this choice of scales?

11

Spruce budworm infestations

Pine and spruce forests in North America are susceptible to occasional outbreaks of infestation by various insects, notably the spruce budworm and the forest tent caterpillar. Plagues of these insects can occur at regular intervals, with a typical period on the order of thirty to forty years. In view of the significance of such outbreaks to commercial logging operations (an outbreak can cause $\$10^8$ worth of damage to the timber industry), an understanding of the dynamics of the tree/insect ecosystem is of some importance.

The spruce budworm infestations provide a good example of how different models arise in practice. Holling, Jones, and Clark (1979) developed a simulation model for use by the Canadian Forestry Service. This model consisted of some 30,000 difference equations that essentially described a discretized tree population density as a function of age class and spatial location, both of which are functionally essential variables for practical forestry management. As such, this model is designed for evaluation of different pest management strategies by simulation. However, if we are more immediately concerned with the *scientific* problem of *why* budworm infestations occur in an oscillatory manner, then a simpler model may in fact be more relevant in obtaining insight.

In contrast to the Holling simulation model, the Ludwig–Jones–Holling model (from a 1978 paper) consists of only three ordinary differential equations and is designed to investigate the problem of how and why the oscillations occur; as we shall see, the model is capable of giving strong analytic insights into the behavior of the system. The initial task is to identify the relevant variables, which must be included in any sensible model. The first is the density of budworm larvae, B, measured in units of larvae per acre (of forest). In the presence of healthy tree foliage, the budworm population will undergo the usual exponential growth. There are two factors that act to limit this growth. The first is the predation of budworm larvae by birds and other small mammals, and the other is the denudation of foliage by the budworms themselves. The first will depend essentially on the population B, but the second requires a second variable that measures the health of the foliage, for example, in foliage units (fu), which are taken as proportional to needles per branch. This second variable is denoted as E. Deforestation occurs when trees are denuded for several years in a row; they then wither and die. We thus require a variable to describe tree density. Specifically, we choose S to represent the branch surface area per acre (of forest), which takes account not only of the number density of healthy trees but also their age and is the physically useful measure of tree density.

Exponential growth of budworm at low densities corresponds to the equation

$$\frac{dB}{dt} = r_B B, \tag{11.1}$$

where r_B is the natural fecundity. Populations are in practice limited by finite resources, and the simplest realistic model of this is the logistic model

$$\frac{dB}{dt} = r_B B(1 - B/K_B), \tag{11.2}$$

where K_B represents the saturation population density. From what has already been said, this density (larvae per acre) will depend both on the available branch surface area and on the foliage health. It is obvious to choose

$$K_B = KS, \tag{11.3}$$

so that K is the carrying capacity in units of larvae per branch surface area. K itself must depend on E, the foliage health variable, such that when $E \to 0$ (total denudation) $K \to 0$, whereas when E is 'large,' K tends to a constant limit K'. The simplest model is the logistic curve $K = K'E/(T_E + E)$, but we also wish to take into account the possibility of budworm competition for resources when E is very low. That is, the logistic curve assumes $K \sim E$ when $E \to 0$: surviving larvae use up what resources are available, and the remainder are content to die. But in reality, larvae will compete for low levels of resources, and the sustainable population is likely to be lower than $O(E)$, because fewer will be able to maintain an adequate intake of nutrient. The simplest model that represents such specific competition is the sigmoidal curve

$$K = \frac{K'E^2}{T_E^2 + E^2}. \tag{11.4}$$

Predation, for example, by birds, is represented by a sink term in the equation for B,

$$\frac{dB}{dt} = \ldots - g(B). \tag{11.5}$$

At high values of B, we expect a finite bird population to be glutted with larvae, and then $g \to \beta$ as $B \to \infty$, say. Again, at low values of B, $g \to 0$, but competition will act to reduce predation below $O(B)$, because birds will simply seek other sources of food, rather than scour branches religiously. Thus we pose

$$g = \frac{\beta B^2}{\alpha^2 + B^2}, \tag{11.6}$$

where in addition we choose

$$\alpha = \alpha' S, \tag{11.7}$$

in keeping with the fact that the variable that really affects predation tendencies must be B/S (larvae per branch area), because it is this variable that birds directly *see*. The equation for B is thus

$$\frac{dB}{dt} = r_B B \left[1 - \frac{B}{K'S} \frac{(T_E^2 + E^2)}{E^2} \right] - \frac{\beta B^2}{(\alpha' S)^2 + B^2}. \tag{11.8}$$

Growth of S from seedlings will be exponential for low values of S, but competition for light and space will also lead to a saturation value of S. Moreover, the saturation value will depend on foliage health E, because when E is low, die-back will occur, leading to more stunted trees with fewer branches. We thus write

$$\frac{dS}{dt} = r_s S \left[1 - \frac{S}{E} \frac{K_E}{K_S} \right], \tag{11.9}$$

where K_E and K_S are saturation values of E and S; the saturation value of K_E reflects a logistic growth process for E:

$$\frac{dE}{dt} = r_E E \left(1 - \frac{E}{K_E} \right) \ldots, \tag{11.10}$$

which reflects the idea that fully healthy trees will have a limited foliage capacity. Exponential growth at low values of E is analogous to that of the other variables, though with less obvious justification. Equation (11.10) simply follows Ludwig, Jones, and Holling's (1978) assumption. However, if one associates the 'health' variable with healthy needles per branch, one might in fact expect that at $E = 0$, growth is at a constant rate. Then we would suggest $dE/dt \approx r_E(1 - E/K_E) \ldots$. However, E is intended to measure the overall health of a tree, and when denudation occurs, there is an associated decrease in the vitality of the tree. For this reason, it is sensible to take the regeneration rate as tending to zero as $E \to 0$, thus corroborating Eq. (11.10).

In addition, defoliation by budworm consumption gives a sink term

$$\frac{dE}{dt} = \ldots - P(B/S), \tag{11.11}$$

where B/S is the relevant budworm density, in units of larvae per branch surface area. This takes account of the fact that the 'energy' variable E is an *intrinsic* variable (in thermodynamic terms) that provides a relative measure of health per unit branch area. The coefficient P is the consumption rate (the denudation rate per budworm). Evidently $P \to 0$ as $E \to 0$ (there must be something to eat), and we can expect the previously mentioned effects of competition to give a faster than linear decay; thus $P \sim E^2$ as $E \to 0$ (this reflects the fact that when needles are scarce, budworm compete for the limited resources). Furthermore, we suppose P saturates as E becomes large – so long as not too many budworm are present, they can eat as much of a healthy tree as they want to. In fact, the competition and saturation alluded to here is just that included in Eq. (11.4); thus we choose the same form for P:

$$P = \frac{P' E^2}{T_E^2 + E^2}. \tag{11.12}$$

Gathering the three equations together, we have the model

$$\begin{aligned}\frac{dB}{dt} &= r_B B\left[1 - \frac{B}{K'S}\frac{(T_E^2 + E^2)}{E^2}\right] - \beta\frac{B^2}{(\alpha' S)^2 + B^2},\\ \frac{dS}{dt} &= r_S S\left[1 - \frac{S}{E}\frac{K_E}{K_S}\right],\\ \frac{dE}{dt} &= r_E E\left(1 - \frac{E}{K_E}\right) - P'\frac{B}{S}\frac{E^2}{(T_E^2 + E^2)},\end{aligned} \tag{11.13}$$

and as for the BZ reaction, we are interested in the possibility of sustained (relaxation) oscillations.

11.1 Nondimensionalization and scale analysis

Extensive field work reported in Morris (1963) enabled reasonable estimates to be made for most of the parameters, and these are as follows:

$$\begin{aligned} r_B &\sim 1.5\,\text{year}^{-1},\\ K' &\sim 355\,\text{larvae branch}^{-1},\\ T_E &\sim ?\,(\ll 1)\ (\text{units of fu br}^{-1}),\\ \beta &\sim 43200\,\text{lv acre}^{-1}\text{y}^{-1},\\ \alpha' &\sim 1.1\,\text{lv br}^{-1},\\ r_S &\sim .1\,\text{y}^{-1},\\ K_S &\sim 25440\,\text{br acre}^{-1},\\ K_E &\sim 1\ \text{fu br}^{-1},\\ r_E &\sim .9\,\text{y}^{-1},\\ P' &\sim .002\ \text{fu lv}^{-1}\ \text{y}^{-1}.\end{aligned} \tag{11.14}$$

The only uncertainty lies in the transition value T_E, which measures the 'health' variable at which budworm begin to have enough resources. Ludwig et al. (1978) analyzed the model with $T_E = 0$. However, in this case, if the predation rate P' is high enough, E can be depleted to zero; in which case $S \to 0$, and the population dies out. Thus in order to model oscillations, we need to retain $T_E \neq 0$. We shall generally suppose $T_E \ll 1$ (as we expect, i.e., regeneration does not take place till $E < T_E$ is very low). The choice of $K_E = 1$ simply reflects the choice of E as a relative measure of health.

Our first task is to nondimensionalize the model in a useful way. The longest exponential growth time is r_S^{-1}, which we choose as a timescale. Morris (1963) records outbreaks in New Brunswick in 1770, 1806, 1878, 1912, and 1949, suggesting a periodicity of 35 years (missing 1842). This is consistent with a timescale of ~10 years ($= r_S^{-1}$), which suggests that this is a sensible scaling. It is simplest to suppose the other variable scales to be the saturation values; thus we write

$$B = K'K_S B^*, \qquad S = K_S S^*, \qquad E = K_E E^*, \qquad t = r_S^{-1} t^*, \tag{11.15}$$

as (by now) usual, and drop the asterisks. The resultant dimensionless model is then

$$\varepsilon_1 \frac{dB}{dt} = B\left[1 - \frac{B}{S}\left(\frac{\delta^2 + E^2}{E^2}\right)\right] - \frac{\lambda B^2}{(\nu^2 S^2 + B^2)}, \tag{11.16a}$$

$$\frac{dS}{dt} = S[1 - S/E], \tag{11.16b}$$

$$\varepsilon_2 \frac{dE}{dt} = E(1 - E) - \frac{\gamma B}{S}\left(\frac{E^2}{\delta^2 + E^2}\right), \tag{11.16c}$$

where

$$\begin{aligned} \varepsilon_1 &= r_S/r_B \sim 1/15, \\ \lambda &= \beta/r_B K' K_S \sim .4 \times 10^{-2}, \\ \nu &= \alpha'/K' \sim .3 \times 10^{-2}, \\ \delta &= T_E/K_E \ll 1, \\ \varepsilon_2 &= r_S/r_E \sim 1/9, \\ \gamma &= P'K'/r_E K_E \sim 0.7. \end{aligned} \tag{11.17}$$

We see that the scaling is *consistent*, insofar as all the parameters are $< O(1)$, and we can begin trying to understand whether the model, *with these parameter values*, will oscillate.

As with the Belousov reaction, it is possible to do stability analysis on steady states, but as we did there, we will now simply attempt a direct asymptotic solution of the system. On a fast timescale ($\sim$1 year), that is, $t \sim \varepsilon_1, \varepsilon_2$, we expect B and E to reach quasi-equilibrium given by

$$B \sim S, \qquad E(1 - E) \sim \gamma, \tag{11.18}$$

where we take λ, ν, $\delta \ll 1$. Equation (11.18) has two solutions, of which that for larger E is (quasi-) stable (see Fig. 11.1). Thus if $\gamma < 1/4$, we expect E, B (and thus S) to evolve toward a steady state given by the larger root of $E(1 - E) = \gamma$, namely, $E = [1 + (1 - 4\gamma)^{1/2}]/2$. We see, however, that $\gamma \approx 0.7$. Thus we must ask what happens if $\gamma > 1/4$.

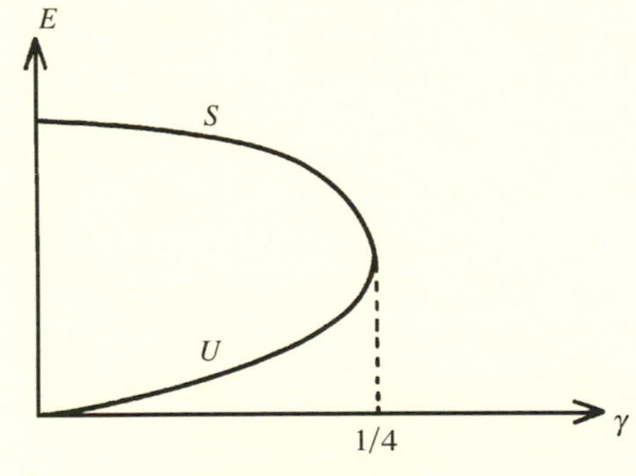

Fig. 11.1. Equilibria for E if $\gamma < 1/4$

11.2 Ludwig–Jones–Holling (LJH) analysis

The fast equations for $B/S = \theta$ and E in Eq. (11.16) can be written in the approximate form

$$\varepsilon_1 \frac{d\theta}{dt} \approx \theta[1-\theta] - (\lambda/S)\left(\frac{\theta^2}{\nu^2+\theta^2}\right),$$
$$\varepsilon_2 \frac{dE}{dt} \approx E(1-E) - \gamma\theta, \tag{11.19}$$

so long as $E \gg \delta$. As before, $\theta \approx 1$ is a stable quasi-equilibrium, but then E (and so also S) decreases unmercifully if $\gamma > 1/4$. Let us therefore consider the equilibria of Eq. (11.19) more carefully. These are given by

$$(\nu S/\lambda)(1-\theta) = \left[\frac{\theta/\nu}{1+(\theta/\nu)^2}\right], \tag{11.20}$$

as illustrated in Fig. 11.2. For $S > \lambda/2\nu$, the large $O(1)$ value is unique and stable. With $\gamma > 1/4$, E decreases, and thus also S decreases. When S becomes of $O(\lambda)$, then Eq. (11.20), written in the form

$$1-\theta = \frac{\lambda}{S}\left(\frac{\theta}{\nu^2+\theta^2}\right), \tag{11.21}$$

has the approximate roots

$$\theta \approx \frac{1}{2} \pm \frac{1}{2}(1-4\lambda/S)^{1/2}, \tag{11.22}$$

which coalesce and disappear when $S = 4\lambda$. At this point θ would jump (on the basis of Eq. (11.19)) to the lower approximate equilibrium

$$\theta \approx \nu^2 S/\lambda \; (= O(\nu^2)); \tag{11.23}$$

this immediately lowers the level of foliage removal to a very small value, and allows for slow regeneration of E according to

$$\varepsilon_2 \frac{dE}{dt} \approx E - \gamma\nu^2 S/\lambda \tag{11.24}$$

over a long timescale. Thus after the sudden jump down in B (i.e., θ), the intermediate

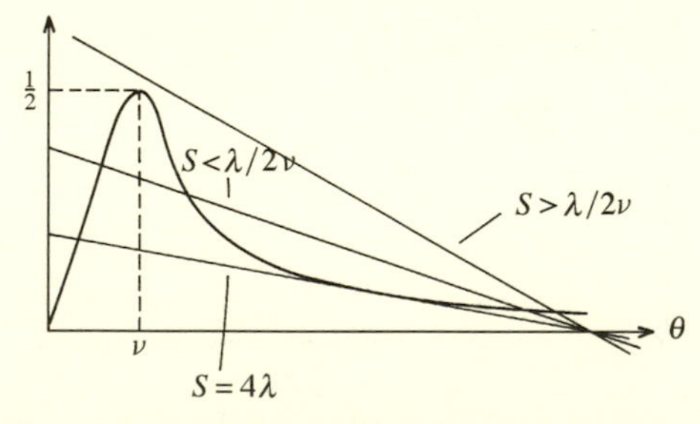

Fig. 11.2. Equilibria of Eq. (11.20) for various S

rate equation for S will carry S toward E (in time $O(1)$), and then E evolves from its very low value exponentially, according to

$$\frac{dE}{dt} \approx \frac{1}{\varepsilon_2}\left[1 - \frac{\gamma\nu^2}{\lambda}\right]E \approx E/\varepsilon_2; \tag{11.25}$$

although the apparent timescale is $O(\varepsilon_2)$, the very low value of E will require a long time to recover. In fact, for this to be $> O(1)$, we require E to reach exponentially low levels.

The final part of the oscillation then occurs as E increases toward one, because S also increases, and when $S \approx \lambda/2\nu\,(<1)$, the two lower roots of Eq. (11.21) coalesce; at this point, the budworm population rapidly jumps to the high outbreak quasi-equilibrium again, and the cycle repeats itself.

11.3 Summary

From our discussion, we have indicated that, provided δ is small enough (i.e., if $E \gg \delta$), the model may be able to undergo relaxation oscillations, even if we approximate $(E^2 + \delta^2)/E^2$ by 1; this is in contrast to Ludwig et al.'s (1978) statements on the basis of a numerical solution: we comment on this further below. Neglecting δ, the model can be written

$$\begin{aligned} \varepsilon_1\frac{dB}{dt} &= B\left(1 - \frac{B}{S}\right) - \frac{\lambda B^2}{(\nu^2S^2 + B^2)}, \\ \frac{dS}{dt} &= S\left(1 - \frac{S}{E}\right), \\ \varepsilon_2\frac{dE}{dt} &= E(1 - E) - \gamma B/S. \end{aligned} \tag{11.26}$$

We now trace the schematic sequence of events during an oscillation.

Stage I: Foliage recovery

Here E grows from a low value, followed by S. B is approximately in equilibrium given by

$$B \approx \frac{\lambda}{2} - \left\{\frac{\lambda^2}{4} - \nu^2S^2\right\}^{1/2}, \tag{11.27}$$

and because S grows more slowly than E, we have $S \ll E$; thus although $E \sim 1$, $S \ll 1$ and

$$B \sim \nu S^2/\lambda; \tag{11.28}$$

$S \sim e^t$ and E grows rapidly to equilibrium $E \approx 1$ according to

$$\varepsilon_2\frac{dE}{dt} = E(1 - E) \tag{11.29}$$

(because $\gamma B/S \sim \nu S/\lambda \sim S \ll 1$).

Stage II: Tree recovery; outbreak

E reaches 1, and then S recovers, and approaches 1 via its own logistic growth curve. B grows according to Eq. (11.20), and the situation would lead to a steady state with low budworm population if $\lambda > 2\nu$. With the values cited, however, $\lambda/2\nu = 2/3$; and so when S reaches $\lambda/2\nu$, B jumps rapidly from $\lambda/2$ ($\ll 1$) to the outbreak value $B \sim S$. As this jump takes place (on a timescale $O(\varepsilon_1)$), E begins to decline rapidly (on a timescale $O(\varepsilon_2)$) according to

$$\varepsilon_2 \frac{dE}{dt} = E(1 - E) - \gamma; \tag{11.30}$$

if $\gamma < 1/4$, E would reach a new lower equilibrium, and a constant high value of B could be maintained indefinitely – permanent outbreak. However, our estimates give $\gamma = 0.7$, so that Eq. (11.30) leads to a rapid decay toward zero; and as $E \to 0$, we have $E \sim \gamma|t|/\varepsilon_2$. It then follows that S begins to follow E down toward zero, and although there is a slight delay, this is followed by a precipitate decrease, so that $S \sim [A + (\varepsilon_2/\gamma)\ln(1/|t|)]^{-1}$ (where A is constant). Thus S and E hurtle toward zero at the same time, and B follows S down according to Eq. (11.22): $B \sim \frac{S}{2}[1 + (1 - 4\lambda/S)^{1/2}]$.

Stage III: Termination

The outbreak can be terminated by S decreasing to reach 4λ, at which point B jumps rapidly (in time $O(\varepsilon_1)$) to the post-outbreak value $B \sim \nu^2 S^2/\lambda = O(\lambda\nu^2)$. This is about $O(10^{-7})$ of the $O(1)$ values prevalent during an outbreak, which suggests that this quantitative analysis may not apply ($O(10^{-3})$ is more realistic). This sudden drop in predation rate will cause a rapid stabilization of foliage. Now we might expect $S \sim \lambda$ when $|t| \sim \exp(-1/\lambda\varepsilon_2)$, and then $E \sim \frac{1}{\varepsilon_2}\exp(-1/\lambda\varepsilon_2)$. For any reasonable values of λ and ε_2, this is ridiculously small, and would be much less than any practical value of δ. Therefore, it seems unlikely that one can seriously avoid the incorporation of δ into the analysis. Furthermore, the decline in E can only be arrested if B drops rapidly; that is, we would probably need $\varepsilon_1 \ll \varepsilon_2$, which is not in fact the case.

The following recovery occurs as in stage I. E begins to recover exponentially, but it is vital in the model that this be enabled by a sufficiently realistic prescription of regeneration and consumption. In order to estimate such quantities as the period, we would estimate the recovery time from a minimum value of $E_{\min}$ to $O(1)$ as $t \sim \varepsilon_2 \ln(1/E_{\min}) \sim 1/\lambda$. This is indeed long, but too long, and these several foregoing inadequacies suggest that we try to seek a satisfactory description when E does in fact reach $O(\delta)$.

11.4 Finite saturation foliage health, revisited

The LJH analysis was really predicated on the basis that the S and E equations were both slow, so that the slow variation of S and E could cause the budworm population to shuttle back and forth between the multiple steady states of Eq. (11.16a). But in reality, both B and E are fast variables, and moreover, we cannot legitimately ignore

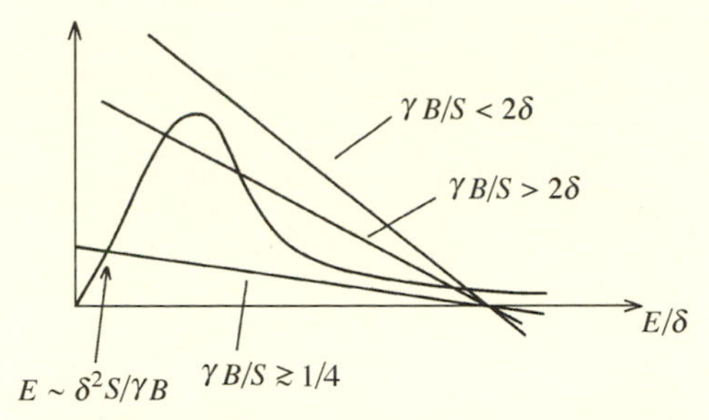

Fig. 11.3. Equilibria of Eq. (11.31) for various B

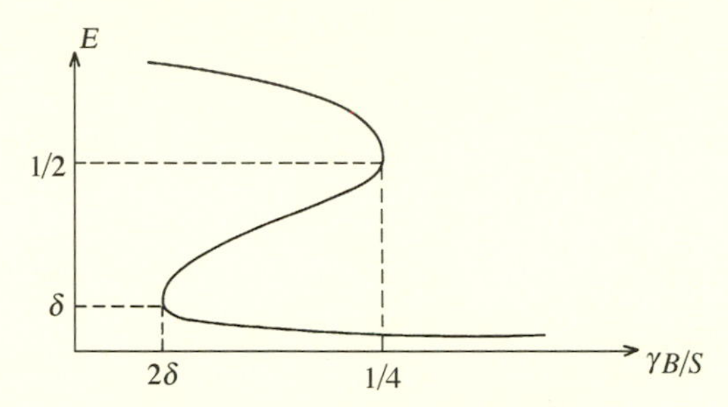

Fig. 11.4. Variation of E with B/S

the saturating effect of a nonzero δ. In this section, we therefore examine the possible multiple equilibria of B *and* E.

Equilibria of E satisfy

$$(1-E)/\{\gamma B/\delta S\} = \frac{E/\delta}{1+(E/\delta)^2}, \tag{11.31}$$

and as shown in Fig. 11.3, there are one or two *stable* steady states, depending on whether $\gamma B/S \gtrless 1/4$, $\gamma B/S \gtrless 2\delta$. There is evidently hysteresis as $\gamma B/S$ increases through $1/4$ and as $\gamma B/S$ decreases through 2δ (approximately). The lower stable state is approximately $E \sim (\gamma B/2S) - \{(\gamma B/2S)^2 - \delta^2\}^{1/2}$, with the turning point at $\gamma B/S \sim 2\delta$, whereas the upper branch is approximately $E \sim \frac{1}{2} + (\frac{1}{4} - \frac{\gamma B}{S})^{1/2}$, with the knee at $\gamma B/S \sim 1/4$; see Fig. 11.4.

Equilibria of B are given by zeros of Eq. (11.16a); thus

$$\frac{\nu S}{\lambda}\left[1-\nu\left(\frac{\delta^2+E^2}{E^2}\right)\frac{B}{\nu S}\right] = \frac{(B/\nu S)}{1+(B/\nu S)^2}. \tag{11.32}$$

The values of $B/\nu S$ are determined much as for E, but in addition depend on E. There are two essential cases: when $E \sim 1$, $(\delta^2+E^2)/E^2 \sim 1$, and when $E < \delta$, $(\delta^2+E^2)/E^2 \approx \delta^2/E^2$. We consider the two cases separately, although both can be represented schematically, as in Fig. 11.5. These give rise to the hysteretic curves in Fig. 11.6, which differ only in the position of the left-hand nose of the curve. Figures 11.4 and 11.6 are redrawn in Fig. 11.7 for convenience.

We can now describe the evolution of B, S, and E on the basis that the fast variables B and E will be located on stable branches of these curves, with rapid transition between them at the turning points. The slow time variation is caused by

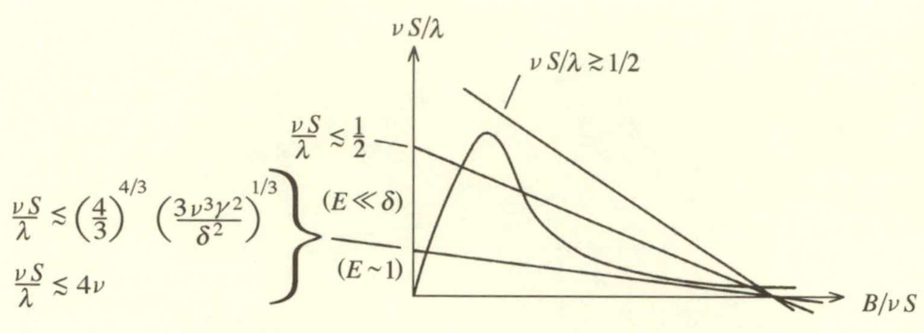

Fig. 11.5. Equilibria of Eq. (11.32) for various S

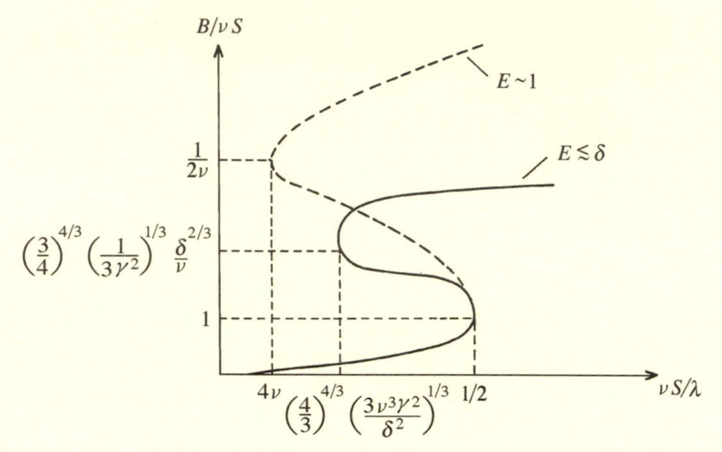

Fig. 11.6. B/S versus S, for large or small E

S trying to reach E. Let us consider what happens when starting from an initial healthy state, where B is low, but S and E are high (a pre-outbreak stage). This corresponds to point e on Fig. 11.7(i), f on Fig. 11.7(ii). If $S \neq E$, then S tends toward E. Suppose $S < E$; then S increases (fg in Fig. 11.7(ii)), so B/S increases and E decreases. A healthy, stable equilibrium then exists if there is a value $S = E$ such that $E > 1/2$ and $S < \lambda/2\nu \ll 1/4\gamma$; in practice, $E > 1/2$ is assured, because on fg, $B/S < \nu \ll 1/4\gamma$, and therefore such a stable equilibrium exists if

$$S = E \sim \frac{1}{2} + \left(\frac{1}{4} - \frac{\gamma B}{S}\right)^{1/2}, \qquad \frac{B}{S} \sim \frac{\lambda}{2S} - \left\{\left(\frac{\lambda}{2S}\right)^2 - \nu^2\right\}^{1/2} \tag{11.33}$$

has a root, that is,

$$S = \frac{1}{2} + \left[\frac{1}{4} - \frac{\lambda\gamma}{2S} + \gamma\left\{\left(\frac{\lambda}{2S}\right)^2 - \nu^2\right\}^{1/2}\right]^{1/2}. \tag{11.34}$$

Because $\lambda \sim \nu \ll 1$, such a root has $S \sim 1$ ($>1/2$), hence approximately $S \approx 1$, provided $(\lambda/2S) > \nu$, that is, provided $\lambda/2\nu > 1$. Thus if

$$\lambda > 2\nu, \tag{11.35}$$

there exists a stable healthy state in which $S = E \approx 1$ and $B \approx (\lambda/2) - \{(\lambda/2)^2 - \nu^2\}^{1/2}$. In our case, $\lambda/2\nu = 2/3$, so this is not the case.

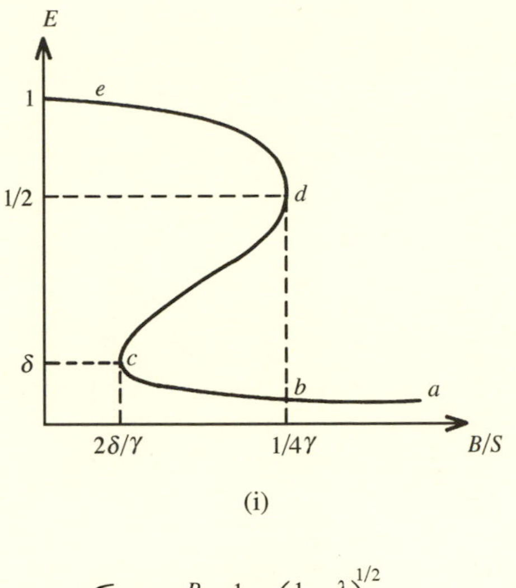

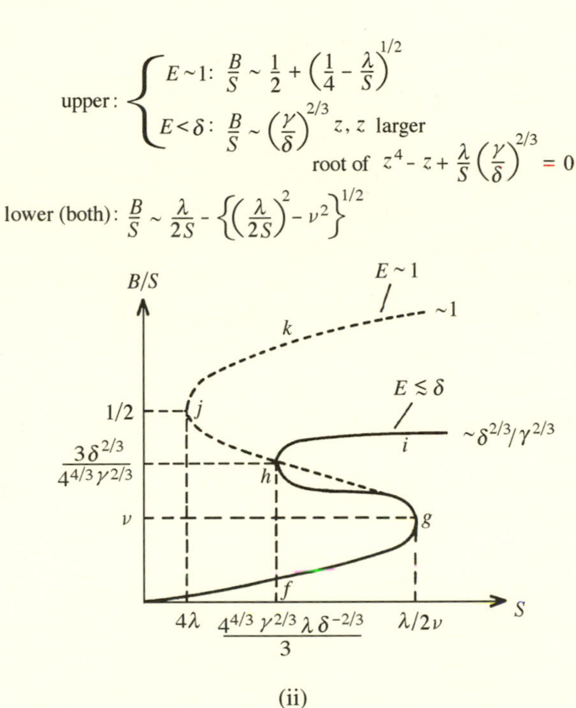

Fig. 11.7. (i) E, B/S and (ii) B/S, S equilibrium curves

S now increases to $\lambda/2\nu$, where it jumps to the upper portion of the dotted curve (because $E \sim 1$), on which $B/S \sim 1$. At the same time, E rapidly follows the upper branch *ed* of Fig. 11.7(i). If $1/4\gamma > 1$, then E remains $O(1)$, and a stable defoliated state will exist. Thus if

$$\gamma < 1/4, \tag{11.36}$$

a stable pestilent state exists in which $B/S \sim 1$, $S = E \sim \frac{1}{2} + (\frac{1}{4} - \gamma)^{1/2}$. In our case, however, $\gamma \approx 0.7$, so this possibility is not allowed either. In fact, when B/S

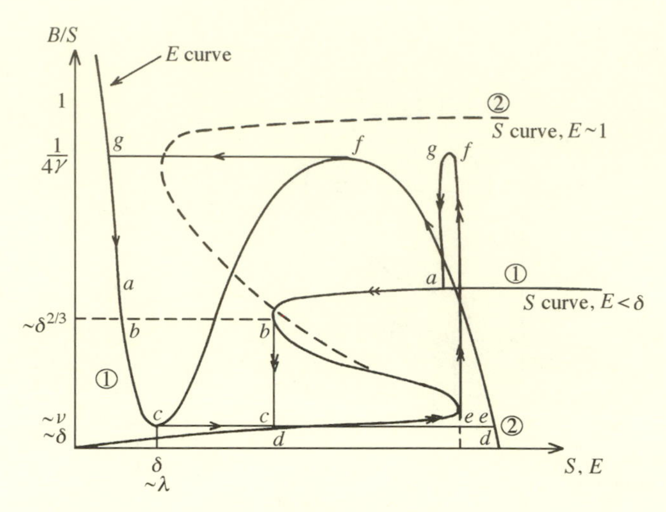

Fig. 11.8. Phase plot of an oscillatory solution

jumps to $O(1)$ in Fig. 11.7(ii), it simultaneously falls off the $E/(B/S)$ curve at e in Fig. 11.7(i), and E collapses to the lower branch at b. In consequence, B/S collapses in Fig. 11.7(ii) to the upper branch of the lower ($E < \delta$) curve. Thus in the outbreak, B jumps from $O(\nu)$ to $O(1)$ and back to $O(\delta^{2/3})$, whereas E falls from $O(1)$ to $o(\delta)$ (in fact, $O(\delta^2)$). Because now $S > E$, S decreases along the upper branch ih of the $E < \delta$ curve until h, where $S \sim \lambda\delta^{-2/3}$, which we suppose $> \delta > E$ (if $\lambda > \delta^{5/3}$; otherwise a stable state could exist on hi, with $E < \delta$, $S < \delta$, $B \sim \delta^{5/3}$, i.e., a very weak forest with quite low budworm density). Then B/S drops off the upper branch at h to f; if the resulting value of B/S, that is, $O(\nu)$, is less than $O(\delta)$ (i.e., if $\nu < \delta$), then E is driven along bc to c where there is again a rapid transition to $E \sim 1$. The cycle now repeats.

The sequence of events is illustrated in Fig. 11.8, where the B/S versus S and B/S versus E curves are superimposed. We understand how to 'operate' Fig. 11.8 as follows. For a given S, measured on the abscissa, we take the corresponding ordinate on the S curves to give the value B/S; we choose curve 1 for $E \sim 1$, that is, corresponding to the right-hand branch of the E curve, and curve 2 for $E < \delta$, that is, corresponding to the left-hand branch. For given B/S on the S curves 1 or 2, we find the value of E on the E curve by taking the same ordinate B/S and finding the corresponding abscissa. This process is illustrated in Fig. 11.9 for two pairs (S_1, E_1) on curve 1 and (S_2, E_2) on curve 2. Then if $S < E$, S increases (slowly), whereas if $S > E$, S decreases. In Fig. 11.8, we show the course of the oscillation described above. Corresponding points a, b, c, etc., are marked on each curve. The outbreak occurs from e to f but is rapidly suppressed again from g to a, by which time E has decreased dramatically. Notice the intermediate threshold ab, where S decreases slowly and B is moderate, which is followed by a rapid decrease in B and consequent refoliation. A time series of the oscillation is shown in Fig. 11.10. It compares favorably to numerical results of Ludwig et al. (1978). In particular, S decreases to a minimum of $O(\lambda/\delta^{2/3})$, while B decreases to $O(\nu S)$. Thus if $\lambda = O(\delta^{2/3})$, S remains $O(1)$ throughout. The period between outbreaks is

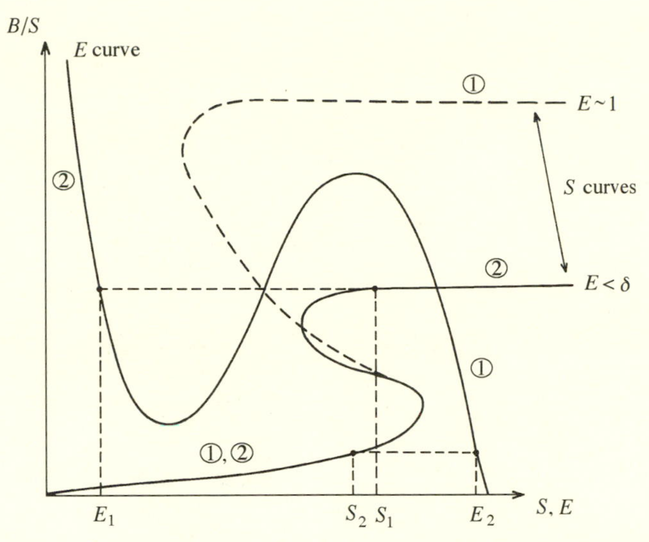

Fig. 11.9. Evolution of the oscillation in terms of (S, E) pairs on the two equilibrium curves

governed to leading order by the time spent between d and e (because the passage from a to b is relatively rapid, as $E \ll 1$). At d, $S \sim (4^{4/3}/3)\lambda(\gamma/\delta)^{2/3}$, whereas at e, $S \sim \lambda/2\nu$. Approximately, $E \approx 1$. The time t_{de} between d and e is thus

$$t_{de} \approx \int_{\frac{4^{4/3}\lambda}{3}(\gamma/\delta)^{2/3}}^{\lambda/2\nu} \frac{dS}{S(1-S)},$$

that is,

$$t_{de} \approx \ln\left[\frac{\{3 - 4^{4/3}\lambda(\gamma/\delta)^{2/3}\}}{(2\nu - \lambda)4^{4/3}(\gamma/\delta)^{2/3}}\right]. \tag{11.37}$$

We can use this to get a plausible estimate for δ. If we take $t_{de} = 3$ (30 years, allowing 5 years for the other parts of the period), and with other values given by Eq. (11.17), we find, for what it is worth, $\delta^{2/3} \approx 7.4 \times 10^{-2}$, $\delta \approx .02$, and the minimum value of S is 0.09. Actually, the time from e to f and from c to d may not be insignificant, as it involves essentially exponential growth from a very low value. Exponential growth of B from $O(\nu)$ to $O(1)$ would take $t \sim \varepsilon_1 \ln(1/\nu)$, whereas that of E from $O(\delta)$ to $O(1)$ is $t \sim \varepsilon_2 \ln(1/\delta)$. The first is $\sim 1/3$ (3 years), whereas the second would be $\sim .4$ (4 years) if δ were 0.02. If we take $t_{de} = 2.5$, we get $\delta^{2/3} \approx 4.7 \times 10^{-2}$, $\delta \approx .01$, and the minimum value of S is 0.14.

11.5 Synopsis

The analysis given here is simple in principle, if complicated in practice. How can one generally hope to find a way through the maze of possibilities available? When the brain fails, the computer may take over, but it is unlikely that numerical computations could come close to producing the detailed qualitative information we have obtained here. In such cases, numerical computation is best used as a verificational tool, and *always*, one needs to know what question one is asking. But the general methodology

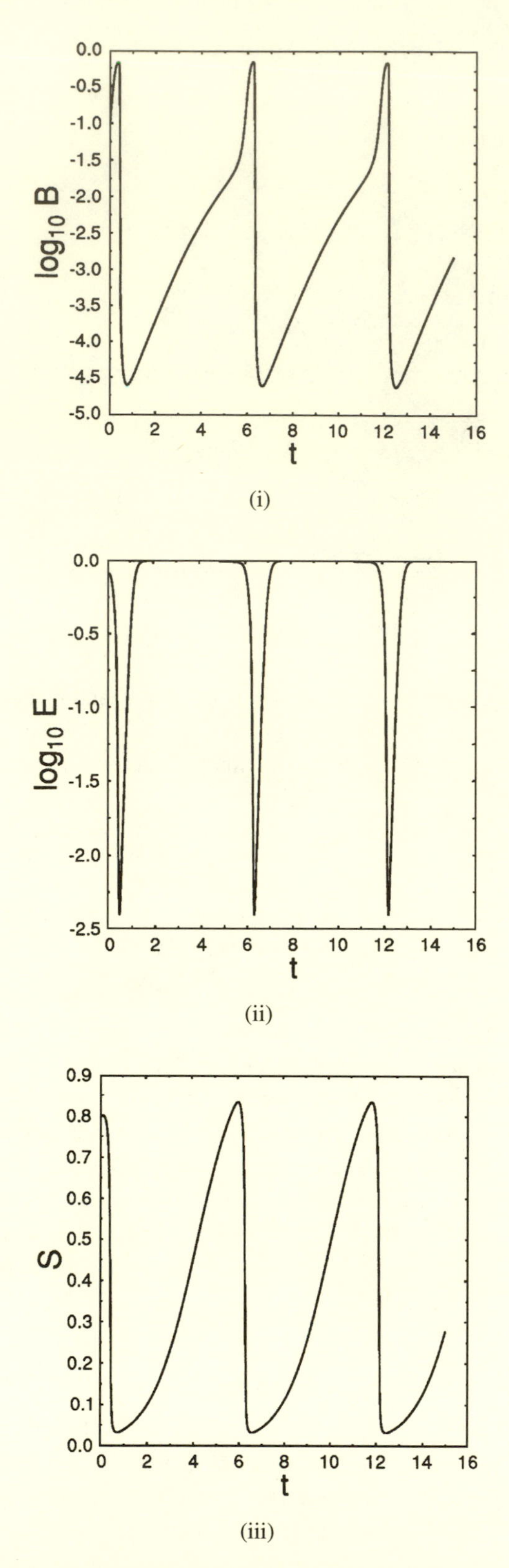

Fig. 11.10. Time series of a numerical solution of the budworm model for (i) B, (ii) E, and (iii) S. Parameter values used are $\lambda = 0.004, \nu = 0.003, \delta = 0.02, \gamma = 0.7, \epsilon_1 = 0.066$, and $\epsilon_2 = 0.11$. Note the logarithmic scales on B and E.

of singular perturbation analysis is exemplified by our approach, where we have been led by plausible analyses to develop a description that is internally consistent, insofar as it goes. Although the analysis may seem complicated, it actually contains very little other than first year calculus and analytic geometry!

11.6 Notes and references

The material here is based largely on the seminal paper by Ludwig et al. (1978). There is also an informative undergraduate teaching module, 'Man in competition with the spruce budworm: an application of differential equations,' prepared under the auspices of Professors Beltrami and Frauenthal at S.U.N.Y., Stony Brook (Department of Applied Mathematics and Statistics). Much of the data used in designing the large-scale simulation and subsequently the three-variable LJH model was collected in the forests of New Brunswick and published under the aegis of R.F. Morris (1963), and provides the parameter estimates given in the text. A summary of the then-current state of the spruce budworm problem is given in the series of nine short articles by Belyea et al. (1975). Other relevant work is in the papers by McNamee, McLeod, and Holling (1981), who compare prototype models for various different insect-forest systems; Stedinger (1984), who discusses improvements of the original site model (Jones, 1979); and Horowitz and Ioinovici (1985), who analyze the Ludwig et al. (1978) system as a feedback control system, where the controls are spraying (an extra term $-B[1-\exp(-m)]$ in the B equation) and logging (a term $-lS$ in the S equation) – compare Exercise 2 below. Régnière and You (1991) discuss a simulation model, with particular emphasis on the foliage dynamics.

Budworm outbreaks are common in Michigan, Maine, Minnesota, and the eastern provinces of Canada. It is only in New Brunswick that the oscillatory character of the outbreaks is manifested. Elsewhere fire is the primary killer of mature forests. In New Brunswick, insecticide spraying has been used since the 1950s to control the outbreaks. The result is that budworm infestation is rife, with most of New Brunswick in a fragile condition of incipient outbreak. The economic costs of the spraying program have continued to rise, but the result of its cessation would be an immediate and disastrous outbreak.

The simulation model A part of the province of New Brunswick is divided into 393 sites, each being a rectangle of 66 square miles; 95% of the province is covered by these sites. In each site, seventy-eight variables are monitored, these being (i) budworm density, (ii) quantity of new (current year's) foliage, (iii) quantity of old foliage, and seventy-five age classes of trees (or, more specifically, the land fraction of trees of the different classes); these classes are, roughly, (1) age ≤ 20 years, (2–74) ages 21–93 years, and (75) age ≥ 94 years. Another input representing variable weather conditions is incorporated. The model then computes the values of the $78 \times 393 = 30{,}654$ state variables from one year to the next: it is essentially a 3×10^4 dimensional map.

A description of the model is given by Jones (1979) (see also Holling et al. (1979)). Although we may have given the impression that the LJH model is realistic, it falls far short of the more detailed budworm site model, which includes detailed description of different larval stages, budworm survival under conditions of migration, predation and parasitism, different consumption of new and old foliage, and so on. Two examples will illustrate the radically different possibilities for modeling various terms.

We modeled the defoliation rate as

$$d_1 = \frac{P' N_L E^2}{T_E^2 + E^2}, \tag{11.38}$$

where N_L $(= B/S)$ is the larval density per branch surface area. By contrast, Jones (1979) puts

$$d_1 = E[1 - \exp(-d_0 N_L / E)], \tag{11.39}$$

and in fact this refers to the consumption rate of new foliage (and E is the new foliage density): Consideration is also given to the (different) rate of consumption of old foliage. The point is that Eq. (11.38) emphasizes the consumption rate per larva as a function of foliage density, whereas Eq. (11.39) focuses on the consumption of foliage as a function of larval density.

A second term that is treated in a more methodical way by Jones (1979) is the budworm reproduction rate. We use a reproduction rate

$$r = r_B \left[1 - \frac{N_2}{K'} \left(\frac{T_E^2 + E^2}{E^2}\right)\right]. \tag{11.40}$$

Jones treats reproduction through the composition of fecundity, taken as a nonlinear function of pupal weight, itself proportional to foliage consumption rate d_1 above, and survival, which is itself a function of E. The resulting functional forms would be rather different.

Dispersal The simulation model includes a description of egg dispersal. Adult female moths can be transported a typical distance of 25 miles in a single night. One natural way to describe dispersal is by a diffusion term, providing there is no bias in direction. (However, prevailing winds may imply that an advective term is more realistic.) With the inclusion of spatial variation arises the possibility (just as for the Belousov reaction) of propagating waves of outbreak. Ludwig et al. (1979) studied this possibility, in particular with regard to two questions: what is the critical size of a patch of forest that can support an outbreak? and what is the width of an effective barrier to the spread of an outbreak?

Exercises

1. Let $\delta, \lambda, \nu \to 0$ and $B, E, S \sim O(1)$ in the dimensionless spruce budworm model. Show that a pair of nontrivial steady states exist if $\gamma < 1/4$, and examine their linear stability, without assuming ε_1 and ε_2 are small. Examine the limit in which ε_1 and ε_2 *are* small. (Note: the criterion for the roots of

a cubic $\lambda^3 + a_1\lambda^2 + a_2\lambda + a_3 = 0$ to have all roots with Re $\lambda < 0$ are that $a_i > 0$ for each i and $a_1a_2 > a_3$. More generally, one has the Routh–Hurwitz conditions, given in Appendix 2 of the book by Murray (1989).)

Use a phase-plane analysis to describe fast trajectories (on a timescale $t \sim \varepsilon_i$) in the (B, E) plane in the cases $\gamma > 1/4$, $\gamma < 1/4$.

2. One way of modeling the use of pesticide spraying is to add a term $-pB$ to the right-hand side of Eq. (11.8), where p is a measure of spraying intensity. Justify, or improve, the use of such a term, and analyze its effect on the dynamics of the model. Consider also variants, such as if $p = H(B - B_c)$ (H being the Heaviside step function), to indicate spraying only if B exceeds a critically monitored value.

3. We have indicated that periodic outbreaks occur if $\mu = \lambda/\nu < 2$, $\gamma > 1/4$. What happens in other parts of the (μ, γ) plane? By examining the definitions of μ and γ, consider which of the following nontoxic defense strategies might work: introduce other, budworm-resistant tree species; introduce a budworm-specific predator. Can you think of any other strategies?

4. Suppose that budworm larvae disperse (via adult moth migration) through a random-walk mechanism, where the net distance traveled in a time interval Δt is the random variable d. It can be shown that the average result of this for a population of dispersers is a diffusion term with diffusion coefficient D equal to half the variance of d; specifically, we can estimate D via the ensemble statistic (estimated variance)

$$D \approx \frac{1}{2N\Delta t}\sum_{i=1}^{N} d_i^2,$$

where d_i is the distance traveled by the ith random walker: then $D = \overline{d^2}/2\Delta t$. Consider the simplified (dimensionless) budworm model with diffusion and an appropriate length scale, where we take $S = E$ and suppose $\varepsilon_1 \ll \varepsilon_2$. Show that an outbreak can exist as a traveling wave of speed c, where c is approximately the wave speed of the Fisher equation $u_t = u_{xx} + u(1-u)$. (The Fisher equation has traveling waves for all $c \geq 2$, but $c = 2$ is what is observed in practice.) Show that the outbreak has a top hat profile for $\theta = B/S$, which travels into healthy forest and leaves destruction in its wake, and that the approximate width of the outbreak zone (where $\theta \approx 1$) is

$$X \approx \frac{8\varepsilon_2}{(4\gamma - 1)^{1/2}}\sin^{-1}(1/2\sqrt{\gamma}).$$

Hence give the dimensional width X_d, and estimate typical values for a range of values of the diffusion coefficient. What are the corresponding (dimensional) wave speeds?

5. Rather than lumping the branch area into a single variable S, we wish to consider the foliage variable F (for E) as a function of age class a (compare the discussion of the simulation model in Section 11.6). Discuss the merits

of the following model as a suitable generalization of Eq. (11.13):

$$\frac{\partial F}{\partial t} + \frac{\partial F}{\partial a} = r_F F\left[1 - \frac{F}{K_F}\right] - P'\frac{B}{S}\frac{F^2}{\left(T_F^2 + F^2\right)},$$

$$\frac{\partial h}{\partial t} + \frac{\partial h}{\partial a} = r_F F h\left[1 - \frac{hK_F}{K_h F}\right],$$

$$\frac{\partial N}{\partial t} + \frac{\partial N}{\partial a} = -\mu(F)N,$$

$$\frac{\partial B}{\partial t} = r_B B\left[1 - \frac{B}{Ks}\right] - \frac{\beta B^2}{(\alpha' s)^2 + B^2},$$

where F, h, and N are the age distributions of tree foliage, tree height, and tree density as functions of age a and time t, so that

$$n = \int_0^\infty N(a, t)\, da, \qquad f = \int_0^\infty N(a, t)F(a, t)\, da$$

are the number of trees per acre of forest and the quantity of foliage (e.g., needles per branch per acre of forest). Also the branch surface area distribution is

$$S = \sigma h^2,$$

where σ is a shape factor, and the total branch surface area per acre of forest is

$$s = \int_0^\infty N(a, t)S(a, t)\, da.$$

B is the budworm density (larvae per acre of forest), and the carrying capacity K (larvae per branch area) is given by

$$K = \frac{K' f^2}{T_F^2 + f^2}.$$

The model is supported by initial conditions for B, F, N, and h on $t = 0$, and for E, h on $a = 0$, and also

$$N(0, t) = \rho[N^* - n].$$

12

Chemical reactors

Chemical reactions typically involve the release or absorption of heat. Formation of a compound from its constituent molecules requires energy, or heat of formation, and this heat is drawn from the surroundings. Thus we can assign a quantity ΔH to any reaction, which is the heat of formation in units of heat per unit mass, normally taken as calories per mole. Thus the reduction of wüstite (FeO) by coke (C) occurs via the net reaction

$$\mathrm{FeO + C \rightarrow Fe + CO}, \tag{12.1}$$

which requires 37,884 cal/mole FeO to take place. Reactions such as this that absorb heat are called *endothermic*, whereas reactions that release heat are called *exothermic*. Thus temperature is affected by the reactions that occur, and in many industrial processes, this coupling is extremely important. The reason is that as well as releasing or absorbing heat as they occur, reactions inevitably proceed at rates that are functions of temperature. This is more or less obvious, because a reaction consists of an alteration in the energy state of the molecules involved, and where these have a higher internal energy (hence temperature), the reaction can proceed more easily.

These combined couplings lead to negative feedback (for an endothermic reaction) or positive feedback (for an exothermic reaction); for example, in the latter case, a higher temperature leads to higher reaction rates, and hence higher heat releases. This self-enhancement can lead to a phenomenon known as *thermal runaway*, in which the temperature increases without limit, and in fact does so in finite time. In practice, this can cause melting, or (in the case of gaseous products) explosion. Such phenomena are not desirable in industrial plants, so that some understanding of the interaction of temperature and reactions is required.

Many industrial plants are concerned with the production of metals from their ores. An iron blast furnace serves to produce iron from iron ore, which may consist, for example, of various iron oxides. Coke and iron ore pellets are fed in at the top of the furnace (see Fig. 12.1) in alternate layers. The coke acts as a catalyst. At high temperatures it is oxidized to CO, which then reacts with the oxides to reduce them to iron. Hot air to enable these reactions to proceed is blasted in through *tuyères*. Near the tuyères, the principal reactions are typically

$$\begin{aligned} \mathrm{C + O_2} &\rightarrow \mathrm{CO_2}, \\ \mathrm{C + CO_2} &\rightarrow \mathrm{2CO}, \end{aligned} \tag{12.2}$$

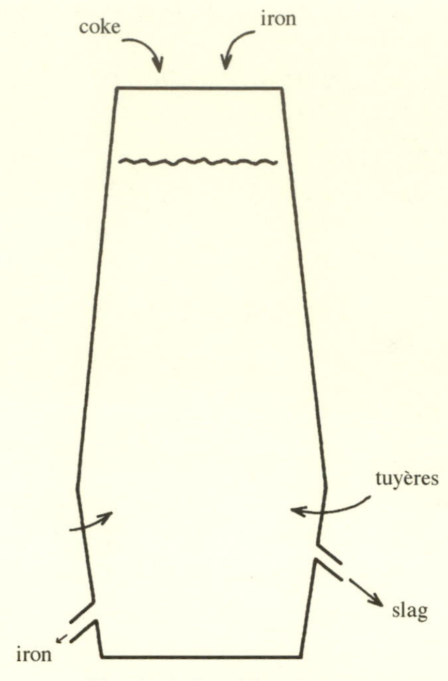

Fig. 12.1. Iron blast furnace

whereas further up, the reduction step occurs via

$$\begin{aligned} FeO + CO &\rightarrow Fe + CO_2, \\ CO_2 + C &\rightarrow 2CO. \end{aligned} \tag{12.3}$$

The overall result of Eqs. (12.2) and (12.3) is Eq. (12.1). Blast furnaces are massive objects, typically 40 meters high and 10 meters in diameter, and useful modeling of the coupled reaction heat transfer problem is of some interest (and difficulty).

A different type of reactor is the *fluidized bed roaster* (see Fig. 12.2), which is used in roasting sulphides. Here a typical reaction involves the reduction of pyrite (FeS_2):

$$4FeS_2 + 11O_2 \rightarrow 2Fe_2O_3 + 8SO_2, \tag{12.4}$$

producing iron oxides (here hematite) for subsequent reduction to iron. Notice the sulphur dioxide emission, which leads to acid rain. It is not only industrially important, but also environmentally important to understand the reaction dynamics. The reaction (12.4) is *exothermic*, with $\Delta H = -2 \times 10^5$cal/mole FeS_2. This is one good reason why roasting is done in a fluidized bed: the necessity for conducting the reaction briskly leads to high temperature in the solid pellets, and the fluidized bed enables this heat to be dissipated efficiently. Reaction in a packed bed (with pellet to pellet contact) could lead to damage via sintering or even melting.

If we wish to model the processes in such reactors, we must realize that the reactions take place on the surface of solid pellets or within the interior if they are porous. Particularly in the sulphide roasting operation, any large-scale model should

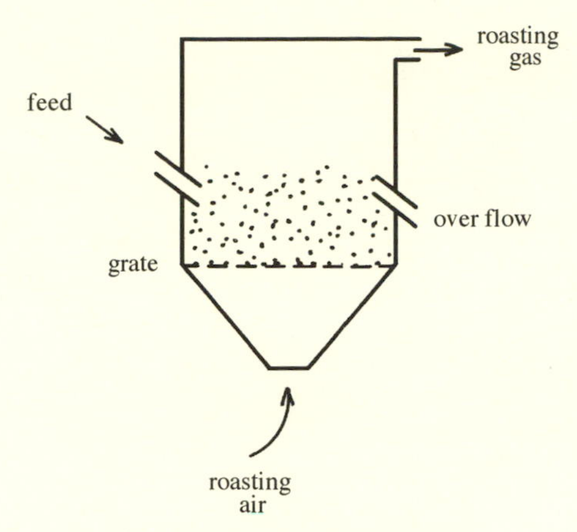

Fig. 12.2. Fluidized bed roaster

treat solid and gas as having locally different temperatures, and the gas-to-particle heat transfer must be controlled by the rate at which the reaction proceeds within the pellet. It thus becomes necessary to separate the modeling effort into two parts: the macroscopic, reactor-scale problem, and a local (microscopic) description of the reaction and heat transfer within a single particle. The first requires the second, and thus we begin with a description of reaction processes for a single pellet.

12.1 Mathematical modeling

The reaction process can be separated into several separate stages (see Fig. 12.3). As the gaseous reactant flows past the pellets, the reactant diffuses to the surface of the pellet. Equally, gaseous products from the reaction can diffuse to the gas flow. The rate of diffusion depends on the differential gas concentration and the local mass transfer coefficient. If we suppose that the pellet is porous, then the gaseous reactant will diffuse through the pore space, as shown in Fig. 12.3(ii). Depending on the nature of the reaction, the next step may involve solid-state diffusion through a solid crust (the 'ash' or residue produced by the reaction) to the unreacted solid; alternatively, if only gaseous products occur, the reactant is adsorbed (and desorbed) on the solid catalyst surface (Fig. 12.3(iii)). On the surface, the reaction occurs; and as a consequence, heat transfer through the pellet takes place.

There are possible complications in this description; in particular, these depend on whether the pellet is porous or not and on whether it shrinks as it reacts or not. As a consequence, a wide variety of different models may be appropriate, depending on circumstances. The book by Szekely, Sohn, and Evans (1976) gives a thorough discussion of many such models from the point of view of the chemical engineer. We shall only scour the surface of this complicated subject here.

An important notion in building a model is the idea of a *rate-controlling step*. If we think, for example, of a sequence of reactions (as in the Belousov reaction), then the rate-controlling reaction is the slowest one. Similarly, the rate-controlling process amongst those described above will be the slowest one. If the others are much faster,

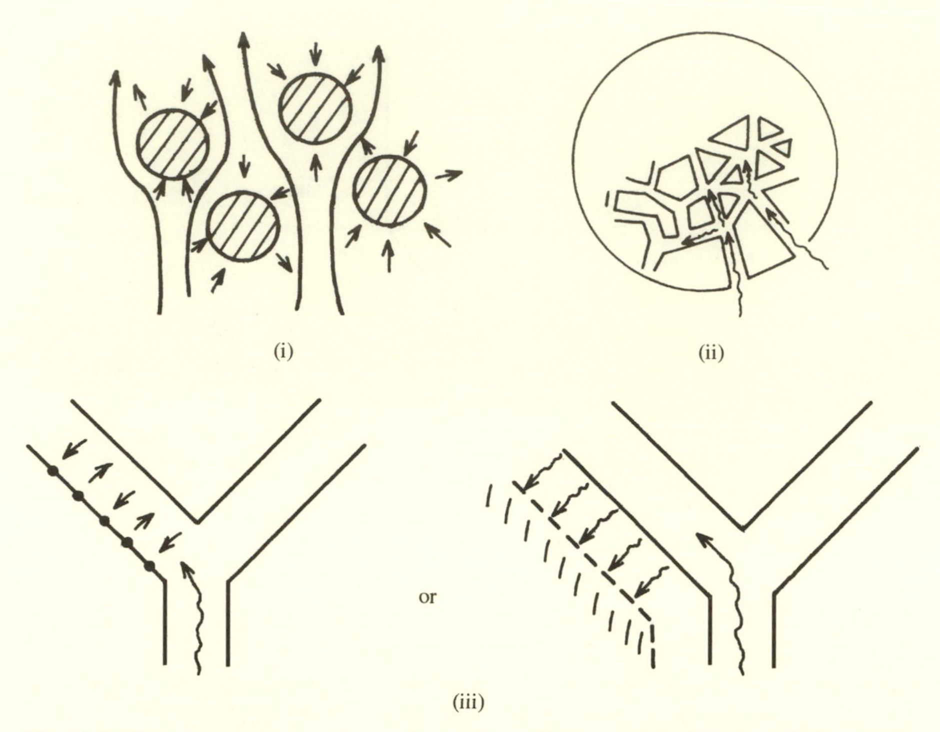

Fig. 12.3. Reactant transport to the catalyst surface: (i) gas-particle diffusion; (ii) intra-pellet diffusion; (iii) interfacial reaction

we can suppose them to be in (quasi-) equilibrium, so that only the rate-controlling mechanism appears in the overall process model as an evolution equation. The other processes essentially determine constitutive relations between relevant variables.

In what follows, we will consider a porous pellet subject to gaseous diffusion through the pores, and we will consider three nonequilibrium processes within the pellet. These are the conductive transfer of heat, (Fickian) diffusion of reactant, and the surface chemical reaction itself. In principle, this supposes that adsorption and desorption are rapid processes. In this case, we are led to various so-called 'isotherms' that give the locally equilibrated proportion of surface sites where gaseous reactant is adsorbed. One then finds that the reaction rate depends on this proportion, as well as on temperature. The dependence is generally complicated, but a simple representative reaction of the form

$$A \ldots \xrightarrow{r} B \ldots \tag{12.5}$$

has a rate r that will depend on the concentration of A and also on the absolute temperature T; a common supposition is the *Arrhenius term*

$$r \propto \exp[-E/RT], \tag{12.6}$$

where R is the gas constant, $R = 8.3\,\mathrm{J\,mole^{-1}\,K^{-1}}$, and E is the *activation energy*. In order to model the behavior of the pellet, we *average* by supposing that the surface concentration of the gaseous reactant A is proportional to its volumetric concentration.

So let us now consider a simple first-order reaction of a gaseous reactant with concentration c (moles per liter) with a solid catalyst pellet, such that the rate of reaction is (moles per liter per second)

$$r = k_0 c \exp[-E/RT], \tag{12.7}$$

and the reaction *absorbs* heat at a rate ΔH (cal per mole), which may be positive (an endothermic reaction) or negative (an exothermic reaction). We have seen (for example in Chapters 9 and 10) that reaction rates typically involve products of concentrations. First-order reaction rates such as Eq. (12.7) can then be derived in limiting cases, as for example in the Michaelis–Menten reaction, which is effectively first order. This reduction is effected through the existence of a rate-controlling reaction. We take an arbitrary, fixed volume V of the pellet. Conservation of reactant within the pellet can then be modeled by the conservation law

$$\frac{d}{dt}\int_V c\,dV = -\int_S [-D\boldsymbol{\nabla}c].\mathbf{n}\,dS - \int_V r\,dV, \tag{12.8}$$

where the second term arises through the diffusive flux $-D\boldsymbol{\nabla}c$ (with diffusion coefficient D), and the third represents the volumetric sink due to the reaction. Applying the divergence theorem, assuming the resulting integrands are continuous (i.e., $c \in C^2$) and using the fact that V is arbitrary, implies that c satisfies

$$c_t = \boldsymbol{\nabla}.[D\boldsymbol{\nabla}c] - r. \tag{12.9}$$

In an entirely analogous manner, the energy conservation law is

$$\frac{\partial}{\partial t}(\rho c_p T) = \boldsymbol{\nabla}.[k\boldsymbol{\nabla}T] - r\Delta H, \tag{12.10}$$

where ρ is density, c_p is specific heat, T is temperature, and k is the thermal conductivity. If we take ρ, c_p, and k as constant, then Eq. (12.10) can be written in terms of the thermal diffusivity

$$\kappa = k/\rho c_p; \tag{12.11}$$

thus

$$T_t = \kappa\nabla^2 T - r\Delta H/\rho c_p. \tag{12.12}$$

The model equations (12.9) and (12.12), with r given by Eq. (12.7), give two (parabolic) equations for T and c. For a spherical pellet with boundary S, we could prescribe

$$T = T_0, \qquad c = c_0 \quad \text{on } S \tag{12.13}$$

(depending in reality on heat and mass transfer coefficients from the external gas flow).

This is a formidable problem in itself, but some analysis may be possible. It is first necessary to nondimensionalize the model. We do this as follows. We expect that

$c \leq c_0$, so $c \sim c_0$ is feasible. However, the temperature *variation* ΔT is really determined by $r\Delta H$, and thus we can suppose $T - T_0 \sim \Delta T$, with ΔT being determined from a balance of, say, thermal conduction and heat release (or absorption). If the pellet is of radius a, then we choose $\mathbf{x} \sim a$, and a balance of T_t with $\kappa \nabla^2 T$ gives the *thermal timescale* $t \sim a^2/\kappa$.

We thus define

$$T = T_0 + (\Delta T)T^*, \qquad c = c_0 c^*, \qquad \mathbf{x} = a\mathbf{x}^*, \qquad t = (a^2/\kappa)t^*, \tag{12.14}$$

substitute into the equations, and drop the asterisks. The reaction rate is

$$r = r_0 r^*, \tag{12.15}$$

where we define

$$r_0 = k_0 c_0 \exp[-E/RT_0]. \tag{12.16}$$

One then finds that, dimensionlessly,

$$\begin{aligned} r &= c \exp\left[\frac{\beta T}{1+\delta T}\right], \\ T_t &= \nabla^2 T \pm r, \\ c_t &= \frac{1}{Le}\nabla^2 c - \mu r, \end{aligned} \tag{12.17}$$

where $\pm$ signifies exothermic (+) or endothermic (−) and the parameters are defined by

$$\begin{aligned} Le &= \kappa/D \quad \text{(Lewis number)}, \\ \mu &= r_0 a^2/c_0\kappa, \\ \beta &= \frac{E\Delta T}{RT_0^2}, \\ \delta &= \frac{\Delta T}{T_0}, \\ \Delta T &= a^2 r_0 |\Delta H|/k. \end{aligned} \tag{12.18}$$

In different circumstances, the values of these parameters can take widely varying values, but it is realistic to consider the situation where

$$\delta \ll 1, \qquad \beta \gg 1, \qquad \mu < 1, \qquad Le > 1. \tag{12.19}$$

The assumption $\delta \ll 1$ (more importantly, $\delta \ll \beta$, i.e., $E/RT_0 \gg 1$) yields the *Frank–Kamenetskii* approximation

$$r \approx c \exp(\beta T), \tag{12.20}$$

which enables some analysis to be carried out more easily. We now consider a particular case.

12.2 Thermal runaway

Let us suppose $\Delta H < 0$ (exothermic reaction), $Le \gg 1$, $\mu \ll 1$, $\beta \gg 1$, and $\delta \ll 1$. Then $1/Le \ll 1$, and $r \ll 1$ (e.g., initially, while $\beta T < O(1)$), so we can suppose that $c_t \ll 1$, and thus c varies slowly from its initial value $c = 1$. Then the temperature reacts 'rapidly,' with $c \approx 1$ (more generally, $c = c(\tau)$, where τ is some slow time variable). Thus T satisfies

$$T_t = \nabla^2 T + c \exp\left[\frac{\beta T}{1 + \delta T}\right]. \tag{12.21}$$

Adopting the Frank–Kamenetskii approximation, we also write

$$T = \theta/\beta, \qquad \lambda = \beta c, \tag{12.22}$$

so that the steady-state version of Eq. (12.21) is

$$\nabla^2 \theta + \lambda e^{\theta} = 0, \tag{12.23}$$

with boundary condition $\theta = 0$ on S.

In one spatial dimension, we solve

$$\theta_{xx} + \lambda e^{\theta} = 0, \qquad \theta = 0 \quad \text{on } x = \pm 1. \tag{12.24}$$

The solution can be obtained exactly and is

$$\theta = \theta_0 - 2 \ln \cosh\left\{\sqrt{\frac{\lambda}{2} e^{\theta_0}}\, x\right\}. \tag{12.25}$$

In this equation, the (maximum) temperature θ_0 is that at $x = 0$ (by symmetry), and is determined by satisfying the boundary condition at $x = 1$, whence

$$e^{\theta_0/2} = \cosh\left\{\sqrt{\frac{\lambda}{2}} e^{\theta_0/2}\right\}. \tag{12.26}$$

The solutions of this transcendental equation are easily studied graphically (see Fig. 12.4): We see that there are 2, 1, or 0 solutions depending on whether $\lambda < \lambda_c$, $\lambda = \lambda_c$, or $\lambda > \lambda_c$, where λ_c is the value given by

$$1 = \sqrt{\frac{\lambda}{2}} \sinh\left\{\left(\frac{\lambda + 2}{2}\right)^{1/2}\right\}, \tag{12.27}$$

whence $\lambda_c \approx .878$.

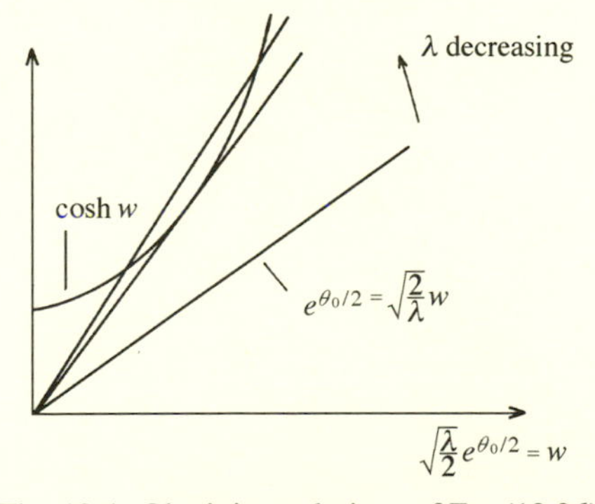

Fig. 12.4. Obtaining solutions of Eq. (12.26)

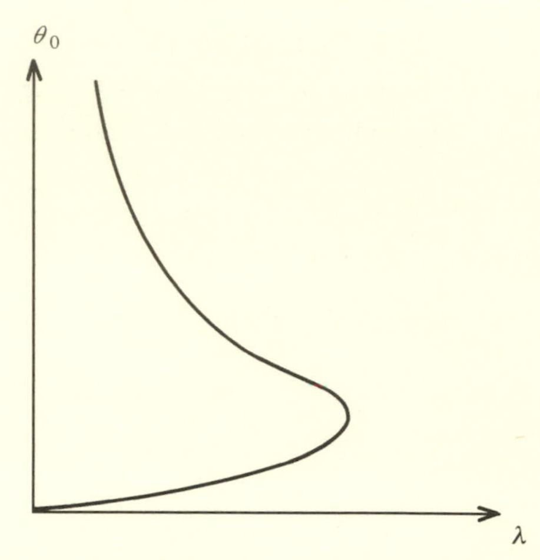

Fig. 12.5. Response diagram for θ_0 as a function of λ

We thus obtain the response curve in Fig. 12.5. This is the celebrated phenomenon of *thermal runaway*. The multiplicity of solutions for $\lambda < \lambda_c$, and the nonexistence of any steady solution for $\lambda > \lambda_c$, is associated with the following facts. For $\lambda < \lambda_c$, the lower branch is stable, but the upper branch is unstable. For initial values of θ that are large enough (for example, the values of θ need to be larger than the value on the top branch at some point), and also for *any* initial θ when $\lambda > \lambda_c$, the solution to the time-dependent problem blows up in finite time. That is to say, $\theta \to \infty$ at some x as $t \to t_c$. The basic reason for this is the positive feedback associated with the convex source term e^{θ}. Indeed, the space-independent version of the time-dependent equation, that is, $\theta_t = \lambda e^{\theta}$, has the exact solution $\theta = \theta(0) + \ln[1/(1 - t/t_c)]$, which blows up (everywhere) as $t \to t_c$. When diffusion is included, blowup is only guaranteed if λ is large enough, and then it occurs at a single point.

Reality intervenes in two ways: mathematical and physical. Physical reality does not permit infinite temperatures; something else happens first. This could be a severe explosion or else sudden melting. More typically, the rapid depletion of reactant concentration at high temperatures will cause the reaction to terminate. In either event, thermal runaway may be viewed as a real catastrophe. The mathematical limitation lies in the approximation: the Frank–Kamenetskii approximation breaks down when θ is too large, and the saturating effect of the Arrhenius term becomes important once more.

In order to study this further, we write

$$\varepsilon = \delta/\beta = (E/RT_0)^{-1} \ll 1, \tag{12.28}$$

and consider the temperature equation in the form

$$\theta_t = \nabla^2\theta + \lambda \exp\left[\frac{\theta}{1 + \varepsilon\theta}\right], \tag{12.29}$$

with $\theta = 0$ on S. We again consider one-dimensional steady solutions, but now we suppose $\theta_0 \sim O(1/\varepsilon)$, so that the Frank-Kamenetskii approximation breaks down.

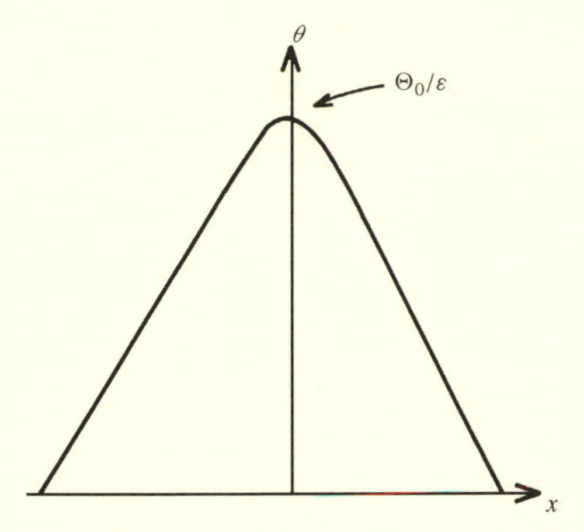

Fig. 12.6. Temperature profile on the upper branch of Fig. 12.4

This is on the upper branch, when λ is very small. We write

$$\theta = \Theta/\varepsilon, \tag{12.30}$$

so that

$$\Theta_{xx} + \varepsilon\lambda \exp\left[\frac{1}{\varepsilon}\left(\frac{\Theta}{1+\Theta}\right)\right] = 0. \tag{12.31}$$

Let us suppose that $\Theta \approx \Theta_0$ at $x = 0$ is the maximum value. *Relative* to this value, the exponential term will be very small for $\Theta < \Theta_0$. This suggests that Θ will be approximately linear ($\Theta_{xx} \approx 0$) away from $x = 0$, with a sharp turning region near $x = 0$ where the curvature is important (see Fig. 12.6). In order to get a sensible exponential term, we write

$$\Theta = \Theta_0 + \varepsilon\phi, \tag{12.32}$$

and then matching to the presumed linear profile in $x \gtrless 0$ requires the local rescaling to be

$$x = \varepsilon X, \tag{12.33}$$

so that the dynamics near $x = 0$ is given by

$$\phi_{XX} + \varepsilon^2\lambda \exp\left[\frac{1}{\varepsilon}\left\{\frac{\Theta_0 + \varepsilon\phi}{1+\Theta_0+\varepsilon\phi}\right\}\right] = 0. \tag{12.34}$$

In order to obtain a balance, we now suppose

$$\lambda = \frac{1}{\varepsilon^2}\exp[-p/\varepsilon], \qquad p = O(1) \tag{12.35}$$

and *choose*

$$\Theta_0 = \frac{p}{1-p}. \tag{12.36}$$

At leading order, we then have

$$\Theta \sim \Theta_0(1-x) \quad \text{in } x > 0, \tag{12.37}$$

and Eq. (12.34) is, to leading order,

$$\phi_{XX} + \exp\left[\frac{\phi}{(1+\Theta_0)^2}\right] = 0 \tag{12.38}$$

with

$$\phi_X = 0 \quad \text{on } X = 0, \qquad \phi_X \to -\Theta_0 \quad \text{as } X \to \infty. \tag{12.39}$$

Thus

$$\frac{1}{2}\phi_X^2 + (1+\Theta_0)^2 \exp\left[\frac{\phi}{(1+\Theta_0)^2}\right] = \frac{1}{2}\Theta_0^2, \tag{12.40}$$

whence $\phi = \phi_0$ at $X = 0$ is given by

$$\begin{aligned}\phi_0 &= (1+\Theta_0)^2 \ln\left[\frac{\Theta_0^2}{2(1+\Theta_0)^2}\right] \\ &= \frac{1}{(1-p)^2}\ln[p^2/2].\end{aligned} \tag{12.41}$$

Thus the maximum value of θ is given by

$$\theta_0 \sim \frac{1}{\varepsilon}\left(\frac{p}{1-p}\right) + \frac{1}{(1-p)^2}\ln(p^2/2), \tag{12.42}$$

where

$$p = \varepsilon \ln(1/\varepsilon^2 \lambda). \tag{12.43}$$

In turn, this estimate breaks down when $p \to 1$, specifically when $1 - p = O(\varepsilon)$ (and $\theta_0 \sim 1/\varepsilon^2$) or $\lambda \approx \frac{1}{\varepsilon^2}\exp(-1/\varepsilon)$.

In order to continue the analysis when $\theta \sim 1/\varepsilon^2$, we seek a further distinguished limit that can describe this *hot branch*. Evidently, if $\theta \sim 1/\varepsilon^2$, then the Arrhenius term begins to saturate. We now expect that θ may be more or less quadratic over most of the domain, as suggested in Fig. 12.7. On the basis of the breakdown exhibited by

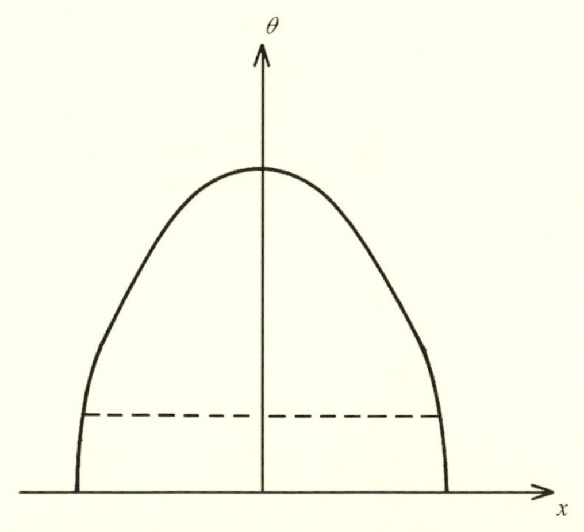

Fig. 12.7. Temperature profile on the hot branch

Eq. (12.42), a rescaling of the equations is suggested via

$$\theta = \psi/\varepsilon^2, \qquad \lambda = \frac{1}{\varepsilon^2}\exp(-1/\varepsilon)\Lambda, \tag{12.44}$$

but with $x = O(1)$. Substitution of these new variables leads to the exact equation

$$\psi'' + \Lambda \exp\left[\frac{-1}{(\psi + \varepsilon)}\right] = 0, \tag{12.45}$$

with $\psi = 0$ on $x = \pm 1$. The solution is now approximated by Eq. (12.45) with $\varepsilon = 0$, with the first integral

$$\frac{1}{2}\psi'^2 + \Lambda \int_{\psi_0}^{\psi} e^{-1/u}\, du = 0, \tag{12.46}$$

where $\psi = \psi_0$ at $x = 0$. Although the reduced problem has a singularity at $\varepsilon = 0$ when $\psi \to 0$, this should not affect the leading order solution. We thus integrate Eq. (12.46) with $\psi = \psi_0$ at $x = 0$ and $\psi = 0$ at $x = 1$, whence

$$\int_{\psi_0}^{\psi} \frac{dv}{\left\{\int_v^{\psi_0} e^{-1/u}\, du\right\}^{1/2}} = -(2\Lambda)^{1/2} x, \tag{12.47}$$

and ψ_0 is determined from

$$(2\Lambda)^{1/2} = \int_0^{\psi_0} \frac{dv}{\left\{\int_v^{\psi_0} e^{-1/u}\, du\right\}^{1/2}}. \tag{12.48}$$

This determines ψ_0 as a monotonically *increasing* function of Λ. For small values of ψ_0, application of Laplace's method to Eq. (12.48) gives

$$\psi_0 \sim \frac{1}{\ln(2\Lambda)}, \tag{12.49}$$

which matches to Eq. (12.42). For large values of Λ (and thus ψ_0), we obtain the approximate result

$$\psi_0 \sim \Lambda/2. \tag{12.50}$$

We thus finally obtain the hot branch solution, and as shown in Fig. 12.8, the complete

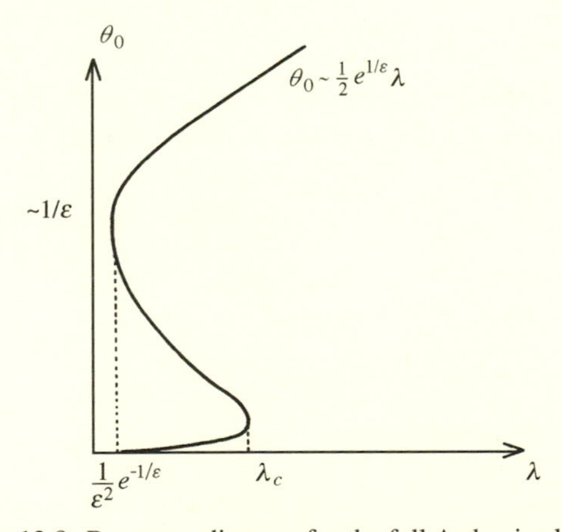

Fig. 12.8. Response diagram for the full Arrhenius kinetics

response curve is S-shaped, the typical hysteresis being associated with stable upper and lower branches with the intermediate branch being unstable. As we have indicated, the hot branch is unlikely to exist in practice, owing to catastrophic failure of the system.

12.3 More realistic models: heat and mass transfer

We return to Eq. (12.17), which we write in the form (for an exothermic reaction)

$$\begin{aligned} \theta_t &= \nabla^2\theta + \beta c \exp\left[\frac{\theta}{1+\varepsilon\theta}\right], \\ c_t &= \frac{1}{Le}\nabla^2 c - \mu c \exp\left[\frac{\theta}{1+\varepsilon\theta}\right]. \end{aligned} \tag{12.51}$$

We recall the definitions of β and μ:

$$\begin{aligned} \beta &= \frac{Ea^2|\Delta H|k_0 c_0}{RkT_0^2}\exp(-E/RT_0), \\ \mu &= \frac{a^2 k_0}{\kappa}\exp(-E/RT_0). \end{aligned} \tag{12.52}$$

Notice β depends on c_0, the surface value of reactant concentration, and both β and μ depend strongly on T_0, the surface temperature. Now, whereas the prescription of T_0 is feasible in an isothermal bath, it is more appropriate in a furnace to consider a gas stream at some ambient tempreature T_∞ and with reactant concentration c_∞. The surface temperature and concentration are then determined by appropriate heat and mass transfer '*correlations*,' that is, we prescribe heat and mass transfer coefficients h_T and h_c, such that the boundary conditions at the particle surface S are

$$\begin{aligned} -D\frac{\partial c}{\partial n} &= h_c(c - c_\infty), \\ -k\frac{\partial T}{\partial n} &= h_T(T - T_\infty), \end{aligned} \tag{12.53}$$

where $\mathbf{n}$ is the unit outward normal to the surface. Because T_0 and c_0 are no longer prescribed, we now define the dimensionless variables in terms of T_∞ and c_∞ (equivalently: write the gas mainstream temperature and concentration as T_0 and c_0); we then find the same equations as above, but with the boundary conditions

$$\begin{aligned} N_{Sh}^{-1}\frac{\partial c}{\partial n} &= 1 - c, \\ -N_{Nu}^{-1}\frac{\partial \theta}{\partial n} &= \theta, \end{aligned} \tag{12.54}$$

where the Sherwood and Nusselt numbers are defined by

$$N_{Sh} = ah_c/D, \qquad N_{Nu} = ah_T/k. \tag{12.55}$$

Correlations for N_{Sh} and N_{Nu} are discussed by Szekely et al. (1976). Typical values are $N_{Sh} > 10$, $N_{Nu} > 10$, and we see that in this case it is reasonable to take the isothermal boundary conditions $\theta = 0$, $c = 1$ as approximations to those prescribed in (12.54).

We can now ask what conditions on the other parameters are likely to pertain in reality. We can expect $\varepsilon \ll 1$, and then our previous analysis suggests that we require T_0 to be low enough that $\beta < 1$, unless also $\mu \gg 1$. In order to make headway, we define the controlling temperature

$$T_c = \left\{ \frac{E|\Delta H| c_0}{R\rho c_p} \right\}^{1/2}. \tag{12.56}$$

For plausible values $E \sim |\Delta H| \sim 8000$ cal mole^{-1}, $c_0 = 10^{-5}$ mole cm^{-3}, $R = 2$ cal mole^{-1} K^{-1}, $\rho = 1$ gm cm^{-3}, and $c_p = .3$ cal gm^{-1} K^{-1}, we have $T_c \sim 330$ K, and

$$\frac{\beta}{\mu} = \left(\frac{T_c}{T_0} \right)^2. \tag{12.57}$$

If we vary T_0, then μ (and thus also β) will vary dramatically, so that cases $\mu \gg 1$, $\mu \sim 1$, $\mu \ll 1$, are all of interest. However, Eq. (12.57) shows that it is plausible to take $\beta \sim \mu$ or $\beta \ll \mu$ (e.g., $\beta/\mu \sim 10^{-1}$). Because μ varies so strongly with T_0, it makes sense to consider $\beta \sim \mu$; thus we define

$$\beta = \mu\gamma, \qquad \gamma = (T_c/T_0)^2 \sim O(1). \tag{12.58}$$

12.4 The case $Le \gg 1$, $\varepsilon \ll 1$, and $\gamma = O(1)$

There are three choices: $\mu \ll 1$ (slow reaction), $\mu \sim 1$ (intermediate), and $\mu \gg 1$ (fast). If $\mu \ll 1$, the reaction proceeds slowly (on a timescale $t \sim 1/\mu$). Temperature equilibrates rapidly, thus $\theta \sim \beta$, and

$$\begin{aligned} \nabla^2\theta + \beta c e^\theta &\approx 0, \\ \frac{1}{\mu} c_t &\approx \frac{1}{\mu Le} \nabla^2 c - c. \end{aligned} \tag{12.59}$$

If $\mu\, Le \ll 1$, then diffusion is dominant, and $c = 1 - O(\mu\, Le)$. On the other hand, if $\mu\, Le \gg 1$, then diffusion is irrelevant, and c decreases exponentially with time.

Now suppose $\mu \sim 1$. The governing equations can be approximated as

$$\begin{aligned} \theta_t &= \nabla^2\theta + \gamma\mu c e^\theta, \\ c_t &= -\mu c e^\theta, \end{aligned} \tag{12.60}$$

but must be solved numerically. If, however, $\gamma \ll 1$, then $\theta = O(\gamma)$ and $c = \exp[-\mu t]$, and a perturbation solution for θ can be written. A boundary layer in reactant concentration forms near the pellet surface.

Finally, suppose $\mu \gg 1$. In a sense, this is ideal, as the reaction proceeds most rapidly. With $\beta \gg 1$, the temperature is, however, prone to runaway, but this may be offset by the rapid decline in c. Ignoring diffusive terms, we have

$$\theta_t \approx \beta c e^\theta, \qquad c_t \approx -\mu c e^\theta, \tag{12.61}$$

whence

$$\mu\theta + \beta c = \beta, \tag{12.62}$$

and thus

$$c \approx 1 - \theta/\gamma, \tag{12.63}$$

and

$$\theta_t \approx \mu(\gamma - \theta)e^{\theta}. \tag{12.64}$$

There is thus an 'explosive' initial runaway, during which the reaction proceeds rapidly; but the temperature quickly saturates as the reactant runs out, with $\theta \to \gamma$. The picture is complicated, however, by thermal boundary layers at the pellet surface, where θ jumps from 0 to γ.

12.5 Nonporous pellet

As an example of how other models may be appropriate in other situations, we consider the case where the pellet is not porous. In this case, the reaction proceeds at the pellet surface at a rate $\mathcal{R}$, related to the speed v at which interface is consumed, by

$$\mathcal{R} = \rho_s v, \tag{12.65}$$

where ρ_s is the molar density (moles per liter). (This assumes a single mole of solid is involved in the reaction step (12.5).) Thus $\mathcal{R}$ has units of moles per area per time, as v is a velocity. In order to relate this to the previous discussion, note that in writing Eq. (12.7) for a porous solid, we tacitly suppose that there is an internal specific surface area S_s with units cm^2/cm^3, or just cm^{-1}: Roughly, $S_s \sim d_p^{-1}$, where d_p is pore spacing (or grain size). Then $r \sim \mathcal{R}S_s = vcS_s$, where $v = v(T)$; and the reference rate is v_0, where $k_0 = v_0 S_s$.

For a nonporous pellet, v is the rate of erosion of the pellet. The heat released at the surface is then $\rho_s v(-\Delta H)$. For example, consider a spherical pellet $r = S(t)$. Then

$$\dot{S} = -v, \qquad v = v_0 \exp[-E/RT]. \tag{12.66}$$

The heat loss to the gas stream is $h_T(T - T_\infty)$, so that, including the jump due to the heat source at $r = S$, we have

$$-k\frac{\partial T}{\partial r} = -\rho_s v(-\Delta H) + h_T(T - T_\infty). \tag{12.67}$$

The temperature of the pellet satisfies the diffusion equation.

To nondimensionalize, we put (as before)

$$\begin{aligned} r = ar^*, &\qquad t = (a^2/\kappa)t^*, \\ T = T_\infty(1 + \varepsilon\theta), &\qquad v = v_0 \exp[-E/RT_\infty]v^*, \end{aligned} \tag{12.68}$$

where

$$\varepsilon = RT_\infty/E, \tag{12.69}$$

to obtain (on dropping asterisks)

$$\theta_t = \nabla^2\theta, \tag{12.70}$$

with

$$\begin{aligned} v &= \exp\left[\frac{\theta}{1+\varepsilon\theta}\right], \\ \dot{S} &= -\nu v, \\ -\frac{1}{N_{Nu}}\frac{\partial\theta}{\partial r} &= -\Gamma\nu v + \theta \quad \text{on } r = S, \end{aligned} \tag{12.71}$$

where

$$\begin{aligned} \nu &= \frac{av_0}{\kappa}\exp[-E/RT_\infty], \\ \Gamma &= (T_r/T_\infty)^2, \\ T_r &= \left\{\frac{E(-\Delta H)\rho_s}{R\rho c_p N_{Nu}}\right\}^{1/2}, \\ N_{Nu} &= h_T a/k. \end{aligned} \tag{12.72}$$

We have here a complicated moving boundary problem. As before, however, some insight can be gained from examination of various limits. Let us suppose $N_{Nu} \gg 1$, $\varepsilon \ll 1$.

If $\nu \ll 1$, $\Gamma < 1$, then the reaction is slow and the pellet temperature equilibrates; thus (if $\theta = O(1)$)

$$\theta \approx \Gamma\nu e^\theta \sim \Gamma\nu, \qquad \dot{S} \sim -\nu. \tag{12.73}$$

If $\nu = O(1)$, the problem must be solved numerically, unless $\Gamma \ll 1$, in which case $\theta \approx \Gamma\nu$ on $r = S$, $\dot{S} \approx -\nu$, and $\theta_t = \nabla^2\theta$ with $\theta \approx 0$ on $r = 1 - \nu t$. If $\Gamma, \nu \sim 1$, and $N_{Nu} \gg 1$, then we have

$$\theta \approx \Gamma\nu\exp\left[\frac{\theta}{1+\varepsilon\theta}\right] \quad \text{at } r = S. \tag{12.74}$$

Solutions of this relation are portrayed in Fig. 12.9. We see the familiar hysteretic curve as ν varies. In particular, θ jumps to a hot branch ($\theta \sim \Gamma\nu\exp(1/\varepsilon)$) if $\Gamma\nu > e^{-1}$. Thus for $\Gamma\nu > e^{-1}$, thermal runaway would in effect occur.

Some simplification can be made in this case via the formal limit $\nu \gg 1$ (but $\Gamma \ll 1$, so $\Gamma\nu \sim O(1)$). Then we put

$$t = \tau/\nu, \qquad r = S - s/\nu^{1/2}, \tag{12.75}$$

so that (approximately)

$$\theta_\tau = \theta_{ss}, \tag{12.76}$$

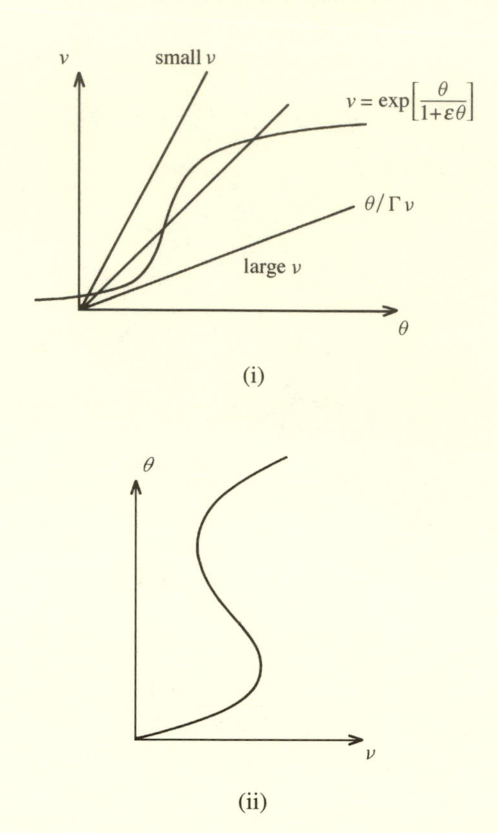

Fig. 12.9. Multiple solutions for Eq. (12.74): (i) ν and $\theta/\Gamma\nu$ versus θ for various ν; (ii) θ versus ν.

with

$$\theta \to 0 \quad \text{as } s \to \infty, \tag{12.77a}$$

$$S' = -\exp\left[\frac{\theta}{1+\varepsilon\theta}\right], \tag{12.77b}$$

$$(\nu^{1/2}/N_{Nu})\frac{\partial\theta}{\partial s} = -\Gamma\nu\exp\left[\frac{\theta}{1+\varepsilon\theta}\right] + \theta \quad \text{on } s = 0. \tag{12.77c}$$

Thus S uncouples, and Eq. (12.77c) is approximately

$$\theta = \alpha e^{\theta} + \beta\frac{\partial\theta}{\partial s} \quad \text{on } s = 0, \qquad \alpha = \Gamma\nu, \qquad \beta = \nu^{1/2}/N_{Nu}. \tag{12.78}$$

Numerical solution of Eqs. (12.76) and (12.77) with α, $\beta = O(1)$, and $\theta = 0$ at $\tau = 0$ is required to ascertain parameter values $\alpha = \alpha_c(\beta)$ beyond which blowup occurs. Evidently, $\alpha_c(0) = e^{-1}$. Because for $\alpha > e^{-1}, \alpha e^{\theta} > \theta$, we expect $\alpha_c > e^{-1}$ for $\beta > 0$ (because then $\partial\theta/\partial s < 0$). When β is large, then $\theta = O(\tau^{1/2}/\beta)$ (from a similarity solution). Then $S' \sim -1$ for $\tau = O(1)$, and $S \to 0$ at $\tau = 1$, so that in this case the temperature rise is only $O(1/\beta)$.

12.6 Macroscopic modeling

The problems discussed above are of interest in their own right. Nevertheless, the problem of large-scale interest is that of describing the heat and mass transfer in a reactor with a packed or fluidized bed of catalyst particles. In this case, it is appropriate to treat the bed as a porous medium through which gaseous reactant flows turbulently. There will be heat and mass transfer between solid and gas, and the results of the pellet models can in principle be applied to parameterize these local processes.

Consider, for example, a packed bed of nonporous pellets undergoing a gas–solid reaction

$$A_s + B_g \to P_g, \tag{12.79}$$

where A_s is the solid reactant and B_g the gaseous reactant. If the particles have typical (initial) diameter d_p, then the specific surface area per unit volume is

$$a \sim C(1-\phi)/d_p\xi, \tag{12.80}$$

where the shape factor $C = 6$ for spherical particles, ϕ is the porosity (void fraction) of the bed, and $d_p\xi$ ($\xi \leq 1$) is the particle radius. From Eq. (12.65), the reaction rate is $\mathcal{R} = \rho_s v$ (moles of A_s or B_g per unit area of particle per unit time). Therefore the volumetric reaction rate is

$$r = a\mathcal{R} \tag{12.81}$$

(moles of A_s or B_g per unit volume per unit time). Taking into account the variation of particle size with time, then

$$r = \frac{C(1-\phi)\rho_s v}{d_p\xi} = \frac{-\rho_s C(1-\phi)\xi_t}{\xi}, \tag{12.82}$$

where ξ is described via Eq. (12.66) (with $S = d_p\xi$).

Suppose the gas flow is directed axially. If the gas velocity is v_g and the solid velocity is v_s, then conservation of mass of the reactant is

$$\frac{\partial}{\partial t}(\phi c) + \frac{\partial}{\partial z}(\phi v_g c) = -h_D a(c - c_s), \tag{12.83}$$

where c_s is the concentration at the particle surfaces of the reactant B_g. The term on the right is the rate of supply of reactant to the particles, where h_D is a mass transfer coefficient. The supply is equal to the rate of reaction, so that

$$\mathcal{R} = h_D(c - c_s), \qquad r = h_D a(c - c_s), \tag{12.84}$$

which determines c_s.

Conservation of mass of solid reactant is

$$\frac{\partial}{\partial t}[\rho_s(1-\phi)] + \frac{\partial}{\partial z}[\rho_s(1-\phi)v_s] = -r. \tag{12.85}$$

Note that both A_s and B_g are consumed in the reaction. This is balanced by production of the gaseous product P_g. In general, we may expect a settling velocity v_s (<0)

as the particles are consumed. If particles settle without compaction, that is, no rearrangement occurs other than that due to shrinkage, then

$$\rho_s(1-\phi)\frac{\partial v_s}{\partial z} = -r \tag{12.86}$$

determines v_s, and ϕ is constant.

The flow in the reactor is usually turbulent. For simplicity, we assume that the gas volume flux $U = \phi v_g$ is constant (and prescribed) and that the reactor operates in a steady state. Thus

$$\begin{aligned} U\frac{\partial c}{\partial z} &= -r, \\ r &= -\frac{-\rho_s C(1-\phi)\xi_t}{\xi}, \end{aligned} \tag{12.87}$$

where (using Eq. (12.66))

$$\xi_t = -(v_0/d_p)\exp[-E/RT_s], \tag{12.88}$$

and T_s is the surface temperature of the particles. If T_g is the gas temperature, then heat transfer from the particles gives Eq. (12.67)

$$-k\frac{\partial T}{\partial r} = -\rho_s v_0 \exp[-E/RT_s](-\Delta H) + h_T(T_s - T_g) \tag{12.89}$$

on $r = d_p\xi$, whereas for turbulent flow, the gas temperature satisfies

$$\rho_g U c_{pg}\frac{\partial T_g}{\partial z} = h_T(T_s - T_g)a, \tag{12.90}$$

where ρ_g and c_{pg} are suitably defined gas density and specific heat, assumed constant, and a is the specific surface area given by Eq. (12.80).

Our problem is thus to solve Eq. (12.87) for c and Eq. (12.90) for T_g, while in so doing determining T_s and ξ. The latter is determined in terms of T_s by Eq. (12.88), whereas T_s is determined by the extra condition (12.89) used in solving

$$T_t = \kappa\nabla^2 T \tag{12.91}$$

in $0 < r < d_p\xi$. Note that, properly, ξ_t should be the time derivative following the particles, so that, in a steady state, Eq. (12.87) is

$$\xi r = -\rho_s C(1-\phi)v_s\frac{\partial \xi}{\partial z}, \tag{12.92}$$

and v_s is determined from Eq. (12.86).

Let the height of a reactor be l. The natural length scale is then l, and the residence time of the solids is, from Eq. (12.86),

$$t_r \sim l/v_s \sim \rho_s(1-\phi)/r. \tag{12.93}$$

On the other hand, if the reaction time is t_R, then we can assume that l is chosen so that the reaction is approximately completed by the time the solids reach the base. Hence Eq. (12.87) gives

$$t_R \sim \rho_s C(1-\phi)/r, \tag{12.94}$$

and we see that $t_R \sim t_r$. From Eq. (12.88), we choose

$$t_R = d_p/\bar{v}, \tag{12.95}$$

where

$$\bar{v} = v_0 \exp(-E/R\bar{T}), \tag{12.96}$$

and $\bar{T}$ is a typical solid temperature.

Next, we suppose the gas concentration is chosen so that $c \approx 0$ at the outlet. In fact, in this simple model, c is uncoupled and need not be considered further. Combining Eqs. (12.92) and (12.86), we have

$$v_s = v_s^{ex}\xi^C, \tag{12.97}$$

where v_s^{ex} is the prescribed (negative) velocity at the outlet $z = l$, determined from the pellet feed rate.

Now the thermal conduction time in the particles is $t_c \sim d_p^2/\kappa$, where κ is the thermal diffusivity. For example, if $d_p \sim 1$ cm, $\kappa \sim 10^{-2}$ cm^2 s^{-1}, then $t_c \sim 100$ s. If the residence time is larger than this, that is, $t_c \ll t_R$, then the particles are approximately in equilibrium, and heat loss is governed by

$$\rho_s c_{ps} V \dot{T}_s = -k \left.\frac{\partial T}{\partial r}\right|_{d_p\xi} A, \tag{12.98}$$

where c_{ps} is the solid specific heat, and V and A are the particle volume and area, respectively. Then $V/A = d_p\xi/C$; thus, using Eq. (12.89),

$$\rho_s c_{ps} \frac{d_p\xi}{C} v_s \frac{\partial T_s}{\partial z} = -\rho_s \bar{v} \exp\left(\frac{E}{R\bar{T}} - \frac{E}{RT_s}\right)(-\Delta H) + h_T(T_s - T_g). \tag{12.99}$$

We have to solve Eqs. (12.90) and (12.99) for T_s and T_g with v_s given by Eqs. (12.97), and ξ given by (12.88), in the form

$$v_s \frac{\partial \xi}{\partial z} = -\frac{\bar{v}}{d_p} \exp\left(\frac{E}{R\bar{T}} - \frac{E}{RT_s}\right). \tag{12.100}$$

We have boundary conditions

$$\begin{aligned} \xi = 1, \qquad T_s = T_s^{ex} \quad &\text{at } z = l, \\ T_g = T_g^{in} \quad &\text{at } z = 0. \end{aligned} \tag{12.101}$$

To make this model dimensionless, choose $\bar{T} = T_s^{ex}$ and define scales

$$v_s \sim \left|v_s^{ex}\right|, \qquad t \sim t_R = d_p/\bar{v}, \tag{12.102}$$

and from our previous discussion, we *assume* l is such that $t_R \sim l/v_s$; specifically, we choose to scale z as

$$z \sim \left|v_s^{ex}\right| t_R, \tag{12.103}$$

and define

$$\lambda = l/\left|v_s^{ex}\right| t_R, \tag{12.104}$$

assumed to be $O(1)$.

We define

$$T_s = \bar{T} + (\Delta T)\theta_s,$$
$$T_g = \bar{T} + (\Delta T)\theta_g, \tag{12.105}$$

where

$$\Delta T = T_s^{ex} - T_g^{in}. \tag{12.106}$$

Then the dimensionless equations and boundary conditions are

$$\xi^C \frac{\partial \xi}{\partial z} = \exp\left[\frac{\theta_s}{\varepsilon(1+\delta\theta_s)}\right],$$
$$-\xi^{C+1}\frac{\partial \theta_s}{\partial z} = \beta \exp\left[\frac{\theta_s}{\varepsilon(1+\delta\theta_s)}\right] + \gamma(\theta_s - \theta_g), \tag{12.107}$$
$$\xi \frac{\partial \theta_g}{\partial z} = s\gamma(\theta_s - \theta_g),$$

where we use conservation of mass in the form $\rho_g U = \rho_s |v_s^{ex}|(1-\phi)$, and we have

$$\delta = \Delta T/\bar{T}, \qquad \varepsilon = R\bar{T}^2/E\Delta T,$$
$$\beta = \frac{C(-\Delta H)}{c_{ps}\Delta T}, \qquad \gamma = \frac{Ch_T t_R}{\rho_s c_{ps} d_p}, \tag{12.108}$$
$$s = \frac{c_{ps}}{c_{pg}},$$

and

$$\xi = 1, \qquad \theta_s = 0 \quad \text{on } z = \lambda,$$
$$\theta_g = -1 \quad \text{on } z = 0.$$

It is possible to gain approximate solutions in certain asymptotic limits, in the (by now) usual way. A natural extension is to suppose that $T_g = T_s$ at $z = l$, that is, that ΔT is unknown *a priori* but must be chosen such that $\theta_g = 0$ at $z = \lambda$. This suggests that $\gamma \sim O(1)$ (also $s \sim O(1)$). If $\varepsilon \ll 1$, then we can use Eq. (12.96) to give an estimate for $\bar{T}$ as

$$\bar{T} \sim \frac{E}{R \ln\left\{\dfrac{\rho_s c_{ps} v_0}{Ch_T}\right\}}. \tag{12.109}$$

Further analysis is possible if $\varepsilon \ll 1$ and also (because $C = 6$ for spherical particles) $C \gg 1$.

12.7 Notes and references

A very useful description of the industrial processes involved in modeling gas–solid reactions is given by Szekely et al. (1976). From the mathematical point of view, it is perhaps hampered by a plethora of detail, which sometimes obscures the basic message. A later review is that by Ramichandran and Doraiswamy (1982). A browse through, for example, the AIChE journal, or *Chemical Engineering and*

Science, shows that reactor modeling is a vigorous subject with plenty of industrial applications: See, for example Silveston et al. (1994), Chan and McElwain (1994), Cao, Varma, and Strieder (1993), Kilpinen (1988), Hagan, Hirshowitz, and Pirkle (1988), and Brooks, Balakotaiah, and Luss (1988). Mathematical analysis of single-pellet reactions is given in the two volumes by Aris (1975). Combustion is treated by Buckmaster and Ludford (1982). Both of these are rather wordy accounts.

The Gel'fand equation The Frank–Kamenetskii approximation that leads to Eq. (12.23) was enunciated by Frank–Kamenetskii (1955). The equation itself is sometimes called the Gel'fand equation after a study by Gel'fand (1963), but its history is much older, dating back to Liouville (1853); a brief review is given by Stuart (1967). It is possible to find the general solution in two dimensions by using complex variables $z = x + iy$ and $\bar{z} = x - iy$ when Eq. (12.23) becomes $4\theta_{z\bar{z}} + \lambda e^{\theta} = 0$, and the general solution (for $\lambda > 0$) is

$$\theta = \ln\left[\frac{8}{\lambda}\frac{|w'|^2}{[1+|w|^2]}\right], \tag{12.110}$$

where $w(z)$ is an arbitrary analytic function. However, the difficulty of fitting boundary conditions makes this less than useful. Some exact solutions are possible in three dimensions (see Exercise 4), but more generally an ingenious method due to Gel'fand can be used to convert the radially symmetric form of the Liouville equation (12.23) to a dynamic system in two or three dimensions. Specifically (see, for example, Wake and Hood, 1993), the equation is

$$\theta'' + \frac{(n-1)}{r}\theta' + \lambda e^{\theta} = 0 \tag{12.111}$$

in $\mathbf{R}^n$, with $\theta' = d\theta/dr$ and r as the radius. Putting

$$p = \lambda r^2 e^{\theta}, \qquad q = 2 + ru', \qquad r = e^{-t}, \tag{12.112}$$

we have

$$\begin{aligned} \dot{p} &= -pq, \\ \dot{q} &= p + (n-2)q - 2(n-2). \end{aligned} \tag{12.113}$$

It can be shown (see Exercise 4) that for $n = 3$, an unbounded multiplicity of solutions exists for λ in the neighborhood of a critical value λ_c, if we require $\theta = 0$ on $r = 1$.

Exercises

1. If M_A, M_B, and M_P are the molecular weights of A_s, B_g, and P_g in the reaction (see Eq. (12.79))

$$A_s + B_g \to P_g,$$

deduce that $M_A + M_B = M_P$; and by writing an appropriate conservation law for the molar density p of P_g, show that conservation of mass applies with a density $\rho = \phi(M_B c + M_P p) + (1-\phi)\rho_s M_A$ and mass flux $\rho u = \phi v_g(M_B c + M_P p) + (1-\phi)\rho_s M_A v_s$. Show also that the gas specific heat per unit volume is defined by (see Eq. (12.90))

$$\rho_g c_p = M_B c c_B + M_P p c_P,$$

where c_B and c_P are the specific heats of B_g and P_g, respectively.

2. Devise a numerical method to solve Eqs. (12.76) and (12.78), to wit,

$$\begin{aligned} \theta_\tau &= \theta_{ss} \quad \text{on } (0, \infty), \\ \theta &= \alpha e^\theta + \beta \frac{\partial \theta}{\partial s} \quad \text{on } s = 0, \\ \theta &\to 0 \quad \text{as } s \to \infty. \end{aligned}$$

Find the function $\alpha_c(\beta)$ such that thermal runaway occurs (for $S > 0$) for $\alpha > \alpha_c$; is your result consistent with $\alpha_c(0) = e^{-1}$?

3. A numerical method to explore Eq. (12.107) is to solve the equations backward from $z = 0$ (i.e., in $z < 0$), with $\theta_g = \theta_s = 0$, $\xi = 1$ on $z = 0$, and then define $\lambda = -z$ when $\theta_g = -1$ (if this exists; ξ may reach zero first). Do this, and compute λ as a function of ε, holding δ, C, γ, β fixed and $O(1)$. Compute also the profiles of ξ, θ_s, and θ_g. Show that if ε is small, one needs β to be small in order for a solution to exist; and in this case, use the numerical solution to suggest a limiting asymptotic structure. Explore both cases $\beta > 0$ and $\beta < 0$.

4. Solve explicitly the equation

$$\nabla^2\theta + \lambda e^\theta = 0, \qquad \theta = 0 \quad \text{on } r = 1,$$

in a cylindrical geometry, and show that there are no solutions if $\lambda > \lambda_c$, and calculate λ_c.

By rewriting the Gel'fand equation (for the sphere) in the form Eq. (12.113), use a phase plane analysis to show that, if $\theta = 0$ on $r = 1$ (and $\theta_r = 0$ on $r = 0$), then multiple solutions exist if λ is near a critical value λ_c, and that when $\lambda = \lambda_c$, infinitely many solutions exist. Show that $\lambda_c = 2$. Compute or sketch the bifurcation diagram of $\theta(0)$ versus λ.

5. If $Le = 1$ in Eq. (12.51), show that the Shvab–Zeldovich variable $c \pm \mu T$ is constant, and derive the resultant single equation for T. Deduce (for an exothermic reaction) that if $\mu \gg 1$, $T = O(1/\mu)$. Further deduce, that if $\beta \gg \mu$ (and $\beta \ll \delta$), then the reaction will proceed slowly but finish with a flourish.

6. For an exothermic reaction governed by Eq. (12.51), if $\beta \gg 1$, $\delta \ll \beta$, give reasons why $c < 1/\beta$ for times of $O(1)$ if thermal runaway is to be prevented. What does this imply about μ? Suppose now $\mu \gg 1$ and $Le \gg 1$. Show that $c \approx \exp[-\mu t e^\theta]$, and hence, by writing $\tilde{t} = \mu t$ and considering $\tilde{t} = O(1)$, show that $\theta \sim 1/\mu$ and that if $\theta = \Theta/\mu$, Θ satisfies (approximately) $\Theta_{\tilde{t}} \approx e^{-\tilde{t}}$. Hence find the maximum temperature rise.

7. Prove that the (steady) temperature field generated by a slow, *endothermic* reaction ($\Delta H > 0$) is globally stable (and unique) if T is prescribed on the boundary.

8. Prove that the lower branch of Fig. 12.5 is linearly stable and that the upper branch is linearly unstable.

13

Groundwater flow

The theory of groundwater movement is one that has immense practical importance in the day-to-day management of reservoirs, flood prediction, description of water table fluctuations, and the like. Although there are numerous complicating effects of soil physics and chemistry that can be important in certain cases, at its simplest, the theory of groundwater flow is conceptually very easy to understand.

Groundwater is water that lies below the surface of the Earth. Below a piezometric (constant pressure) surface called the *water table*, the soil or rock is *saturated*, which is to say, the *pore space* is completely full of water. Above this surface, the soil is unsaturated, and the pore space contains both water and air. In exceptional circumstances, soil can become desiccated, but there is usually some water present. Following precipitation, water infiltrates the subsoil and causes a local rise in the water table. The excess hydrostatic pressure thus produced leads to groundwater flow toward rivers, lakes, etc.

In order to characterize this flow, the soil (or rock) is thought of as a porous medium. For example, soil consists of an aggregation of variously sized mineral particles, and the pore space between the particles can admit a water flux when the saturated soil is subjected to a pressure gradient. Our first problem is therefore to relate the flow rate of water in a porous medium to the pressure gradient.

As an idealized case, suppose the pores consist of uniform cylindrical tubes of radius a. If a is small enough that the Reynolds number corresponds to laminar flow in the tubes, then Poiseuille flow in each tube leads to a volume flux

$$Q = \frac{\pi a^4}{8\mu} |\nabla p|, \tag{13.1}$$

where μ is the liquid viscosity and ∇p is the pressure gradient along the tube. A more realistic porous medium is *isotropic*, which is to say that if the pores have this tubular shape, the tubules will be arranged randomly and form an interconnected network. However, between nodes of this network, Eq. (13.1) will still apply, and an appropriate generalization is to suppose that the volume flux vector is given by

$$\mathbf{Q} \approx -\frac{a^4}{\mu X} \nabla p, \tag{13.2}$$

where the approximation takes account of small interactions at the nodes; the numerical factor $X \gtrsim 1$ takes some account of the arrangement of the pipes.

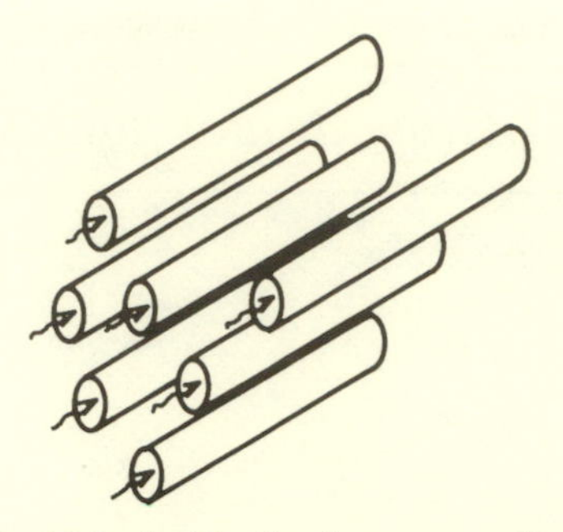

Fig. 13.1. An idealized porous medium

A porous medium is often characterized by its *porosity* ϕ (i.e., the pore volume fraction) and its *grain size* d. The latter characterizes the 'coarseness' of the medium and could be the typical dimension of sand grains in a pile of sand, or the typical spacing between (or length of) tubules in a connected network. In terms of the latter, we would have

$$\phi \sim a^2/d^2, \tag{13.3}$$

so that $\mathbf{Q}/d^2 \sim -(\phi^2 d^2/\mu)\boldsymbol{\nabla} p$. We define the volume flux per unit area (having units of velocity) as the discharge $\mathbf{u}$. Darcy's law then relates this to an applied pressure gradient by the relation

$$\mathbf{u} = -\frac{k}{\mu}\boldsymbol{\nabla} p, \tag{13.4}$$

where k is an empirically determined parameter called the *permeability*.

Although the validity of Darcy's law can be motivated theoretically, it ultimately relies on experimental measurements for its accuracy. The permeability k has dimensions of $(\text{length})^2$, which as we have seen is related to the mean 'grain size.' If we write $k = d^2 C$, then the number C depends on the pore configuration. For a tubular network (in three dimensions), one finds $C \sim \phi^2$, specifically, $C \approx \phi^2/72\pi$ (so long as ϕ is relatively small). An often used relation is that of Carman and Kozeny, which applies to pseudo-spherical grains (for example sand grains); this is

$$C \approx \frac{\phi^3}{180(1-\phi)^2}. \tag{13.5}$$

The factor $(1-\phi)^2$ takes some account of the fact that as ϕ increases toward one, the resistance to motion becomes negligible. In fact, for media consisting of uncemented (i.e., separate) grains, there is a critical value of ϕ beyond which the medium as a whole will deform. Depending on the grain-size distribution, this value is about 0.5 to 0.6. When the medium as a whole deforms, the description of the intergranular fluid flow can still be taken to be given by Darcy's law, but this now constitutes a particular choice of the interactive drag term in a two-phase flow model.

In the case of soils, empirical power laws of the form

$$C \sim \phi^m \tag{13.6}$$

Table 13.1. *Rock and soil permeabilities*

$k(\mathrm{m}^2)$	Material
10^{-8}	Gravel
10^{-10}	Sand
10^{-12}	Fractured igneous rock
10^{-13}	Sandstone
10^{-14}	Silt
10^{-18}	Clay
10^{-20}	Granite

are often used, with much higher values of the exponent (e.g., $m = 8$). Such behavior reflects the (chemically-derived) ability of clay-rich soils to retain a high fraction of water even at large suction. Table 13.1 gives typical values of the permeability of several common rock and soil types, ranging from coarse gravel and sand to finer silt and clay.

Under gravity, the hydraulic head gradient is ρg, and the discharge is given by the Darcy flux

$$u = k\rho g/\mu. \tag{13.7}$$

This value is known as the hydraulic conductivity. It has (evidently) units of velocity but is of less general use insofar as it depends on the properties of water, whereas k depends only on the porous medium. Nevertheless, the hydraulic conductivity is of intuitive value. It is related to the permeability by a factor of approximately 10^5 (in cgs units), that is, $k = 10^{-8}$ cm^2 if $u = 10^{-3}$ cm s^{-1}. The unit of permeability known as the *darcy* is defined as 1 darcy $= 10^{-8}$ cm^2.

Going back to the theoretical derivation of Darcy's law, we can calculate the (particle) Reynolds number for the porous flow. If v is the velocity in a pore, then we define a Reynolds number based on grain size as

$$Re_{gr} = \frac{\rho v a}{\mu} \sim \frac{\rho Q}{\mu a} \sim \frac{\rho|\mathbf{u}|d}{\mu\sqrt{\phi}}. \tag{13.8}$$

If we take $\phi \sim O(1)$, $k \approx 10^{-2}d^2$, and $\mu/\rho = 10^{-2}$ cm^2 s^{-1}, then we have in cgs units the hydraulic conductivity $u = 10^5 k = 10^3 d^2$; thus

$$Re_{gr} \approx 10^5 d^3, \tag{13.9}$$

with d in centimeters. Thus the flow is laminar for $d < .3$ cm, that is, $k < 10^{-3}$ cm^2, corresponding to a gravel. Only for very coarse gravel could the flow become turbulent, but for water percolation in rocks and soils, we invariably have slow flow.

In other situations, and notably for gas stream flow in fluidized beds or in packed catalyst reactor beds, the flow is rapid and turbulent. In this case, the Poiseuille flow balance $-\boldsymbol{\nabla} p = \mu\mathbf{u}/k$ can be replaced by the *Ergun equation*

$$-\boldsymbol{\nabla} p = \rho|\mathbf{u}|\mathbf{u}/k' \tag{13.10}$$

(or more generally, a sum of the two terms), which reflects the fact that turbulent flow in a pipe is resisted by *Reynolds stresses*, which are generated by the fluctuation of the inertial terms in the momentum equation. Just as for the laminar case, the parameter k', having units of length, depends both on the grain size d and on ϕ. We will have $k' \propto d$, $k' \to 0$ as ϕ as $\phi \to 0$. A Carman-Kozeny type equation is

$$k' = \frac{\phi^3 d}{175(1-\phi)}. \tag{13.11}$$

13.1 Basic groundwater flow

Darcy's equation is supplemented by an equation for the conservation of the fluid phase (or phases, for example in oil recovery, where these may be oil and water). For a single phase, this equation is of the simple conservation form

$$\frac{\partial}{\partial t}(\rho\phi) + \nabla.(\rho\mathbf{u}) = 0, \tag{13.12}$$

supposing there are no sources or sinks within the medium. In this equation, ρ is the material density, that is, mass per unit volume *of the fluid*. A term ϕ is not present in the divergence term, because $\mathbf{u}$ has already been written as a volume flux (i.e., the ϕ has already been included in it).

Eliminating $\mathbf{u}$, we have the parabolic equation

$$\frac{\partial}{\partial t}(\rho\phi) = \nabla.\left[\frac{k}{\mu}\rho\nabla p\right], \tag{13.13}$$

and we need a further equation of state (or two) to complete the model. The simplest assumption corresponds to incompressible groundwater flowing through a rigid porous medium. In this case, ρ and ϕ are constant, and the governing equation reduces (if also k is constant) to Laplace's equation

$$\nabla^2 p = 0. \tag{13.14}$$

This simple equation will form the basis for most of the discussion, but we mention two further variants of interest.

The first concerns the flow of a compressible fluid (e.g., a gas) in a nondeformable medium. Then ϕ is constant (so k is constant), but ρ is determined by pressure and temperature. If we ignore the effects of temperature (but note this is inappropriate in, for example, hydrothermal circulation; see Chapter 14), then we can assume $p = p(\rho)$ with $p'(\rho) > 0$, and

$$\rho_t = \frac{k}{\mu\phi}\nabla.[\rho p'(\rho)\nabla\rho], \tag{13.15}$$

which is a nonlinear diffusion equation for ρ. If $p \propto \rho^\gamma$, $\gamma > 0$, this is degenerate when $\rho = 0$, and the equation has interesting properties (see below).

A different variant occurs in *consolidation theory*, where the medium compacts under its own weight and in so doing expels the pore fluid. If we take ρ as constant, we have to postulate a relation between the porosity ϕ and the pore pressure p. In practice, it is found that soils (for example), when compressed, obey a (nonreversible) relation between ϕ and the *effective pressure*

$$p_e = P - p, \tag{13.16}$$

where P is the overburden pressure.

The concept of effective pressure, or more generally effective stress, is an extremely important one. The idea is that the total imposed pressure (e.g., the overburden pressure due to the weight of the rock or soil) is borne by both the pore fluid and the porous medium. The pore fluid is typically at a lower pressure than the overburden, and the extra stress (the effective stress) is that which is applied through grain-to-grain contacts. Thus the effective pressure is that which is transmitted through the porous medium, and it is in consequence of this that the medium responds to the effective stress; in particular, the *characteristic relation* between ϕ and p_e represents the nonlinear pseudo-elastic effect of compression.

As p_e increases, so ϕ decreases; thus we can write (neglecting hysteresis)

$$p_e = p_e(\phi), \qquad p_e'(\phi) < 0. \tag{13.17}$$

With ρ equal to a constant, we obtain the nonlinear diffusion equation

$$\phi_t = \nabla . \left[\frac{k(\phi)}{\mu} \left| p_e'(\phi) \right| \nabla \phi \right] \tag{13.18}$$

if we take P as constant. A similar relation applies for unsaturated soils. Here the soil matrix can be immobile, but if the soil is dried by suction, the retained moisture fraction ϕS ($< \phi$, the porosity) depends on p_e. The suction characteristic curves are similar to the consolidation curves due to capillary action and exhibit the same hysteresis phenomenon. The same equation (13.18) therefore applies (providing P is constant).

13.2 Dam seepage

The boundary conditions we apply to Laplace's equation (13.14) are that at a solid (impermeable) surface, the normal velocity is zero:

$$\mathbf{u}.\mathbf{n} = 0, \tag{13.19}$$

where $\mathbf{n}$ is the unit outward normal. At a permeable surface (for example, the Earth's surface, or more generally at the level of saturation in the ground, i.e., the water table, the pressure could be constant, and so

$$p = 0 \tag{13.20}$$

(by subtracting off atmospheric pressure)).

Consider now the problem of determining the rate of leakage through an earthfill dam built on an impermeable foundation. The configuration is as shown in Fig. 13.2, where we have illustrated the (unrealistic) case of a dam with vertical walls: In reality the cross section would be trapezoidal. A reservoir of height h_0 abuts a dam of width l.

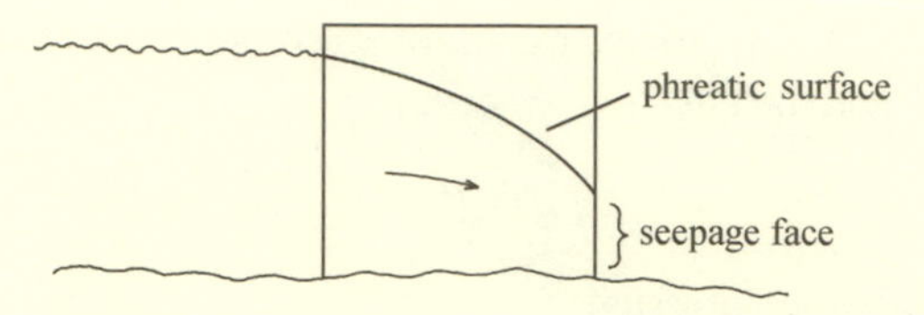

Fig. 13.2. Geometry of dam seepage problem

Water flows through the dam between the base $y = 0$ and a free surface (known as the *phreatic* surface) $y = h$, below which the dam is saturated and above which it is unsaturated. As a simple assumption, we suppose that this free surface also provides an upper limit to the region of groundwater flow. We therefore neglect the flow in the unsaturated region, and the free boundary must be determined by a kinematic boundary condition, which is that

$$\frac{d}{dt}(h - y) = 0, \qquad \frac{d}{dt} = \frac{\partial}{\partial t} + (\mathbf{u}/\phi).\nabla, \tag{13.21}$$

where d/dt is a material derivative for the fluid flow. The boundary condition (13.21) expresses the idea that the free surface is *defined* by the fluid elements that constitute it, so that the fluid velocity at $y = h$ is the same as the velocity of the interface itself.

In the two-dimensional configuration shown in Fig. 13.2, we therefore have to solve

$$\begin{gathered} u = -\frac{k}{\mu} p_x, \qquad v = -\frac{k}{\mu}(p_y + \rho g), \\ u_x + v_y = 0, \end{gathered} \tag{13.22}$$

where $\mathbf{u} = (u, v)$, with boundary conditions that

$$\begin{gathered} v = 0 \quad \text{on } y = 0, \\ p = 0, \qquad v = \phi h_t + u h_x \quad \text{on } y = h, \\ p = \rho g(h_0 - y) \quad \text{on } x = 0, \\ p = 0 \quad \text{on } x = l. \end{gathered} \tag{13.23}$$

These conditions describe the impermeable base at $y = 0$, the free surface at $y = h$, hydrostatic pressure on $x = 0$, and atmospheric pressure at $x = l$ (the seepage face). The free boundary is to be determined as part of the solution.

13.3 Dupuit approximation

We nondimensionalize the variables by scaling as follows:

$$\begin{gathered} x \sim l, \qquad y \sim h_0, \qquad p \sim \rho g h_0, \qquad u \sim k\rho g h_0/\mu l, \\ v \sim k\rho g/\mu, \qquad t \sim \phi h_0 \mu / k\rho g, \end{gathered} \tag{13.24}$$

all in order to obtain various obvious balances in the equations and boundary conditions. The Dupuit–Forchheimer approximation is based on the idea that $h_0 \ll l$, so that the dam is relatively long and thin. To this end, we define

$$\delta = h_0/l; \tag{13.25}$$

the equations are then

$$
\begin{gathered}
u = -p_x, \qquad v = -(p_y + 1), \\
p_{yy} + \delta^2 p_{xx} = 0,
\end{gathered} \tag{13.26}
$$

with

$$
\begin{gathered}
p_y = -1 \quad \text{on } y = 0, \\
p = 0, \qquad h_t = -(p_y + 1) + \delta^2 p_x h_x \quad \text{on } y = h, \\
p = 1 - y \quad \text{on } x = 0, \\
p = 0 \quad \text{on } x = 1.
\end{gathered} \tag{13.27}
$$

We now proceed (if $\delta \ll 1$) by expanding $p = p_0 + \delta^2 p_1 + \ldots$, etc. It is clear that the leading order approximation for p is just

$$
p \approx h - y. \tag{13.28}
$$

This fails to satisfy the condition at $x = 1$, where a boundary layer is necessary to bring back the x derivatives of p, unless there is no seepage face, that is, $h(1) = 0$. However, we also note that with $p_y + 1 = O(\delta^2)$, then $h_t = O(\delta^2)$, which suggests $h = h_0 + \delta^2 h_1 + \ldots$ and $h_0 =$ constant. Alternatively, we realize that $h_t = O(\delta^2)$ simply indicates that the timescale of relevance to transient problems is longer than our initial guess $O(\phi h_0 \mu / k\rho g)$, so that we *rescale t* with $1/\delta^2$. Putting $t = \bar{t}/\delta^2$ (and subsequently omitting the overbar), we rewrite the kinematic boundary condition as

$$
\delta^2 h_t = -(p_y + 1) + \delta^2 p_x h_x. \tag{13.29}
$$

We now seek expansions

$$
\begin{gathered}
p = p_0 + \delta^2 p_1 + \ldots, \\
h = h_0 + \ldots,
\end{gathered} \tag{13.30}
$$

and we find successively

$$
p_0 = h_0 - y, \tag{13.31}
$$

and

$$
p_{1yy} = -h_{0xx}, \qquad p_{1y} = 0 \quad \text{on } y = 0, \tag{13.32}
$$

whence

$$
p_{1y} = -y h_{0xx}, \tag{13.33}
$$

so that Eq. (13.29) gives

$$
h_{0t} = h_0 h_{0xx} + h_{0x}^2; \tag{13.34}
$$

dropping the subscript, we obtain the nonlinear diffusion equation

$$
h_t = (h h_x)_x. \tag{13.35}
$$

Notice that this equation is not valid to $x = 1$, because we require $p = 0$ at $x = 1$, in contradiction to Eq. (13.31). We therefore expect a boundary layer there where p

changes rapidly. Equation (13.35) is a second-order equation requiring two boundary conditions. One is that

$$h = 1 \quad \text{on } x = 0, \tag{13.36}$$

but it is not so clear what the other is. It can be determined by means of the following trick. Define

$$U = \int_0^h p\,dy, \tag{13.37}$$

and note that the flux q is given by

$$q = \int_0^h u\,dy = -\int_0^h p_x\,dy = -\frac{\partial U}{\partial x}, \tag{13.38}$$

because $p = 0$ on $y = h$. Furthermore,

$$\begin{aligned} U &= 1/2 \quad \text{at } x = 0, \\ U &= 0 \quad \text{at } x = 1, \end{aligned} \tag{13.39}$$

and therefore we have the exact result

$$\int_0^1 q\,dx = 1/2. \tag{13.40}$$

In a steady state, $h_t = -q_x = 0$, so q is constant, and therefore

$$q = 1/2 = -hh_x. \tag{13.41}$$

The steady solution (away from $x = 1$) is therefore

$$h = (1-x)^{1/2}, \tag{13.42}$$

and there is (to leading order) no seepage face at $x = 1$.

In fact, the derivation of Eq. (13.41) applies for unsteady problems also. If we suppose that q does not jump rapidly near $x = 1$, then we can use the Dupuit approximation $q = -hh_x$ in Eq. (13.40), and an integration yields

$$h = 0 \quad \text{at } x = 1 \tag{13.43}$$

as the general condition.

The boundary layer *structure* near $x = 1$ can be described as follows: Near $x = 1$, we have $h \sim (1-x)^{1/2}$, $p \sim h - y$, and so we put

$$x = 1 - \varepsilon X, \qquad h = \varepsilon^{1/2} H, \qquad y = \varepsilon^{1/2} Y, \qquad p = \varepsilon^{1/2} P, \tag{13.44}$$

and if we choose

$$\varepsilon = \delta^2 \tag{13.45}$$

to bring back the x derivatives in Laplace's equation, we get

$$P_{XX} + P_{YY} = 0, \tag{13.46}$$

with

$$\begin{aligned} &P_Y = -1 \quad \text{on } Y = 0, \\ P = 0, \qquad &P_Y - P_X H_X = -1 \quad \text{on } Y = H, \\ P \sim H - Y, \qquad &H \sim X^{1/2} \quad \text{as } X \to \infty, \\ &P = 0 \quad \text{on } X = 0. \end{aligned} \tag{13.47}$$

Exact solutions of this problem can be found using complex variables, but for many purposes the outer (Dupuit) approximation is sufficient, together with a consistently scaled boundary layer problem (which at least suggests that the asymptotic procedure makes sense).

Subsurface flow: similarity solutions

It is left as an exercise for the reader to show that the Dupuit approximation applies in the more general context of a slowly varying water table. Because the water table more or less follows the surface topography, this means slowly varying topography, that is, small hillslopes: this is normally the case. Moreover, in three spatial dimensions, if the water table is at a height h, which is a slowly varying function of x, y (the horizontal variables) as well as t, then the generalization of the Dupuit–Forchheimer equation is

$$h_t = \boldsymbol{\nabla}.[h\boldsymbol{\nabla}h], \tag{13.48}$$

and the (vector) flux in the horizontal direction is

$$\mathbf{q} = -h\boldsymbol{\nabla}h; \tag{13.49}$$

here $\boldsymbol{\nabla} = (\partial/\partial x, \partial/\partial y)$. This equation arises in many other contexts, and in general it must be solved numerically. This is straightforward if $h \neq 0$, when it behaves like an ordinary diffusion equation. However, the degeneracy of the equation when $h \to 0$ gives solutions a distinctive property, which we now illustrate by discussing some particular exact solutions.

For simplicity, we consider the equation in one dimension again, and we look for a similarity solution that represents the release of a unit quantity of groundwater at $t = 0$. Thus we solve

$$\begin{aligned} &h_t = (hh_x)_x, \\ &h = 0 \quad \text{at } t = 0,\ x \neq 0, \\ &\int_{-\infty}^{\infty} h\,dx = 1 \end{aligned} \tag{13.50}$$

(i.e., $h = \delta(x)$ at $t = 0$). A similarity solution to this problem exists in the form

$$h = t^{-1/3} g(\xi), \qquad \xi = x/t^{1/3}, \tag{13.51}$$

where g satisfies

$$(gg')' + \frac{1}{3}(\xi g)' = 0. \tag{13.52}$$

Because $g \to 0$ as $\xi \to \pm\infty$, we have

$$g\left(g' + \frac{1}{3}\xi\right) = 0, \tag{13.53}$$

with solutions

$$g = 0 \quad \text{or} \quad g = g_0 - \frac{1}{6}\xi^2. \tag{13.54}$$

Neither solution alone can satisfy both the constraints

$$g(\infty) = 0, \qquad \int_{-\infty}^{\infty} g(\xi)\,d\xi = 1, \tag{13.55}$$

and we are forced to allow g to be piecewise smooth. Thus

$$\begin{aligned} g &= 0, \quad \xi > \xi_0, \\ g &= \frac{1}{6}(\xi_0^2 - \xi^2), \quad \xi < \xi_0, \end{aligned} \tag{13.56}$$

where $\int_0^{\xi_0} \frac{1}{6}(\xi_0^2 - \xi^2)\,d\xi = 1$, that is, $\xi_0 = 3^{2/3}$. Although g'' and g' do not exist at ξ_0, it is clear that Eq. (13.52) is satisfied (in a generalized sense) everywhere. The rule is that where the diffusion coefficient is zero, one can have discontinuities in slope, because there is no diffusion to smooth them out.

This also makes physical sense, because the conservation law (13.48) is originally derived in integral form,

$$\frac{\partial}{\partial t}\int_S h\,ds = -\int_C [-h\nabla h].\mathbf{n}\,ds, \tag{13.57}$$

where C is the boundary of an arbitrary surface S. It is only the *assumption* that h is twice continuously differentiable ($h \in C^2$) that implies the point form of the equation. If we write $h\nabla h = \nabla(h^2/2)$, then we see that if ∇h has jump discontinuities where $h = 0$, then $h^2/2$ is differentiable everywhere, and Eq. (13.56) does indeed make sense. By taking a 'pillbox' contour that contains such a moving front, we find that the normal velocity of the front is given by

$$v_n = \left[-\frac{\partial}{\partial n}(h^2/2)\right]_-^+ \Big/ [h]_-^+, \tag{13.58}$$

where $\pm$ denotes limiting values ahead of and behind the front. Deferring the limit process, we find

$$v_n = -\left.\frac{\partial h}{\partial n}\right|_-, \tag{13.59}$$

which agrees with Eqs. (13.51) and (13.56), for example. More generally, if $\mathbf{q} = \nabla[Q(h)]$,

$$v_n = -\frac{\partial}{\partial n}\left[\frac{\partial Q}{\partial h}\right]_-; \tag{13.60}$$

for example, if $Q = h^{m+1}/(m+1)$, $m > 0$, then $v_n = -\partial(h^m)/\partial n\,|_-$, and we must have $h \sim (n_f - n)^{1/m}$ near the front, where n is the normal coordinate and $n_f(t)$ is the front position. In particular, this shows that for the equation $h_t = \nabla.[h^m \nabla h]$, with an initial condition having compact support and such that $|\nabla h|$ is finite at $h = 0$, the solution has a stationary front until the profile becomes singular at $h = 0$. This is the 'waiting-time' property of such nonlinear diffusion equations.

13.4 Consolidation

Let us now return to Eq. (13.13), where we assume that the pore fluid has constant density ρ. If we include gravity, then we have

$$\frac{\partial \phi}{\partial t} = \nabla.\left[\frac{k}{\mu}\nabla(p + \rho g y)\right], \tag{13.61}$$

where y is the vertical coordinate; as previously discussed, ϕ depends on the effective pressure $p_e = P - p$, where P is overburden pressure, as indicated in Fig. 13.3. The dependence is, however, nontrivial and involves hysteresis. Specifically, a soil follows the *normal consolidation line*, providing consolidation is occurring, that is, $\dot{p}_e > 0$. However, if at some point the effective pressure is reduced, only a partial recovery of ϕ takes place. When p_e is increased again, ϕ more or less retraces its path to the normal consolidation line and then resumes its normal consolidation path. A crude representation of this behavior is

$$\phi = f\left[\max_{\tau < t} p_e(\tau)\right], \tag{13.62}$$

where f represents the normal consolidation line. Here we will ignore effects of hysteresis, and suppose $\phi = f(p_e)$, or equivalently $p_e = p_e(\phi)$. Thus

$$p = P - p_e(\phi), \qquad k = k(\phi), \tag{13.63}$$

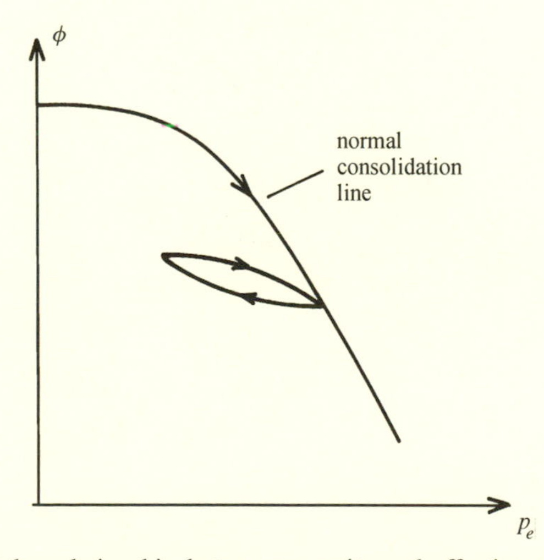

Fig. 13.3. Form of the relationship between porosity and effective pressure. A hysteretic decompression-reconsolidation loop is indicated. In soil mechanics this relationship is often written in terms of the *void ratio* $e = \phi/(1 - \phi)$, and specifically $e = e_0 - C_c \log p_e$, where C_c is the *compression index*.

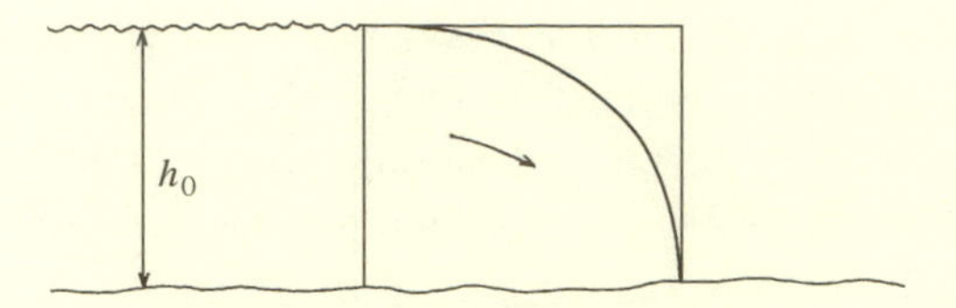

Fig. 13.4. Idealized dam geometry

so that

$$\phi_t = \nabla.\left[\frac{k(\phi)}{\mu}\{|p'_e(\phi)|\nabla\phi + \nabla P + \rho g y\}\right]. \tag{13.64}$$

This is a nonlinear diffusion equation for ϕ, which generally requires numerical solution.

As an example of a method of problem formulation, let us reconsider the problem of dam seepage. As shown in Fig. 13.4, suppose the dam is of height h_0, and the top is level with the water surface. We have

$$P = \rho_s g(h_0 - y), \tag{13.65}$$

and in the Dupuit approximation

$$p \approx \rho g(h - y), \tag{13.66}$$

where h is the free surface. Thus

$$p_e = \Delta\rho g(h_0 - y) + \rho g(h_0 - h), \tag{13.67}$$

where $\Delta\rho = \rho_s - \rho$.

For purposes of illustration, we suppose (not very realistically, for ϕ)

$$\begin{aligned} k &= k_0\phi^m, \\ \phi &= \phi_0 \exp[-p_e/\tau], \end{aligned} \tag{13.68}$$

whence

$$\phi = \phi_0 \exp\left[-\frac{1}{\tau}\{\Delta\rho g(h_0 - y) + \rho g(h_0 - h)\}\right]. \tag{13.69}$$

With u being the horizontal velocity, conservation of mass can be written as

$$\frac{\partial}{\partial t}\int_0^h \phi\, dy + \frac{\partial}{\partial x}\int_0^h u\, dy = 0, \tag{13.70}$$

where

$$u = -\frac{k}{\mu}p_x \approx -\frac{k\rho g}{\mu}h_x; \tag{13.71}$$

hence

$$\frac{\partial}{\partial t}\int_0^h \phi\, dy \approx \frac{\rho g}{\mu}\frac{\partial}{\partial x}\left[\left\{\int_0^h k\, dy\right\}h_x\right], \tag{13.72}$$

which gives the appropriate (dimensional!) generalization of Eq. (13.35) for variable ϕ and k. We scale the variables as follows:

$$x \sim l, \qquad h \sim h_0, \tag{13.73}$$

and after some algebra, we find the dimensionless equation

$$\frac{\partial}{\partial t}[e^{\lambda rh} - e^{\lambda h}] = \frac{\partial}{\partial x}[\{e^{\Lambda rh} - e^{\Lambda h}\}h_x], \tag{13.74}$$

where

$$\lambda = \rho g h_0/\tau, \qquad r = \rho_s/\rho, \qquad \Lambda = m\rho g h_0/\tau, \tag{13.75}$$

and where we scale t as

$$t \sim \frac{ml^2\mu}{k_0\phi_0^{m-1}\rho g h_0}\exp[(\Lambda - \lambda)r]. \tag{13.76}$$

With typical values $\rho = 2 \times 10^3$ kg m^{-3}, $g = 10$ m s^{-2}, and $h_0 = 10$ m, we have $\rho g h_0 \sim 2$ bars, whereas typically $\tau > 10$ bars. If $\lambda, \Lambda \ll 1$, then approximately,

$$\lambda h_t = \frac{\partial}{\partial x}[\Lambda h h_x] \tag{13.77}$$

as before. If, on the other hand, $r > 1$ and $\lambda, \Lambda \gg 1$, then approximately

$$\frac{\partial}{\partial t}e^{\lambda rh} \approx \frac{\partial}{\partial x}[e^{\Lambda rh}h_x], \tag{13.78}$$

and by rescaling t and putting $w = \exp(\lambda rh)$, this can be written in the form

$$w_t = (w^m)_{xx}. \tag{13.79}$$

As before, the second-order equation (13.74) requires two boundary conditions. One of these is that $h = 1$ at $x = 0$. The other follows from a trick similar to that used previously. Define

$$K(p) = h_0\phi_0^m\frac{\tau}{m}\exp\left[-\frac{m}{\tau}P\right]\{e^{mp/\tau} - 1\}, \tag{13.80}$$

so that $\partial K/\partial p = k$, and $K = 0$ when $p = 0$ (at $y = h$). Then

$$-\frac{\partial}{\partial x}\int_0^h \frac{1}{\mu}K(p)\,dy = \int_0^h -\frac{k}{\mu}p_x\,dy = \int_0^h u\,dy = q,$$

say, and thus

$$\int_0^l q\,dx = \int_0^{h_0}\frac{1}{\mu}K_0\,dy, \tag{13.81}$$

because $K = 0$ at $x = l$ ($p = 0$), and $K = K_0(y)$ at $x = 0$ is known. This leads to the second condition for h. Carrying out the calculation, we find

$$h = 0 \quad \text{at } x = 1, \tag{13.82}$$

as before. The steady solution of Eq. (5.74) is thus

$$\frac{1}{\Lambda r}e^{\Lambda r h}-\frac{1}{\Lambda}e^{\Lambda h}=\left[\frac{1}{\Lambda r}e^{\Lambda r}-\frac{1}{\Lambda}e^{\Lambda}\right](1-x). \tag{13.83}$$

If $\Lambda \gg 1$, this gives (for $x < 1$)

$$h \approx 1+\frac{1}{\Lambda r}\ln(1-x), \tag{13.84}$$

and thus the effect of Λ is to steepen the profile, with a sharp drop near $x = 1$.

13.5 Solute dispersivity

Much of the interest in modeling groundwater flow lies in the prediction of solute transport, in particular understanding how pollutants will disperse: For example, how do nitrates used for agricultural purposes disperse via the local groundwater system? Most simply, one would simply add a diffusion term to the advection of the solute concentration c:

$$c_t+\mathbf{u}.\nabla c=\nabla.[D\nabla c]. \tag{13.85}$$

Referring back to Table 13.1, a porous sand has a hydraulic conductivity of order 10^{-1} cm s^{-1}. The degree of diffusive spreading of a sharp front is of the order of D/u, that is (with $D \sim 10^{-6}$ cm^2 s^{-1}) 10^{-5} cm, and it would seem that solutes simply advect with the groundwater. That this is not so is due to a remarkable phenomenon called *Taylor dispersion* after a seminal paper by Taylor (1953), who first described it.

Taylor's paper considers the diffusion of a solute in a tube of circular cross section through which a Poiseuille flow passes. If the mean velocity is U and the tube is of radius a, then the velocity is $2U(1-r^2/a^2)$, and the concentration satisfies the equation

$$c_t+2U(1-r^2/a^2)c_x=D\left(c_{rr}+\frac{1}{r}c_r+c_{xx}\right), \tag{13.86}$$

where x is measured along the tube and r is the radial coordinate. Taylor showed, rather ingenuously, that when the Péclet number $Pe = aU/D$ is large, then the effect of the diffusion term in Eq. (13.86) is to *disperse* the mean solute concentration diffusively about the position of its center of mass, $x = Ut$, with a dispersion coefficient of $a^2U^2/48D$. Aris (1956) later improved this to

$$D_T=\frac{a^2U^2}{48D}+D, \tag{13.87}$$

which is valid for $aU/D \gtrsim 1$ (see Exercise 2). The dispersive mechanism is due to the radial variation of the velocity profile, which can disperse the solute even if the diffusion coefficient is very small.

Typically, this is generalized for porous media by writing the dispersion coefficient as

$$D_T=D^*+D_l, \tag{13.88}$$

where D^* represents molecular diffusion and D_l longitudinal dispersion. The tortuosity of the flow paths and the possibility of adsorption on to the solid causes D^* to be less than D, and ratios D^*/D between 0.01 and 0.5 are commonly observed. In porous media, remixing at pore junctions causes the dependence of D_l on the flow velocity to be less than quadratic, and a relation of the form

$$D_l = \alpha v^m, \tag{13.89}$$

where v is flow velocity, fits experimental data reasonably well for values $1 < m < 1.2$. Mixing at junctions also causes transverse dispersion to occur, with a coefficient D_t that is measured to be less than D_l by a factor of order 10^2 when $Pe \gg 1$. Dispersion is thus a tensor property.

Pollution leakage

Consider leakage of a pollutant of concentration c (mass per unit volume of liquid) from the reservoir in Section 13.2. Conservation of solute is described by

$$\frac{\partial}{\partial t}(\phi c) + \boldsymbol{\nabla}.(c\mathbf{u}) = \boldsymbol{\nabla}.\{\mathbf{D}_T : \boldsymbol{\nabla} c\}, \tag{13.90}$$

where $\mathbf{D}_T$ is the dispersion tensor. We ignore transverse dispersion and take

$$\mathbf{D}_T = \beta d\mathbf{u}, \tag{13.91}$$

where d is grain size, and β is an $O(1)$ coefficient. Because $\boldsymbol{\nabla}.\mathbf{u} = 0$ (see Eq. (13.22)), we have (with ϕ constant)

$$\phi c_t + \mathbf{u}.\boldsymbol{\nabla} c = \beta d\left(u\frac{\partial^2 c}{\partial x^2} + v\frac{\partial^2 c}{\partial y^2}\right). \tag{13.92}$$

Boundary conditions for this equation are (in appropriate units), for the geometry illustrated in Fig. 13.2,

$$\begin{aligned} c = 1 &\quad \text{on } x = 0, \\ \frac{\partial c}{\partial y} = 0 &\quad \text{on } y = 0, \\ (\mathbf{u} : \boldsymbol{\nabla} c).\mathbf{n} = 0 &\quad \text{on } y = h, \\ \frac{\partial c}{\partial x} = 0 &\quad \text{on } x = l, \end{aligned} \tag{13.93}$$

where we assume no flux of water from the unsaturated zone $y > h$ and continuity of c across the seepage face.

Using the nondimensionalization in Eq. (13.24) (but with $t \sim l^2\mu/k\rho g h_0$), we have the dimensionless equation

$$\delta^2\phi c_t + \delta^2 u c_x + v c_y = \frac{\beta d}{h_0}\{c_{yy} + \delta^3 c_{xx}\}, \tag{13.94}$$

together with

$$
\begin{aligned}
c &= 1 \quad \text{on } x = 0, \\
\frac{\partial c}{\partial y} &= 0 \quad \text{on } y = 0, \\
\frac{\partial c}{\partial x} &= 0 \quad \text{on } x = 1, \\
-\delta^3 u c_x h_x + v c_y &= 0 \quad \text{on } y = h.
\end{aligned}
\tag{13.95}
$$

Note from Eqs. (13.26), (13.31), and (13.33) that

$$v = \delta^2 V, \qquad u \approx -h_x, \qquad V \approx y h_{xx}; \tag{13.96}$$

thus

$$\phi c_t + u c_x + V c_y = \gamma \{ c_{yy} + \delta c_{xx} \}, \tag{13.97}$$

with

$$
\begin{aligned}
c = 1 \quad \text{on } x = 0, &\qquad \frac{\partial c}{\partial y} = 0 \quad \text{on } y = 0, \\
\frac{\partial c}{\partial x} = 0 \quad \text{on } x = 1, &\qquad -\delta u c_x h_x + V c_y = 0 \quad \text{on } y = h.
\end{aligned}
\tag{13.98}
$$

Dispersion is thus controlled by the dispersion parameter

$$\gamma = \frac{\beta d l^2}{h_0^3} \tag{13.99}$$

and is significant if l is large enough. For example, if $d = 10^{-3}$ m, $h = 10$ m, then $\gamma \gtrsim O(1)$ if $l > 10^3$ m.

A problem of this type only makes sense in a time-dependent situation, because obviously the steady solution is just $c = 1$. To model the propagation of a front from $x = 0$ toward $x = 1$, notice that $c \approx c(x, t)$ satisfies the boundary conditions, and then

$$\phi c_t + u c_x = \nu c_{xx}, \tag{13.100}$$

where from Eq. (13.42)

$$u = \frac{1}{2}(1 - x)^{-1/2}, \qquad \nu = \frac{\beta d l}{h_0^2}. \tag{13.101}$$

The front from $x = 0$ propagates at the linear liquid velocity u/ϕ and spreads out with a width $\Delta x \sim (\nu t)^{1/2}$ (see Exercise 3).

13.6 Heterogeneous porous media

In large-scale flow fields, the permeability may vary over many orders of magnitude owing to the quasi-random stacking of different geologic layers of material. For example, in sedimentary basins one may be interested in flow over scales of kilometers, whereas the permeability, measured at intervals of ten centimeters on a sedimentary core, may appear completely noisy. The question then arises, what value to choose for the permeability?

The theory of random fields addresses this as follows. We suppose (based on observations) that the hydraulic conductivity K has a lognormal distribution and write

$$\ln K = \ln K_l + f, \qquad E(f) = 0, \tag{13.102}$$

where $E(f)$ is the expected, or mean value of f. If f is stationary, then

$$E[f(\mathbf{x}+\boldsymbol{\xi})f(\mathbf{x})] = R_{ff}(\boldsymbol{\xi}) \tag{13.103}$$

defines the autocovariance function. In addition, $f(\mathbf{x})$ has the spectral representation as a Fourier–Stieltjes integral,

$$f(\mathbf{x}) = \int e^{i\mathbf{k}.\mathbf{x}} dZ_f(\mathbf{k}), \tag{13.104}$$

where Z_f is a complex-valued stochastic process. Then the power spectral density S_{ff} is defined through

$$\begin{aligned} E[dZ_f(\mathbf{k})\overline{dZ}_f(\mathbf{k}')] &= 0, \quad \mathbf{k} \neq \mathbf{k}', \\ E[dZ_f(\mathbf{k})\overline{dZ}_f(\mathbf{k}')] &= S_{ff}(\mathbf{k})\, d\mathbf{k}, \quad \mathbf{k} = \mathbf{k}', \end{aligned} \tag{13.105}$$

where the overbar denotes the complex conjugate. The Wiener–Khintchin theorem is then that

$$R_{ff}(\boldsymbol{\xi}) = \int e^{i\mathbf{k}.\boldsymbol{\xi}} S_{ff}(\mathbf{k})\, d\mathbf{k}, \tag{13.106}$$

where $d\mathbf{k}$ denotes the volume element in $\mathbf{k}$ space. For example, a common form of autocovariance function is

$$R_{ff}(\boldsymbol{\xi}) = \sigma_f^2 \exp[-|\boldsymbol{\xi}|/\lambda], \tag{13.107}$$

where σ_f^2 is the variance of f and λ is called the correlation length. This corresponds to an autocorrelated red noise process and has corresponding power spectrum

$$S_{ff}(\mathbf{k}) = \frac{\sigma_f^2 \lambda^3}{\pi^2(1+\lambda^2|\mathbf{k}|^2)^2}. \tag{13.108}$$

Now we return to the problem of determining an effective permeability. If ϕ is the hydraulic head (usually, $\phi = p + \rho g z$, where z is altitude), then for an incompressible flow,

$$\mathbf{q} = -K\boldsymbol{\nabla}\phi, \qquad \boldsymbol{\nabla}.\mathbf{q} = 0. \tag{13.109}$$

Thus $K\nabla^2\phi + \boldsymbol{\nabla}K.\boldsymbol{\nabla}\phi = 0$, or $\nabla^2\phi + \boldsymbol{\nabla}(\ln K).\boldsymbol{\nabla}\phi = 0$. Thus, if we write $\phi = H + h$, where $E(h) = 0$, then (with K_l constant)

$$\nabla^2 H + \nabla^2 h + \boldsymbol{\nabla}f.\boldsymbol{\nabla}H + \boldsymbol{\nabla}f.\boldsymbol{\nabla}h = 0. \tag{13.110}$$

Taking the mean value,

$$\nabla^2 H + E(\boldsymbol{\nabla}f.\boldsymbol{\nabla}h) = 0; \tag{13.111}$$

hence

$$\nabla^2 h + \boldsymbol{\nabla} f . \boldsymbol{\nabla} H + (\boldsymbol{\nabla} f . \boldsymbol{\nabla} h)' = 0, \tag{13.112}$$

where the prime denotes the fluctuating part.

We simplify (or close) the model by neglecting the quadratic interaction terms, thus

$$\nabla^2 h \approx \mathbf{J} . \boldsymbol{\nabla} f, \qquad \mathbf{J} = -\boldsymbol{\nabla} H, \tag{13.113}$$

and in terms of their power spectra (see Eq. (13.104)),

$$dZ_h \approx -\frac{J_i k_i}{k^2} \, dZ_f, \tag{13.114}$$

where summation over the i components is implied and $k = |\mathbf{k}|$. From this we deduce

$$S_{hh}(\mathbf{k}) = \frac{J_i J_j k_i k_j}{k^4} S_{ff}(\mathbf{k}), \tag{13.115}$$

which relates the known f to the unknown h.

Now we consider the mean flow $\mathbf{Q} = E(\mathbf{q})$. This has components

$$\begin{aligned} Q_i = K_l E\left[e^f\left(J_i - \frac{\partial h}{\partial x_i}\right)\right] &= K_l E\left[\left(1 + f + \frac{f^2}{2} + \ldots\right)\left(J_i - \frac{\partial h}{\partial x_i}\right)\right] \\ &\approx K_l\left[\left(1 + \frac{\sigma_f^2}{2}\right) J_i - E\left(f \frac{\partial h}{\partial x_i}\right)\right], \end{aligned} \tag{13.116}$$

where we neglect terms higher than quadratic. To compute the second term, we use the spectral representations to obtain

$$E\left(f \frac{\partial h}{\partial x_i}\right) = \int -i k_i E(dZ_f \overline{dZ_h}) = \left\{\int \frac{k_i k_j}{k^2} S_{ff}(\mathbf{k}) \, d\mathbf{k}\right\} J_j. \tag{13.117}$$

We define

$$F_{ij} = \int \frac{k_i k_j}{k^2} S_{ff}(\mathbf{k}) \, d\mathbf{k}; \tag{13.118}$$

then the effective form of Darcy's law relating the mean flow $\mathbf{Q}$ to the mean head H is

$$Q_i = -\bar{K}_{ij} \frac{\partial H}{\partial x_j}, \tag{13.119}$$

where the mean hydraulic conductivity tensor is

$$\bar{K}_{ij} = K_l\left[\left(1 + \frac{\sigma_f^2}{2}\right)\delta_{ij} - F_{ij}\right]. \tag{13.120}$$

The important points to note are that the variance of the distribution of K distorts the mean conductivity and that the distortion is generally anisotropic. For example, if we consider a layered medium, then the conductivities parallel to and perpendicular to the layering are found to be

$$\begin{aligned} K_{\parallel} &\approx K_l\big(1 + \sigma_f^2/2\big), \\ K_{\perp} &\approx K_l\big(1 - \sigma_f^2/2\big). \end{aligned} \tag{13.121}$$

Obviously, these results are valid for sufficiently small σ_f and are meaningless for $\sigma_f^2 > 2$. A better approximation in the latter case is to take $K_\parallel, K_\perp = \exp(\pm\sigma_f^2/2)$.

13.7 Notes and references

Flow in porous media is described in the books by Bear (1972) and Dullien (1979). More recent books, for example, by Bear and Bachmat (1990), give more theoretical, deductive treatments based on homogenization (see below) or averaging (see Chapter 16), with a concomitant loss of readability. The classic geologists' book on groundwater is by Freeze and Cherry (1979), and the classic engineering text is by Polubarinova-Kochina (1962). A short introduction, of geographic style, is by Price (1985). A more mathematical survey, with a variety of applications, is by Bear and Verruijt (1987). The book edited by Cushman (1990) contains a wealth of articles on topics of varied and current interest, including dispersion, homogenization, averaging, dual porosity models, multigrid methods, and heterogeneous porous media. Further information on the concepts of soil mechanics can be found in Lambe and Whitman (1979).

Darcy's law and homogenization The 'derivation' of Darcy's law can be carried out in a more formal way using the method of homogenization. This is essentially an application of the method of multiple (space) scales to problems with microstructure. Usually (for analytic reasons) one assumes that the microstructure is periodic, although this is probably not strictly necessary (so long as local averages can be defined).

Consider the Stokes flow equations for a viscous fluid in a medium of macroscopic length l, subject to a pressure gradient of order $\Delta p/l$. If the microscopic (e.g., grain size) length scale is d, and $\varepsilon = d/l$, then if we scale velocity with $d^2\Delta p/l\mu$ (appropriate for local Poiseuille-type flow), length with l, and pressure with Δp, the Navier–Stokes equations can be written in the form

$$\begin{aligned} \boldsymbol{\nabla}.\mathbf{u} &= 0, \\ 0 &= -\boldsymbol{\nabla} p + \varepsilon^2\nabla^2\mathbf{u}, \end{aligned} \tag{13.122}$$

together with the no-slip boundary condition,

$$\mathbf{u} = 0 \text{ on } S: f(\mathbf{x}/\varepsilon) = 0, \tag{13.123}$$

where S is the interfacial surface. We put $\mathbf{x} = \varepsilon\boldsymbol{\xi}$ and seek solutions in the form

$$\begin{aligned} \mathbf{u} &= \mathbf{u}^{(0)}(\mathbf{x}, \boldsymbol{\xi}) + \varepsilon\mathbf{u}^{(1)}(\mathbf{x}, \boldsymbol{\xi}) \ldots \\ p &= p^{(0)}(\mathbf{x}, \boldsymbol{\xi}) + \varepsilon p^{(1)}(\mathbf{x}, \boldsymbol{\xi}) \ldots . \end{aligned} \tag{13.124}$$

We then have $p^{(0)} = p^{(0)}(\mathbf{x})$, and $\mathbf{u}^{(0)}$ satisfies

$$\begin{aligned} \boldsymbol{\nabla}_\xi.\mathbf{u}^{(0)} &= 0, \\ 0 &= -\boldsymbol{\nabla}_\xi p^{(1)} + \nabla_\xi^2\mathbf{u}^{(0)} - \boldsymbol{\nabla}_x p^{(0)}, \end{aligned} \tag{13.125}$$

equivalent to Stokes' equation for $\mathbf{u}^{(0)}$ with a forcing term $-\nabla_x p^{(0)}$. If $\mathbf{w}^j$ is the velocity field that (uniquely) solves

$$\begin{aligned} \nabla_\xi . \mathbf{w}^j &= 0, \\ 0 &= -\nabla_\xi P + \nabla_\xi^2 \mathbf{w}^j + \mathbf{e}_j, \end{aligned} \tag{13.126}$$

with periodic (in $\boldsymbol{\xi}$) boundary conditions and $\mathbf{u} = 0$ on $f(\boldsymbol{\xi}) = 0$, where $\mathbf{e}_j$ is the unit vector in the ξ_j direction, then (because the equation is linear) we have (summing over j)

$$\mathbf{u}^{(0)} = -\frac{\partial p^{(0)}}{\partial x_j} \mathbf{w}^j. \tag{13.127}$$

We define the average velocity

$$\langle \mathbf{u} \rangle = \frac{1}{V} \int_V \mathbf{u}^{(0)} \, dV, \tag{13.128}$$

where V is the volume over which S is periodic. Averaging Eq. (13.127) then gives

$$\langle \mathbf{u} \rangle = -\mathbf{K} : \nabla p, \tag{13.129}$$

where the (dimensionless) hydraulic conductivity tensor is defined by

$$K_{ij} = \left\langle w_i^j \right\rangle. \tag{13.130}$$

See, for example, Ene's article in the book edited by Cushman (1990).

Unsaturated flow In writing Darcy's law, we assumed saturated flow; that is, the pore space is filled with liquid. In normal groundwater flow, there is a region above the saturated zone where water only partially occupies the pore space. This region is called unsaturated and is described by the moisture constant θ, which is the volume percentage of liquid. If ϕ is the porosity, we also write $\theta = \phi S$ and call S the saturation (thus $0 < S < 1$). The permeability of fluid flow depends on the saturation and is often written as $kk_r(S)$, where k is the permeability of the saturated soil and k_r is the relative permeability; a typical dependence is of the form $k_r \approx S^3$.

As water is sucked from a soil, the saturation is reduced, and the dependence of θ on the suction pressure is called the suction characteristic. It is a hysteretic relationship, as shown in Fig. 13.5, owing to the hysteresis associated with varying contact angles during advance or retreat of fluid (see Section 6.2). In Fig. 13.5, we plot θ as a function of the *effective pressure* (equivalent to the suction). Typically, $p_e \to \infty$ as $\theta \to 0$, and some moisture is always retained. As $\theta \to \phi$, $p_e \to p_a$, termed the air entry pressure head. This can be interpreted as a cohesion, because if $p_e = 0$, we would expect the soil to lose its strength.

The seepage face Despite the Dupuit result in Section 12.3, a seepage face always exists in reality (otherwise the fluid velocity would be infinite at the outlet). Thus the boundary of the domain is itself divided by a free contour. Numerical solution of such problems is discussed by Crank (1984); see also Tayler (1986). The use of a

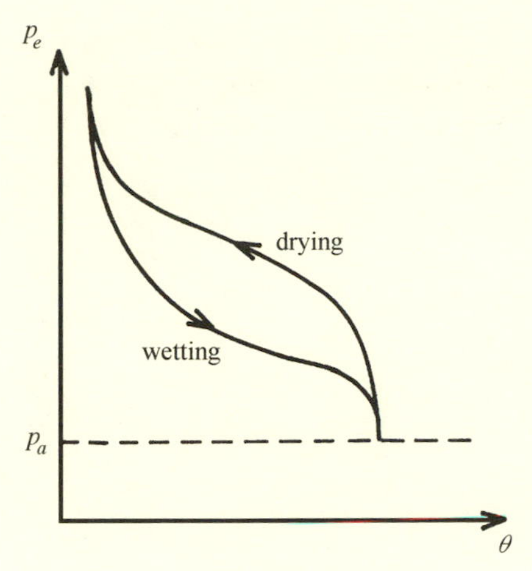

Fig. 13.5. Suction characteristic

Baiocchi transformation enables reformulation as a variational inequality, which is useful both for formulating numerical solutions and for proving such solutions really exist. An example of a problem of this sort would be mixed groundwater/overland flow down a hillslope.

Piping Many dams are built of concrete, and in this case the problems associated with seepage do not arise, owing to the virtual impermeability of concrete. Earth and rockfill dams do exist, however, and are liable to failure by a mechanism called *piping*. The Darcy flow through the porous dam causes channels to form by eroding away fine particles. The resultant channelization concentrates the flow, increasing the force exerted by the flow on the medium and thus increasing the erosion/collapse rate of the channel wall. We can write Darcy's law as a force balance on the liquid phase,

$$0 = -\phi \nabla p - \frac{\phi \mu}{k} \mathbf{v}_l - \phi \rho_l g \mathbf{k} \tag{13.131}$$

($\mathbf{k}$ being vertically upward), and $\phi \mu \mathbf{v}_l / k$ is an *interactive drag term* (see also Chapter 16); then the corresponding force balance for the solid phase is

$$0 = -(1 - \phi) \nabla p_s + \frac{\phi \mu}{k} \mathbf{v}_l - (1 - \phi) \rho_s g \mathbf{k}, \tag{13.132}$$

where p_s is the pressure in the solid. For a granular solid, we can expect grain motion to occur if the interactive force is large enough to overcome friction and cohesion; the typical kind of criterion is that the shear stress τ satisfy

$$\tau \geq c + p_e \tan \phi \tag{13.133}$$

(see Chapter 7), but in view of the large confining pressure and the necessity of dilatancy for soil deformation, the piping criterion will in practice be satisfied at the toe of the dam (i.e., the front), and piping channels will eat their way back into the dam, in much the same way that river drainage channels eat their way into a hillslope.

A simpler criterion at the toe then follows from the necessity that the effective pressure on the grains be positive. A lucid discussion by Bear and Bachmat (1990, p. 153) indicates that the solid pressure is related to the *effective* pressure p_e, which controls grain deformation, by

$$p_e = (1-\phi)(p_s - p), \tag{13.134}$$

and in this case the piping criterion at the toe is that $p_e < 0$ in the soil there, or $\partial p_e/\partial z > 0$. From Eqs. (13.131), (13.132), and (13.134), this implies piping if

$$\frac{\mu v}{k} > (\rho_s - \rho_l)(1-\phi)g, \tag{13.135}$$

where v is the vertical component of $\mathbf{v}_l$. This criterion is given by Bear (1972). More generally, piping can be expected to occur if p_e reaches 0 in the soil interior (ignoring cohesion). Sellmeijer and Koenders (1991) developed a model for piping.

Heterogeneous porous media The first-order geostatistical theory described in this chapter is reviewed by Gelhar (1984). Solute transport in heterogeneous media is treated by Dagan (1984). A paper by Hewett (1986) considers the case of a medium with a fractal distribution of permeability: in this case, the geostatistical method fails, and a different computational procedure must be adopted.

Exercises

1. Subsurface groundwater flow occurs in three dimensions (x, y, z) and may be thought of as being due to gravity and the variation of the height of the *water table* $h(x, y, t)$ above some reference state $z=0$. Motivate the Dupuit–Forchheimer approximation $\partial/\partial z \gg \partial/\partial x, \partial/\partial y$ for a typical drainage basin, and derive the corresponding (dimensionless) nonlinear diffusion equation for h. By writing the equation in cylindrical coordinates, show that the radial spreading of a fixed mass of water at the origin is described by a similarity solution, and find the rate of spread of the front.
   How is the equation changed if a source term owing to rainfall is included (units of volume per unit area per unit time)? If the boundaries for the flow are lakes or rivers at height $z = h_0$, what are appropriate boundary conditions for the equation for h? What happens if $h > h_g$ anywhere, where $z = h_g$ is the ground level?
2. Derive the Taylor–Aris dispersion coefficient by solving Eq. (13.86). Before doing so study Taylor's (1953) original paper: is his derivation of the result to your taste? Now follow the following procedure. (You will need to be able to follow multiple scales methods for partial differential equations; see for example Kevorkian and Cole's (1982, pp. 510 f.) derivation of the Korteweg–de Vries equation.)

   (i) Scale Eq. (13.86) by writing $r \sim a$, $x \sim l$, $t \sim l/U$, and define $\varepsilon = a/l$, $Pe = aU/D$.

(ii) Assume $\varepsilon \ll 1$, $Pe \sim 1$ (or $Pe \gg 1$), and write c as an expansion in powers of $\varepsilon Pe \ll 1$.
(iii) Solve for $c^{(0)}$, $c^{(1)}$, and $c^{(2)}$, applying the boundary conditions $\partial c/\partial r = 0$ at $r = 0, 1$. Hence show that $c^{(0)} = c^{(0)}(x - t)$ and secular terms are produced at $O(\{\varepsilon Pe\}^2)$.
(iv) Redo the analysis, introducing the mass-centered coordinate $\xi = x - t$ and the slow time $\tau = \varepsilon Pet$, and seek a multiple timescale expansion $c = c^{(0)}(\xi, t, \tau) + \dots$.
(v) Derive the solution up to $O(\{\varepsilon Pe\}^2)$, and show that in order to remove secular terms, $c^{(0)} = c^{(0)}(\xi, \tau)$ must satisfy

$$c_\tau^{(0)} = \left[\frac{1}{48} + \frac{1}{Pe^2}\right] c_{\xi\xi}^{(0)}.$$

Deduce that the Taylor–Aris dispersion coefficient relative to the mean flow is

$$\frac{a^2U^2}{48D} + D.$$

Note that $aU/D \gg 1$ is not required, only that $a \ll l$ (or equivalently $t \gg a/U$).

3. By suitable transformations, show that the dispersion equation (13.100) can be transformed to the form

$$c_\tau = 3\nu W \frac{\partial}{\partial \xi}\left(W \frac{\partial c}{\partial \xi}\right),$$

where ξ is a coordinate centered on a point moving at velocity u/ϕ and $W = (1 - \xi - \tau)^{1/3}$. Deduce that if $\nu \ll 1$, then an initial step function profile $c = 1, \xi < 0$; $c = 0, \xi > 0$, evolves approximately as

$$c \approx \frac{1}{2}[1 + \operatorname{erfc} \zeta], \tag{13.136}$$

where

$$\zeta = \frac{\xi}{2\left\{\frac{9\nu}{5}[1 - (1 - \tau)^{5/3}]\right\}^{1/2}}. \tag{13.137}$$

4. *The Buckley-Leverett equation*
Suppose two fluids (for example, water and air, or oil and water) flow under gravity in a porous medium of porosity ϕ. Write down the equations of mass conservation and Darcy flow for the two fluids having pressures p_i, fluxes $\mathbf{q}_i$, and saturations S_i. If fluid 2 is the *wetting fluid* (that is to say, its contact angle with the solid phase is acute), then the capillary relation is given by

$$p_1 - p_2 = p_c(S),$$

where $S_2 = S$ (and $S_1 = 1 - S$). Explain the physical basis of the capillary pressure, and illustrate the expected form of $p_c(S)$.

If flow is purely vertical, and z is the vertical coordinate, show that $q = q_1 + q_2 = q(t)$ (which may be taken as given), and hence deduce that S satisfies

$$\phi\frac{\partial S}{\partial t} + \frac{\partial}{\partial z}\left[M_{eff}\left\{q/M_1 + \frac{\partial p_c}{\partial z} + (\rho_1 - \rho_2)g\right\}\right] = 0,$$

where the phase mobilities are defined by

$$M_i = k_i/\mu_i$$

and depend on S through the relative permeability, and $M_{eff} = (M_1^{-1} + M_2^{-1})^{-1}$. Show that this is well-posed if $p_c'(S) < 0$. The Buckley–Leverett equation follows by neglecting p_c in this equation. Show that in this case, the equation is hyperbolic, and analyze the nature of shock formation if the relative permeabilities are given by $k_{1r} = S^3$, $k_{2r} = (1 - S)^3$.

5. A sedimentary basin compacts vertically by expelling porewater. Show that if u^s and u^l are solid and liquid fluxes, then conservation of mass for the two phases, together with Darcy's law for the relative flux $u^l - u^s$, leads to the equation

$$\frac{\partial \phi}{\partial t} = \frac{\partial}{\partial z}\left[(1-\phi)\frac{k}{\mu}\left\{\frac{\partial p}{\partial z} + \rho_l g\right\}\right],$$

where ϕ is porosity, p is pore fluid pressure, and it is assumed that $u^s = u^l = 0$ at $z = 0$. If the total pressure is $P = p_0 - \rho_s g z$, and the effective pressure $p_e = P - p$ is a function of ϕ, deduce that ϕ satisfies a nonlinear diffusion equation, and explain the basis of the boundary conditions

$$\phi = \phi_i, \qquad \dot{h} = \dot{m}_s + u^s \quad \text{on } z = h(t),$$

where $\dot{m}_s$ is the sedimentation rate (the rate at which sediment is deposited at the ocean floor).

By scaling k with k_0, p_e with $[p]$, z with $[p]/(\rho_s - \rho_l)g$, and t with $z/\dot{m}_s$, show that the problem depends on the single parameter

$$\lambda = \frac{k_0(\rho_s - \rho_l)g}{\mu \dot{m}_s}.$$

By using a perturbation expansion in $1/\lambda$, show that if $\lambda \gg 1$, then h is approximately given by

$$\dot{h} = \frac{1 - \phi_0}{1 - \phi_0 e^{-h}},$$

and solve the equation. Show also that if $k = h_0(\phi/\phi_0)^m$ with $m \gg 1$ (a value of $m \approx 8$ is sometimes quoted), this solution is valid provided $h < \Pi = \frac{1}{m}\ln\lambda$ and that this is satisfied for $t < t_0 = (\Pi - \phi_0 + \phi_0 e^{-\Pi})/(1 - \phi_0)$. What happens in this case if $t > t_0$?

6. Show that if we take the normal consolidation line (13.62) to be determined by the function $f(p_e)$, $f'(p_e) = -c$, then (ignoring gravity) the equation for $\psi = p_e$ may be written in dimensionless form as

$$\nabla^2\psi = H(\psi_t)\psi_t,$$

where $H(u)$ is the Heaviside step function. By defining $u = \psi_t$, show that the equation can be written in the form

$$\frac{\partial h}{\partial t} = \nabla^2 u,$$

for a suitable function $h(u)$, and devise a numerical method to solve this equation.

7. *Dual porosity models*
In a fractured permeable medium, one conceptualizes the flow as being either slow through matrix blocks M, or rapid through widely spaced fractures that separate the blocks. Each system has an effective permeability, k_m (matrix), and k_f (fractures), and one can write

$$\mathbf{v}_m = -\frac{k_m}{\mu}\boldsymbol{\nabla} p_m, \qquad \mathbf{v}_f = -\frac{k_f}{\mu\Phi}\boldsymbol{\nabla} p_f,$$

for the Darcy flow in each, where Φ is the fracture porosity (volume fraction of fractures). In particular, $\mathbf{v}_f$ is the local linear velocity on the fracture scale, so that if fractures are of width h, then $k_f/\Phi \sim h^2$. The total flux is

$$\mathbf{u} = \Phi\langle\mathbf{v}_f\rangle + (1-\Phi)\langle\mathbf{v}_m\rangle,$$

where $\langle\ \rangle$ denotes an average over many matrix blocks. In order to calculate $\langle\mathbf{v}_m\rangle$ and $\langle\mathbf{v}_f\rangle$, let L be the macroscopic length scale and let l be the matrix block dimension (thus $\Phi \sim h/l$ for planar fractures). Scale lengths with L, and then suppose that the fracture pressure $p_f = p_f(\mathbf{x}, \mathbf{x}/\varepsilon)$, where $\varepsilon = l/L$. Show that if the fractures are thin, so that $\partial p_f/\partial n \approx 0$ on the block boundaries ∂M (why?), then we must have

$$p_f = \bar{p}(\mathbf{x}) + \varepsilon\tilde{p}_f(\mathbf{x}/\varepsilon),$$

where $\bar{p}$ is the local mean pressure, and for periodic blocks we could take $\tilde{p}_f$ to be periodic. In particular, show that, on ∂M,

$$\frac{\partial\tilde{p}_f}{\partial n} = -\frac{\partial\bar{p}}{\partial x_i}\frac{\partial y_i}{\partial n},$$

where $\mathbf{y} = \mathbf{x}/\varepsilon$ is a local block coordinate. Show that if $p_m = \bar{p}(\mathbf{x}) + \varepsilon\tilde{p}_m(\mathbf{y})$, with $\tilde{p}_m$ periodic, then

$$\nabla_y^2\tilde{p}_m = 0 \quad \text{in } M, \qquad \tilde{p}_m = \tilde{p}_f \quad \text{on } \partial M,$$

where $\boldsymbol{\nabla}_y$ represents the gradient with respect to $\mathbf{y}$. Show also that

$$\langle\mathbf{v}_m\rangle \approx -\frac{k_m}{\mu}\boldsymbol{\nabla}_x\bar{p},$$

and that

$$\langle \mathbf{v}_f \rangle = -\frac{k_f}{\mu\Phi}[\boldsymbol{\nabla}_x \bar{p} + \langle \boldsymbol{\nabla}_y \tilde{p}_f \rangle_{\partial M}],$$

where the average is essentially over the block boundaries ∂M. Use the expressions for $\partial \tilde{p}_f/\partial n$ and the problem definition for $\tilde{p}_m$ to show that

$$\langle \boldsymbol{\nabla}_y \tilde{p}_f \rangle_{\partial M} = -\left\langle \mathbf{n} \left\{ \frac{\partial \tilde{p}_m}{\partial n} + \frac{\partial \bar{p}}{\partial x_j} n_j \right\} \right\rangle_{\partial M}.$$

If d is grain size, so that $k_m \sim d^2$ and also $d \ll h \ll l \ll L$, show that in order of magnitude,

$$\frac{|\boldsymbol{\nabla}_y \tilde{p}_m|}{|\nabla_x \bar{p}|} \sim \frac{u_m}{u_f} \cdot \frac{h^3}{d^2 l},$$

where $u_m \sim \langle \mathbf{v}_m \rangle$, $u_f \sim \Phi \langle \mathbf{v}_f \rangle$ are the matrix and fracture fluxes. Use plausible values of h, d, and l to show that commonly, $\delta = h^3/d^2 l \ll 1$, and deduce in this case that if u_f is significant ($u_f \gtrsim u_m$), then $\partial \tilde{p}_m/\partial n \ll \partial \bar{p}/\partial x_j$. Deduce that an approximate expression for the total flux $\mathbf{u}$ is determined by

$$u_i = -\frac{k_{ij}}{\mu} \frac{\partial \bar{p}}{\partial x_j},$$

where the permeability tensor is given by

$$k_{ij} = \{k_f + (1 - \Phi) k_m\} \delta_{ij} - k_f \langle n_i n_j \rangle_{\partial M}.$$

(For further information on double-porosity models, see the articles by Douglas and Arbogast and by Ene, in the book edited by Cushman (1990); the classic early paper is by Barenblatt, Zheltov, and Kochina (1960).)

14

Convection in a porous medium

14.1 Introduction

Much of the literature of fluid mechanics and its applications is concerned with the idea of *stability*: given some (usually simple) flow state, what happens if we perturb it by a small amount? The physical import of this notion is that in nature, small fluctuations are always present, so that any solution of a mathematical model for a physical system will only have a practical meaning if it is stable to small perturbations. If it is unstable, then naturally present small fluctuations will drive the system elsewhere. Thus, unstable steady states will not be physically observable in nature.

A simple example of the distinction between stable and unstable steady states is in the behavior of the simple pendulum, governed by the equation (suitably non-dimensional)

$$\ddot{\theta} + \sin\theta = 0. \tag{14.1}$$

The lower rest state, $\theta = 0$, is stable, whereas the upper rest state, $\theta = \pi$, is unstable: This corresponds to our common observation.

14.2 Linear stability

There are various mathematical frameworks within which one may study stability, and these can be classed as either *linear* or *nonlinear*. To begin, let us approach this theory in fairly general terms. Suppose we have a system governed (for example) by a nonlinear evolution equation

$$u_t = N(u), \tag{14.2}$$

where N is a nonlinear operator, often (but not necessarily) a differential operator, and u may be a scalar, or more often a vector. Suppose there is a steady state u_0 (i.e., independent of time) that thus satisfies

$$N(u_0) = 0 \tag{14.3}$$

together with any associated constraints (e.g., boundary conditions). We wish to examine the evolution of *perturbations* to this steady state by writing $u = u_0 + v$, so that

$$v_t = N[u_0 + v] - N[u_0], \tag{14.4}$$

which is a nonlinear equation for v (assuming u_0 is known).

The method of *linear stability analysis* is predicated upon the assumption that the perturbation v is *small*, so that Eq. (14.4) can be *linearized*. Simplistically, we expand the *operator* N as a Taylor series in v; thus

$$N(u_0 + v) = N(u_0) + \mathcal{L}[u_0]v + O(v^2), \tag{14.5}$$

where $\mathcal{L}[u_0]$ is a linear operator acting on v. Usually, this expansion is straightforward. Formally, we are doing functional analysis, and $\mathcal{L}$ is called the Fréchet derivative of N. Such terminology is not necessary to understand the procedure, however.

The linearized equation for v follows from neglecting the small quadratic terms, so that v satisfies the *linear* equation

$$v_t = \mathcal{L}v. \tag{14.6}$$

For an autonomous system (N does not depend explicitly on t), $\mathcal{L}$ is also independent of t, and separable solutions $v = e^{\lambda t}\phi$ will exist, whence the *eigenfunction* ϕ satisfies

$$\mathcal{L}\phi = \lambda\phi. \tag{14.7}$$

This is an *eigenvalue* problem, and, depending on the problem, the typical sort of behavior we can expect is that there are an infinite set of solutions $\{\phi, \lambda\}$ to Eq. (14.7). For a finite-dimensional problem, for example, a boundary value problem in a finite domain, we obtain a denumerable set of eigenvalues $\lambda_1, \lambda_2, \ldots$, which may be ordered in decreasing values of $\mathrm{Re}\,\lambda$. For an infinite-dimensional problem in an infinite domain, we might expect a continuous spectrum of λs. In either case, the general solution of Eq. (14.7) will be a superposition of these different *modes*, either as a sum or an integral. We then say that the system is linearly unstable if $\mathrm{Re}\,\lambda > 0$ for any of the eigenmodes, because the corresponding solution will grow exponentially in time.

As an example, consider the pendulum equation (14.1). It can be written as a system of the form (14.2) for $(u_1, u_2)^T = (\theta, \dot{\theta})^T$, or we can simply barge straight in and linearize. Denoting the steady state of interest as θ_0, so $\theta_0 = 0$ or π, put $\theta = \theta_0 + v$, thus

$$\begin{aligned} \sin(\theta_0 + v) &= \sin\theta_0 \cos v + \cos\theta_0 \sin v \\ &\approx v\cos\theta_0 + O(v^3), \end{aligned} \tag{14.8}$$

because $\sin\theta_0 = 0$. Thus, linearizing, we have

$$\ddot{v} + (\cos\theta_0)v = 0, \tag{14.9}$$

and solutions are

$$v = \exp[\lambda t], \qquad \lambda = \pm[-\cos\theta_0]^{1/2}. \tag{14.10}$$

Thus for $\theta_0 = 0$, $\lambda = \pm i$, and the perturbed solutions oscillate about zero. This situation is termed *neutral stability*. For $\theta_0 = \pi$, $\lambda = \pm 1$, and the positive root means that this state is unstable.

An example that incorporates a spatially distributed system is that of a nonlinear diffusion equation

$$u_t = [D(u)u_x]_x, \tag{14.11}$$

for example in $-\infty < x < \infty$ with $u \to 0$ as $x \to \pm\infty$. The linearized system about the steady state $u = 0$ is just

$$v_t = D_0 v_{xx}. \tag{14.12}$$

It is because the model is autonomous in x (translation invariant) and the basic state is as well, that solutions to Eq. (14.12) are separable in space as well as time, thus

$$v = \exp[\sigma t + ikx] \tag{14.13}$$

will satisfy Eq. (14.12), provided *a dispersion relation*

$$\sigma = -k^2 D_0 \tag{14.14}$$

is satisfied. (More commonly, dispersion relations are reserved to describe systems where $\sigma = i\omega$, so that the modes are neutrally stable, but the main point is that the growth rate σ, or *frequency* ω, is a function of *wavenumber* k.) A general solution of Eq. (14.12) then consists of a Fourier superposition of modes (14.13) integrated over wavenumber space,

$$v = \int_{-\infty}^{\infty} a(k) \exp[-k^2 D_0 t + ikx]\, dk. \tag{14.15}$$

Notice that v can satisfy the zero conditions at infinity, even though the individual modes (14.13) do not.

14.3 Nonlinear stability

The neutral stability result for the pendulum points to the need for fuller information: we cannot be sure that the effects of nonlinearity, however small, will not cause perturbations ultimately to grow or decay. For example, consider the stability of the equilibrium $A = 0$ of the equation

$$\frac{dA}{dt} = i\omega A \pm |A|^2. \tag{14.16}$$

In order to assess such issues, *nonlinear stability analysis* is necessary. Lyapounov functionals can provide global stability results, as can energy methods, but most practical nonlinear results use modified perturbation methods for the case where v is small but not infinitesimal. In the main, such methods are only necessary when the linear stability analysis gives a degenerate result, and in applications this mostly arises at *bifurcation points* (see below).

Bifurcation parameters

There are a variety of different methods of doing nonlinear stability analysis, most of which boil down to the same thing, and here we will illustrate the ideas using the method of multiple scales. First, we need the idea of *bifurcations* as a parameter varies. Suppose that a system depends on a forcing parameter, which can be (for example) externally controlled: denote this as μ. Then a given steady state will have a spectrum of eigenvalues in its linear stability analysis that depend on μ. In particular, the eigenvalue σ with maximal real part (which controls the stability) may have $\lambda = \text{Re}\,\sigma > 0$ for some values of μ but $\lambda < 0$ for others. To be definite, suppose that for $\mu < 0$, $\text{Re}\,\lambda < 0$, and for $\mu > 0$, $\text{Re}\,\lambda > 0$. Then $\mu = 0$ is called a *bifurcation point*, where the stability of the basic state changes. At $\mu = 0$, and in fact for $|\mu| \ll 1$, linear stability is not sufficient to determine the behavior of the system, and a nonlinear analysis is necessary.

Multiple-scale analysis

The details of the procedure vary from case to case, but a standard type goes as follows. We define

$$\mu = s\varepsilon^2, \tag{14.17}$$

with $s = \pm 1$ to indicate a *subcritical* ($s = -1$) or *supercritical* ($s = +1$) situation, and $\varepsilon \ll 1$ is a parameter that represents the size of the perturbation we study. The choice of Eq. (14.17) thus represents a *distinguished limit* whose choice is dictated by the analysis, as we shall see. The idea is that with $\mu \ll 1$, the growth (or decay) rate $\text{Re}\,\lambda$ is in general $\sim\mu$, so that growth or decay occurs on the long timescale $t \sim 1/\mu \sim 1/\varepsilon^2$. This suggests the explicit consideration of a slow timescale

$$\tau = \varepsilon^2 t. \tag{14.18}$$

However, if (for example) the neutrally stable mode at $\mu = 0$ is oscillatory, for example, $\lambda = \pm i\omega$ at $\mu = 0$, there will in addition be a fast timescale of behavior, $t = O(1)$. The two cases are referred to as *exchange of stability* ($\text{Im}\,\lambda = 0$) and *Hopf bifurcation* ($\text{Im}\,\lambda \neq 0$). In what follows, we consider the first case, and then we need only introduce the slow timescale. In oscillatory cases, both timescales need to be introduced, hence the title of multiple-scale methods.

We put

$$v = \varepsilon v_1(\tau) + \varepsilon^2 v_2(\tau) + \ldots \tag{14.19}$$

into Eq. (14.4). The expansion of Eq. (14.5) can be extended as

$$N(u_0 + v) = N(u_0) + Lv + L_2(v, v) + L_3(v, v, v) + \ldots, \tag{14.20}$$

where $L_2(v, v)$ is quadratic (i.e., $L_2(v, w)$ is bilinear), $L_3(v, v, v)$ is cubic ($L_3(u, v, w)$ is trilinear), and so on. Thus we have

$$\begin{aligned}\varepsilon^2 \frac{d}{d\tau}[\varepsilon v_1 + \varepsilon^2 v_2 + \ldots] = {} & L[\varepsilon v_1 + \varepsilon^2 v_2 + \ldots] \\ & + L_2[\varepsilon v_1 + \ldots, \varepsilon v_1 + \ldots] \\ & + L_3[\varepsilon v_1 + \ldots, \varepsilon v_1 + \ldots, \varepsilon v_1 + \ldots] \ldots \\ & + \varepsilon^2 s \frac{\partial L}{\partial \mu}[\varepsilon v_1 + \ldots] \ldots\end{aligned} \tag{14.21}$$

where the operators L_i are evaluated at $\mu = 0$. We are assuming that the linear eigenvalue problem $L\phi = \lambda\phi$ has a zero eigenvalue (and nontrivial ϕ) when $\mu = 0$. Then we expect $v \sim \varepsilon\phi + \ldots$ for $\mu \neq 0$.

Equating powers of ε, we have

$$\text{at } O(\varepsilon): \quad L(v_1) = 0; \tag{14.22a}$$

$$\text{at } O(\varepsilon^2): \quad L(v_2) = -L_2(v_1, v_1); \tag{14.22b}$$

$$\begin{aligned}\text{at } O(\varepsilon^3): \quad L(v_3) &= \dot{v}_1 - s\frac{\partial L}{\partial \mu}v_1 - L_3(v_1, v_1, v_1) \\ &\quad - \{L_2(v_1, v_2) + L_2(v_2, v_1)\}.\end{aligned} \tag{14.22c}$$

From $O(\varepsilon)$,

$$v_1 = A(\tau)\phi, \tag{14.23}$$

where ϕ is the eigenfunction of L corresponding to $\lambda = 0$. The higher-order equations will therefore have no solution unless their right-hand sides are orthogonal to eigenfunctions of the adjoint problem $L^*\eta = 0$. These solvability conditions at higher order determine the amplitude functions $A(\tau)$, etc., sequentially.

A typical example is when $\phi = e^{ikx}$ is the eigenfunction and L is self-adjoint. If N and v are real, then $\bar{\phi} = e^{-ikx}$ must also be an eigenfunction. Then Eq. (14.23) is replaced by

$$v_1 = A\phi + \bar{A}\bar{\phi}, \tag{14.24}$$

where $\bar{A}$ must be the complex conjugate of A in order that v_1 be real, and thus

$$\begin{aligned} L_2(v_1, v_1) &= A^2 L_2(\phi, \phi) + |A|^2\{L_2(\phi, \bar{\phi}) + L_2(\bar{\phi}, \phi)\} + \bar{A}^2 L_2(\bar{\phi}, \bar{\phi}) \\ &= l_{22}A^2 e^{2ikx} + l_{20}|A|^2 + \bar{l}_{22}\bar{A}^2 e^{-2ikx}. \end{aligned} \tag{14.25}$$

Because this contains no terms proportional to $e^{\pm ikx}$, v_2 can be solved for. It is because of the lack of *secular terms* (which require a nontrivial solvability condition to be satisfied) that we take $t \sim 1/\varepsilon^2$ and $\mu \sim \varepsilon^2$ rather than $\mu \sim \varepsilon$.

In a similar way, the right-hand side of Eq. (14.22c) is

$$e^{ikx}\left[\dot{A} - sl_1'A - |A|^2 A\{l_{31} + l_{21}\}\right] + \text{cc} + \text{other terms}, \tag{14.26}$$

where cc means the complex conjugate of the e^{ikx} term and other terms are not secular. Now v_3 will in general only have a solution if the terms in $e^{\pm ikx}$ are absent, and from this we derive the amplitude equation for A,

$$\dot{A} = sl_1'A + k_3|A|^2 A, \tag{14.27}$$

known as the Ginzburg–Landau equation.

Depending on the sign of l_1' and k_3, Eq. (14.27) shows that for $\mu > 0$ or $\mu < 0$, there will exist a nontrivial steady solution of small amplitude, whose stability is opposite to that of the basic state. If the basic state is unstable for $\mu > 0$ (thus $l_1' > 0$), and if $k_3 < 0$, then a stable nonzero solution exists for $\mu > 0$: see Fig. 14.1.

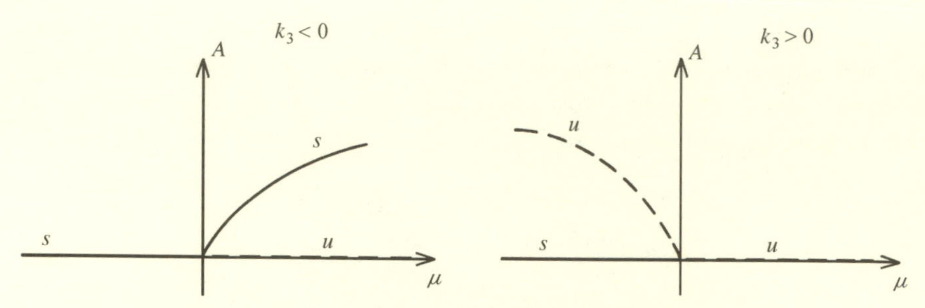

Fig. 14.1. Supercritical (left) and subcritical (right) bifurcations

14.4 Convection

A physical example of the above ideas occurs when a layer of fluid is heated from below or from within. The propensity of fluid to expand when heated causes a density inversion to occur, and if the heating is strong enough, a circulatory motion ensues, termed *convection*. Convection of a fluid layer is a well-studied phenomenon and occurs in many natural settings: in the atmosphere, in the Earth's mantle, and in many industrial situations, for example, involving solidification. Convection will also occur in a porous medium, and it is this problem that we consider in this chapter, mainly because the model is simpler (Darcy's law replaces the Navier Stokes equation).

There are several applications of porous medium convection. An interesting example is supplied by the relatively recent discovery of 'black smokers' on the ocean floor. These extraordinary edifices are vents that pour forth very hot water, massively contaminated with sulphides and other minerals, hence the (often) black color. Measured temperatures of the ejected fluids are up to 300°C.

They are observed at mid-ocean ridges, where upwelling in the mantle below leads to the partial melting of rock and the existence of magma chambers. The rock between these chambers and the ocean flow is extensively fractured, permeated by seawater, and strongly heated by the magma below. Consequently, strong thermal convection occurs, and the water passing nearest to the magma chamber is able to dissolve minerals with ease. The upwelling water is concentrated into fracture zones, where it rises rapidly. We shall see an explanation for this concentration when we study strongly nonlinear convection later on.

Another striking example of porous medium convection is afforded by hot springs and geysers, such as those in Yellowstone National Park. Here meteoric groundwater is heated by a subterranean magma chamber, again leading to strong thermal convection concentrated on the way up into fissures. However, whereas the ocean hydrostatic pressure prevents boiling from occurring, this is not the case for geysers, and boiling of the water causes the periodic eruption of steam and water that is familiar to tourists. A similar phenomenon can occur in domestic hot water systems, usually if they are incorrectly constructed. The onset of boiling leads to a rapid surge upward, followed by a compensating influx of cold water, and subcooled convection again. Because of the enormous density difference between steam and water, it is the boiling that really controls the whole process.

14.5 A mathematical model

Let us consider a permeable aquifer bounded above and below by impermeable boundaries and heated from below. Because convection is driven by thermally induced buoyancy, we assume that the density of liquid in the aquifer is given by

$$\rho = \rho_0[1 - \alpha(T - T_0)], \tag{14.28}$$

where T_0 is a reference temperature, which we take to be that prescribed at the top of the layer. Let the porosity, or void fraction, of the rock be ϕ. The conservation of fluid mass in the rock gives

$$(\rho\phi)_t + \nabla.[\rho\mathbf{u}] = 0, \tag{14.29}$$

where $\mathbf{u}$ is the fluid flux (equal to porosity times velocity). Darcy's law is

$$\mathbf{u} = -\frac{k}{\mu}[\nabla p + \rho g\mathbf{j}], \tag{14.30}$$

where $\mathbf{j}$ is a unit vector vertically upward, k is the permeability, μ the liquid viscosity, p the pressure, and g is gravity. The energy equation is

$$\frac{\partial}{\partial t}[\{\rho\phi c_l + \rho_r(1-\phi)c_r\}T] + \nabla.[\rho c_l\mathbf{u}T] = \nabla.[k_T\nabla T], \tag{14.31}$$

where c_l, c_r are specific heats of liquid and rock, ρ_r is the density of rock, T is temperature, and k_T is the average thermal conductivity.

We prescribe boundary conditions

$$\begin{aligned} T &= T_0 + \Delta T, \qquad u_n = 0 \quad \text{at } y = 0,\\ T &= T_0, \qquad u_n = 0 \quad \text{at } y = d; \end{aligned} \tag{14.32}$$

thus ΔT is the prescribed temperature drop and d is the layer depth.

Thermal conductivity

For ease of presentation, it is convenient to assume specific heats are equal and constant and also that k_T is constant, and we will later make these assumptions. As a side issue in the *modeling*, however, it is interesting to study the prescription of k_T, because this arises through surface averaging and hence depends on the microstructure of the medium. Two particular types of averaging might be appropriate, as is easily seen from Fig. 14.2. If the medium consists of a stacked array of plates, then it is easy from first principles to show that the average thermal conductivities parallel and perpendicular to the plates, denoted by $k_{\parallel}$ and $k_{\perp}$ respectively, are given by

$$k_{\parallel} = \phi k_l + (1-\phi)k_r, \tag{14.33a}$$

$$k_{\perp}^{-1} = \phi k_l^{-1} + (1-\phi)k_r^{-1}, \tag{14.33b}$$

where k_l and k_r are the thermal conductivities of liquid and rock, respectively; and although these are the same if $k_l = k_r$, they may be very different if k_l/k_r is small or large.

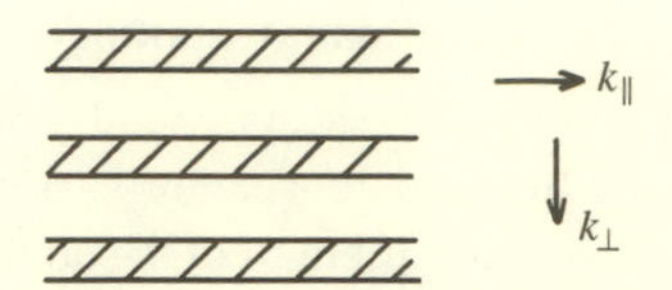

Fig. 14.2. A layered porous medium

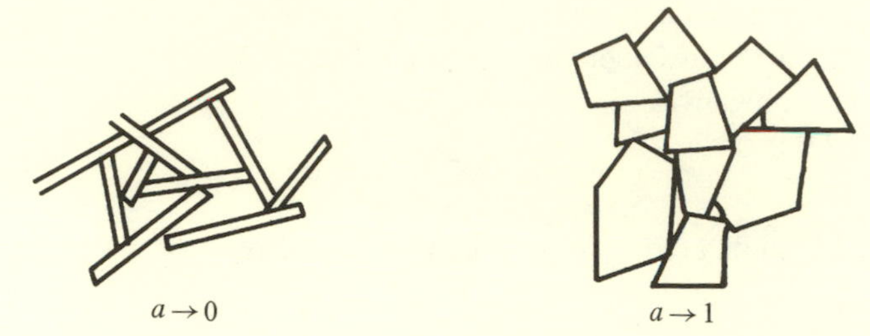

Fig. 14.3. Claylike structure (left) and rocklike structure (right)

Two particular cases that are more representative of real soil or rock than that in Fig. 14.2 arise when the *specific contact surface area* a tends to zero or one. We define the specific contact surface area to be the contact area between rock grains as a fraction of total surface area. For unconsolidated clays, we can expect $a \to 0$, whereas for well-cemented, low-permeability rocks, we can expect $a \to 1$.
To some extent, a depends on ϕ, such that $a \to 0$ is associated with high values of ϕ, whereas $a \to 1$ is associated with low values of ϕ. It is difficult to draw much distinction between the two cases of Eq. (14.33) if $k_l \approx k_r$. Suppose, therefore, for example, that the rock is more conductive, so that $k_r \gg k_l$; then Eq. (14.33) gives

$$\begin{aligned} k_{\parallel} &\approx (1-\phi)k_r \quad (\text{unless}\, \phi \to 1), \\ k_{\perp} &\approx k_l/\phi \quad (\text{unless}\, \phi \to 0). \end{aligned} \tag{14.34}$$

How do these compare to the sort of structure indicated in Fig. 14.3? If $k_r \gg k_l$ and $a \to 1$, then we expect conduction to be largely through the solid phase, and

$$k \approx (1-\phi)k_r, \quad a \to 0. \tag{14.35}$$

If, on the other hand, $a \to 0$, then $\phi \to 1$; typically, the solid phase will be configured as thin platelets, with effective conductivity $(1-\phi)k_r$. In addition, the liquid will contribute ϕk_l, the two systems hardly interacting. Consequently, we could suppose

$$k \approx \phi k_l + (1-\phi)k_r, \quad a \to 0. \tag{14.36}$$

For both cases, one could then suggest that the parallel result, Eq. (14.33a) is more realistic. Partly, this is a reflection of the three-dimensional nature of rocks: no one constituent truly acts as a barrier, at least at the microscopic level.

14.6 Nondimensionalization

The Boussinesq approximation

A typical value of α is $\sim 10^{-5}$ K^{-1}, so that even if $\Delta T \sim 10^3$ K, as is likely for hydrothermal convection, we have the *Boussinesq number*[1]

$$B = \alpha \Delta T \ll 1. \tag{14.37}$$

It then follows that where ρ appears algebraically in the equations, we can at leading order put $\rho \approx \rho_0$. However, where ρ appears as a derivative, we retain the term in B. Although this seems to be nowhere, in fact the density is present in this way in Darcy's law, because the hydrostatic term $\rho_0 g$ merely determines the vertical variation of pressure. This can be seen by eliminating the pressure from the equations.

Nondimensionalization

In keeping with the above, we write

$$\begin{aligned} T &= T_0 + (\Delta T)T^*, \\ \rho &= \rho_0[1 - BT^*], \\ p &= p_0 - \rho_0 g y + [p]p^*, \\ \mathbf{u} &= [u]\mathbf{u}^*, \\ (x, y) &= d(x^*, y^*), \\ t &= [t], \end{aligned} \tag{14.38}$$

where the pressure, velocity, and timescales $[p]$, $[u]$, and $[t]$ are yet to be chosen. As threatened, we will take the specific heats c_l and c_r as equal to a constant c and take the thermal conductivity k_T as constant. The equations become, dropping the asterisks,

$$\begin{aligned} [1 - BT]_t + \left\{\frac{[t][u]}{\phi d}\right\}\boldsymbol{\nabla}.[(1 - BT)\mathbf{u}] &= 0, \\ \mathbf{u} = -\frac{k}{\mu[u]}\left[\frac{[p]}{d}\boldsymbol{\nabla} p - B\rho_0 g T\mathbf{j}\right]&, \\ \left[\{\rho_0(1 - BT)\phi + \rho_r(1 - \phi)\}\frac{c\Delta T}{[t]}\right]\frac{\partial T}{\partial t} + \frac{\rho_0[u]c\Delta T}{d}(1 - BT)\mathbf{u}.\boldsymbol{\nabla} T & \\ = \frac{k_T \Delta T}{d^2}\nabla^2 T&, \end{aligned} \tag{14.39}$$

with boundary conditions

$$\begin{aligned} T = 1, &\quad u_n = 0 \quad \text{on } y = 0, \\ T = 0, &\quad u_n = 0 \quad \text{on } y = 1. \end{aligned} \tag{14.40}$$

[1] There are two different numbers called the Boussinesq number (Massey, 1986). Our definition here puts $B = 1/\beta$, where β is the second of these definitions.

We choose $[p]$, $[u]$, and $[t]$ as follows to balance various terms in Eq. (14.39):

$$\begin{aligned} [u] &= \kappa_T/d, \qquad \kappa_T = k_T/\rho_0 c, \\ [t] &= d^2/\kappa_m, \qquad \kappa_m = k_T/\{\rho_0\phi + \rho_r(1-\phi)\}c, \\ [p] &= \frac{\mu d[u]}{k} = \frac{\mu\kappa_T}{k}, \end{aligned} \tag{14.41}$$

and then Eq. (14.39) become

$$-B(\rho_0\phi/\rho_m)T_t + \nabla.[(1-BT)\mathbf{u}] = 0, \tag{14.42}$$

where

$$\rho_m = \rho_0\phi + \rho_r(1-\phi), \tag{14.43}$$

$$\mathbf{u} = -\nabla p + RT\mathbf{j}, \tag{14.44}$$

where

$$R = \frac{\alpha\Delta T\rho_0 g\, dk}{\mu\kappa_T}, \tag{14.45}$$

and

$$\{1 - B(\rho_0\phi/\rho_m)T\}\frac{\partial T}{\partial t} + (1-BT)\mathbf{u}.\nabla T = \nabla^2 T. \tag{14.46}$$

We now make the (regular) Boussinesq approximation by letting $B \to 0$ (note that $\rho_0\phi/\rho_m < 1$). Consequently,

$$\begin{aligned} \nabla.\mathbf{u} &= 0, \\ \mathbf{u} &= -\nabla p + RT\mathbf{j}, \\ T_t + \mathbf{u}.\nabla T &= \nabla^2 T, \end{aligned} \tag{14.47}$$

together with the boundary conditions (14.40).

14.7 Stability analysis

Steady state

The basic conductive steady state is one of no motion, thus

$$\begin{aligned} \mathbf{u} &= \mathbf{0}, \\ T &= 1 - y, \\ p &= -\frac{R}{2}(1-y)^2. \end{aligned} \tag{14.48}$$

The pressure profile is merely a quadratic correction to the linear hydrostatic profile subtracted off in Eq. (14.38). Obviously, a reference pressure is supplied; here the assumption being that $p = 0$ at $y = 1$. However, in a closed system this is of secondary importance.

Linear stability

The single dimensionless parameter in the reduced model is the Rayleigh number, R. It is a measure of the thermal forcing, and we can expect that if it is large enough, the conductive steady state will be unstable. To examine this, we consider a two-dimensional situation, where the coordinates are (x, y). Then $\mathbf{u} = (u, v)$, and we define a stream function ψ by

$$u = \psi_y, \qquad v = -\psi_x, \tag{14.49}$$

to satisfy the continuity equation. Put

$$T = 1 - y + \theta, \tag{14.50}$$

then

$$\begin{aligned} \nabla^2\psi &= -R\theta_x, \\ \theta_t + \psi_x + (\psi_y\theta_x - \psi_x\theta_y) &= \nabla^2\theta, \end{aligned} \tag{14.51}$$

where we have eliminated p by cross differentiation. The linearized system is obtained by neglecting the quadratic terms in the energy equation, thus

$$\begin{aligned} \nabla^2\psi &= -R\theta_x, \\ \nabla^2\theta &= \psi_x + \theta_t, \end{aligned} \tag{14.52}$$

with

$$\theta = \psi = 0 \quad \text{on } y = 0, 1. \tag{14.53}$$

Dispersion relation

The equations are autonomous in x, y, and t. The domain is unbounded in x and t coordinates. Therefore there are separable solutions

$$\begin{aligned} \psi &= f(y)e^{\sigma t + ikx}, \\ \theta &= g(y)e^{\sigma t + ikx}, \end{aligned} \tag{14.54}$$

whence f and g satisfy

$$f'' - k^2 f = -ikRg, \tag{14.55a}$$

$$g'' - k^2 g = ikf + \sigma g, \tag{14.55b}$$

with $f = g = 0$ on $y = 0, 1$. Solutions are clearly exponentials in y. In fact, because there are no first derivatives present and because of the homogeneous boundary conditions, we can see that the relevant solution for f must be

$$f = \sin n\pi y, \tag{14.56}$$

whence

$$g = \left(\frac{n^2\pi^2 + k^2}{ikR}\right) \sin n\pi y \tag{14.57}$$

from Eq. (14.55a), and the second equation gives the consistency condition for these to be valid, which is the dispersion relation for the eigenvalue σ. It is

$$-\frac{(n^2\pi^2+k^2)^2}{ikR}=ik+\sigma\left(\frac{n^2\pi^2+k^2}{ikR}\right), \tag{14.58}$$

whence

$$\sigma=\frac{Rk^2}{n^2\pi^2+k^2}-(n^2\pi^2+k^2). \tag{14.59}$$

We see that there is a denumerable set of eigenvalues for a given k, given by $n=1,2,\ldots$. The most general solution for (f,g) would consist of a Fourier integral over k, together with a Fourier series in n.

For the purposes of stability, however, we only need to know about the value of σ with greatest real part. Here we see that σ is real. Moreover $\partial\sigma/\partial n<0$, so that for given k, the largest value of σ is obtained from $n=1$; this value is

$$\sigma=\frac{Rk^2}{\pi^2+k^2}-(\pi^2+k^2). \tag{14.60}$$

As a function of wavenumber k (or in fact k^2), σ is as shown in Fig. 14.4. As $k^2\to\infty$, $\sigma\to-\infty$, whereas at $k^2=0$, $\sigma=-\pi^2$.

The derivative is given by

$$\frac{\partial\sigma}{\partial k^2}=\frac{\pi^2R}{(\pi^2+k^2)}-1, \tag{14.61}$$

so that σ has the shape shown in Fig. 14.4 if $R>1$. In this case, there is a maximum at

$$\pi^2+k^2=(\pi^2R)^{1/2}, \tag{14.62}$$

where

$$\sigma=R-2(\pi^2R)^{1/2}. \tag{14.63}$$

We therefore see that instability sets in when σ given by Eq. (14.63) becomes positive, which is at a critical value

$$R=R_c=4\pi^2\approx 39.48 \tag{14.64}$$

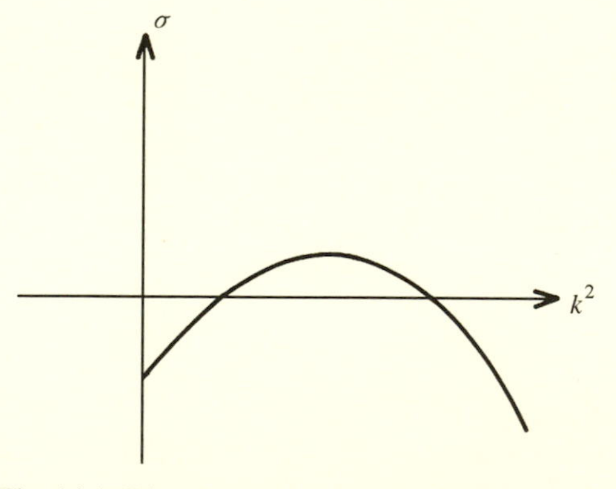

Fig. 14.4. Dispersion relation given by Eq. (14.60)

and at a critical wavenumber

$$k = k_c = \pi,$$

corresponding to an aspect ratio for a single cell of one.

Thus the onset of convection occurs as an exchange of stability. We now turn to the nonlinear problem of establishing the stability of the convective state for $R > R_c$.

14.8 Nonlinear stability analysis

We revert to the nonlinear equations (14.51):

$$\begin{aligned} \nabla^2\psi &= -R\theta_x, \\ \theta_t + \psi_x + \psi_y\theta_x - \psi_x\theta_y &= \nabla^2\theta. \end{aligned} \tag{14.65}$$

In keeping with the earlier discussion, we put

$$\begin{aligned} R &= R_c + s\varepsilon^2, \\ t &= \varepsilon^2\tau, \\ \psi &= \varepsilon\psi_1 + \varepsilon^2\psi_2 + \dots \\ \theta &= \varepsilon\theta_1 + \varepsilon^2\theta_2 + \dots, \end{aligned} \tag{14.66}$$

so that

$$\begin{aligned} \nabla^2[\varepsilon\psi_1 + \dots] &= -(R_c + s\varepsilon^2)[\varepsilon\theta_1 + \dots]_x, \\ \varepsilon^2[\varepsilon\theta_1 + \dots]_\tau &+ [\varepsilon\psi_1 + \dots]_x + [\varepsilon\psi_1 + \dots]_y[\varepsilon\theta_1 + \dots]_x \\ &- [\varepsilon\psi_1 + \dots]_x[\varepsilon\theta_1 + \dots]_y \\ &= \nabla^2[\varepsilon\theta_1 + \dots], \end{aligned} \tag{14.67}$$

and we obtain the following sequence of problems on equating powers of ε:

$$\text{at } O(\varepsilon)\text{: } \begin{aligned} \nabla^2\psi_1 + R_c\theta_{1x} &= 0, \\ \nabla^2\theta_1 - \psi_{1x} &= 0; \end{aligned} \tag{14.68}$$

$$\text{at } O(\varepsilon^2)\text{: } \begin{aligned} \nabla^2\psi_2 + R_c\theta_{2x} &= 0, \\ \nabla^2\theta_2 - \psi_{2x} &= \psi_{1y}\theta_{1x} - \psi_{1x}\theta_{1y}; \end{aligned} \tag{14.69}$$

$$\text{at } O(\varepsilon^3)\text{: } \begin{aligned} \nabla^2\psi_3 + R_c\theta_{3x} &= -s\theta_{1x}, \\ \nabla^2\theta_3 - \psi_{3x} &= \theta_{1\tau} + \{\psi_{2y}\theta_{1x} + \psi_{1y}\theta_{2x} - \psi_{2x}\theta_{1y} - \psi_{1x}\theta_{2y}\}. \end{aligned} \tag{14.70}$$

To simplify matters, we will restrict attention to a domain $0 < x < 1$ (corresponding to the critical wavenumber $k_c = \pi$) on which we enforce the boundary conditions

$$\theta_x = \psi = 0 \quad \text{at } x = 0, 1. \tag{14.71}$$

This restriction is one of convenience only. It forces the convection cells to be of uniform width. More generally, we can expect slight variation of cell width in an infinite domain, which can be accommodated by allowing the solutions to depend also on a slow space scale $X = x/\varepsilon$. The reason for expecting a length scale $1/\varepsilon$ rather than $1/\varepsilon^2$ is that the *neutral stability* curve (Eq. (14.60)) where $\sigma = 0$ is

$$R = \frac{(\pi^2 + k^2)^2}{k^2}, \tag{14.72}$$

which has a quadratic minimum at k_c. Thus if $k - k_c = O(\varepsilon)$, $R - R_c = O(\varepsilon^2)$, and a variation in k of $O(\varepsilon)$ suggests a space scale $O(1/\varepsilon)$.

At $O(\varepsilon)$, the convective solution can be written, from Eqs. (14.54)–(14.57),

$$\begin{pmatrix} \psi_1 \\ \theta_1 \end{pmatrix} = Ae^{i\pi x} \begin{pmatrix} 2\pi i \\ 1 \end{pmatrix} \sin \pi y + (\text{cc}), \tag{14.73}$$

where $A(\tau)$ is an amplitude function, which can be taken as real.

If we write $v = (\psi, \theta)^T$, then Eq. (14.68) can be written as

$$Lv_1 = 0, \tag{14.74}$$

where

$$L = \begin{pmatrix} \nabla^2 & R_c \frac{\partial}{\partial x} \\ -\frac{\partial}{\partial x} & \nabla^2 \end{pmatrix}. \tag{14.75}$$

Note that if

$$v = e^{ip\pi x} \begin{pmatrix} a \\ b \end{pmatrix} \sin q\pi y, \tag{14.76}$$

then

$$Lv = \begin{pmatrix} -(p^2 + q^2)\pi^2 & ip\pi R_c \\ -ip\pi & -(p^2 + q^2)\pi^2 \end{pmatrix} v, \tag{14.77}$$

so that for v of the form (14.76), we can identify L with the matrix multiplication given by Eq. (14.77). Note also that (the matrix) L is invertible unless $p = q = 1$ (corresponding to Eq. (14.73)). We denote the matrices in Eq. (14.77) as L_{pq}.

The $O(\varepsilon)$ solutions can be written

$$\begin{aligned} \psi_1 &= -4\pi A \sin \pi x \sin \pi y, \\ \theta_1 &= 2A \cos \pi x \sin \pi y, \end{aligned} \tag{14.78}$$

whence

$$\psi_{1y}\theta_{1x} - \psi_{1x}\theta_{1y} = 4\pi^3 A^2 \sin 2\pi y, \tag{14.79}$$

so that Eq. (14.69) is

$$L \begin{pmatrix} \psi_2 \\ \theta_2 \end{pmatrix} = A^2 \begin{pmatrix} 0 \\ 4\pi^3 \end{pmatrix} \sin 2\pi y. \tag{14.80}$$

It follows that the particular integral is given by

$$\begin{pmatrix} \psi_2 \\ \theta_2 \end{pmatrix} = A^2 \begin{pmatrix} a_2 \\ b_2 \end{pmatrix} \sin 2\pi y, \tag{14.81}$$

where

$$\begin{pmatrix} a_2 \\ b_2 \end{pmatrix} = L_{02}^{-1} \begin{pmatrix} 0 \\ 4\pi^3 \end{pmatrix}. \tag{14.82}$$

(The complementary function proportional to $(\psi_1, \theta_1)^T$ need not be included, as it does not affect the determination of A.)

From Eq. (14.77),

$$L_{02}^{-1} = -\frac{1}{4\pi^2} \begin{pmatrix} 1 & 0 \\ 0 & 1 \end{pmatrix}, \tag{14.83}$$

therefore

$$\begin{pmatrix} a_2 \\ b_2 \end{pmatrix} = \begin{pmatrix} 0 \\ -\pi \end{pmatrix}, \tag{14.84}$$

so that

$$\begin{aligned} \psi_2 &= 0, \\ \theta_2 &= -\pi A^2 \sin 2\pi y. \end{aligned} \tag{14.85}$$

Thus we have

$$\begin{aligned} -s\theta_{1x} &= 2\pi s A \sin \pi x \sin \pi y, \\ \theta_{1\tau} &= 2\dot{A} \cos \pi x \sin \pi y, \\ \{\psi_{2y}\theta_{1x} + \psi_{1y}\theta_{2x} - \psi_{2x}\theta_{1y} - \psi_{1x}\theta_{2y}\} &= -\psi_{1x}\theta_{2y} \\ &= 8\pi^4 A^3 \cos \pi x[\sin \pi y - \sin 3\pi y], \end{aligned} \tag{14.86}$$

whence

$$L \begin{pmatrix} \psi_3 \\ \theta_3 \end{pmatrix} = e^{i\pi x} \begin{pmatrix} -i\pi s A \\ \dot{A} + 4\pi^4 A^3 \end{pmatrix} \sin \pi y + (\text{cc} + \text{higher harmonics}). \tag{14.87}$$

It is only the fundamental term $e^{i\pi x} \sin \pi y$ that causes a solvability problem, because the higher harmonic (here $e^{i\pi x} \sin 3\pi y$) can be solved via Eq. (14.77) by inverting L_{13}.

Solvability condition

In order to find a (bounded) solution of the form $(a, b)^T e^{i\pi x} \sin \pi y$ to Eq. (14.87), we require

$$L_{11} \begin{pmatrix} a \\ b \end{pmatrix} = \begin{pmatrix} -i\pi s A \\ \dot{A} + 4\pi^4 A^3 \end{pmatrix}. \tag{14.88}$$

Because L_{11} is singular, this equation has a solution if and only if the right-hand side is orthogonal to the null vector of the adjoint L_{11}^* of L_{11}, given by

$$L_{11}^* = \begin{pmatrix} -2\pi^2 & i\pi \\ -4i\pi^3 & -2\pi^2 \end{pmatrix}; \tag{14.89}$$

this vector is $\eta = (i, 2\pi)^T$. Therefore the solvability condition for Eq. (14.87) is (because the inner product $\langle u, v\rangle$ is $\bar{u}^T v$)

$$\overline{(i, 2\pi)} \begin{pmatrix} -i\pi s A \\ \dot{A} + 4\pi^4 A^3 \end{pmatrix} = 0,$$

that is,

$$\dot{A} = \frac{1}{2} s A - 4\pi^4 A^3, \tag{14.90}$$

and this finally gives the amplitude equation governing A. We see that the bifurcation is supercritical: when $s = 1$ ($R = R_c + \varepsilon^2$), then the equilibrium amplitude A is

$$A_{eq} = \frac{1}{2\sqrt{2}\pi^2}, \tag{14.91}$$

or in terms of ψ,

$$\psi \approx -\frac{1}{\pi}[2(R - R_c)]^{1/2} \sin \pi x \sin \pi y, \tag{14.92}$$

the negative sign indicating a clockwise rotation in $0 < x < 1$. The finite amplitude convective motion is stable.

The Fredholm alternative

The solvability condition derived above is an example of a more general statement deriving from the theory of the Fredholm alternative in Hilbert space. If u, v are functions in a Hilbert space (meaning, pertinently, that there is an *inner product* $\langle u, v\rangle$) then the idea is as follows. Suppose L is a linear operator acting on the functions in a Hilbert space. A standard definition of inner product for functions in L_2 (square Lebesgue integrable) is

$$\langle u, v\rangle = \int \bar{u}^T v, \tag{14.93}$$

where u, v may be complex-valued vector functions. We will use this definition for illustration here.

Suppose

$$Lu = 0 \tag{14.94}$$

has nontrivial solutions; then the adjoint L^* will also, that is,

$$L^*\eta = 0, \tag{14.95}$$

where the adjoint L^* is defined by

$$\langle u, L^*v\rangle = \langle Lu, v\rangle. \tag{14.96}$$

For the inner product we have defined, L^* is determined in practice by integrating by parts; that is, if V is the domain of interest, and S is its boundary, then

$$\int_V [\bar{v}^T L^*\eta - \eta^T \overline{Lv}]\, dV = \int_S [\text{boundary terms}]\, dS = 0 \tag{14.97}$$

for homogeneous boundary conditions. It follows from this that if $L^*\eta = 0$, then

$$Lv = f \tag{14.98}$$

can only have a solution if

$$\langle \eta, v \rangle = \int_V \bar{\eta}^T f \, dV = 0 \tag{14.99}$$

for any η in the null space of L^*. That this condition is also sufficient follows from the Fredholm alternative.

In the present example, a comparison of Eqs. (14.75), (14.77), and (14.89) suggests that

$$L^* = \begin{pmatrix} \nabla^2 & \frac{\partial}{\partial x} \\ -R_c \frac{\partial}{\partial x} & \nabla^2 \end{pmatrix}. \tag{14.100}$$

Then we find that if $v = (v_1, v_2)^T$, $\eta = (\eta_1, \eta_2)^T$,

$$\begin{aligned} \bar{v}^T L^* \eta - \eta^T \overline{Lv} = {} & \boldsymbol{\nabla}.[\bar{v}_1 \boldsymbol{\nabla} \eta - \eta_1 \boldsymbol{\nabla} \bar{v}_1 + \bar{v}_2 \boldsymbol{\nabla} \eta_2 - \eta_2 \boldsymbol{\nabla} \bar{v}_2] \\ & + \frac{\partial}{\partial x}[\bar{v}_1 \eta_2 - R_c \eta_1 \bar{v}_2], \end{aligned} \tag{14.101}$$

and thus Eq. (14.97) is satisfied for homogeneous boundary conditions.

14.9 Boundary layer theory

Nonlinear stability theory, or weakly nonlinear theory, tells us what happens when R is close to critical. The theory has the advantage of being clean and easy (in principle – messy in practice) but has the disadvantage that the net result is the sign of only one coefficient (which determines whether the bifurcation is subcritical or supercritical), for which the theory is cost-ineffective, and more importantly, real life is rarely near criticality. Bifurcation analysis is very useful to tell us what solutions can be produced, but as a quantitative predictor, it is less useful. Nevertheless, it enjoys enormous popularity (particularly, these days, with mathematical physicists) because of its conceptual simplicity – and yet, when generalized to include spatial dependence, it can lead to the description of complicated, nontrivial solutions, including chaos, which are very difficult to analyze by other means.

Fortunately, there are other tricks. If R is close to R_c, or indeed $R - R_c = O(1)$, small amplitude theory is entirely satisfactory. When $R \gg 1$, however, it is inadequate. In this case, singular perturbation theory can be used. The idea is that if $R \gg 1$, then we can expect motion to be *vigorous*, in the sense that $\psi \gg 1$. Then advection of heat is large, heat conduction is small, and we can expect thermal boundary layers at the edges of the cell (where the thermal boundary conditions are prescribed).

In what follows, we analyze convection in a closed system as described above. The analysis is based on work by Booker (1981), though the results here are somewhat different.

At high R, if the motion becomes more rapid, we can expect the thermal gradient at the top and at the base to become larger. Because it is the lateral temperature gradients

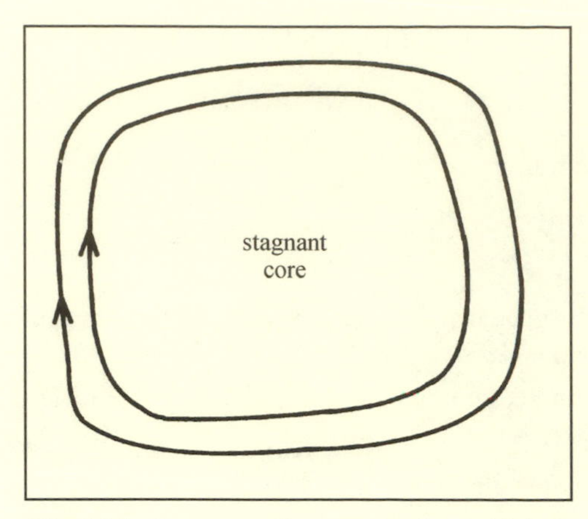

Fig. 14.5. Porous medium convection at high R

that induce flow, we can then expect the region of upwelling and downwelling flow to become concentrated, as shown in Fig. 14.5. In fact, Booker describes an experiment using a Hele–Shaw cell, which suggests that when R is large, one develops a relatively stagnant core, with the motion and thermal gradients concentrated on the periphery. We now attempt to substantiate this picture.

Firstly, recall the Eq. (14.51), which we can write in the form

$$\begin{aligned} \nabla^2 \psi &= -RT_x, \\ \psi_y T_x - \psi_x T_y &= \nabla^2 T. \end{aligned} \tag{14.102}$$

We expect $\psi \gg 1$, so we *rescale* $\psi \sim \Lambda \gg 1$ (formally, put $\psi = \Lambda \tilde{\psi}$, substitute in, drop overtildes...); we get

$$\nabla^2 \psi = -\frac{R}{\Lambda} T_x, \tag{14.103a}$$

$$\psi_y T_x - \psi_x T_y = \frac{1}{\Lambda} \nabla^2 T. \tag{14.103b}$$

The magnitude of Λ is to be determined, so that the rescaled ψ is $O(1)$ within the cell. Boundary conditions are

$$\begin{aligned} \psi = 0, \qquad T = 1/2 \quad &\text{on } y = 0, \\ \psi = 0, \qquad T = -1/2 \quad &\text{on } y = 1, \\ \psi = T_x = 0 \quad &\text{on } x = 0, 1. \end{aligned} \tag{14.104}$$

We have changed the top and bottom temperatures for convenience.

Core flow

If we assume that $\Lambda \ll R$, then to leading order we find from Eq. (14.103) that

$$\begin{aligned} T_x &\approx 0, \\ \psi_y T_x - \psi_x T_y &\approx 0, \end{aligned} \tag{14.105}$$

whence

$$T \approx T_c(y),\\ \psi \approx \psi_c(y), \tag{14.106}$$

where the core values T_c and ψ_c $(= O(1))$ are to be determined. Booker's experiment suggests $\psi_c =$ constant, in fact. It should be noted that this is not the only possibility. If $\mathbf{u} \neq \mathbf{0}$, then to all orders of $1/\Lambda$, Eq. (14.103b) implies (via the Prandtl–Batchelor theorem) $T = 0$, so that in fact $\nabla^2\psi = 0$. However, it is difficult to find a convincing asymptotic structure based on this possibility.

Thermal plume

Suppose the plume (at $x = 0$, say) is of thickness δ_p and the temperature jump across it is of order θ_p. The appropriate rescaling is given by

$$T \sim \theta_p, \qquad T_c \sim \theta_p, \qquad \psi \sim 1, \qquad x \sim \delta_p, \tag{14.107}$$

and we balance conduction in Eq. (14.103b) and buoyancy in Eq. (14.103a) by choosing

$$\frac{\Lambda}{R\delta_p^2} = \frac{\theta_p}{\delta_p}, \qquad \frac{1}{\delta_p} = \frac{1}{\Lambda\delta_p^2}. \tag{14.108}$$

The leading order plume equations for the rescaled variables are

$$\psi_{XX} = -T_X, \tag{14.109a}$$

$$\psi_y T_X - \psi_X T_y = T_{XX}, \tag{14.109b}$$

whence

$$\psi_X = T_c - T, \qquad \psi = \int_0^X (T_c - T)\,dX, \tag{14.110}$$

and thus

$$\psi_c = \int_0^\infty (T_c - T)\,dX. \tag{14.111}$$

Introduction of Von Mises variables y, ψ enables Eq. (14.109b) to be written as

$$\frac{\partial T}{\partial y} = \frac{\partial}{\partial \psi}\left[(T - T_c)\frac{\partial T}{\partial \psi}\right], \tag{14.112}$$

with $T_\psi = 0$, and on $\psi = 0$, $T = T_c$ on $\psi = \psi_c$. Integration of Eq. (14.112) yields

$$\frac{d}{dy}\int_0^{\psi_c} T\,d\psi = 0, \tag{14.113}$$

whence

$$\int_0^{\psi_c} T\,d\psi = C, \tag{14.114}$$

a constant.

Thermal boundary layer

We require a boundary layer at $y = 0$ (and $y = 1$) to enable T to jump to the boundary value and also for ψ to jump to zero. It is impossible to do both of these at once, and there are therefore two nested boundary layers.

(i) Buoyancy

The outermost of these layers balances buoyancy with vorticity; appropriate scales (near $y = 0$) are

$$y \sim \delta_h, \qquad \psi \sim 1, \qquad T \sim \theta_p, \tag{14.115}$$

and we choose

$$\frac{1}{\delta_h^2} = \frac{R}{\Lambda}\theta_p. \tag{14.116}$$

Assuming $\Lambda\delta_h \gg 1$, then at leading order the rescaled variables satisfy

$$T \approx T(\psi), \qquad \psi_{YY} \approx -T'(\psi)\psi_x, \tag{14.117}$$

a nonlinear diffusion equation for ψ, with $\psi = 0$ on $Y = 0$, and $\psi \to \psi_c(0)$ as $Y \to \infty$.

(ii) Heat conduction

Nearer the boundary, we have

$$y \sim \delta_\theta, \qquad \psi \sim \delta_\theta/\delta_h, \qquad T \sim 1. \tag{14.118}$$

We choose

$$\frac{1}{\delta_h} = \frac{1}{\Lambda\delta_\theta^2} \tag{14.119}$$

(note $\delta_\theta \ll \delta_h$ if $\Lambda\delta_h \gg 1$), and then the rescaled equations give

$$\Psi \sim -u_0(x)Y, \qquad \Psi_Y T_x - \Psi_x T_Y = T_{YY}, \tag{14.120}$$

where u_0 is determined from the buoyancy layer.

The final relation arises through conserving buoyancy $\int T\,d\psi$ as the thermal boundary layer turns the corner. This requires

$$\theta_p = \delta_\theta/\delta_h. \tag{14.121}$$

Thus, finally, we obtain the values

$$\Lambda = 1/\varepsilon^4, \qquad \theta_p = \varepsilon, \qquad \delta_h = \varepsilon^2, \qquad \delta_\theta = \varepsilon^3, \qquad \delta_p = \varepsilon^4, \tag{14.122}$$

where

$$\varepsilon = R^{-1/9}. \tag{14.123}$$

We see that $\Lambda\delta_h = 1/\varepsilon^2 \gg 1$ as assumed and $\Lambda/R = \varepsilon^5 \ll 1$ also. In Fig. 14.6, we summarize the rescaled equations and the location and rescalings of the major boundary layers.

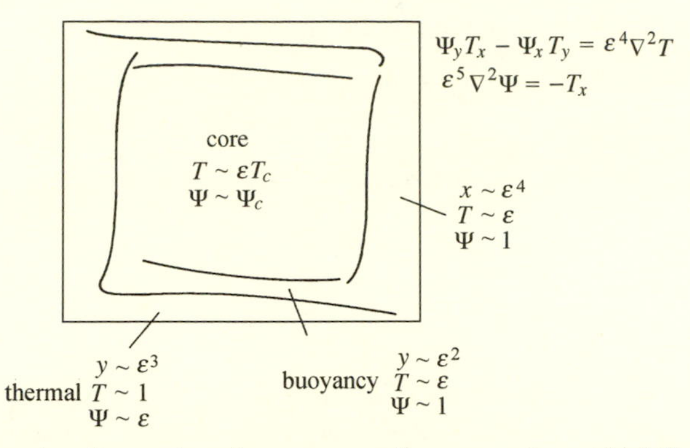

Fig. 14.6. Different scaling regions for porous medium convection at high Rayleigh number

Solution procedure

Without going into the details of matching through the corners, we sketch the structure of the solution. In the plume, we have to satisfy

$$\begin{aligned} T_y &= [(T - T_c)T_\psi]_\psi, \\ T_\psi &= 0 \quad \text{on } \psi = 0, \\ T &= T_c \quad \text{on } \psi = \psi_c, \end{aligned} \tag{14.124}$$

with an initial condition determined from the thermal boundary layer equation. Then ψ_c is determined from

$$\psi = \int_0^X (T_c - T)\, dX, \qquad \psi_c = \int_0^\infty (T_c - T)\, dX. \tag{14.125}$$

From Eq. (14.124), we determine $T_p(\psi) = T$ at $y = 1$. Equivalently, $T = -T_p(\psi)$ at the base of the cold plume, and in the basal buoyancy layer we solve

$$\begin{aligned} \psi_{YY} &= -T_p'(\psi)\psi_x, \\ \psi &= 0 \quad \text{on } Y = 0, \\ \psi &\to \psi_c \quad \text{as } Y \to \infty, \end{aligned} \tag{14.126}$$

with an initial condition by matching to the corner region. From this, we determine $u_0(x)$, and then the thermal boundary layer has

$$\frac{\partial T}{\partial x} = -u_0 \frac{\partial^2 T}{\partial \Psi^2}, \tag{14.127}$$

in Von Mises coordinates x, Ψ. From this, we determine the initial condition for the plume, in particular $\int_0^{\psi_c} T\, d\psi$.

This reduction is thus self-consistent, provided T_c can be determined. Now Eq. (14.124) is a degenerate diffusion equation, and we can expect that the degeneracy enables T_c to be found. For example, we have $\int_0^{\psi_c} T\, d\psi = C$; given C and ψ_c, we expect T_c to be determinable. Then Eq. (14.125) gives ψ_c, and C is determined through Eq. (14.127).

If we write $T = T_c + \theta$, then the problem for θ is

$$\begin{aligned} \theta_y &= (\theta\theta_\psi)_\psi - T_c', \\ \theta_\psi &= 0 \quad \text{on } \psi = 0, \\ \theta &= 0 \quad \text{on } \psi = \psi_c, \end{aligned} \tag{14.128}$$

together with an appropriate initial condition. Then also

$$\psi_X = -\theta, \qquad \psi_c = -\int_0^\infty \theta\, dX. \tag{14.129}$$

Booker suggests $\psi_c =$ constant and $T_c' =$ constant. However, if $\psi_c =$ constant, then $\theta \sim (\psi_c - \psi)^{1/2}$ near ψ_c, so that ψ reaches ψ_c in a finite value of X. In order that $\theta \sim \psi_c - \psi$ as $\psi \to \psi_c$, ψ_c must vary with y. In fact, a similarity solution can be found, and this is also appropriate to the initial condition at $y = 0$. It is of the form

$$\theta = h(y)F(\eta), \quad \eta = \psi/h(y), \tag{14.130}$$

where $h(y) = T_c(y) +$ constant. Then F satisfies

$$\begin{aligned} F - \alpha\eta F' &= (FF')' - b, \\ F'(0) = 0, &\qquad F(-\beta) = 0, \end{aligned} \tag{14.131}$$

where h satisfies

$$h^{2\alpha-1} = by + \text{constant}, \tag{14.132}$$

and $\psi_c = -\beta h(y)$; thus

$$\psi_X = -h(y)F[\psi/h(y)], \quad \psi(0) = 0, \tag{14.133}$$

whence

$$\psi = -h(y)\phi(X), \qquad X = \int_0^\phi \frac{d\phi}{F(\phi)}, \tag{14.134}$$

and thus $\phi(\infty) = \beta$, that is, $F(\beta) = 0$ (with $F'(\beta)$ finite). This is automatically satisfied because from Eq. (14.131), F is an even function. The choice $\alpha = 1$ gives a linear temperature profile. In general, Eq. (14.131) requires a numerical solution. Further details are not pursued here.

14.10 Notes and references

Two major books on stability of fluid flows are those by Drazin and Reid (1981) and Chandrasekhar (1961). Of these, the former is largely (but not entirely) concerned with instability in shear flows, whereas the latter considers mainly convective instability: the basic Rayleigh–Bénard instability of a heated fluid layer, together with the modifications induced by rotation and magnetic fields, the motivation being applications in astrophysics.

Instabilities in fluids can be broadly classified according to whether viscosity (or some other dissipative mechanism) is relevant or not. Examples of diffusive instabilities, where viscosity is important, are those in shear flows, convection, and other more

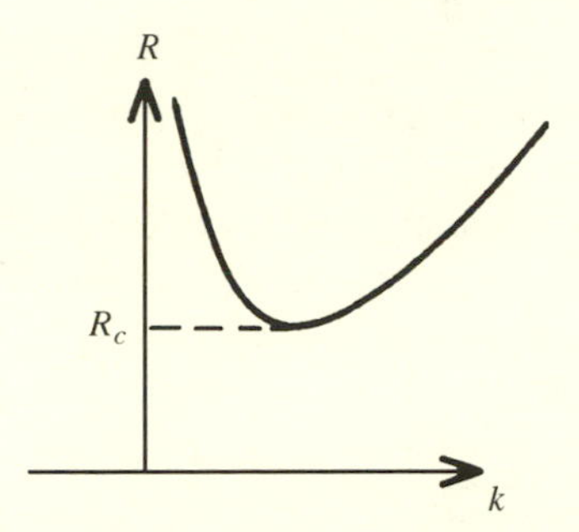

Fig. 14.7. A typical critical curve for instability

exotic situations, for example, double-diffusive instability (Turner, 1973). Examples of essentially inviscid instabilities are the Rayleigh–Taylor instability (heavy fluid overlying light fluid), Kelvin–Helmholtz instability (wind generating water waves), and the baroclinic instability of zonal flows in meteorology.

In general, the distinction is manifested through the dispersion relation. If, for example, modes $\exp[i(kx + \omega t)]$ are excited, then $\sigma = i\omega = \sigma(k, R)$ is a function both of wavenumber k and forcing parameter R (be this Rayleigh, Reynolds, or some other number). The dispersion relation for porous medium convection, Eqs. (14.59) or (14.60), displays a typical form of such relations: σ is concave ($\partial^2\sigma/\partial k^2 < 0$), with $\sigma < 0$ for $k = 0$ and $k \to \infty$ and $\partial\sigma/\partial R > 0$. Thus, for sufficiently large R, a band of wavenumbers is unstable (see Fig. 14.4), and thus there is a critical curve $R(k)$ that separates stability ($R < R_c = \min_k R$) from instability ($R > R_c$). This is illustrated in Fig. 14.7. The main distinction between viscous and inviscid, or between *dissipative* and *dispersive* instabilities (see Gibbon and McGuinness (1981), who also consider the nonlinear stability of such systems), is that dispersive systems have real dispersion relations of the form $l(\omega, k) = 0$, whose roots $\omega(k)$ therefore occur as complex conjugates; complex roots ω cause instability, whereas real values of ω lead to neutrally stable waves. Dissipative systems have a complex dispersion relation, and the roots do not generally occur as conjugates. In both cases, the critical curve is given by $\mathrm{Re}\,\sigma = 0$. However, in the dissipative case, this separates regions $\mathrm{Re}\,\sigma > 0$ and $\mathrm{Re}\,\sigma < 0$; whereas in the dispersive case, it separates regions where $\mathrm{Re}\,\sigma > 0$ and $\mathrm{Re}\,\sigma = 0$. An example of a dispersive instability is the Kelvin–Helmholz instability, in which fluid (e.g., air) of density ρ_a flows over fluid (e.g., water) of density ρ_w with speed U. One can show that σ is given by

$$\rho_w[kg + \sigma^2] = \rho_a[kg - (\sigma + ikU)^2], \tag{14.135}$$

with a dispersive instability if the shear U is greater than a critical value U_c given by

$$U_c = \left[\frac{g\left(\rho_w^2 - \rho_a^2\right)}{k\rho_w\rho_a}\right]^{1/2}. \tag{14.136}$$

(Note that for this case, all sufficiently short wavelength disturbances are unstable. Surface tension removes the instability at large k; see Exercise 1.)

Boundary layer theory For ordinary Rayleigh–Bénard convection (see also Exercise 3), boundary layer theory is well-developed; see, for example, Roberts (1979) and Jimenez and Zufiria (1987). It is surprising that for the apparently easier porous medium convection problem, the theory is less clear. The discussion here is related to the analysis by Booker (1981). Booker's analysis was based on a Hele–Shaw analog experiment, in which thermal boundary layers developed surrounding a stagnant thermally stratified core. He used experimental observation to close his model and predicted that the Nusselt number $Nu \sim R^{2/5}$. His values are $\delta_\theta \sim R^{-2/5}, \delta_p \sim R^{-1/5}$, as compared to our values $\delta_\theta \sim R^{-1/3}, \delta_p \sim R^{-4/9}$. Robinson and O'Sullivan (1976) suggested $Nu \sim R^{2/3}$, based on computational results, though these were at relatively low values of $R \gtrsim 2000$.

Exercises

1. Fluid of density ρ_1 flows with horizontal speed U past stationary fluid of density ρ_2. The surface tension at the interface is T. Assuming each fluid is of infinite extent, find the dispersion relation corresponding to small disturbances of the interface, and deduce that the flow is unstable if

$$U^2 > \frac{(\rho_2 + \rho_1)}{\rho_1 \rho_2} \left[\frac{g}{k}(\rho_2 - \rho_1) + Tk \right].$$

Hence show that the critical shear for instability is

$$U_c = \left[\frac{2(\rho_2 + \rho_1)}{\rho_1 \rho_2} \{Tg(\rho_2 - \rho_1)\}^{1/2} \right]^{1/2},$$

and that if $\rho_2 \gg \rho_1$, this is approximately $(2/\rho_1)^{1/2}(\rho_2 g T)^{1/4}$. Find values appropriate to wind over water, and compute the corresponding critical shear.

2. *Lapwood convection*
Consider the onset of convection in a porous medium, where the base is impermeable and the top is permeable (for example, convection in sea-floor crustal rock at mid-ocean ridges). Specify the appropriate boundary conditions for this case, and show that the appropriate form of the stream function $f(y)$ in Eq. (14.54), when $\sigma = 0$, can be taken as

$$f(y) = \sinh \alpha_+ \sin \alpha_- y + \sin \alpha_- \sinh \alpha_+ y,$$

where (with $R > k^2$)

$$\alpha_+ = [k^2 + k\sqrt{R}]^{1/2}, \qquad \alpha_- = [k\sqrt{R} - k^2]^{1/2}.$$

Hence show that the dispersion relation is given by

$$\frac{\tanh \alpha_+}{\alpha_+} + \frac{\tan \alpha_-}{\alpha_-} = 0.$$

Denoting $F(\alpha) = [\tanh \alpha]/\alpha$, indicate graphically the form of $F(\alpha)$ and thus also $F^{-1}(\theta)$ (for $0 < \theta < 1$); by plotting $[\tan \alpha]/\alpha$, show that (for

$\alpha > 0$) $F^{-1}[(-\tan\alpha)/\alpha] = G(\alpha)$ is defined as a positive monotone increasing function on strips $[\alpha_n, n\pi)$, where $(2n-1)\pi/2 < \alpha_n < n\pi$, $n = 1, 2, \ldots$, and $G(\alpha_n) = 0$, $G(n\pi) = \infty$. Deduce that for given k, there are an infinite number of critical values of R where $\sigma = 0$, given by the intersections in the first quadrant of the hyperbola $\alpha_+^2 - \alpha_-^2 = 2k^2$ with $\alpha_+ = G(\alpha_-)$. If the intersection values are denoted by $(\alpha_-^{(n)}, \alpha_+^{(n)})$, with $\alpha_-^{(1)} < \alpha_-^{(2)} < \ldots$, show that $\alpha_+^{(1)} < \alpha_+^{(2)} < \ldots$, $\alpha_+^{(n)} \approx \alpha_-^{(n)} \approx n\pi$ as $n \to \infty$ and that the corresponding values of R, denoted R_n, satisfy $R_1 < R_2 < \ldots$. If $\alpha_1^* < \alpha_2^* < \ldots$ denote the roots of $\alpha = G(\alpha)$, show that $\alpha_n < \alpha_n^* < n\pi$, $\alpha_n^* \to n\pi$ as $n \to \infty$ and that $\alpha_-^{(n)}$ increases monotonically with k, with $\alpha_-^{(n)}(0) = \alpha_n^*$, $\alpha_-^{(n)}(\infty) = n\pi$. Deduce that each critical curve $R_n(k)$ is convex, $R_n \to \infty$ as $k \to 0$, and $k \to \infty$, so that the critical value for the onset of convection is $R_c = \min_k R_1(k)$. Show that

$$\left[\frac{\alpha_1^{*2}}{k} + k\right]^2 < R_1(k) < \left[\frac{\pi^2}{k} + k\right]^2,$$

and deduce that $4\alpha_1^{*2} < R_c < 4\pi^2$. Devise a numerical method to compute R_c (the actual value is 27.1). This problem was studied by Lapwood (1948).

3. The Boussinesq equations for convection in a fluid layer heated from below can be written in the dimensionless form (see Section 2.3)

$$\begin{aligned} \boldsymbol{\nabla}.\mathbf{u} &= 0, \\ \frac{1}{\sigma}\frac{d\mathbf{u}}{dt} &= -\boldsymbol{\nabla} p + \nabla^2\mathbf{u} + Ra\,\theta\mathbf{k}, \\ \frac{d\theta}{dt} &= \nabla^2\theta, \end{aligned}$$

where $\theta = 1$ on $z = 0$, $\theta = 0$ on $z = 1$, and where the *Prandtl* and *Rayleigh* numbers are

$$\sigma = \frac{\mu}{\rho_0\kappa}, \qquad Ra = \frac{\alpha g d^3 \rho_0 \Delta T}{\mu\kappa}.$$

Write down the steady state with $\mathbf{u} = \mathbf{0}$, and show that for free slip conditions ($\mathbf{u}.\mathbf{n} = \partial(\mathbf{u}.\mathbf{t})/\partial n = 0$ on the boundary), this is unstable to convection if

$$Ra > Ra_c = 27\pi^4/4 \approx 657.5$$

and the critical wavenumber is $k_c = \pi/\sqrt{2}$.

4. By considering a multiple-scale expansion in terms of a slow timescale $\tau = \varepsilon^2 t$ and a slow space scale $X = \varepsilon x$, where $Ra = Ra_c + \varepsilon^2$ in the notation of Exercise 3, show that convective perturbations to the basic state of the form $\varepsilon A e^{ik_c x} + \ldots$ will evolve according to the evolution equation

$$\frac{\partial A}{\partial \tau} = kA - l|A|^2 A + m\frac{\partial^2 A}{\partial X^2}.$$

Comment on plausible boundary conditions for, and likely behavior of, this equation. (See Newell and Whitehead, 1969.)

5. When Ra is large and σ is very large, derive a rescaled version of the Rayleigh–Bénard convection model of Exercise 3 by defining $\delta = Ra^{-1/3} \ll 1$ and then rescaling the variables. It is convenient, for two-dimensional flows, to define a stream function via $\mathbf{u} = (\psi_z, 0, -\psi_x)$ and the vorticity ω (actually its negative) as $\omega = \nabla^2\psi$; then rescale $\psi, \omega \sim 1/\delta^2$. Show that, in the limit $\delta \to 0$, $\theta \approx \theta_c(\psi)$; and by integrating the temperature equation round a (closed) streamline, show that $\theta_c =$ constant ($= 1/2$, by symmetry) (this is an example of the Prandtl–Batchelor theorem). Deduce that thermal boundary layers exist at the boundary, and show that for free slip boundary conditions, these are of thickness $O(\delta)$. What happens if no-slip boundary conditions are applied?

6. *Convection in a fluid loop*
A domestic back boiler is configured as shown in Fig. 14.8. Show that in a Boussinesq model of the flow, the mean fluid velocity (across a cross section) v is a function only of t, $v = v(t)$, and explain why v satisfies

$$\rho_0 \dot{v} = -\frac{\partial p}{\partial s} - \rho g \frac{\partial z}{\partial s} - \frac{8\mu}{a^2} v,$$

where we assume the pipes are circular, of radius a, s is the distance along the pipes, and z is vertical height. By assuming $\rho = \rho_0[1 - \alpha(T - T_0)]$, derive the equation in the form

$$\rho_0 \dot{v} = -\frac{\partial \Pi}{\partial s} + \alpha \rho_0 g (T - T_0) \frac{\partial z}{\partial s} - \frac{8\mu}{a^2} v,$$

where $p + \rho_0 g z = \Pi$; and by integrating round the closed loop C of length L, show that v satisfies

$$\rho_0 L \dot{v} = \alpha \rho_0 g \oint_C (T - T_0)\, dz - \frac{8\mu L}{a^2} v.$$

If heat losses are neglected in the pipe, show that T satisfies

$$T_t + v T_s = 0.$$

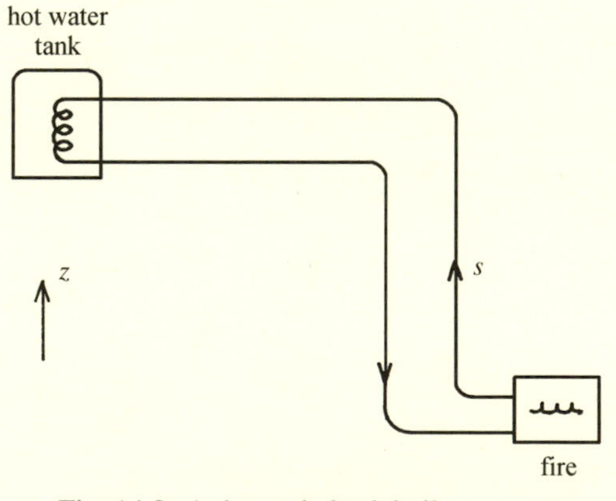

Fig. 14.8. A domestic back boiler system

Further, show that if the fire at $s = 0$ delivers heat at a rate q_H (e.g., in kilowatts), then a suitable jump condition for the temperature across $s = 0$ is

$$v \Delta T = \frac{q_H}{\rho_0 c_p A},$$

where $A = \pi a^2$ is the tube cross-sectional area, and c_p is the specific heat. Similarly, show that the jump in T across the hot water tank at $s = L/2$ is

$$v \Delta T = -\frac{q_c}{\rho_0 c_p A},$$

where q_c is the cooling rate. What is the final boundary condition?

By choosing suitable scales, nondimensionalize the model and show that it depends on the single dimensionless parameter

$$\delta = \frac{\rho_0 a^2}{8 \mu L} \left(\frac{q_H \alpha g}{8 \mu \pi c_p} \right)^{1/2}$$

(you should assume $q_H = q_c$). By considering the particular case where the connecting pipes are vertical, show that there is a unique steady state (up to an additive constant for the temperature) and examine its stability. What happens if the pipe contour $z(s)$ is modified? Also analyze the model when the cooling rate is given by $q_c = h(T|_{\frac{1}{2}-} - T_0)$. (This is the celebrated Welander loop, studied by Welander (1967), Keller (1966), and, more recently, Herrero and Velazquez (1990).)

15

River flow

15.1 The role of fluid mechanics

Fluid dynamics is pervasive in modeling many natural phenomena. Indeed, one school of thought has it that fluid dynamics and applied mathematics are synonymous. But in practice, it is often the case that 'classical' fluid mechanics is of little practical use. One reason for this is that at Reynolds numbers in excess of about 10^3, fluid flow becomes *turbulent* and consists of a spatially and temporally chaotic flow, consisting of eddies of many different scales. To describe such flows, one must use models that are semiempirical, and (in fact) simpler than the Navier–Stokes equations; and as often as not, the models are specific to the situation under consideration. In this chapter we give one example of such a situation, that of river flow, and in Chapter 16 we consider another, that of two-phase flow.

15.2 The mechanics of drainage basins

Rainwater that falls in a catchment area of a particular river basin makes its way back to the ocean (or sometimes to an inland lake) by seepage into the ground and then, through groundwater flow, to outlet streams and rivers. In severe storm conditions, the rainfall intensity may exceed the soil infiltration capacity, and then direct runoff to discharge streams can occur as overland flow. Depending on local topography, soil cover, and vegetation, one or more transport process may be the norm. Overland flow can also occur if the soil becomes saturated.

River flow itself occurs on river beds that are typically quasi-one-dimensional, sinuous channels with variable and rough cross section. Moreover, if channel cross sectional area is A (m^2) and discharge is Q ($m^3\ s^{-1}$), then an appropriate Reynolds number for the flow is

$$Re = Q/\nu A^{1/2}, \tag{15.1}$$

where ν is the kinematic viscosity. If $A^{1/2} = 10$ m, $\nu = 10^{-6}\ m^2\ s^{-1}$, and $Q = 1$ $m^3\ s^{-1}$ (a small value), then $Re \sim 10^5$. Inevitably, river flow is turbulent for all but the smallest rivulets. Thus, to model river flow and to explain the response of river discharge to storm conditions, as measured on flood hydrographs, for instance, one must model a flow that is essentially turbulent and that exists in a rough, irregular channel.

15.3 Mathematical model

Our starting point is that the flow is essentially one-dimensional: or at least, we focus on this aspect of it. In addition to the cross-sectional area (of the *flow*) A and discharge Q, we introduce a longitudinal, curvilinear distance coordinate s, and we assume that the river axis changes direction slowly with s. Then conservation of mass is, in its simplest form,

$$\frac{\partial A}{\partial t} + \frac{\partial Q}{\partial s} = 0. \tag{15.2}$$

This ignores, for the moment, source terms owing to infiltration seepage and overland flow from the catchment.

Eq. (15.2) must be supplemented by an equation for Q as a function of A, and this arises through consideration of momentum conservation. There are three levels at which one may do this: by exact specification, as in the Navier–Stokes momentum equation; by including inertia but averaging, as in Darcy's law; and most simply, by ignoring inertia and applying a force balance using a semiempirical friction factor. We begin by opting for the last choice, which should apply for sufficiently 'slow' (in some sense) flow. Later we will consider more complicated models.

We have already defined the Reynolds number *Re* in terms of Q and A, or equivalently, a mean velocity $u = Q/A$ and a channel depth $d \sim A^{1/2}$. 'Slow' here means a small *Froude number*, defined by

$$Fr = \frac{u}{(gd)^{1/2}} = \frac{Q}{g^{1/2}A^{5/4}}. \tag{15.3}$$

If $Fr < 1$, the flow is *tranquil*; if $Fr > 1$, it is *rapid*. Gravity is of relevance, because the flow is ultimately due to gravity.

Now let l be the perimeter of a cross section, and let τ be the mean shear stress exerted at the bed (longitudinally) by the flow. If the downstream slope is α, then a force balance gives

$$l\tau = \rho g A \sin\alpha, \tag{15.4}$$

where ρ is density. For turbulent flow, the shear stress is given by the friction law

$$\tau = f\rho u^2, \tag{15.5}$$

where the friction factor f may depend on the Reynolds number. Because

$$u = Q/A, \tag{15.6}$$

and defining the hydraulic radius

$$R = A/l, \tag{15.7}$$

we derive the relations

$$u = (g/f)^{1/2} R^{1/2} S^{1/2}, \tag{15.8}$$

where

$$S = \sin\alpha, \tag{15.9}$$

and

$$Q = \left(\frac{g}{fl}\right)^{1/2} A^{3/2} S^{1/2}. \tag{15.10}$$

For wide, shallow rivers, l is essentially the width. For a more circular cross section, then $l \sim A^{1/2}$, and

$$Q = (g/f)^{1/2} A^{5/4} S^{1/2}. \tag{15.11}$$

The relation (15.8) is the Chezy velocity formula, and $C = (g/f)^{1/2}$ is the Chezy roughness coefficient. Notice that the Froude number, in terms of the hydraulic radius, is

$$Fr = \frac{u}{(gR)^{1/2}} = (S/f)^{1/2}, \tag{15.12}$$

and tranquillity (at least in uniform flow) is basically due to slope.

Alternative friction correlations exist. That due to Manning is an empirical formula to fit measured stream velocities and is of the form

$$u = R^{2/3} S^{1/2}/n, \tag{15.13}$$

where Manning's roughness coefficient n takes typical values in the range .01–.1 $\mathrm{m}^{-1/3}$ s, depending on stream depth, roughness, etc. For Manning's formula, we have

$$\begin{aligned} Q &\sim A^{4/3} \quad \text{if } R \sim A^{1/2}, \\ Q &\sim A^{5/3} \quad \text{if } l \text{ is width,} \qquad R = A/l \sim A. \end{aligned} \tag{15.14}$$

Thus we see that for a variety of stream types and velocity laws, we can pose a relation between discharge and area of the form

$$Q \sim A^{m+1}, \quad m > 0, \tag{15.15}$$

with typical values $m = 1/4 - 2/3$. In practice, for a given stream, one could attempt to fit a law of the form (15.15) by direct measurement.

15.4 The flood hydrograph

We can nondimensionalize the equation for A so that it becomes

$$\frac{\partial A}{\partial t} + A^m \frac{\partial A}{\partial s} = 0, \tag{15.16}$$

a first-order nonlinear hyperbolic equation, also known as a kinematic wave equation, whose solution can be written down. Specifically, take initial data parameterized as

$$A = A_0(\sigma), \qquad s = \sigma > 0, \qquad t = 0. \tag{15.17}$$

Then the characteristic equations are

$$\begin{aligned} \frac{dA}{dt} &= 0, \\ \frac{ds}{dt} &= A^m, \end{aligned} \tag{15.18}$$

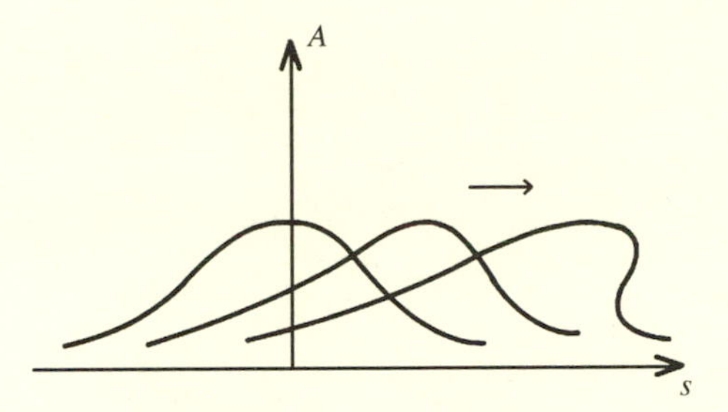

Fig. 15.1. Formation of a shock wave

whence

$$A = A_0(\sigma), \qquad s = \sigma + A_0^m t; \tag{15.19}$$

thus

$$A = A_0[s - A^m t] \tag{15.20}$$

determines A implicitly. It is a familiar fact that humped initial conditions $A_0(\sigma)$ will lead to propagation of a kinematic wave and then to shock formation, as shown in Fig. 15.1, when $\partial A/\partial s$ reaches infinity. Because $\partial A/\partial s = A_s = A_0'[s - A^m t](1 + mt\,A^{m-1}A_s)$, a shock first forms on the characteristic through σ that maximizes $-(A_0^m)'$ and at a time t that is the inverse of this maximum. Thereafter a shock exists at a point $s_d(t)$, and propagates at a rate given, by consideration of the integral conservation law

$$\frac{\partial}{\partial t}\int_{s_1}^{s_2} A\,ds = -[Q]_{s_1}^{s_2}, \tag{15.21}$$

by

$$\dot{s}_d = \frac{[Q]_{s_d-}^{s_d+}}{[A]_{s_d-}^{s_d+}}. \tag{15.22}$$

As an application, we consider the flood hydrograph, which measures discharge at a fixed value of s as a function of time. As an idealization of a flood, we consider an initial condition

$$A \approx A_0\delta(s) \quad \text{at } t = 0, \tag{15.23}$$

where $\delta(s)$ is the delta function, representing the overland flow of a short period of localized rainfall. Because $A = f(s - A^m t)$, it follows that $A \approx 0$ except where $s = A^m t$. The humped initial condition causes a shock to form at $s_d(t)$, with $s_d(0) = 0$, and we have

$$\begin{aligned} A &= 0, \qquad s > s_d, \\ A &= (s/t)^{1/m}, \qquad s < s_d, \end{aligned} \tag{15.24}$$

as shown in Fig. 15.2. The shock speed is given by

$$\dot{s}_d = (Q/A)|_{s_d-} = \left.\frac{A^m}{m+1}\right|_{s_d-} = \frac{s_d}{(m+1)t}, \tag{15.25}$$

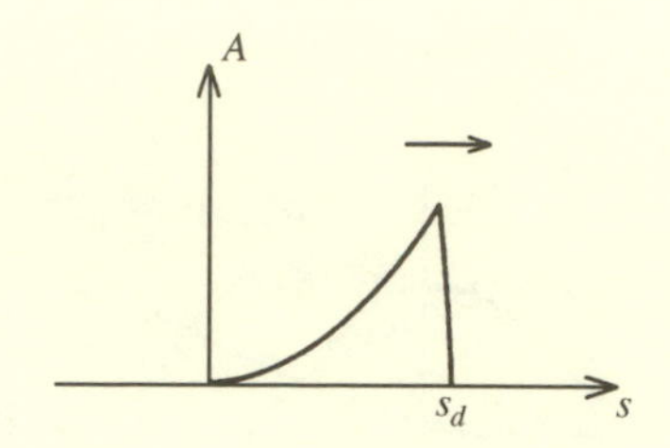

Fig. 15.2. Propagation of a shock front

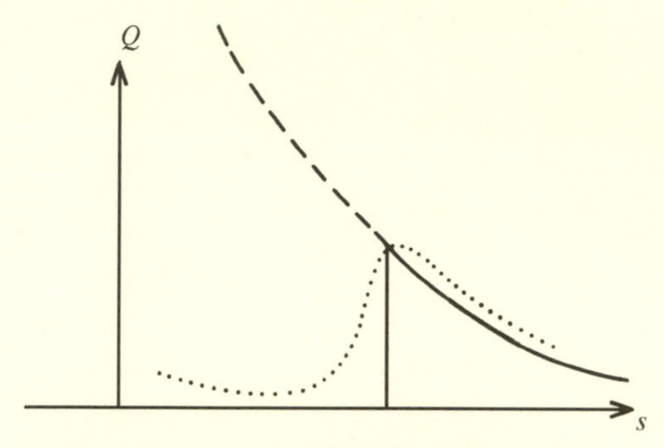

Fig. 15.3. Ideal and observed flood hydographs

whence $s_d \propto t^{1/(m+1)}$. To calculate the coefficient of proportionality, we use conservation of mass in the form

$$\int_0^{s_d} A\,ds = A_0, \tag{15.26}$$

whence, in fact,

$$s_d = [(m+1)A_0/m]^{m/(m+1)} t^{1/(m+1)}. \tag{15.27}$$

Denoting $b = [(m+1)A_0/m]^{m/(m+1)}$, the flood hydrograph at a fixed station $s = s^*$ is then as follows. For $t < t^*$, where

$$t^* = (s^*/b)^{m+1}, \tag{15.28}$$

$Q = 0$. For $t > t^*$, $A = (s^*/t)^{1/m}$, and thus

$$Q = \frac{s^{*(m+1)/m}}{(m+1)} t^{-(m+1)/m}. \tag{15.29}$$

This result is illustrated in Fig. 15.3, together with a typical observed hydrograph. The smoothed observation can be explained by the fact that a more realistic initial condition would have delivery of the storm flow over an interval of space and time. Also, one can expect that a more realistic model will allow diffusive effects.

15.5 Acceleration: stability and waves

We now reexamine the momentum equation, which we previously assumed to be described by a force balance. Again, consider the equations in dimensional form.

Conservation of mass is written in the form

$$\frac{\partial A}{\partial t} + \frac{\partial}{\partial s}(Au) = 0, \tag{15.30}$$

and then conservation of momentum (from first principles) leads to the equation (adopting the friction law (15.5))

$$\rho \frac{\partial (Au)}{\partial t} + \rho \frac{\partial}{\partial s}(Au^2) = \rho g A \sin\alpha - \rho l f u^2 - \frac{\partial}{\partial s}(A\bar{p}), \tag{15.31}$$

where $\bar{p}$ is the mean pressure. Now the pressure is approximately hydrostatic; thus $p \approx \rho g z$, where z is depth. Then $\bar{p}A \approx \int \frac{1}{2}\rho g h^2\,dx$, where h is total depth and x is width; and supposing $\partial h/\partial s$ is independent of x and that the width does not vary with s, we find

$$\frac{\partial}{\partial s}(A\bar{p}) = \rho g A \frac{\partial \bar{h}}{\partial s}, \tag{15.32}$$

where $\bar{h}$ is the mean depth. Using Eq. (15.30), Eq. (15.31) reduces to

$$u_t + u u_s = g \sin\alpha - f l u^2/A - g\frac{\partial \bar{h}}{\partial s}. \tag{15.33}$$

Eq. (15.30) and (15.33) are known as the St. Venant equations.

Nondimensionalization

We choose scales for $u = Q/A$, t, s, A, R (the hydraulic radius, $= A/l$), and $\bar{h}$ as follows, in keeping with the assumed balances adopted earlier:

$$\begin{aligned} Au &\sim Q, \\ g \sin\alpha &\sim l f u^2/A = f u^2/R, \\ t &\sim s/u, \\ s &\sim d/\sin\alpha, \\ \bar{h}, R &\sim d, \end{aligned} \tag{15.34}$$

where we can suppose Q is a typical observed discharge and d is a typical observed depth. Explicitly, the scales are

$$\begin{aligned} &[\bar{h}], [R] = d, \qquad [s] = d/\sin\alpha, \\ &[u] = (gd \sin\alpha/f)^{1/2}, \qquad [t] = (fd/g \sin^3\alpha)^{1/2}, \\ &[A] = Q(f/gd \sin\alpha)^{1/2}, \end{aligned} \tag{15.35}$$

and we put $u = [u]u^*$, etc., and drop asterisks. The resulting equations are

$$\begin{aligned} A_t + (Au)_s &= 0, \\ F^2[u_t + u u_s] &= 1 - u^2/R - h_s, \end{aligned} \tag{15.36}$$

where we choose $h \sim R \sim A$ for a wide channel and $h \sim R \sim A^{1/2}$ for a rounded channel. For simplicity, here we choose $h = R = A$. The momentum equation can then be written

$$F^2(u_t + u u_s) = 1 - u^2/A - A_s. \tag{15.37}$$

The Froude number F is given by

$$F = \frac{[u]}{(gd)^{1/2}} = (\sin\alpha/f)^{1/2}. \tag{15.38}$$

Waves

Note that the length scale $d/\sin\alpha$ is long, but not as long as the total river length, which is of the order $D/\sin\alpha$, where D is the elevation drop and $D \gg d$, usually. Thus the theory of the preceding sections is implicitly concerned with disturbances on longer length scales (and timescales). For example, if we take a river of length $L \sim 100$ km, slope $\sin\alpha = 10^{-3}$, velocity $u = 30$ cm s^{-1}, and depth $d = 3$ m, then the drainage response time for storm flow is $\sim L/u \sim 4 \times 10^5$ s ~ 3 days, which is about right. The timescale for response in Eq. (15.37) is then calculated in units of $d/u \sin\alpha \sim 10^4$ s, which is 4 hours; although for these values, $F^2 \sim .003$, so the actual response time is $t \sim F^2$, corresponding to about a minute. Evidently, the quasi-steady theory is satisfactory for many purposes.

Locally, we can take the basic river flow as constant,

$$u = A = 1, \tag{15.39}$$

and we examine its stability by writing

$$u = 1 + v, \qquad A = 1 + a \tag{15.40}$$

and linearizing. We obtain the linear system

$$\begin{aligned} a_t + a_s + v_s &= 0, \\ F^2(v_t + v_s) &= -2v + a - a_s, \end{aligned} \tag{15.41}$$

whence

$$F^2\left(\frac{\partial}{\partial t} + \frac{\partial}{\partial s}\right)^2 v = -2\left(\frac{\partial}{\partial t} + \frac{\partial}{\partial s}\right)v - v_s + v_{ss}. \tag{15.42}$$

Solutions $v = \exp[iks + \sigma t]$ exist, provided σ satisfies

$$F^2(\sigma + ik)^2 + 2(\sigma + ik) + ik + k^2 = 0 \tag{15.43}$$

or

$$\tilde{\sigma} = -i\tilde{k} - 1 \pm [1 - i\tilde{k} - \tilde{k}^2/F^2]^{1/2}, \tag{15.44}$$

where we write

$$\sigma = \tilde{\sigma}/F^2, \qquad k = \tilde{k}/F^2. \tag{15.45}$$

There are thus two wave-like disturbances. The possibility of instability exists if either value of $\tilde{\sigma}$ has positive real part. We define the positive square root in Eq. (15.44) to be that with positive real part. From Fig. 15.4, we see that as $F \to \infty$, $p = [1 - i\tilde{k} - c\tilde{k}^2/F^2]^{1/2}$ has real part greater than one, so that the corresponding value of $\tilde{\sigma}$ has Re $\tilde{\sigma} > 0$, indicating instability. As F decreases, the positive square

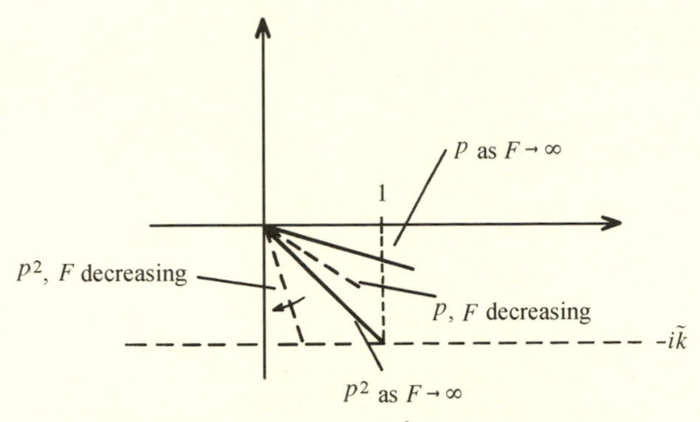

Fig. 15.4. Representation of p^2 and p for varying F

root rotates (in complex p space) clockwise, so that there is a definite value F_c such that instability exists for $F > F_c$. This is given by the condition that

$$\left[1 - i\tilde{k} - \tilde{k}^2/F_c^2\right]^{1/2} = 1 - i\beta \tag{15.46}$$

for some β, whence we find

$$F_c = 2. \tag{15.47}$$

Thus for tranquil flow, $F < O(1)$, the flow is stable. For rapid flow, $F > O(1)$, it is unstable. For all values of F, we have $p = \alpha - i\beta$, where $\beta > 0, \alpha > 0$, so that

$$\tilde{\sigma} = -1 \pm \alpha - i\tilde{k} \mp i\beta, \tag{15.48}$$

whence the solutions are of the form

$$\begin{aligned} &\exp[\{(-1+\alpha)t + i(\tilde{k}s - (\tilde{k}+\beta)t)\}/F^2], \\ &\exp[\{-(1-\alpha)t + i(\tilde{k}s + (\beta - \tilde{k})t\}/F^2]. \end{aligned} \tag{15.49}$$

Thus the wave that goes unstable (when $\alpha = 1$) propagates downstream, whereas the other wave, always stable, propagates downstream unless $\beta > \tilde{k}$, which is if and only if $\alpha < 1/2$, that is,

$$F < F_- = \frac{2\tilde{k}}{(3 + 4\tilde{k}^2)^{1/2}}. \tag{15.50}$$

Note that F_- depends on $\tilde{k}$ and that $0 < F_- < 1$; we therefore have three distinct ranges for F:

$F > 2$: two waves downstream, one unstable;
$1 < F < 2$: two waves downstream, both stable;
$F < 1$: stable waves can propagate upstream and downstream.

To go further than this requires a study of the nonlinear system (15.36). We see that the transition at $F = 1$ is associated with the ability of waves to propagate upstream. The transition at $F = 2$ is sometimes called Vedernikov instability and is associated with the formation of downstream propagating *roll waves*.

15.6 Nonlinear waves

In keeping with the linear theory, we may expect that if F is large enough, linear disturbances will grow, and we might then expect shocks to form. To examine such shocks, we rescale s and t with F^2 in Eq. (15.36), and put

$$\gamma = 1/F^2. \tag{15.51}$$

The equations are then

$$\begin{aligned} A_t + (Au)_s &= 0, \\ u_t + uu_s &= 1 - u^2/A - \gamma A_s, \end{aligned} \tag{15.52}$$

and they can be written in the form

$$\frac{\partial}{\partial t}\begin{pmatrix} A \\ u \end{pmatrix} + \begin{pmatrix} u & A \\ \gamma & u \end{pmatrix}\frac{\partial}{\partial s}\begin{pmatrix} A \\ u \end{pmatrix} = \begin{pmatrix} 0 \\ 1 - u^2/A \end{pmatrix}. \tag{15.53}$$

Characteristics

The analysis of characteristics for systems of hyperbolic equations is as follows. Suppose a state vector ψ satisfies the system

$$P\psi_t + Q\psi_s = f(\psi), \tag{15.54}$$

where P, Q are matrices. Let $P^{-1}Q$ have left eigenvector $v(\psi)$ and eigenvalues $\lambda(\psi)$, specifically

$$v^T Q = \lambda v^T P. \tag{15.55}$$

Then for each such eigenpair, we have

$$(Pv)^T[\psi_t + \lambda\psi_s] = v^T f. \tag{15.56}$$

Hence on the characteristics defined by

$$\frac{ds}{dt} = \lambda, \tag{15.57}$$

ψ satisfies

$$(Pv)^T\frac{d\psi}{dt} = v^T f, \tag{15.58}$$

which provides a method of integrating the system in principle.

In the present case, P is the identity, and we seek the eigenvalues of $Q = \begin{pmatrix} u & A \\ \gamma & u \end{pmatrix}$, which are given simply by

$$\lambda = u \pm (A\gamma)^{1/2}. \tag{15.59}$$

Thus nonlinear waves propagate downstream if $\gamma < u^2/A$, but one will propagate upstream if $\gamma > u^2/A$. This is consistent with the preceding linear theory. Because Eq. (15.52) are of second order, simple shock wave formation analysis is probably only possible (if at all) in simple cases (see Exercises 3 and 4). Equations (15.52) are very similar to those of gas dynamics, or shallow water equations, and we expect the equations will support the existence of propagating shocks in a similar way.

15.7 Sediment transport

An important constituent of river dynamics is the degree to which the flow transports sediments. Through processes of erosion and deposition, rivers can alter their own bedforms and courses, and this is of concern to hydraulic engineers, as well as being of interest in its own right.

Sediment is transported in one of two basic ways in alluvial rivers (meaning those that flow in valleys of erodible sediments, be they sand, gravel, clay, or silt). The larger particles, such as sand and gravel, roll or jump (the latter process is known as saltation) at the river bed and constitute the bedload transport. Smaller particles – clays and silts – are transported as suspended load. No transport at all occurs unless a critical shear stress is exceeded, which can be defined as

$$\tau_c = \mu \Delta\rho g D_s, \tag{15.60}$$

and is called the Shields stress. Here D_s is particle grain size, $\Delta\rho$ is the density difference between particle and water densities ρ_s and ρ_w, and μ is a dimensionless coefficient, which varies somewhat with D_s, but is roughly equal to 0.05. When the shear stress exerted by the river flow exceeds the Shields stress, then bedload transport occurs. Various empirical relationships have been proposed for bedload Q_b, for example, the Meyer-Peter/Müller relationship

$$Q_b = \frac{K\rho_s l}{\rho_w^{1/2}\Delta\rho g}[\tau - \tau_c]_+^{3/2}, \tag{15.61}$$

where $[\tau]_+ = \max(\tau, 0)$, $K \approx 10$ and l is the stream width. The units of Q_b are mass per unit time.

Suspended sediment exists in a dynamic balance between erosion and deposition. Erosion occurs via entrainment of particles by turbulent eddies at the bed and is thus related to the shear stress. One proposed relationship is that the erosion rate is

$$\dot{E} \approx 0.007 u_*^3 / v_s^2, \tag{15.62}$$

where $u_* = (\tau/\rho_w)^{1/2}$ is the turbulent friction velocity, representing the typical mean velocity of turbulent eddies, and v_s is the particle-setting velocity. An expression for the deposition rate is given by

$$\dot{D} \approx 13 v_s^2 c / u_*, \tag{15.63}$$

where c is the mean concentration of suspended sediment, measured as a volume fraction. Each of these expressions is a velocity. The particle-settling velocity depends on c and on the particle Reynolds number R_p. If both of these are small, then

$$v_s = \frac{\Delta\rho g D_s^2}{18\eta}, \tag{15.64}$$

where η is the viscosity. Empirical modifications to this relationship are available for larger values of c and R_p. We see that v_s depends strongly on D_s; thus the erosion rate decreases dramatically as D_s increases (and in reality, we might have $\dot{E} = 0$ for $\tau < \tau_c$), and the deposition rate increases. A balance of $\dot{E}$ with $\dot{D}$ thus yields an equilibrium suspended sediment concentration given by

$$c \approx 5 \times 10^{-4} \left[\frac{\tau}{\rho_w v_s^2} \right]^2, \tag{15.65}$$

and is thus proportional to D_s^{-8}. It seems clear that for a basal sediment having a distribution of grain sizes, theoretical descriptions are likely to be much more complicated and ultimately are strongly constrained by experimental data.

Dune formation

The erosion and transport of sediments from river beds causes these beds to develop a variety of shapes, depending on the flow rate. At low (tranquil, i.e., Froude number $Fr < 1$) flows, ripples are formed; these are relatively flat corrugations in the bed. At higher velocities, their amplitude becomes larger, and they also become more asymmetric, with the steep (lee) faces being downstream to the flow: see Fig. 15.5. As the Froude number increases through one, the bed becomes flat again and then develops antidunes – regular corrugations that are (unlike dunes) in phase with the

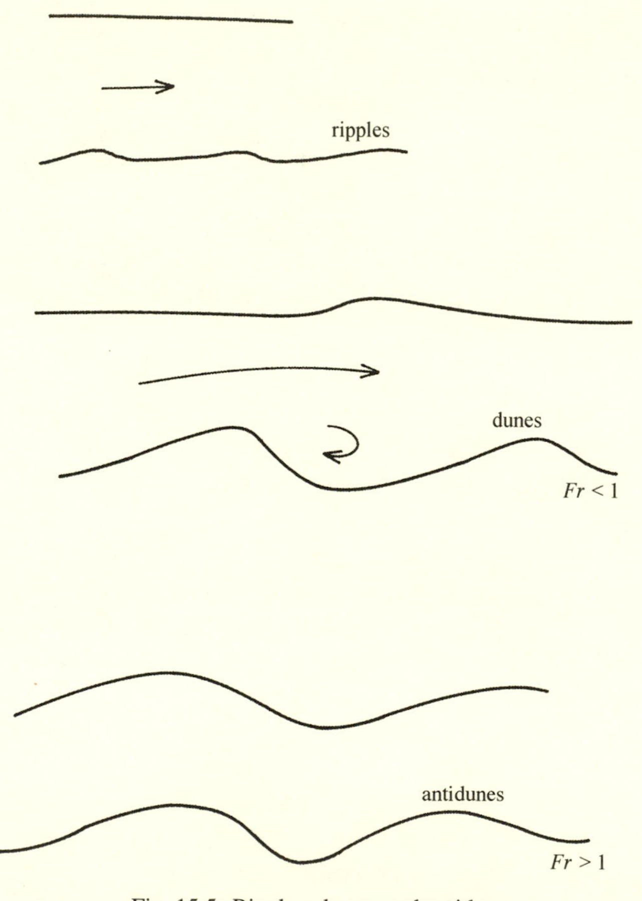

Fig. 15.5. Ripples, dunes, and antidunes

water surface. Dunes are also subject to three-dimensional instability, forming arcuate structures, which are familiar in desert environments (being formed by the same mechanism).

River meanders

River channels are themselves subject to instabilities; in particular, they commonly evolve a sinuous shape, known as a meander. The mechanism depends on the erosion of sediments from the banks: if a bend begins to develop, then the flow is slightly higher at the outside, and erosion will be greater there. Hence the bend develops further. Models of channel stability involve downstream and lateral sediment transport, both of which are strongly affected in the vicinity of banks by the presence of secondary flow induced by channel curvature. This secondary circulation aids in the transport of sediment scoured from banks to the high-velocity thread in the deeper part of the stream, and thus plays a role in the maintenance of cross-sectional shape. If the channel is too shallow (or too wide), however, the pattern of streamwise and transverse flow may split into multiple cells, resulting in the formation of midchannel bars. Under the right conditions, these bars can become exposed, leading to the development of what is known as a braided channel.

15.8 Drainage networks

The other morphological feature of river flow is that river channels typically form arterial networks. These are reticulate structures (see Fig. 15.6), whose characteristics are described by various measures such as drainage basin area A, length of the main-

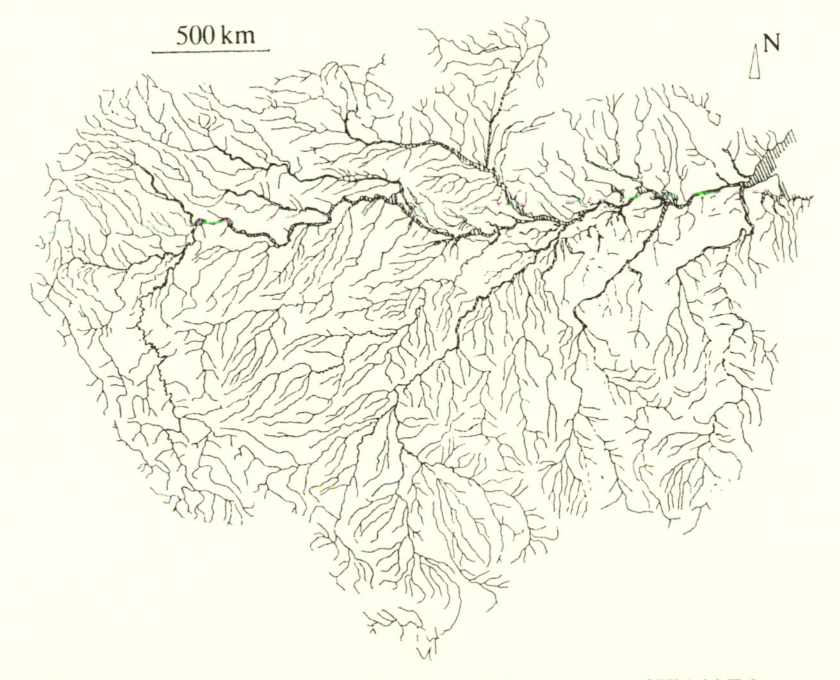

Fig. 15.6. The Amazon River Basin. Reproduced by courtesy of Hideki Takayusu

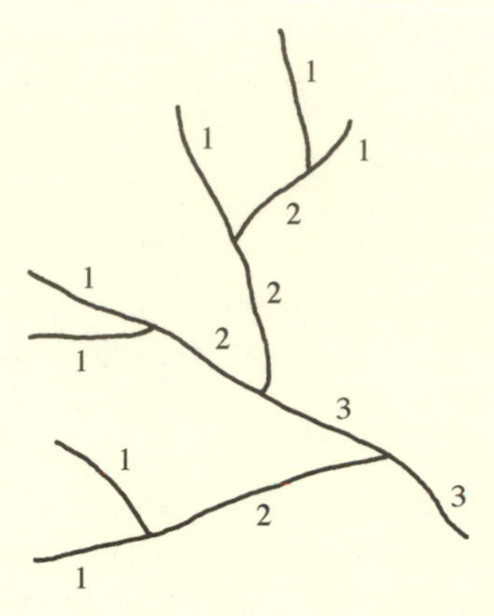

Fig. 15.7. The Strahler ordering system

stream L, channel density D_d, – equal to the total channel length divided by the basin area – and the Strahler order number, defined as follows: exterior links (i.e., headwater streams) have order one. If two Strahler streams of order Ω_1 and Ω_2 join, the resultant link has order $\max(\Omega_1, \Omega_2)$, unless $\Omega_1 = \Omega_2$, in which case the new order is $\Omega_1 + 1$. Figure 15.7 shows a typical example.

Various power law relationships exist between these quantities for a variety of river basins. For example, *Hack's law* is

$$L \approx 1.9A^{0.6}, \tag{15.66}$$

where L is measured in kilometers and A in square kilometers. Another relation is between the mean length P of rivers of a given order and the number N of such rivers. With P measured in kilometers, we have

$$N \approx 5.3 \times 10^6 P^{-1.83}. \tag{15.67}$$

Power laws like these are indicative of *fractal structure*, and indeed the networks themselves form approximately fractal curves. The Amazon river system shown in Fig. 15.6 has a fractal dimension of 1.85, for example.

Network formation

The problem of explaining and predicting the formation of drainage networks is one that has potential applications in other situations: the formation of underground cave systems, for example. The dominant processes are erosion of the substrate and the transport of the sediments. A conceptually simple model assumes that all water flow takes place surficially as overland flow. If the water depth is h and the ground topography is given by $z = H$ (where z points vertically upwards), then a plausible mathematical model, due to Kramer and Marder, is the following:

$$\frac{\partial H}{\partial t} = -\frac{P}{P_0}|\mathbf{v}|[1 + |\nabla H^2|]^{1/2} + \nu\nabla.[|\mathbf{v}|h\nabla H], \tag{15.68a}$$

$$\frac{\partial h}{\partial t} + \nabla.[h\mathbf{v}] = f_p. \tag{15.68b}$$

The second equation describes conservation of mass of water, where $\mathbf{v}$ is the water velocity and f_p is a source term owing to rainfall. The first equation describes erosion of the topography and incorporates two processes: erosion (and removal) of sediment, modeled as proportional to the product of flow velocity $\mathbf{v}$ and water pressure P; as we shall see, $P \sim h$, so this term represents the physically plausible dependence of erosion rate on increasing velocity and shear stress. The second term in (15.68a) represents transport of sediment, proportional to water flux and directed down the hillslope. Again, the functional dependence only aims to be qualitatively appropriate.

We require two constitutive relations for $\mathbf{v}$ and P. We choose them to satisfy Bernoulli's law and also require $\mathbf{v} \propto -\boldsymbol{\nabla} s$, where $s = H + h$. A specific prescription is

$$\mathbf{v} = -\frac{(2gh)^{1/2}\boldsymbol{\nabla} s}{[1+|\boldsymbol{\nabla} s|^2]^{1/2}}, \qquad P = \frac{\rho g h}{1+|\boldsymbol{\nabla} s|^2}, \tag{15.69}$$

which also has $|\mathbf{v}| \sim (2gh)^{1/2}$ for large $|\boldsymbol{\nabla} s|$, corresponding to free fall velocities.

Nondimensionalization

We choose scales

$$h \sim [h], \qquad H, s \sim [H], \qquad \mathbf{x} \sim l, \qquad t \sim [t], \qquad \mathbf{v} \sim [v], \qquad f_p = [f_p]\phi, \tag{15.70}$$

where $[H]$, l, $[f_p]$ are prescribed, and the rest are to be chosen. We choose

$$[t] = [h]/[f_p], \qquad [h][v] = l[f_p], \qquad [v] = \{2g[h]\}^{1/2}[H]/l, \tag{15.71}$$

which implies

$$[h] = \left\{\frac{l^2[f_p]}{(2g)^{1/2}[H]}\right\}^{2/3}, \tag{15.72}$$

and the other scales can be computed from Eq. (15.71). The model equations can now be written as

$$\begin{aligned}
\frac{\partial H}{\partial t} &= -\frac{\beta h^{3/2}|\boldsymbol{\nabla} s|[1+\delta^2|\boldsymbol{\nabla} H|^2]^{1/2}}{[1+\delta^2|\boldsymbol{\nabla} s|^2]^{3/2}} + \gamma\boldsymbol{\nabla}.\left\{\frac{h^{3/2}|\boldsymbol{\nabla} s|\boldsymbol{\nabla} H}{[1+\delta^2|\boldsymbol{\nabla} s|^2]^{1/2}}\right\},\\
\frac{\partial h}{\partial t} &= \boldsymbol{\nabla}.\left\{\frac{h^{3/2}\boldsymbol{\nabla} s}{[1+\delta^2|\boldsymbol{\nabla} s|^2]^{1/2}}\right\} + \phi,\\
s &= H + \kappa h,
\end{aligned} \tag{15.73}$$

wherein

$$\beta = \frac{\rho g[f_p][t]l}{P_0[H]}, \qquad \gamma = \frac{\nu[f_p][t]}{l}, \qquad \delta = \frac{[H]}{l}, \qquad \kappa = \frac{[h]}{[H]}. \tag{15.74}$$

To estimate the parameters, we choose values consistent with expectations; thus

$$[f_p] \sim 1 \text{ m y}^{-1}, \qquad l \sim 10^3 \text{ m}, \qquad [H] \sim 20 \text{ m}, \tag{15.75}$$

corresponding to typical rainfall and hillslope topography. We then find $[h] \sim 5$ mm, which thus represents a typical film thickness for distributed overland flow. The other parameters are (choosing $\nu = 0.1$ and $P_0 = 3 \times 10^{10}$ kg m^{-1} s^{-2}, corresponding to a basic erosion rate of 1 mm y^{-1})

$$\beta \sim 2 \times 10^{-7}, \qquad \gamma \sim 10^{-6}, \qquad \delta \sim 2 \times 10^{-2}, \qquad \kappa \sim 5 \times 10^{-4}, \tag{15.76}$$

all of them small, so that we can make an asymptotic reduction of Eq. (15.73) to

$$\frac{\partial H}{\partial t} = -\beta h^{3/2}|\nabla s| + \gamma \nabla . [h^{3/2}|\nabla s|\nabla H], \tag{15.77a}$$

$$\frac{\partial h}{\partial t} = \nabla . [h^{3/2}\nabla s] + \phi, \tag{15.77b}$$

$$s = H + \kappa h, \tag{15.77c}$$

where the terms in κ, β, and γ are retained in view of possible singular perturbations.

Channel formation

A general solution, even a numerical one, is not possible for Eq. (15.77), because if a channelized network does form, one would expect it to consist of thin regions (streams) where $h \gg 1$ and the lateral length scale is very small. In effect, we might expect streams to be like shocks for the slowly evolving topography. Here we simply describe a simpler problem, involving the generation of channels from an initially uniform overland flow.

We suppose that $\phi = 0$ and that a prescribed uniform overland flow descends a uniform hillslope in the downstream (y) direction. Specifically, we take this uniform state to be

$$s = -\beta t - y, \qquad h = 1, \tag{15.78}$$

and we consider perturbed solutions in which $h = h(x, t)$, but still $s = -\beta t - y$; thus $H = -y - \beta t - \kappa h(x, t)$ (here, x is a cross-stream variable). Thus the water surface is level, and $|\nabla s| = 1$, so that Eq. (15.77a) becomes, with $\phi = 0$,

$$-\beta - \kappa \frac{\partial h}{\partial t} = -\beta h^{3/2} + \gamma \nabla . [h^{3/2}(-\kappa h_x, -1)]; \tag{15.79}$$

we rescale x and t as

$$t = \frac{\kappa}{\beta}\tau, \qquad x = \left(\frac{\gamma\kappa}{\beta}\right)^{1/2} X, \tag{15.80}$$

to obtain

$$h_\tau = h^{3/2} - 1 + \frac{\partial}{\partial X}[h^{3/2}h_X]. \tag{15.81}$$

Note that with the values indicated in Eq. (15.76), this gives $x \sim 0.05$, $t \sim 2.5 \times 10^3$, corresponding to a lateral length scale of 50 m and a timescale of 12 years. This length scale is within reach of a channel width scale (particularly given uncertainties in the governing scales).

The assumption $h = h(x, t)$, $\boldsymbol{\nabla} s = (0, -1)$ is strictly incompatible with Eq. (15.77b), but we see *a posteriori* that the rescaling induces a scaled form of the equation,

$$\boldsymbol{\nabla}.[h^{3/2}\boldsymbol{\nabla} s] = \gamma h_\tau - (\gamma\kappa/\beta)\phi; \tag{15.82}$$

because $\gamma \ll 1$, $\gamma\kappa/\beta \ll 1$, we see that the assumption $h = h(x, t)$, $\boldsymbol{\nabla} s = (0, -1)$ is in fact a consistent approximation to describe channel development.

If we perturb the steady solution $h = 1$ of Eq. (15.81) by putting $h = 1 + u$, then for small perturbations u, we have

$$u_\tau = \frac{3}{2}u + u_{xx}, \tag{15.83}$$

and solutions of the form $u = \exp[\sigma\tau + ikX]$ exist if $\sigma = (3/2) - k^2$, so that the steady solution is always unstable to the formation of channels. Depending on the precise boundary conditions applied to Eq. (15.81), nonlinear equilibration of Eq. (15.81) may take place. For example (see Exercise 5), if $h \to 0$ as $X \to \pm\infty$, then a steady finite channel exists. This shows that this model is capable of initiating channels, although it is as yet unclear whether it leads to the formation of a fully arterial network.

15.9 Notes and references

Books on hydrology tend to be geographic in nature, describing the processes important in the hydrological cycle. For example, Chorley (1969) or Ward and Robinson (1990) are useful introductions. Books on hydraulics, on the other hand, concentrate on the fluid dynamics of the river flow itself. An example is the book by French (1994); an older classic is that by Chow (1959). A nice book that bridges the gap, and also includes discussion of sediment transport and channel morphology and pattern, is that by Richards (1982). A more detailed account of sediment transport is given by Allen (1985). Flood waves and roll waves have been discussed from the present perspective by Whitham (1974).

Sediment transport The critical Shields stress in Eq. (15.60) was introduced by Shields (1936). Its main problem in applicability is when (as is normal) there is a distribution of grain sizes, and because particles may vary in size from microns (clay) to millimeters (gravel), this is clearly a drawback. Sometimes it is proposed that one use a suitable median grain size: rather a subjective approach, however (e.g., Parker, 1978). Various formulae for bedload transport have been proposed: that due to Meyer–Peter and Müller (1948) has the advantage of simplicity. The erosion rate (15.62) was proposed by Engelund (1970) and the deposition rate (15.63) by Parker (1978).

Bedforms Stability analyses to explain the formation of dunes and anti-dunes have been carried out, relatively successfully, by Kennedy (1963) and Engelund (1970).

The basic model consists of an eddy–viscosity equation for the turbulent mean vorticity, together with an eddy–diffusive suspended sediment transport equation. An appropriate kinematic condition is applied at the bed, and bedload transport is included in this.

Meanders The stability theory of river meanders is also relatively healthy. Ikeda and Parker (1989) present a collection of papers on various aspects of meandering. The basic model here consists of momentum and sediment balance equations, but where balances transverse to the flow are also needed.

Drainage networks The fractal features of river networks are discussed (briefly) by Turcotte (1992) and Takayasu (1990). The mechanistic model of Eq. (15.68) is that of Kramer and Marder (1992). Despite its relative simplicity, the small values of β, γ, and κ in Eq. (15.77) render the system impossible to solve numerically, and future study of this and other models will aim to analyze the equations using singular perturbation techniques.

Another class of mechanistic model is that introduced by Willgoose et al. (1991a, b). The model has a hillslope erosion/transport equation similar to Eq. (15.77a), but the 'channel' equation is replaced by an artificial equation for a channel indicator function Y (so $Y=0$ indicates hillslope, $Y=1$ indicates channel). Although the numerical results do simulate network evolution, this artificiality is a serious drawback to the model. Other models are given by Howard, Dietrich, and Seidl (1994), Tucker and Slingerland (1994), and Kooi and Beaumont (1994).

The study of channel-forming instability may have been initiated by Smith and Bretherton (1972). The model is similar to the Kramer–Marder model but ignores erosion.

More computer-based simulations, such as cellular automata and their variants, have been studied by Masek and Turcotte (1993), Stark (1991), and others.

Exercises

1. Show that for a semicircular channel, $\bar{h}=\pi R/2$, $R=(A/2\pi)^{1/2}$, and derive the corresponding forms of the St. Venant equations. How is the wave theory of Section 15.5 affected in this case?
2. Show that if the bed slope is α, then the surface slope β is approximately given (for small α, β) by $\beta=\alpha-\partial h/\partial s$. Hence show, if $\partial h/\partial s \ll \alpha$, that the resulting modification to the Chezy or Manning model (replacing α by β) acts diffusively in smoothing out shocks in the flood hydrograph. Give an approximate analytic description of the unit hydrograph (cf. Section 15.4) corresponding to a delta function rainfall spike.
3. By first defining $H=A^{1/2}$, show that the St. Venant equations (15.52) can be written in the form

$$\left[\frac{\partial}{\partial t}+\left(u\pm\frac{H}{F}\right)\frac{\partial}{\partial s}\right]\left[u\pm\frac{2H}{F}\right]=1-\frac{u^2}{H^2}.$$

By using a characteristic diagram, describe the evolution of an initial profile given by

$$H = u = H_+, \qquad s > 0,$$
$$H = u = H_-, \qquad s < 0,$$

distinguishing between the cases $F > 1$ and $F < 1$.

4. *The hydraulic jump*

Using the dimensionless form of the mass and momentum equations, show that discontinuities (shocks) in the channel depth travel at a (dimensionless) speed V given by

$$V = \frac{[Au]_-^+}{[A]_-^+} = \frac{\left[F^2 Au^2 + \frac{1}{2}A^2\right]_-^+}{[Au]_-^+},$$

where $\pm$ refer to the values on either side of the jump. Show that a stationary jump at $s = 0$ is possible (this can be seen when a tap is run into a basin) if $Au = Q$ in $s > 0$ and $s < 0$, and

$$\left[\frac{F^2 Q^2}{A} + \frac{A^2}{2}\right]_-^+ = 0.$$

Deduce that for prescribed Q and A_-, a unique choice of $A_+ \neq A_-$ is possible. Show also that the locally defined Froude number is

$$Fr = \frac{FQ}{A^{3/2}},$$

and deduce that the hydraulic jump connects a region of *supercritical* ($Fr > 1$) flow to a *subcritical* ($Fr < 1$) one. (In practice, $A_- < A_+$ if $Q > 0$; if $A_- > A_+$, the discontinuity cannot be maintained.)

5. Show that if $h \to 0$ as $X \to \infty$ in Eq. (15.81), then steady-state solutions exist, with $\max_X h = h_0 = (8/5)^{2/3}$. Show that the channel width is

$$\frac{2^{7/2}5^{-1/3}}{3} B\left(\frac{5}{6}, \frac{1}{2}\right),$$

where $B(a, b)$ is the Euler beta function, $= \int_0^1 v^{a-1}(1-v)^{b-1}\, dv$.

6. Show that if the erosion rate is taken to be time dependent in Eq. (15.78), so that βt is replaced by r, then the appropriate generalization of Eq. (15.81) is

$$h_\tau = -\frac{1}{\beta}\frac{dr}{d\tau} + h^{3/2} + \frac{\partial}{\partial X}\left[h^{3/2}\frac{\partial h}{\partial X}\right].$$

Show that if r is chosen to conserve water $\int_{-L}^{L} h\, dX$ in the finite domain $(-L, L)$, then

$$\frac{dr}{d\tau} = \frac{\beta}{2L}\int_{-L}^{L} h^{3/2}\, dX,$$

whereas if the water flux $\int_{-L}^{L} h^{3/2}\, dX$ is conserved, then

$$\frac{dr}{d\tau} = \beta\left[\frac{\int h^2\, dX - \frac{1}{2}\int hh_X^2\, dX}{\int h^{1/2}\, dX}\right].$$

7. A river flows through a lowland valley. The river level may fluctuate, so that it lies above or below the local groundwater level. Give a *simple* motivation for the model

$$\frac{\partial A}{\partial t} + cA^m \frac{\partial A}{\partial s} = -r(A - B),$$
$$\frac{\partial B}{\partial t} = r(A - B),$$

to describe the variations of river water (A) and groundwater (B), where B is a measure of the amount of groundwater.

Show that small disturbances to the uniform state $A = B = 1$ exist proportional to $\exp[\sigma t + iks]$, and find the dispersion relation relating σ to k. What do these solutions represent?

16

One-dimensional two-phase flow

16.1 Introduction

Two-phase flow occurs in numerous situations in industry, as well as in nature. Two-phase flow refers to the coexistence of two phases of a substance in a flow. Typically (though not always) the phases are liquid and gas, as for example in the common occurrence of steam-water flows.

In boilers, water is heated as it flows through a bank of channels until it starts to boil. This leads to a two-phase flow region until *dryout* occurs, and the flow is of superheated steam. The boundaries that physically divide the various regimes are called the *boiling boundary* and the *superheat boundary*.

A similar situation occurs in nuclear reactors where liquid sodium is commonly used as a coolant. Here it is important that dryout does not occur, because the insulating properties of vapor reduce the cooling efficiency of the flow. Two-phase flow also occurs in condensers, where superheated steam is cooled through the reverse sequence of two-phase and then sub-cooled regions.

Natural examples of two-phase flows include volcanic eruptions, where a variety of such flows can occur, for example, ash flows (solid/liquid) and vesicular eruptions, where dissolved gases are exsolved as the magma rises (and loses pressure), so that the erupting flow is of a gas/liquid mixture.

16.2 Flow regimes

Modeling two-phase flow is complicated by a variety of factors. For a start, the flow is usually turbulent, so that some sort of averaging is necessary to model the mean flow. In addition, the distribution of phases means that averaging must also be done so that average variables such as void fraction can be defined. (This is analogous to the definition of variables such as porosity in permeable media.)

A further complication is that two-phase flows can exist in a variety of regimes, all of which will generally occur in a boiling flow. When boiling commences, small bubbles are nucleated at the wall, detach, and are taken up by the fluid. Initially, the liquid away from the walls may still be *subcooled* (below boiling point), so that heat transfer to the vapor is predominantly at the wall. When the liquid reaches saturation (and in fact becomes slightly superheated), then this regime of *bubbly flow* evolves, by virtue of bubble coalescence and evaporation at bubble interfaces, to

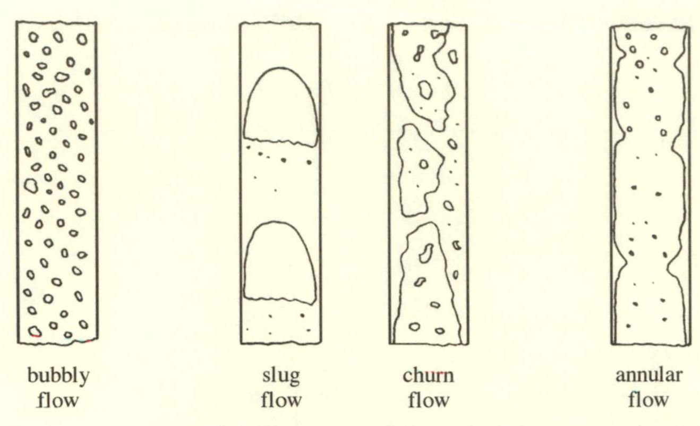

Fig. 16.1. Flow patterns in vertical flow

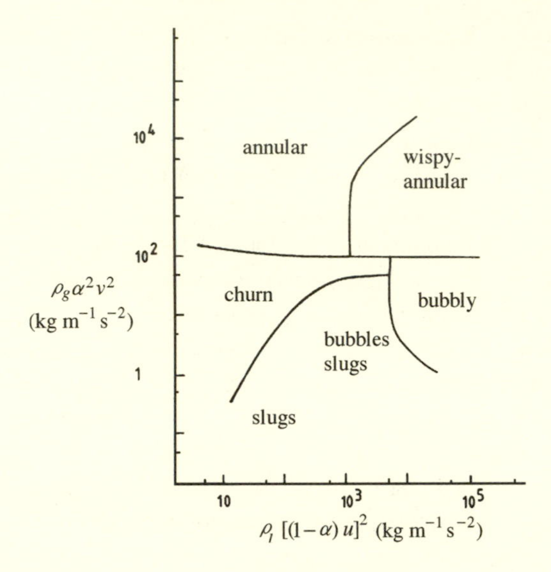

Fig. 16.2. Flow regime map. Redrawn from Hewitt and Roberts (1969).

the regime known as *slug flow*, in which plugs of gas filling the tube alternate with slugs of bubbly fluid. As the evaporation proceeds, the gas plugs become irregular and one gets *churn flow*, which leads finally to *annular flow*, in which the liquid is confined to a film at the tube wall and the gas flows in the core. Shearing between the gas and the liquid causes droplets to be eroded and entrained in the gas. The sequence of flows is portrayed in Fig. 16.1. Various experimentally based laws to determine parametric criteria for the type of regime a particular flow will adopt lead to the construction of *flow regime maps*, an example of which is shown in Fig. 16.2.

16.3 A simple two-fluid model

Two-phase flow equations are averaged in various ways: in time, cross sectionally, and in space. For one-dimensional flow in a tube, we seek relations for cross-sectionally and time-averaged variables representing the two fluids. These variables are the void fraction α, which is the gas volume fraction; u and v, which are the liquid and gas velocities; and averaged pressures p_l and p_g for each phase. We concentrate on steam-water or air-water flow, for example, for which the viscous stresses are manifested through the wall friction, whereas the internal friction is largely due to *Reynolds stresses*; both of these terms must be constituted. In writing the simplest model (in order to examine its structure), we will in fact omit frictional terms for the moment.

Ignoring surface tension, it seems reasonable to take $p_g = p_l = p$, and then equations conserving mass and momentum of each phase (and without change of phase due to boiling or condensation) are

$$\begin{aligned}(\alpha\rho_g)_t + (\alpha\rho_g v)_z &= 0,\\ \{\rho_l(1-\alpha)\}_t + \{\rho_l(1-\alpha)u\}_z &= 0,\\ \rho_g[v_t + vv_z] &= -p_z,\\ \rho_l[u_t + uu_z] &= -p_z.\end{aligned} \tag{16.1}$$

These equations can be derived from first principles in the usual way. They represent four equations for the variables α, u, v, and p, if we suppose that ρ_g and ρ_l, the gas and liquid densities, are given by appropriate equations of state. These are simple generalizations of Euler's equations to the case of two-fluid motion.

For heated flows, one requires also two energy or (more usually) enthalpy equations for the gas and liquid enthalpies h_g and h_l. For *adiabatic* (unheated) flow, these equations are redundant.

Boundary conditions

From a physical consideration of the system, it seems we could prescribe inlet and outlet pressures and the two inlet mass fluxes. The natural boundary conditions for the equations are those of α, u, and v at the inlet and p at the outlet (say). Thus if we solve the system for given α_0 at $z = 0$, as well as u and v, then we will obtain the pressure drop Δp as a functional of α_0, $\Delta p = \Delta p(\alpha_0)$. Inversion of this relation determines the necessary α_0 to obtain the correct pressure drop.

16.4 Other models

There are two other commonly used two-phase flow models. In the homogeneous model, both phases are assumed to move at the same velocity; thus $u = v$, and one considers momentum conservation for the mixture, in the (simplest) form

$$\rho[u_t + uu_z] = -p_z, \tag{16.2}$$

where

$$\rho = \alpha\rho_g + (1-\alpha)\rho_l \tag{16.3}$$

is the mixture density. The homogeneous model should work best for regimes such as bubbly flow, where there may be little relative motion between the phases.

The *drift-flux model* allows a relative motion, but rather than have separate momentum equations, it considers total momentum conservation in the form

$$[\alpha\rho_g v+(1-\alpha)\rho_l u]_t+[\alpha\rho_g v^2+(1-\alpha)\rho_l u^2]_z=-p_z, \tag{16.4}$$

with v being related to u through the drift flux (basically, $v-u$ is constituted as a function of α). This is rather akin to the status of Darcy's law. We will pose a drift-flux model for a device called a thermosyphon later on.

16.5 Characteristics

The form of Eqs. (16.1) suggests that the system should be hyperbolic. It can be written in the form

$$A\psi_t+B\psi_z=0, \tag{16.5}$$

where $\psi=(\alpha,u,v,p)^T$ and, if ρ_g and ρ_l are constant,

$$A=\begin{pmatrix}1&0&0&0\\-1&0&0&0\\0&\rho_l&0&0\\0&0&\rho_g&0\end{pmatrix},\qquad B=\begin{pmatrix}v&0&\alpha&0\\-u&1-\alpha&0&0\\0&\rho_l u&0&1\\0&0&\rho_g v&1\end{pmatrix}. \tag{16.6}$$

We seek characteristics $dz/dt=\lambda$ satisfying $\det(\lambda A-B)=0$; we find that the eigenvalues λ must satisfy

$$\rho_g(1-\alpha)(\lambda-v)^2+\rho_l\alpha(\lambda-u)^2=0, \tag{16.7}$$

whence

$$\lambda=\frac{u\pm isv}{1\pm is},\qquad s=\left[\frac{\rho_g(1-\alpha)}{\rho_l\alpha}\right]^{1/2}. \tag{16.8}$$

It follows that there are two *complex* characteristics unless $u=v$, (the other two are infinite, corresponding to two infinite sound speeds). Consequently, the model is *ill-posed* as it stands. There is thus a fundamental logical inconsistency, and before we go on to consider more complicated models, it is worth pursuing this further. Notice that the ellipticity of the model is not due to the neglect of frictional terms: these are algebraic, and do not affect the characteristics.

To see the practical effect of complex characteristics, consider a uniform state $\psi=\psi_0$, subject to small perturbations proportional to $\exp(\sigma t+ikz)$. Such solutions exist if $\sigma=ik\lambda=\mp k\lambda_I+ik\lambda_R$, where $\lambda=\lambda_R\pm i\lambda_I$ represents Eq. (16.8). Thus if $\lambda_I\neq 0$, there are unstable solutions; moreover, these grow arbitrarily fast at very short wavelengths. These grid scale instabilities are a practical sign of an ill-posed problem.

16.6 More on averaging

In order to resolve this dilemma, let us examine the process of averaging in greater detail. There are many different ways of approaching averaging, and here we follow that outlined by Drew and Wood (1985) or Drew (1983). The idea is to use indicator functions X_k for each phase (labeled by k) such that $X_k(\mathbf{x}, t) = 1$ if $\mathbf{x}$ is in phase k at time t, and $X_k = 0$ otherwise. Averaged equations can then be obtained by multiplying the pointwise conservation laws for phase k by X_k and averaging. In so doing, we consider X_k as a generalized function, which allows us to use integration by parts.

A typical conservation law has the form

$$\frac{\partial}{\partial t}(\rho\psi) + \nabla.(\rho\psi\mathbf{v}) = -\nabla.\mathbf{J} + \rho f. \tag{16.9}$$

Multiplying by X_k and averaging (in practice, this is often a time and/or space average) yields (an overbar denoting the average)

$$\frac{\partial}{\partial t}(\overline{X_k\rho\psi}) + \nabla.[\overline{X_k\rho\psi\mathbf{v}}] = -\nabla.[\overline{X_k\mathbf{J}}] + \overline{X_k\rho f} + \overline{\rho\psi\left\{\frac{\partial X_k}{\partial t} + \mathbf{v}_i.\nabla X_k\right\}} \\ + \overline{\{\rho\psi(\mathbf{v} - \mathbf{v}_i) + \mathbf{J}\}.\nabla X_k}, \tag{16.10}$$

where we assume that $\overline{\nabla f} = \nabla\bar{f}$, $\overline{\partial f/\partial t} = \partial\bar{f}/\partial t$, which will be the case for sufficiently well-behaved f. In Eq. (16.10), $\mathbf{v}_i$ is the average interfacial velocity of the boundary of phase k, and derivatives of X_k are interpreted as generalized functions. In particular, $\partial X_k/\partial t + \mathbf{v}_i.\nabla X_k = 0$, because if ϕ is a smooth test function, then

$$\begin{aligned} &\int\!\!\int \phi\left[\frac{\partial X_k}{\partial t} + \mathbf{v}_i.\nabla X_k\right] dV\,dt \\ &\quad = -\int\!\!\int X_k\left[\frac{\partial\phi}{\partial t} + \nabla.(\phi\mathbf{v}_i)\right] dV\,dt = -\int_{-\infty}^{\infty}\int_{V_k(t)}\left[\frac{\partial\phi}{\partial t} + \nabla.(\phi\mathbf{v}_i)\right] dV\,dt \\ &\quad = -\int_{-\infty}^{\infty}\frac{d}{dt}\int_{V_k(t)}\phi\,dV\,dt = -\left[\int_{V_k(t)}\phi\,dV\right]_{-\infty}^{\infty} = 0, \end{aligned} \tag{16.11}$$

for ϕ vanishing at large values of $|t|$.

The last term in Eq. (16.10) is related to the surface average, because ∇X_k picks out interfacial values. For a smooth test function vanishing at large $\mathbf{x}$, $\mathbf{j}.\nabla X_k$ is defined via

$$\begin{aligned} \int_V \phi\mathbf{j}.\nabla X_k\,dV &= -\int_V X_k\nabla.(\phi\mathbf{j})\,dV \\ &= -\int_{V_k}\nabla.(\phi\mathbf{j})\,dV \\ &= -\int_{S_k}\phi j_n\,dS, \end{aligned} \tag{16.12}$$

where j_n is the normal component of $\mathbf{j}$ at the interface, pointing *away* from phase k. This suggests that $\overline{\mathbf{j}.\nabla X_k}$ can be identified with the *surface* average of $-\mathbf{j}.\mathbf{n}$.

Now put $\psi = 1$, $\mathbf{J} = \mathbf{0}$, and $f = 0$ in Eq. (16.9), corresponding to mass conservation. Then equations of conservation of mass of each phase are, from Eq. (16.10),

$$\frac{\partial}{\partial t}(\overline{X_k\rho}) + \nabla.[\overline{X_k\rho\mathbf{v}}] = \overline{\rho(\mathbf{v} - \mathbf{v}_i).\nabla X_k}. \tag{16.13}$$

The form of Eq. (16.13) suggests that we define the average phase volume, density, and velocity as follows:

$$\alpha_k = \overline{X_k}, \qquad \rho_k = \overline{X_k \rho}/\alpha_k, \qquad \mathbf{v}_k = \overline{X_k \rho \mathbf{v}}/\alpha_k \rho_k, \tag{16.14}$$

so that Eq. (16.13) gives

$$\frac{\partial}{\partial t}(\alpha_k \rho_k) + \boldsymbol{\nabla}.[\alpha_k \rho_k \mathbf{v}_k] = \Gamma_k, \tag{16.15}$$

where $\Gamma_k = \overline{\rho(\mathbf{v} - \mathbf{v}_i).\boldsymbol{\nabla} X_k}$ and represents a mass source due to phase change (without which $\mathbf{v} = \mathbf{v}_i$ at the interface).

Next, consider momentum conservation. With appropriate interpretation of tensor notation, we put

$$\psi = \mathbf{v}, \qquad \mathbf{J} = p\mathbf{I} - \boldsymbol{\tau}, \qquad f = \mathbf{g}, \tag{16.16}$$

where $\boldsymbol{\tau}$ is the deviatoric stress tensor, and $\mathbf{g}$ is gravity. Then

$$\begin{aligned} \frac{\partial}{\partial t}(\overline{X_k \rho \mathbf{v}}) + \boldsymbol{\nabla}.[\overline{X_k \rho \mathbf{v}\mathbf{v}}] = \boldsymbol{\nabla}.[\overline{X_k(-p\mathbf{I} + \boldsymbol{\tau})}] + \overline{X_k \rho \mathbf{g}} \\ + \overline{\{\rho \mathbf{v}(\mathbf{v} - \mathbf{v}_i) + (p\mathbf{I} - \boldsymbol{\tau})\}.\boldsymbol{\nabla} X_k}. \end{aligned} \tag{16.17}$$

Now $\overline{X_k \rho \mathbf{v}} = \alpha_k \rho_k \mathbf{v}_k$, and we would like to have $\overline{X_k \rho \mathbf{v}\mathbf{v}} = \alpha_k \rho_k \mathbf{v}_k \mathbf{v}_k$; but evidently, the latter is not the case. Because the flow is normally turbulent, this can be circumvented by separating $\mathbf{v}$ (and, more generally ψ) into mean and fluctuating parts; thus $\mathbf{v} = \mathbf{v}_k + \mathbf{v}'_k$, so that

$$\overline{X_k \rho \mathbf{v}\mathbf{v}} = \alpha_k \rho_k \mathbf{v}_k \mathbf{v}_k + \overline{X_k \rho \mathbf{v}'_k \mathbf{v}'_k}. \tag{16.18}$$

The second term can be interpreted as the averaged Reynolds stress. The momentum equation can thus be written as

$$\frac{\partial}{\partial t}(\alpha_k \rho_k \mathbf{v}_k) + \boldsymbol{\nabla}.[\alpha_k \rho_k \mathbf{v}_k \mathbf{v}_k] = \boldsymbol{\nabla}.\left[\alpha_k\left(\mathbf{T}_k + \mathbf{T}'_k\right)\right] + \alpha_k \rho_k \mathbf{g} + \mathbf{M}_k + \mathbf{v}^m_{ki} \Gamma_k, \tag{16.19}$$

where

$$\begin{aligned} \alpha_k \mathbf{T}_k &= \overline{X_k(-p\mathbf{I} + \boldsymbol{\tau})}, \\ \alpha_k \mathbf{T}'_k &= \overline{X_k \rho \mathbf{v}'_k \mathbf{v}'_k}, \\ \mathbf{M}_k &= \overline{(p\mathbf{I} - \boldsymbol{\tau}).\boldsymbol{\nabla} X_k}, \\ \mathbf{v}^m_{ki} &= [\overline{\rho \mathbf{v}(\mathbf{v} - \mathbf{v}_i).\boldsymbol{\nabla} X_k}]/[\overline{\rho(\mathbf{v} - \mathbf{v}_i).\boldsymbol{\nabla} X_k}]. \end{aligned} \tag{16.20}$$

Evidently, the average pressure in phase k is $p_k = \overline{X_k p}/\alpha_k$. If we neglect viscous stresses, we can write the interfacial momentum source as

$$\mathbf{M}_k = \overline{p \boldsymbol{\nabla} X_k} = p_{ki} \boldsymbol{\nabla} \alpha_k + \mathbf{M}'_k, \tag{16.21}$$

where

$$\mathbf{M}'_k = \overline{(p - p_{ki}) \boldsymbol{\nabla} X_k}, \tag{16.22}$$

p_{ki} is the average interfacial pressure on phase k, and we use $\overline{\nabla X_k} = \nabla \alpha_k$. Thus the momentum equation can be written as

$$\frac{\partial}{\partial t}(\alpha_k \rho_k \mathbf{v}_k) + \nabla.[\alpha_k \rho_k \mathbf{v}_k \mathbf{v}_k] = -\alpha_k \nabla p_k - (p_k - p_{ki})\nabla \alpha_k + \nabla.\left[\alpha_k \mathbf{T}_k'\right] + \alpha_k \rho_k \mathbf{g} + \mathbf{M}_k' + v_{ki}^m \Gamma_k. \tag{16.23}$$

Commonly, we assume $p_k = p_{ki}$ and the pressure term is explicitly $-\alpha_k \nabla p_k$, an important point to note.

One-dimension: profile coefficients

In one-dimensional turbulent flow, it is common to constitute Reynolds stresses via an eddy viscosity, or equivalently, to define (cf. Eq. (16.18))

$$\overline{X_k \rho \mathbf{v}\mathbf{v}} = D_k \alpha_k \rho_k \mathbf{v}\mathbf{v}, \tag{16.24}$$

where D_k is known as a *profile coefficient*. This coefficient includes the effects of both the Reynolds stresses and also the cross-sectional nonuniformity of the flow. Let us consider the simplest modification to Eq. (16.1) that allows $D_k \neq 1$. A one-dimensional version of Eq. (16.1) is, with ρ_g and ρ_l constant,

$$\alpha_t + (\alpha v)_z = 0, \tag{16.25a}$$

$$-\alpha_t + [(1-\alpha)u]_z = 0, \tag{16.25b}$$

$$\rho_g(\alpha v)_t + \rho_g(D_g \alpha v^2)_z = -\alpha p_z, \tag{16.25c}$$

$$\rho_l[(1-\alpha)u]_t + \rho_l[D_l(1-\alpha)u^2]_z = -(1-\alpha)p_z. \tag{16.25d}$$

It will usually be appropriate to choose $D_g = 1$, but we allow $D_l \neq 1$. Then Eqs. (16.25c) and (16.25d) can be written as

$$\begin{aligned} \rho_g[v_t + v v_z] &= -p_z, \\ \rho_l\left[u_t + (2D_l - 1)u u_z - (D_l - 1)\left[\frac{u^2}{1-\alpha}\right]\alpha_z\right] &= -p_z, \end{aligned} \tag{16.26}$$

and the system can be written as Eq. (16.5) for $\psi = (\alpha, u, v, p)^T$, with

$$A = \begin{pmatrix} 1 & 0 & 0 & 0 \\ -1 & 0 & 0 & 0 \\ 0 & \rho_l & 0 & 0 \\ 0 & 0 & \rho_g & 0 \end{pmatrix}, \quad B = \begin{pmatrix} v & 0 & \alpha & 0 \\ -u & (1-\alpha) & 0 & 0 \\ -\frac{\rho_l(D_l-1)u^2}{(1-\alpha)} & \rho_l(2D_l-1)u & 0 & 1 \\ 0 & 0 & \rho_g v & 1 \end{pmatrix}. \tag{16.27}$$

The characteristics $dz/dt = \lambda$ satisfy $\det(\lambda A - B) = 0$; hence with s defined in Eq. (16.8),

$$(\lambda - u)^2 = \delta[u^2 + 2u(\lambda - u)] - s^2(\lambda - v)^2, \tag{16.28}$$

where $\delta = D_l - 1$. For small values of s and δ, we see that λ is real, provided

$$\delta > s^2(u - v)^2/u^2, \tag{16.29}$$

so that, practically, a very small Reynolds stress (or profile coefficient above one) is sufficient to make the basic system have real characteristics and hence be well-posed. Other possibilities to render the system well-posed can be chosen. In practice, any realistic model will (and *should*) be well-posed, otherwise it will be physically meaningless.

16.7 A simple model for annular flow

In many two-phase boiling flows, annular flows are significant. Because the gas flow is less impeded by the liquid, the flow velocities are high; and consequently, annular flow regions can occupy large parts of the tube. Moreover, because the gas and liquid velocities are very different, a two-fluid model is appropriate. If we denote α, β as gas and liquid volume fractions, u, v as liquid and gas velocities, and h_g and h_l as gas and liquid enthalpies, then a typical one-dimensional set of model equations is given by

$$\begin{aligned}
(\alpha\rho_g)_t + (\alpha\rho_g v)_z &= \Gamma, \\
(\beta\rho_l)_t + (\beta\rho_l u)_z &= -\Gamma, \\
(\beta\rho_l u)_t + (D\beta\rho_l u^2)_z &= -\beta p_z - F_{lw} + F_{li}, \\
(\alpha\rho_g v)_t + (\alpha\rho_g v^2)_z &= -\alpha p_z + F_{gi}, \\
\beta\rho_l[h_{lt} + u h_{lz}] &= \Gamma(h_l - h_{li}) + E_l + Q/A, \\
\alpha\rho_g[h_{gt} + v h_{gz}] &= \Gamma(h_{gi} - h_g) + E_g, \\
E_l + E_g + \Gamma(h_{gi} - h_{li}) &= 0.
\end{aligned} \tag{16.30}$$

The first two of these represent conservation of mass, as before. The next two represent conservation of momentum, and we have chosen to include the wall friction F_{lw} on the liquid and the interfacial friction F_{li} on the liquid (and F_{gi} on the gas, $F_{gi} = -F_{li}$). The next two are enthalpy equations. In an annular flow, boiling takes place at the liquid–gas interface, so that the liquid in particular must be superheated. If the average phasic enthalpy h_k is different from the interfacial value h_{ki}, then there is a convective transfer of enthalpy $\Gamma(h_{ki} - h_k)$ to that phase associated with the phase change term. In addition, there will be a diffusive transport E_k due to heat conduction. Finally, Q is the external heat supply per unit length per unit time, and A is the cross-sectional area. The final relation then represents the volume average of the Stefan condition. There are various other terms that could be included, but they can be treated in the same way as below and are in any case often small.

Nondimensionalization

Our aim is to derive appropriate scaling relationships for the variables and to show how suitable simplifications can often be made. To do this, we use values of the parameters typical to one particular application, that of a steam turbine.

Firstly, we must choose constitutive forms for the various terms that arise through the averaging process. We suppose

$$F_{lw} = \frac{2}{d} f_{lw} \rho_l |u| u, \tag{16.31a}$$

$$F_{li} = -F_{gi} = \frac{2}{d} f_{li} \rho_g |v - \chi u| (v - \chi u) \tag{16.31b}$$

are the friction terms. The numbers f_{lw}, f_{li} are friction factors and are themselves usually considered to be functions of the phasic Reynolds numbers. Here we take them to be constant. The coefficient χ in Eq. (16.31b) represents the speed of interfacial waves; a typical value is $\chi = 2$.

We will assume that the interface is in thermodynamic equilibrium; thus

$$h_{gi} = h_g^{sat}, \qquad h_{li} = h_l^{sat}, \tag{16.32}$$

where these are the saturation values of the enthalpies, and $L = h_g^{sat} - h_l^{sat}$ is the latent heat. Enthalpies are related to temperature by

$$\begin{aligned} h_g &= h_g^{sat} + c_{pg}(T_g - T^{sat}), \\ h_l &= h_l^{sat} + c_{pl}(T_l - T^{sat}), \end{aligned} \tag{16.33}$$

where T^{sat} is the saturation (boiling) temperature, which we take as constant. In terms of temperature, the interfacial heat transfer terms are

$$\begin{aligned} E_g &= H_{gi}(T_{gi} - T_g)/L_s, \\ E_l &= H_{li}(T_{li} - T_l)/L_s, \end{aligned} \tag{16.34}$$

where H_{li} and H_{gi} are heat transfer coefficients, $T_{gi} = T_{li} = T^{sat}$ here, and L_s^{-1} is the (average) surface area per unit volume, which for annular flow we take as

$$L_s^{-1} = 4\alpha^{1/2}/d \approx 4/d. \tag{16.35}$$

In general, H_{ki} are complicated functions of Reynolds number, etc., but we take them as constant.

Now we scale the variables by writing

$$\begin{gathered} z = lz^*, \quad u = Uu^*, \quad v = Vv^*, \quad p = p_0 + Pp^*, \quad \beta = B\beta^*, \\ \Gamma = G\Gamma^*, \quad t = (l/U)t^*, \quad h_g = h_g^{sat} + Lh_g^*, \quad h_l = h_l^{sat} + Lh_l^*; \end{gathered} \tag{16.36}$$

we take l to be the tube length and choose the unknown scales U, V, P, B, and G by effecting the following balances in Eq. (16.30):

$$\begin{gathered} (\alpha\rho_g v)_z \sim \Gamma, \qquad (\beta\rho_l u)_z \sim \Gamma, \qquad F_{lw} \sim F_{li}, \\ \alpha p_z \sim F_{gi}, \qquad \Gamma(h_l - h_{li}) \sim Q/A. \end{gathered} \tag{16.37}$$

We take ρ_g and ρ_l as constants for simplicity, though in reality ρ_g will vary by a reasonable amount. Specifically, we choose (because $\alpha \sim 1$)

$$\begin{gathered} \rho_g V = lG, \qquad \rho_l BU = lG, \qquad U/V = (f_{li}\rho_g/f_{lw}\rho_l)^{1/2}, \\ P = 2lf_{lw}\rho_l U^2/d, \qquad G = Q/AL \end{gathered} \tag{16.38}$$

(if Q is constant, or a typical value if it varies). Substitution of these values into the equations, and omitting the asterisks, yields the following:

$$\begin{aligned}
\alpha &= 1 - c_1\beta, \\
c_2\alpha_t + (\alpha v)_z &= \Gamma, \\
\beta_t + (\beta u)_z &= -\Gamma, \\
c_3[(\beta u)_t + D(\beta u^2)_z] &= -c_1\beta p_z - u^2 + (v - \chi c_2 u)^2, \\
c_4[c_2(\alpha v)_t + (\alpha v^2)_z] &= -\alpha p_z - (v - \chi c_2 u)^2, \\
\beta(h_{lt} + u h_{lz}) &= \Gamma h_l - h_l/c_5 + q, \\
c_2\alpha h_{gt} + \alpha v h_{gz} &= \Gamma h_g - h_g/c_6, \\
\Gamma &= h_g/c_6 + h_l/c_5,
\end{aligned} \tag{16.39}$$

where the parameters are defined by

$$\begin{gathered}
c_1 = B, \qquad c_2 = U/V, \qquad c_3 = Bd/2f_{lw}l, \\
c_4 = d/2f_{li}l, \qquad c_5 = Gc_{pl}/H_{li}L_s^{-1}, \qquad c_6 = Gc_{pg}/H_{gi}L_s^{-1},
\end{gathered} \tag{16.40}$$

and $q = O(1)$ is the dimensionless heat supply.

To estimate the values of the parameters, we choose

$$\begin{gathered}
l = 10\,\text{m}, \qquad \rho_g = 30\,\text{kg m}^{-3}, \qquad \rho_l = 760\,\text{kg m}^{-3}, \qquad f_{lw} = .004, \\
f_{li} = .02, \qquad d = .014\,\text{m}, \qquad A \sim 1.5 \times 10^{-4}\,\text{m}^2, \qquad \dot{m} = .2\,\text{kg s}^{-1}, \\
L_s^{-1} = 4/d, \qquad c_{pl} = 5.2\,\text{kJ kg}^{-1}\,\text{K}^{-1}, \qquad c_{pg} = 4.6\,\text{kJ kg}^{-1}\,\text{K}^{-1}, \\
H_{li} \sim 1.7 \times 10^5\,\text{W m}^{-2}\,\text{K}^{-1}, \qquad H_{gi} \sim .7 \times 10^5\,\text{W m}^{-2}\,\text{K}^{-1},
\end{gathered} \tag{16.41}$$

where $\dot{m}$ is the inlet mass flux; this determines Q via $Ql \sim \dot{m}L$, so that G is equivalently determined from

$$G = \dot{m}/Al. \tag{16.42}$$

These parameters are relevant for steam/water flow at an ambient pressure of 60 bars. We find successively

$$\begin{aligned}
G &\sim 130\,\text{kg m}^{-3}\,\text{s}^{-1}, \\
V &\sim 43\,\text{m s}^{-1}, \\
U &\sim 19\,\text{m s}^{-1}, \\
B &\sim .09, \\
P &\sim 16\,\text{bars}\ (1.6 \times 10^6\,\text{Pa}),
\end{aligned} \tag{16.43}$$

and thus

$$\begin{aligned}
c_1 &\sim .09, \\
c_2 &\sim .44, \\
c_3 &\sim .016, \\
c_4 &\sim .035, \\
c_5 &\sim .014, \\
c_6 &\sim .03.
\end{aligned} \tag{16.44}$$

The fact that all the parameters are less than one indicates that the balances chosen in Eq. (16.40) are correct. In practice, this often has to be done through trial and error.

Analysis

Because all the parameters except c_2 in Eq. (16.44) are in fact small, the system lends itself to an asymptotic reduction. Specifically, $h_g, h_l \ll 1$ (thus the fluids are close to thermodynamic equilibrium); adding the two enthalpy equations then gives

$$\Gamma \approx q. \tag{16.45}$$

Also $\alpha \approx 1$, and the gas momentum equation is just a quadrature for p,

$$0 = -p_z - (v - \chi c_2 u)^2, \tag{16.46}$$

whereas the liquid momentum equation is simply a force balance:

$$u = \lambda v, \tag{16.47}$$

where

$$\lambda = (1 + \chi c_2)^{-1} \sim 0.5. \tag{16.48}$$

The equations thus reduce to

$$\begin{aligned} v_z &\approx q, \\ \beta_t + \lambda(\beta v)_z &= -q, \end{aligned} \tag{16.49}$$

which can even be solved explicitly. Thus, if q is constant,

$$v = v_0 + qz, \tag{16.50}$$

and using characteristics, we find

$$(1 + \lambda\beta) = \frac{(1 + \lambda\beta_0)v_0}{v_0 + qz}, \tag{16.51}$$

if $v = v_0$, $\beta = \beta_0$ at $z = 0$. Thus β decreases and reaches zero at $z = \lambda\beta_0 v_0/q$, corresponding to dryout of the liquid film.

The dramatic collapse we have illustrated here is robust, in the sense that even if a more realistic system is algebraically more intractable (for example, because f_{lw} depends on u, f_{li} depends on v and β, etc.), nevertheless the approximations involved will continue to apply. The result is that numerical computations can be greatly simplified. Simple but accurate models are also of use in establishing parameter dependence of stability characteristics, for example.

In neglecting the parameters c_1, c_3, c_4, c_5, and c_6, we reduce a sixth-order system to a third-order one, two of whose equations are quadratures. All these approximations are potentially singular, as they involve the neglect of small terms, and we must be concerned over whether we also lose the ability to satisfy initial/boundary conditions. Insofar as the system will be hyperbolic, we can afford not to worry about loss of time derivatives, because any resulting rapid transients will be washed out of the system.

Inspecting the reduced system, we see that we can specify v and β at the inlet but not u, as we lose the acceleration terms in the liquid momentum equation. Loss of acceleration in the gas momentum equation does not matter, because the pressure gradient remains and the equation acts as a quadrature for p. Inspection of the enthalpy equations shows that any initial/boundary condition for h_l or h_g is quickly relaxed. Because, in fact, it is natural to prescribe $h_l = h_g = 0$ at the entrance to the annular flow region, there will not actually be any enthalpy boundary layer. The only cause for worry is thus the inability to prescribe u, and we must forgo this luxury. Strictly, an inlet boundary layer analysis is then necessary to complete the solution.

Two important features about this two-phase flow are these: the pressure gradient disappears from the liquid momentum equation because $\beta \ll 1$ (in annular flow); and further, acceleration terms will always be small if $\rho u^2 \ll \Delta p$, which is inevitably the case. Thus it will be a common feature that liquid momentum simply gives a force balance, so that the momentum equations can effectively be taken out of the system.

16.8 Mathematical model of a thermosyphon

A thermosyphon is a device in which fluid flows convectively in a circulation loop under the influence of its own buoyancy. A common example is the back boiler used in domestic hot water appliances. In one part of the fluid loop, heat is applied, and this causes a convective circulation to occur. In the domestic boiler, one does not wish boiling to occur, and this only happens in the event of bad plumbing. In other situations, the heat supplied is sufficient to cause a region of two-phase flow. A natural example is that of *geysering*, as for example in hot springs, where an oscillatory flow is generated, alternating between subcooled, superheated, and boiling phases.

Such oscillations are extreme examples of instabilities that can occur in a steady flow and are of industrial concern in cryogenic air separation processes, where a fluid loop of liquid nitrogen is heated by the external condensation of oxygen. The boiling of the nitrogen gives the buoyancy that generates the flow around the loop, as indicated in Fig. 16.3. Here we model the flow, with parameters relevant to a device known as

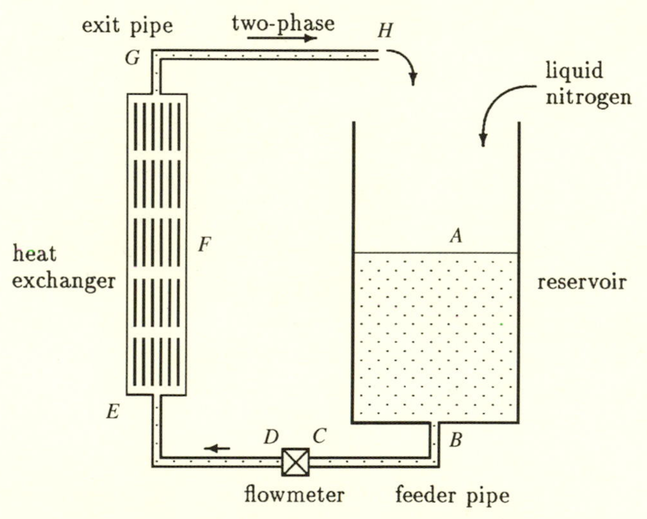

Fig. 16.3. Thermosyphon system. Reproduced from Aldridge and Fowler (1992), by permission of B.G. Teubner, Stuttgart

a plate-fin heat exchanger. A novelty here is that the subcooling of the liquid at the inlet is due to the Clapeyron effect; that is, the excess gravitational head at the inlet to the riser causes the saturation temperature to be higher there. Thus gravity is fundamentally important, both in determining the inlet subcooling and in driving the flow.

In Fig. 16.3, the section AB represents the reservoir, BE is the inlet pipe (with CD being a flowmeter), and GH is the exit pipe. The two-phase section EG consists of a subcooled region EF, $0 < z < r(t)$, where F $(z = r(t))$ denotes the boiling boundary and a two-phase region FG. We model these in turn.

Subcooled region

Suitable averaged equations for the subcooled flow are

$$\rho_t + (\rho u)_z = 0, \tag{16.52a}$$

$$(\rho u)_t + (\rho u^2)_z = -p_z - \rho g - \frac{2}{d} f_{lw} \rho u^2, \tag{16.52b}$$

$$\rho(h_t + u h_z) = \frac{4}{d} q_{lw}, \tag{16.52c}$$

where f_{lw} is the friction factor and q_{lw} is the heat flux delivered to the fluid. With the assumption that the liquid is incompressible, $\rho = \rho_l =$ constant, these equations simplify dramatically. We then have

$$u = u_0(t), \tag{16.53}$$

and the pressure drop from E to F is

$$\Delta p_{EF} = \left[\rho_l u_0' + \rho_l g + \frac{2}{d} f_{lw} \rho_l u_0^2 \right] r, \tag{16.54}$$

where $u_0' = du_0/dt$, provided f_{lw} depends on u_0 only (or is constant). We find h by writing Eq. (16.52c) in characteristic form as

$$\dot{z} = u_0, \qquad \dot{h} = 4q_{lw}/d\rho_l, \tag{16.55}$$

which can be integrated if, for example, q_{lw} is constant (prescribed heat flux conditions). In that case, the solution is just

$$z = \int_s^t u_0(\theta)\, d\theta, \qquad h = h_0 + (4q_{lw}/\rho_l d)(t - s), \tag{16.56}$$

where h_0 is the inlet enthalpy; hence

$$z = \int_{t-\tau}^t u_0(\theta)\, d\theta, \qquad \tau = (\rho_l d/4q_{lw})(h - h_0), \tag{16.57}$$

and, in particular, the position of the boiling boundary is

$$r(t) = \int_{t-\tau_0}^t u_0(\theta)\, d\theta, \tag{16.58}$$

where

$$\tau_0 = \left(\frac{\rho_l d}{4q_{lw}} \right) (h_{sat} - h_0). \tag{16.59}$$

If the heat flux is variable, analytic solutions are not normally available. The commonly used Dittus–Boelter relations, for example, have $q_{lw} \propto u_0^{0.8}\Delta T$, where ΔT is the wall temperature minus the fluid temperature. If this relationship is modified to

$$q_{lw} = q' u_0 (h_w - h), \tag{16.60}$$

however, an analytic solution is again possible, and we find

$$r = r^* = \left(\frac{\rho_l d}{4q'}\right) \ln\left[\frac{h_w - h_0}{h_w - h_{sat}}\right]. \tag{16.61}$$

Thus for prescribed wall temperature, the boiling boundary position is nearly constant; whereas for prescribed heat flux, it responds to variations in the inlet velocity with a distributed delay. The presence of delay in the response contributes toward oscillatory instability. For the more general Dittus–Boelter relation

$$q_{lw} = q' u_0^{0.8} (h_w - h), \tag{16.62}$$

r is given implicitly by the relation (16.58), together with

$$\ln\left[\frac{h_w - h_0}{h_w - h_{sat}}\right] = \frac{4q'}{\rho_l d} \int_{t-\tau_0}^{t} u_0^{0.8}(\theta)\, d\theta, \tag{16.63}$$

which can be usefully approximated by

$$r = r^* u_0^{0.2}, \tag{16.64}$$

where r^* is given by Eq. (16.61), on the basis that 0.2 is small (so that $u_0^{0.2}$ is approximately constant). In this case, the effect of the delay is eradicated.

Two-phase region

The simplest model that takes some account of the different phasic velocities (and hence of the differing flow regimes) is the *drift flux* model developed by Zuber and co-workers. This is essentially a mixture model that includes a phasic drift term. We define α to be the void fraction of steam, and then the mixture defined values of density, velocity, and enthalpy are

$$\begin{aligned} \rho &= \alpha\rho_g + (1-\alpha)\rho_l, \\ u &= [\alpha\rho_g v_g + (1-\alpha)\rho_l v_l]/\rho, \\ h &= [\alpha\rho_g h_g + (1-\alpha)\rho_l h_l]/\rho, \end{aligned} \tag{16.65}$$

where v_g, v_l are the gas and liquid velocities, and h_g, h_l are the gas and liquid enthalpies. The volumetric flux of the mixture is

$$j = \alpha v_g + (1-\alpha) v_l, \tag{16.66}$$

and the drift velocity of the gas, V_j, is defined by

$$V_j = v_g - j = (1-\alpha)(v_g - v_l). \tag{16.67}$$

The drift flux model supplements conservation laws of mass, momentum, and enthalpy with a constitutive law for the drift flux.

Equations for mass of vapor, and mixture mass, momentum, and enthalpy can be written as

$$(\alpha\rho_g)_t + (\alpha\rho_g u)_z = \Gamma - \left[\frac{\alpha\rho_g\rho_l}{\rho}V_j\right]_z, \tag{16.68a}$$

$$\rho_t + (\rho u)_z = 0, \tag{16.68b}$$

$$(\rho u)_t + (\rho u^2)_z + \left[\left(\frac{\rho_l - \rho}{\rho - \rho_g}\right)\frac{\rho_g\rho_l}{\rho}V_j^2\right]_z = -p_z - \rho g - \frac{4}{d}\tau_{mw}, \tag{16.68c}$$

$$(\rho h)_t + (\rho u h)_z + \left[\frac{\alpha\rho_g\rho_l V_j}{\rho}(h_g - h_l)\right]_z = \frac{dp}{dt} + \frac{4}{d}q_{mw}, \tag{16.68d}$$

and in the two-fluid model, these would be supplemented by separate momentum and enthalpy equations for V_j and h_g (for example). Here we assume that $h_g = h_g^{sat}$ is the saturation value and choose V_j via an empirically determined constitutive relationship. In addition, Γ must in general be specified by an averaged Stefan interphase jump condition. Taking our cue from the two-fluid model in Section 16.7, we will suppose that h_l is close to h_l^{sat}, the saturation value. Then h is known in terms of α, and Eq. (16.68a) determines Γ but uncouples from the other equations.

Drift flux correlations

Zuber and Findlay (1965) present a number of expressions for V_j, corresponding to different flow regimes. In the bubbly flow regime, there are two cases. If the motion of a given bubble is influenced by the presence of other bubbles, then the drift velocity necessarily depends on the volume fraction of gas. Zuber and Findlay suggest

(i) for small bubbles (diameter $d_b < .5$ mm)

$$V_j = \frac{g\Delta\rho}{18\mu_l}d_b^2(1-\alpha)^3; \tag{16.69}$$

(ii) for larger bubbles, $d_b < 2$ cm,

$$V_j = 1.53\left(\frac{\sigma g\Delta\rho}{\rho_l^2}\right)^{1/4}(1-\alpha)^{3/2}, \tag{16.70}$$

where σ is surface tension. For larger α, one progresses through a regime of 'churn-turbulent' bubbly flow to slug flow, where it is suggested that

(iii)

$$V_j = .35\left(\frac{g\Delta\rho d}{\rho_l}\right)^{1/2}. \tag{16.71}$$

As a generality, we can take $V_j = V_j(\alpha)$. Typical values are $V_j \sim .1$ m s^{-1}, and V_j decreases as α increases.

Clapeyron relation

The change in saturation temperature due to a change in pressure is given by

$$\Delta T_{sat} = T_{sat}\Delta V\Delta p/L, \tag{16.72}$$

where L is latent heat, $L = h_g^{sat} - h_l^{sat}$, and ΔV is the change in specific volume, $\Delta V = (1/\rho_g) - (1/\rho_l)$. Hence the change in saturation enthalpy is given by

$$\Delta h_l^{sat} = \left(\frac{1}{\rho_g} - \frac{1}{\rho_l}\right)\frac{c_{pl}T_{sat}\Delta p}{L}, \tag{16.73}$$

where c_{pl} is the liquid specific heat. Taking h_g and h_l in the two-phase region to have their respective saturation values, it follows from Eq. (16.65) that

$$h = h_l^{sat} + (\alpha\rho_g L/\rho). \tag{16.74}$$

We take the pipe inlet E in Fig. 16.3 as a reference location, and we suppose

$$h_l = h_0 = h_{sat}^A \tag{16.75}$$

there; that is, we suppose the fluid at A is at saturation. The saturation value of h_l in the pipe is then

$$h_l^{sat} = h_{sat}^A + \left(\frac{1}{\rho_g} - \frac{1}{\rho_l}\right)\frac{c_{pl}T_A}{L}[p - p_E + \Delta p_{AE}], \tag{16.76}$$

where $\Delta p_{AE} = p_E - p_A$ is the pressure head from A to E and is essentially $\Delta p_{AE} = \rho_l g(z_A - z_E)$, where z_A, z_E are the vertical heights of A and E. Thus the inlet subcooling at E is

$$h_l^{sat} - h_0 = \left(\frac{1}{\rho_g} - \frac{1}{\rho_l}\right)\frac{c_{pl}T_A}{L}\Delta p_{AE}. \tag{16.77}$$

Nondimensionalization

We nondimensionalize by writing

$$\begin{aligned} u = [u]u^*, \qquad V_j = [u]V_j^*, \qquad z = lz^*, \qquad p = p_E + \rho_l g l p^*, \\ h = h_0 + (\rho_g L/\rho_l)h^*, \qquad \rho = \rho_l \rho^*, \qquad t = (l/[u])t^*, \end{aligned} \tag{16.78}$$

and we choose $[u]$ to balance gravity with friction in Eq. (16.52b); thus

$$[u] = (gd/2f_{lw})^{1/2}. \tag{16.79}$$

Then the dimensionless equations in the subcooled region can be written (omitting asterisks)

$$\begin{aligned} u &= u_0(t), \\ \epsilon_1 u_0' &= -p_z - 1 - u_0^2, \\ h_t + uh_z &= q_{lw}^*, \end{aligned} \tag{16.80}$$

where

$$\begin{aligned} q_{lw}^* &= 4q_{lw}l/\rho_g L d[u], \\ \epsilon_1 &= [u]^2/gl = d/2lf_{lw}. \end{aligned} \tag{16.81}$$

The two-phase flow equations become

$$\begin{aligned}
\rho_t + (\rho u)_z &= 0,\\
\epsilon_1[(\rho u)_t + (\rho u^2)_z] + \delta\epsilon_1\left[\left(\frac{1-\rho}{\rho-\delta}\right)\frac{1}{\rho}V_j^2\right]_z &= -p_z - \rho - \tau_{mw}^*,\\
\rho &= 1-\alpha+\delta\alpha,\\
h &= a_1(\Delta+p)+\alpha/\rho,\\
\rho(h_t + uh_z) &= q_{mw}^* - (\alpha V_j/\rho)_z + \epsilon_2\, dp/dt,
\end{aligned} \tag{16.82}$$

where

$$\begin{gathered}
\delta = \rho_g/\rho_l, \qquad \Delta = \Delta p_{AE}/\rho_l g l, \qquad a_1 = \frac{\rho_l(\rho_l-\rho_g)c_{pl}T_A g l}{\rho_g^2 L^2},\\
\tau_{mw}^* = 4\tau_{mw}/\rho_l g d, \qquad q_{mw}^* = 4q_{mw}l/\rho_g L d[u],\\
\epsilon_2 = \rho_l g l/\rho_g L.
\end{gathered} \tag{16.83}$$

Appropriate boundary conditions are that

$$\begin{gathered}
h = 0, \qquad p = 0 \quad \text{at } z = 0,\\
p = -\Delta \quad \text{at } z = 1,
\end{gathered} \tag{16.84}$$

if we ignore the pressure drop in the outlet tube GH in Fig. 16.3.

To estimate the values of the parameters in Eqs. (16.80) and (16.82), we use the following estimates, appropriate for nitrogen in an industrial thermosyphon (specifically, a plate-fin heat exchanger):

$$\begin{gathered}
l \sim 2\,\text{m}, \qquad \rho_g \sim 4.6\,\text{kg m}^{-3}, \qquad L \sim 198\,\text{kJ kg}^{-1}, \qquad d \sim 2\,\text{mm},\\
g \sim 9.8\,\text{m s}^{-2}, \qquad f_{lw} \sim 10^{-2}, \qquad \rho_l \sim 808\,\text{kg m}^{-3},\\
c_{pl} \sim 1955\,\text{J kg}^{-1}\,\text{K}^{-1}, \qquad T_A \sim 77\,\text{K}; \qquad q_{lw}, q_{mw} \sim 2000\,\text{W m}^{-2}.
\end{gathered} \tag{16.85}$$

The two-phase stress τ_{mw} is often given by a complicated empirical formula, but we suppose $\tau_{mw} \sim \frac{1}{2}f_{lw}\rho_l u^2 \sim \frac{1}{4}\rho_l g d$, at least in order of magnitude, the latter being the corresponding liquid frictional term. We then find that

$$[u] \sim 1\,\text{m s}^{-1}, \tag{16.86}$$

and hence

$$\begin{gathered}
q_{mw}^*, q_{lw}^* \sim 9, \qquad \epsilon_1 \sim .05, \qquad \delta \sim .005, \qquad \Delta \sim O(1),\\
a_1 \sim 3, \qquad \tau_{mw}^* \sim O(1), \qquad \epsilon_2 \sim .02.
\end{gathered} \tag{16.87}$$

16.9 A reduced model

Let us consider the case where the heat flux is prescribed, so that $q_{mw}^*, q_{lw}^* = q$ (say) are both constant (and equal). We put $\epsilon_1 = 0$ in Eq. (16.80), so that the solution in the subcooled region can be written, following Eq. (16.58), as

$$r(t) = \int_{t-\tau}^{t} u_0(s)\,ds, \tag{16.88}$$

where

$$\tau = \frac{a_1}{q}(\Delta + p_r), \tag{16.89}$$

and p_r is the value of p on $z = r$, which is

$$p_r = -\left(1 + u_0^2\right)r; \tag{16.90}$$

thus

$$\tau = \tau(r, u_0) = \frac{a_1}{q}\left[\Delta - \left(1 + u_0^2\right)r\right], \tag{16.91}$$

and we require a second relation between r and u_0 to complete the system.

In the two-phase region, we put $\epsilon_1 = \delta = \epsilon_2 = 0$, so that

$$\begin{aligned} \rho_t + (\rho u)_z &= 0, \\ [\varepsilon_1 \rho(u_t + u u_z) +]\, p_z &= -\rho - \tau^*_{mw}, \\ \rho(h_t + u h_z) &= q - (\alpha V_j/\rho)_z, \end{aligned} \tag{16.92}$$

where we retain the term proportional to ε_1 temporarily,

$$\rho = 1 - \alpha, \qquad h = a_1(\Delta + p) + (\alpha/\rho), \tag{16.93}$$

and we have

$$\begin{aligned} \alpha = 0, \qquad p = p_r \quad &\text{at } z = r, \\ p = -\Delta \quad &\text{at } z = 1. \end{aligned} \tag{16.94}$$

Characteristics

Our first concern is the equation type of Eq. (16.92). If we define

$$\begin{aligned} \alpha = 1 - \rho, \qquad \nu = 1/\rho, \qquad h = a_1(\Delta + p) + \nu - 1, \\ W(\rho) = \alpha V_j(\alpha)/\rho, \qquad Q(\nu) = -W'(\rho)/\nu, \end{aligned} \tag{16.95}$$

then we can write Eq. (16.92) (with $\varepsilon_1 \neq 0$) in the form

$$A\psi_t + B\psi_z = \mathbf{c}, \tag{16.96}$$

where $\psi = (\nu, u, h)^T$, and

$$A = \begin{pmatrix} -1 & 0 & 0 \\ 0 & a_1\varepsilon_1 & 0 \\ 0 & 0 & 1 \end{pmatrix}, \qquad B = \begin{pmatrix} -u & \nu & 0 \\ -\nu & a_1\varepsilon_1 u & \nu \\ Q & 0 & u \end{pmatrix},$$
$$\mathbf{c} = \begin{pmatrix} 0 \\ -a_1(1 + \nu\tau_{mw}) \\ \nu q \end{pmatrix}. \tag{16.97}$$

This is a third-order system, whose characteristics are given by $dz/dt = \lambda$, where $\det(\lambda A - B) = 0$. Solving this, we have

$$a_1\varepsilon_1(\lambda - u)^3 = \nu^2[\lambda - (u + Q)], \tag{16.98}$$

so that, as $\varepsilon_1 \to 0$, the characteristics are all real and are approximately given by

$$\lambda \approx u + Q, \qquad \lambda \approx \pm\nu/(a_1\varepsilon_1)^{1/2}. \tag{16.99}$$

The latter pair correspond to rapidly propagating sound-like waves. As $\varepsilon_1 \to 0$, they travel rapidly and can be neglected in studying more slowly evolving behavior. Notice that because two speeds are positive and one negative, it is in fact appropriate to have two conditions at $z = r$ and one at $z = 1$ (as in Eq. (16.94)).

The slower speed corresponds to the propagation of density waves through the two-phase region, and because $V_j > 0$, and $V_j'(\alpha) > 0$, then $Q = \alpha V_j'(\alpha) + V_j/(1-\alpha)$ is positive and increases with α.

Steady state

In a steady state, the two-phase region has

$$\begin{aligned} u &= u_0/(1-\alpha), \\ 1-\alpha &= \frac{u_0 + V_j}{u_0 a_1(p_r - p) + (u_0 + V_j) + q(z-r)}, \end{aligned} \tag{16.100}$$

whence $\alpha = \alpha(p, z; p_r, u_0, r)$, and a quadrature gives

$$p_r = \Delta + \int_r^1 \left[1 - \alpha + \tau_{mw}^*\right] dz. \tag{16.101}$$

In principle, then, $p_r = p_r(u_0, r)$. Together with Eq. (16.90), this gives $r = r(u_0)$, so that from Eq. (16.91) $\tau = \tau(u_0)$, and thus

$$r(u_0) = \int_{t-\tau(u_0)}^{t} u_0(s)\, ds; \tag{16.102}$$

in particular, steady solutions satisfy the nonlinear algebraic equation

$$r(u_0) = \tau(u_0)u_0, \tag{16.103}$$

and in general this must be solved numerically. In certain circumstances, nonmonotonicity in the functions r and τ can give rise to multiple (usually three) solutions for u_0, a phenomenon that gives rise to the so-called *Ledinegg* instability of the steady state on the intermediate branch. An example is illustrated in Fig. 16.4.

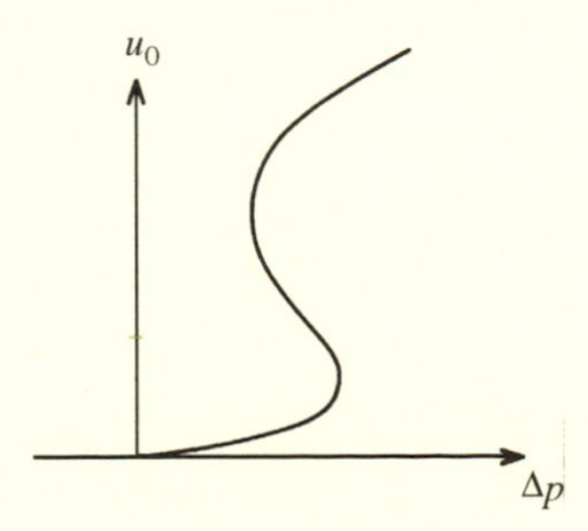

Fig. 16.4. Ledinegg instability

Stability

The derivation of Eq. (16.102) will also apply if the two-phase region responds rapidly to disturbances, in which case the steady solution, Eqs. (16.100) and (16.101), applies quasi-statically. This will be the case if, for example, $Q \gg u_0$, so that disturbances are washed rapidly through the two-phase region. However, it is fairly easy to see in this case that any instability of a steady state is of Ledinegg type.

When $Q \sim u$, the density wave propagation through the two-phase region can interact with the delayed response of the boiling boundary to the inlet velocity and give rise to oscillations. These oscillations are called *density-wave oscillations* and can be analyzed using linear and nonlinear stability methods.

16.10 Notes and references

Two-phase flow has been, and remains, a subject of intense importance for (in particular) the nuclear industry, and there are a good number of references that have been developed through the efforts of various research groups. The books by, or edited by, Wallis (1969), Butterworth and Hewitt (1977), Bergles et al. (1981), Hetsroni (1982), and Whalley (1987) are all useful general sources.

Averaging There are any number of books that deal with this important, but dense, topic. It should be emphasized that the form of the appropriate governing equations remains in itself a subject for research. Good sources are Ishii (1975), Drew (1983), and Drew and Wood (1985).

Instability The Ledinegg instability was described by Ledinegg (1938); it arises when the inlet velocity u_0 is a multiple-valued function of the pressure drop Δp, as in Fig. 16.4. One can in fact prove (with certain assumptions) that if Δp is prescribed, then steady states with $du_0/d\Delta p < 0$ are unstable. Thus if the 'pump characteristic' Δp varies slowly past either turning point of Fig. 16.4, then hysteretic flow 'excursions' will take place. Usually such excursions will also involve a transition in flow regime, and instabilities may also be possible between different regimes (without changing u_0 significantly). It is usually held in engineering circles that 'static instabilities' (i.e., where disturbances $e^{\sigma t}$ exist with σ real and positive) can only exist if $du_0/d\Delta p < 0$, but this is not the case; although normally if $du_0/d\Delta p > 0$, Ledinegg instability is also associated with oscillatory instability.

Oscillatory instabilities include density-wave oscillations, which are associated with the fact that the pressure gradients in single-phase and two-phase regions are different. A perturbation in the inlet velocity $u_0(t)$ leads to a perturbation in the position of the boiling boundary $r(t)$, which causes (generally) a fluctuation in Δp, due both to the change in u_0 and the change in r. However, the change in r is delayed owing to the transit time of the fluid from the inlet, and if this is large enough, then the two effects can oscillate out of phase, leading to an oscillatory instability. Single-channel flows then tend to oscillate periodically, and no further instabilities seem to occur. If the external pressure drop is itself oscillated, then a chaotic response is possible (Dorning, 1989). A recent review is given by Aldridge and Fowler (1996).

Exercises

1. Suppose that in a model of two-phase flow, the pressure drop due to an inlet velocity $u_0(t)$ is represented in the form

$$\Delta p = N[u_0(t)],$$

where N is some nonlinear, autonomous operator. Suppose further that, if $u_0(t) = u^*(1 + \varepsilon e^{\sigma t})$ where u^* is a constant and $\varepsilon \ll 1$, then to leading order, σ satisfies

$$f(\sigma) = 0,$$

where $\Delta p = N(u^*)$ is prescribed constant and

$$\mathcal{L}[u^*]e^{\sigma t} \equiv f(\sigma)e^{\sigma t},$$

where $\mathcal{L}$ is the linearization of N about u^*. Show that if $f(\infty) > 0$, then the existence of a negative slope $d\Delta p/du^* < 0$ implies the existence of a Ledinegg instability.

2. A one-dimensional homogeneous model for flow in a heated tube is given by (cf. Eq. (16.52))

$$\begin{aligned}\rho_t + u\rho_z + \rho u_z &= 0,\\ p_z &= -\kappa\rho u^2,\\ \rho(h_t + uh_z) &= q,\end{aligned}$$

where we neglect gravity and inertia. Assuming ρ $(= \rho_w)$ is constant in the subcooled region $0 < z < r(t)$ and that $h = h_0$, $u = u_0$ on $z = 0$, $h = h_{sat}$ on $z = r(t)$, derive the relations

$$\begin{aligned}z &= \int_{t'}^{t} u_0(s)\,ds,\\ h &= h_0 + \frac{q}{\rho_w}(t - t'),\end{aligned}$$

where q is assumed constant, and deduce that

$$r(t) = \int_{t-\tau}^{t} u_0(s)\,ds,$$

where

$$\tau = \rho_w(h_{sat} - h_0)/q.$$

Show also that the subcooled pressure jump is

$$\Delta p_{sc} = \kappa\rho_w u_0^2 r.$$

Let α be the volume fraction of steam in $z > r$. Show that

$$\begin{aligned}\rho &= \rho_w(1-\alpha) + \rho_s\alpha,\\ \rho h &= \rho_w h_w(1-\alpha) + \rho_s h_s\alpha,\end{aligned}$$

where h_s and h_w ($= h_{sat}$) are steam and water enthalpies, and deduce that

$$h = (\beta/\rho) + \gamma,$$

where

$$\beta = \frac{\rho_w \rho_s L}{\rho_w - \rho_s}, \qquad \gamma = \frac{\rho_w h_w - \rho_s h_s}{\rho_w - \rho_s},$$

L being the latent heat, $h_s - h_w$. Hence show that, in the two-phase region

$$\frac{\partial u}{\partial z} = \frac{q}{\beta},$$

and thus that

$$u = u_0 + \frac{q}{\beta}(z - r)$$

there. Hence show that the two-phase pressure drop between $z = r$ and the end of the tube $z = l$ (assume $\alpha < 1$ there) is

$$\Delta p_{tp} = \kappa \int_r^l \rho u^2 \, dz,$$

where $u = u_0 + q(z - r)/\beta$, and $\rho_t + u\rho_z + \rho u_z = 0$ with $\rho = \rho_w$ on $z = r$.

3. By choosing scales

$$u, u_0 \sim ql/\beta; \qquad z, r \sim l; \qquad t \sim l/u; \qquad \rho \sim \rho_w; \qquad \Delta p \sim \kappa \rho_w u^2 l;$$

show that the dimensionless homogeneous model of Exercise 2 can be written

$$r = \int_{t-\tau^*}^{t} u_0^*(s) \, ds,$$

$$\Delta p^* = u_0^{*2} r + \int_r^1 \rho u^2 \, dz,$$

$$u = u_0^* + (z - r),$$

$$\rho_t + u\rho_z + \rho u_z = 0, \qquad \rho = 1 \quad \text{on } z = r;$$

and the dimensionless quantities are defined by

$$\Delta p^* = \Delta p \left[\frac{\beta^2}{\kappa \rho_w q^2 l^3} \right],$$

$$u_0^* = u_0 \left[\frac{\beta}{ql} \right],$$

$$\tau^* = \rho_w \frac{\Delta h_{sc}}{\beta},$$

where $\Delta h_{sc} = h_{sat} - h_0$ is the inlet subcooling.

Dropping the asterisks, show that in the steady state Δp is given as a function of u_0, by

$$\Delta p = \tau u_0^3 + \left(\frac{1}{2}\tau^2 - \tau \right) u_0^2 + (1 - \tau) u_0 + \frac{1}{2},$$

and find conditions under which Ledinegg instability can occur. Show also that dryout does not occur (i.e., the volume fraction α of steam remains positive), providing

$$1 + (1 - \tau)u_0 < 1/\delta,$$

where $\delta = \rho_s/\rho_w$.

4. Let $\rho, u, \Delta p$, and r be determined by the dimensionless homogeneous model, as in Exercise 3. Show that in the two-phase region, ρ is determined implicitly by

$$\rho = e^{-\zeta},$$
$$z = r(t-\zeta)e^{\zeta} + \int_0^{\zeta} e^{w}[u_0(t-w) - r(t-w)]\,dw.$$

Show that in the steady state, with $u_0 = \bar{u}, r = \bar{r}$,

$$\zeta = \xi = \ln[(\bar{u} + z - \bar{r})/\bar{u}].$$

By putting

$$u_0 = \bar{u} + \varepsilon e^{\sigma t}, \qquad \rho = e^{-\xi} + \varepsilon e^{\sigma t} e^{-\xi} v(\xi)/\bar{u},$$
$$r = \bar{r} + \varepsilon r_1 e^{\sigma t}, \qquad \Delta p = \overline{\Delta p} + \Delta p_1 e^{\sigma t},$$

where $\overline{\Delta p}$ is the steady pressure drop, show that, correct to $O(\varepsilon)$,

$$r_1 = \frac{1}{\sigma}(1 - e^{-\sigma\tau}),$$
$$\Delta p_1 = 2\bar{r} + \bar{u}^2 r_1 (3 - 2e^{\lambda} - (v/\bar{u})|_{\xi=0}) + \int_0^{\lambda} \bar{u}^2 e^{2\xi} v\,d\xi,$$

where

$$\lambda = \ln[(1 + \bar{u} - \bar{r})/\bar{u}],$$

and

$$\sigma v + v_{\xi} = (1 - r_1)e^{-\xi}, \quad \text{with } v = r_1 \text{ on } \xi = 0.$$

Deduce that

$$v = \left(\frac{1 - \sigma r_1}{1 - \sigma}\right) e^{-\sigma\xi} - \left(\frac{1 - r_1}{1 - \sigma}\right) e^{-\xi},$$

and hence show that $\Delta p_1 = 0$ if $f(\sigma) = 0$, where

$$f(\sigma) = 2\bar{r} + (3 - 2e^{\lambda})\bar{u}^2 r_1 - \bar{u} r_1^2$$
$$+ \frac{\bar{u}^2(1 - \sigma r_1)}{(1 - \sigma)(2 - \sigma)}[e^{(2-\sigma)\lambda} - 1] - \frac{\bar{u}^2(1 - r_1)}{(1 - \sigma)}[e^{\lambda} - 1],$$

r_1 being given above. Show that $f(\infty) = 2\bar{r}$, and deduce that a sufficient condition for Ledinegg instability is that $d\overline{\Delta p}/d\bar{u} < 0$.

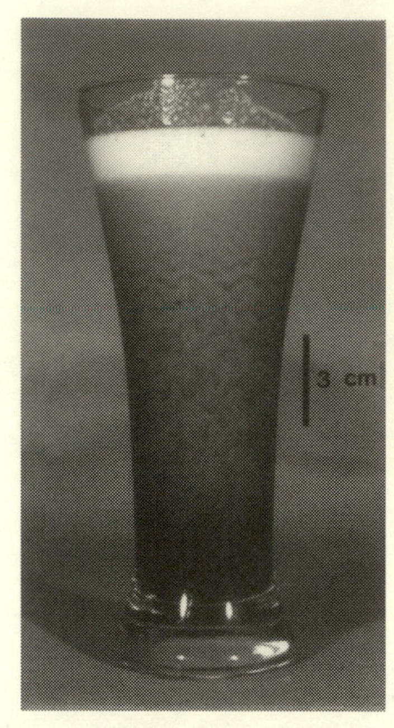

Fig. 16.5. Waves propagating downward in a glass of Guinness. Photograph courtesy of Michael Manga, and reproduced in M. Manga, Waves of bubbles in basaltic magmas and lavas, J. Geophys. Res. **101**, 17,457–17,465, 1996, © The American Geophysical Union.

5. How would you extend the linear analysis of Exercise 4 to a weakly nonlinear analysis? In particular, how could you expand expressions such as $\int_{t-\tau}^{t} u_0(s)\,ds$ if u_0 is considered to be a function of the multiple timescales $t, T = \varepsilon t$? For details of the calculations, see Fowler (1978).
6. When a pint of Guinness is first poured, many tiny bubbles are nucleated. As they rise toward the surface, waves can be seen propagating downward (see Fig. 16.5). To explain this, write down a one-dimensional model for bubbly two-phase flow with variables α (void fraction), u, v (liquid and gas velocities), and p_l, p_g (liquid and gas pressures), that includes gravity and an interactive drag term M. If the bubbles have radius a, show that the assumption of Stokes' drag (e.g., Batchelor 1967, Eq. (4.9.30)) on each bubble leads to

$$M = \frac{\mu D(\alpha)(v-u)}{a^2}, \qquad (*)$$

where $D(\alpha) = 3\alpha$, and explain why this should be appropriate for small α. Conversely, if the bubble density is so large that the liquid drains as in a foam, show that the assumption of Darcy flow (cf. Chapter 13) leads to

$$M \approx \frac{\mu}{k}(1-\alpha)^2(v-u),$$

and for the Carman–Kozeny relation, we again derive $(*)$ with $D(\alpha) \approx 180\alpha^2/(1-\alpha)$. Show that inertia terms can be neglected if $v^2/gh \ll 1$, where h is a typical length scale of the system, and show that, if $(*)$ is generally applicable, with D an increasing function of α, then the model can be reduced to the first-order equation

$$\frac{\partial \alpha}{\partial t} + \frac{\partial Q}{\partial z} = 0,$$

where $Q = \alpha v$ is given by

$$Q = \frac{\Delta \rho g a^2}{\mu} \frac{\alpha^2 (1 - \alpha)^2}{D(\alpha)}.$$

Deduce that voidage waves will travel downward if α is large enough, and show that the speed is of order $\Delta \rho g a^2 / \mu D$. Show also that this is in accord with observation of wave speeds ~ 1 cm s^{-1}, if $D \sim 10$, $\mu / \Delta \rho \sim 10^{-6}$ m^2 s^{-1}, $g \sim 10$ m s^{-2}, and $a \sim 10^{-4}$ m (100 microns). Why do you think these waves form in the first place? (See also Manga (1996).)

Part five

Advanced models

17

Alloy solidification

17.1 Introduction

Loosely speaking, an alloy is a substance consisting of two or more components. Common laboratory examples are aqueous solutions of sodium nitrate, sodium chloride, copper sulphate, ammonium chloride, etc. Industrially important alloys are usually metallic, such as lead–tin, aluminium–tin, etc., whereas a more exotic range of 'alloys' consists of molten igneous rocks – magmas – which may contain many chemically distinct substances, mostly various kinds of silicates. An even more exotic alloy is the Earth's liquid outer core, which is thought to be iron alloyed with a lighter element, often taken to be either sulphur or oxygen.

In many different processes, one is interested in the solidification of such alloys: the formation of the Earth's inner core, the formation of igneous rocks by the crystallization of magma chambers, and the solidification of metal alloys in forming castings. If the resultant solid form were always uniform in composition, there might be little interest in the process, but in fact, alloy solidification often leads to a number of different phenomena that produce *segregation* of the chemical constituents. Such segregation is of both scientific and commercial interest, because in the industrial context, the formation of compositionally nonuniform materials can lead to structural weakness, and hence, ultimately, can cause failure in system components.

One such irregularity is the formation of 'freckles' in metal alloy castings. These consist of columnar 'pipes' whose composition is locally different from that of the surrounding solid. When a cross section is taken, the pipes have the appearance of freckles, hence their name. In this chapter, we focus on this particular phenomenon and will derive and partially analyze a mathematical model of the solidification process that goes some way toward explaining these features.

The equilibrium phase diagram

The first concept of importance that is fundamental to an understanding of the solidification process is the description of thermodynamic equilibrium. In a binary alloy (with which henceforth we shall be exclusively concerned), the chemical composition can be represented by a single concentration c, which we take to be the concentration of the lighter element in the liquid phase. When the liquid alloy is cooled, it begins to freeze at a temperature known as the *liquidus temperature*, which is a function of the

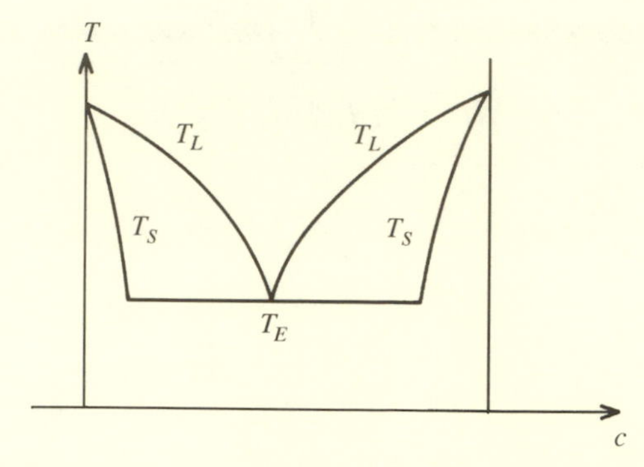

Fig. 17.1. Typical phase diagram for a two-component alloy

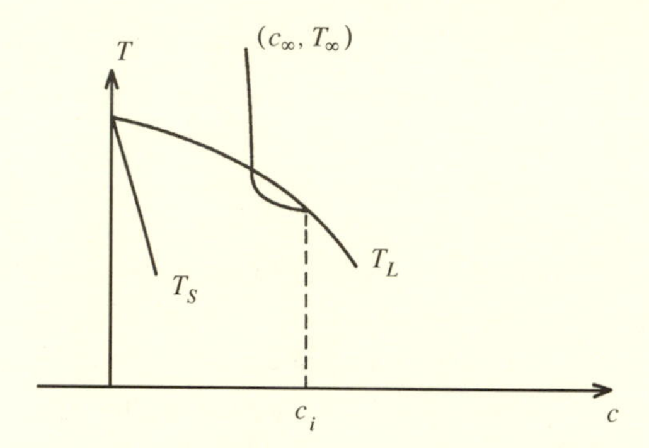

Fig. 17.2. Constitutional supercooling

concentration, $T = T_L(c)$. However, the composition of the solid that forms is not the same. The solid compositional curve that is traced out in (T, c) space is called the *solidus curve*, $T = T_S(c)$ and is illustrated in Fig. 17.1. Typically, the solidus and liquidus curves both decrease away from the values for either pure component, and then the liquidus curves meet at a point called the *eutectic point*. At equilibrium, a solid–liquid interface cannot have a temperature lower than the eutectic temperature T_E; and at this point, a solid of eutectic composition is formed, consisting of a mixture (which may be locally inhomogeneous) of compounds corresponding to both solidus curves.

Constitutional supercooling

The simplest description of solidification is that of an advancing planar interface that separates solid and liquid. The model equations then describe diffusion of temperature and composition, with diffusivities κ (temperature) and D (composition), and it is possible to find (for example) steady traveling-wave solutions. It is instructive to plot the T, c profiles in the liquid on the phase diagram, and this is done in Fig. 17.2. Suppose T_∞, c_∞ represent the temperature and the concentration in the far field of the liquid. Obviously, the liquid is superheated, so $T_\infty > T_L(c_\infty)$. If (for example) $c_\infty < c_E$, then at the interface, the solid that forms has lower concentration than c_∞, and hence the interface must have a value $c = c_i > c_\infty$ in order to conserve mass of the impurity. Now in general, $\kappa \gg D$; hence the composition ahead of the interface

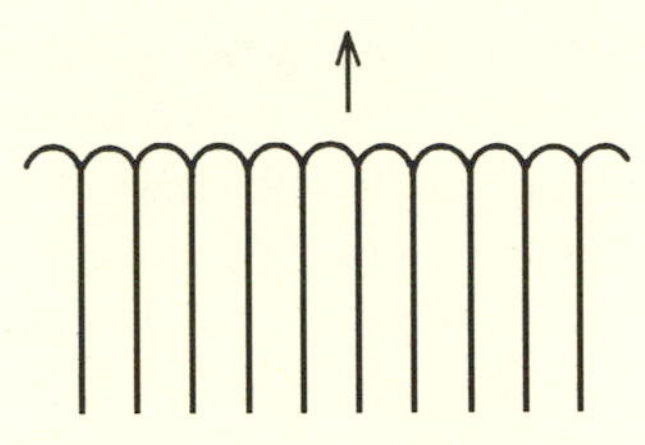

Fig. 17.3. Cellular dendrites

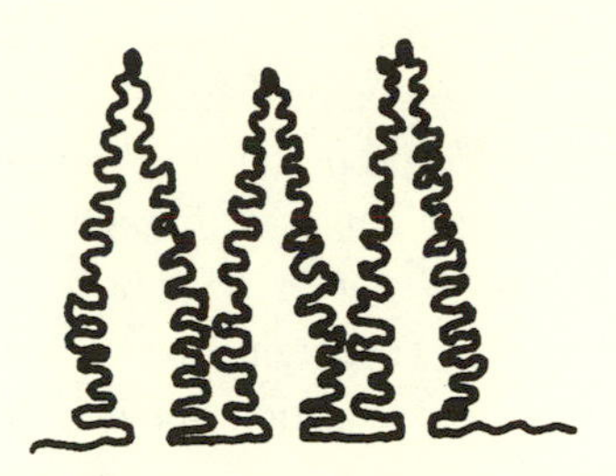

Fig. 17.4. Dendrites

relaxes more rapidly (in a distance $z \sim \sqrt{D}$) from c_i to c_∞ than does the temperature from $T_L(c_i)$ to T_∞ (which is over a distance $z \sim \sqrt{\kappa}$). Hence the (c, T) path in the liquid is more or less horizontal (c changes from c_i to c_∞) and then more or less vertical (T changes from $T_L(c_i)$ to T_∞). As can be seen, the effect of this is that the liquid ahead of the cooling interface has a temperature less than its corresponding liquidus: it is said to be *constitutionally supercooled.*

Interfacial instability

Not surprisingly, the presence of such nonequilibrium supercooling is associated with an instability, first studied by Mullins and Sekerka, and followed up by many other authors. The basic result of the instability is the formation of a corrugated interface, whose shape depends on the magnitude of the instability. For $\kappa \sim D$, when undercooling is moderate, we first obtain sinusoidal wave forms, which for higher amplitude become elongated as shown in Fig. 17.3, and which are known as cellular dendrites.

The most nonlinearly distorted interface occurs in *dendritic* solidification, when the individual cells undergo secondary and tertiary instabilities to form sets of side branches, as shown in Fig. 17.4. Such patterns commonly occur in metallic alloys, and a popular aqueous analog is NH_4Cl (ammonium chloride). Typically, the dendrites are so convoluted and extended that one abandons the idea of following an interface; instead, we consider that when growth is dendritic, then the region of dendrites can be modeled as a porous medium of finite extent, through which the interstitial fluid can flow. Such partially molten regions (and where the solid is connected and immobile) are termed mushy regions, and their modeling gives rise to a distinctive set of equations.

17.2 Modeling mushy layers

Three different approaches to the modeling of mushy zones have been used. One method, which one might call the 'thermodynamic' approach, postulates a sequence of mixture conservation laws with very general forms of constitutive laws for the various fluxes. A second approach uses formal averaging methods, such as were used in Chapter 16, to derive *averaged* conservation laws. These then contain numerous secondary flux terms, arising from averaging nonlinear terms, interfacial jumps, etc., which must be constituted. The final approach, and the one that we adopt here, is based on a fundamental approach from first principles. This is less general than the other two methods, as various approximations are implicitly assumed (e.g., the neglect of viscous dissipation in the pore fluid flow), but generally speaking, the various methods should lead to more or less the same result, providing common sense is used in choosing the various constitutive relations and in estimating their sizes.

We will consider the mush as a rigid porous medium, that is to say, the dendrites form a motionless, nondeforming solid matrix, through which the residual melt flows as in a porous medium. The mush is characterized by a solid fraction ϕ, which is the *mass fraction* of solid phase (note that averaging methods more naturally deal with the volume fraction). The choice of a mass fraction stems from its use in diffusive mixture theory. Let the solid and liquid phases have densities ρ_s and ρ_l, and suppose their (average) concentrations of the light element are ξ^s and ξ^l (where the concentrations are measured as a weight fraction). The mixture density is then defined as

$$\rho = \rho^s + \rho^l, \tag{17.1}$$

where

$$\rho^s = \phi\rho, \qquad \rho^l = \rho(1-\phi) \tag{17.2}$$

are the mass-averaged densities, related to the material properties ρ_s and ρ_l by

$$\rho^l = \rho_l\alpha, \qquad \rho^s = (1-\alpha)\rho_s, \tag{17.3}$$

where α is the liquid volume fraction (i.e., the porosity).

We take the solid velocity $\mathbf{u}^s = \mathbf{0}$, and define the averaged liquid velocity to be $\mathbf{u}^l$. The barycentric mixture velocity is then

$$\mathbf{u} = \chi\mathbf{u}^l, \tag{17.4}$$

where $\chi = 1-\phi$ is the liquid mass fraction.

Conservation of mass

Conservation of mass for each phase in the mush then yields

$$\frac{\partial\rho^s}{\partial t} = m^s, \tag{17.5a}$$

$$\frac{\partial\rho^l}{\partial t} + \boldsymbol{\nabla}.[\rho^l\mathbf{u}^l] = -m^s, \tag{17.5b}$$

where m^s is the local freezing rate, which needs to be determined. Adding these yields the mixture equation

$$\frac{\partial\rho}{\partial t} + \boldsymbol{\nabla}.(\rho\mathbf{u}) = 0. \tag{17.6}$$

Conservation of species

Similarly, we should have two equations for the average species concentration in each phase. However, it is usual to make some simplification to that for the solid. The sum of the two equations, in any case, is the mixture concentration equation

$$\frac{\partial}{\partial t}(\rho\xi) + \boldsymbol{\nabla}.[\rho^l \mathbf{u}^l \xi^l] = \boldsymbol{\nabla}.[\rho^l D_l \boldsymbol{\nabla}\xi^l + \rho^s D_s \boldsymbol{\nabla}\xi^s], \tag{17.7}$$

where the mixture concentration is

$$\xi = \phi\xi^s + (1-\phi)\xi^l, \tag{17.8}$$

and where we suppose the diffusive flux is additive between that in the liquid and that in the solid; D_l and D_s are the respective diffusion coefficients.

One attractive limit is to let D_s (or D_s/D_l) $\to \infty$, so that ξ^s is locally uniform and equal to ξ^{si}, the interfacial value. If in addition, we suppose D_l is large (at least in regard to the microscale) so that the liquid concentration is locally (but not globally) uniform, then $\xi^l = \xi^{li}$ (the average interfacial value of liquid concentration (ξ^{si} is defined similarly)); and if *in addition*, we assume local thermodynamic equilibrium, then ξ^{li} and ξ^{si} are in a ratio determined by the phase diagram, and thus

$$\xi^s/\xi^l = \lambda, \tag{17.9}$$

where λ is known as the partition coefficient.

However, this limit is not at all realistic, and a more appropriate assumption is that $D_s = 0$: diffusion is inoperative in the solid, and the concentration is frozen into the solid through what occurs at the interface. To see what this implies, let us revert to an averaged type of model, as described in Chapter 16, Section 16.6. The general form for the solid concentration equation would be

$$\frac{\partial}{\partial t}(\rho^s\xi^s) + \boldsymbol{\nabla}.[\rho^s\xi^s\mathbf{u}^s] = \boldsymbol{\nabla}.[\rho^s D_s \boldsymbol{\nabla}\xi^s] + d_s + \xi^{si}m^s, \tag{17.10}$$

where d_s is an interfacial solute flux through the solid. (The liquid concentration equation is similar, with s replaced by l and we require $m^l = -m^s$ and the solute interfacial jump condition $d_s + d_l + \xi^{si}m^s + \xi^{li}m^l = 0$.) If $D_s = 0$, then we should specify $d_s = 0$ also, and because $\mathbf{u}^s = \mathbf{0}$, we simply have

$$\frac{\partial}{\partial t}(\rho^s\xi^s) = \xi^{si}m^s. \tag{17.11}$$

Because $\rho\xi = \rho^l\xi^l + \rho^s\xi^s$, then Eq. (17.7) simplifies to

$$\frac{\partial}{\partial t}(\rho^l\xi^l) + \boldsymbol{\nabla}.[\rho^l\xi^l\mathbf{u}^l] = \boldsymbol{\nabla}.[\rho^l D_l \boldsymbol{\nabla}\xi^l] - \lambda\xi^{li}m^s, \tag{17.12}$$

if we use the thermodynamic equilibrium assumption (17.9). Now if $m^s > 0$ (freezing), this is fine, but if $m^s < 0$ (melting), then Eq. (17.9) no longer serves to determine ξ^{si}, which is instead determined by the prior freezing history.

For example, consider a spherical crystal $r = R(t)$, which is surrounded by melt of concentration $\xi^l(t)$. If R first increases to $R_{\max}$ at $t = t_{\max}$ and then melts, we have, for $0 < r < R_{\max}$, $\xi^{si}(r) = \lambda\xi^{li}(\tau)$, where $R(\tau) = r$, $\tau < t_{\max}$. With $\dot{R} = m^s(t)$, this gives

$$\begin{aligned} \xi^{si}(r) &= \lambda\xi^{li}(\tau), \\ \int_0^\tau m^s(t)\,dt &= r; \end{aligned} \tag{17.13}$$

thus determining ξ^{si} implicitly, and in a history-dependent way.

Such a situation should be avoided if at all possible, and two ways are available. Either we assume $m^s > 0$, so that we consider only freezing (and then Eq. (17.12) is appropriate), or we put $\lambda = 0$, which implies that $\xi^s = 0$ anyway. Because the latter approximation is often appropriate, for example for most aqueous solutions and some metallic alloys, we simply take $\lambda = 0$ henceforth, using Eq. (17.12) and ignoring ξ^s altogether.

Conservation of momentum

Again, we would in general propose two momentum equations, one for each phase. For the solid, this is replaced by the assumption $\mathbf{u}^s = \mathbf{0}$. For the liquid, momentum balance is assumed to be given by Darcy's law

$$\alpha\mathbf{u}^l = -\frac{\Pi}{\mu}[\nabla p - \rho_l\mathbf{g}], \tag{17.14}$$

where Π is the permeability (replacing the usual symbol k, which is reserved for the thermal conductivity) and $\mathbf{g}$ is gravity. Darcy's law can in fact be derived from an averaged liquid momentum equation such as Eq. (16.23) by appropriately constituting the interfacial drag $\mathbf{M}_k'$ (and ignoring various terms).

Conservation of energy

Formally, we write two equations for the enthalpy or temperature of each phase, supplemented by an interphasic enthalpy jump condition (the averaged Stefan condition). Except in very rapid flows, for example in gas–solid reactors, local thermal equilibrium can be assumed (not the same as thermodynamic equilibrium), so that only one temperature need be defined, and hence the mixture energy equation suffices. As an enthalpy equation this is

$$\frac{\partial}{\partial t}[\rho^s h_s + \rho^l h_l] + \nabla.[\rho^l h_l \mathbf{u}^l] = \nabla.[k\nabla T], \tag{17.15}$$

where k is an averaged thermal conductivity, for example, volume weighted,

$$k = \alpha k_l + (1-\alpha)k_s, \tag{17.16}$$

and the specific enthalpies are

$$\begin{aligned} h_s &= h_{s0} + c_s T, \\ h_l &= h_{l0} + c_l T, \end{aligned} \tag{17.17}$$

with $c_{s,l}$ being the specific heats; we define the latent heat via

$$L = h_s - h_l, \tag{17.18}$$

when $T = T_l$ is on the liquidus.

Thermodynamic equilibrium

The coexistence of two phases requires us to make some statement about thermodynamic equilibrium, whether the phases are at it or not. The simplest assumption is that of *local equilibrium*, that is, that temperature and concentration fields are locally uniform (on the dendrite spacing scale), and the temperature lies on the liquidus. Thus we prescribe

$$T = T_0 - \Gamma c, \tag{17.19}$$

assuming that $c < c_E$ (so $\Gamma > 0$) and that the liquidus slope is constant (which is not an essential requirement). We will return to the question of the veracity of Eq. (17.19) in Section 17.6.

Model summary

If we define the mass liquid fraction to be

$$\chi = 1 - \phi, \tag{17.20}$$

then the density ρ in Eqs. (17.1)–(17.3) is given by

$$\rho = \frac{\rho_l}{1 - r(1-\chi)}, \tag{17.21}$$

where

$$r = \frac{\rho_s - \rho_l}{\rho_s}. \tag{17.22}$$

Mass conservation is given by Eq. (17.6),

$$\rho_t + \boldsymbol{\nabla}.(\rho\mathbf{u}) = 0. \tag{17.23}$$

Conservation of species follows from Eq. (17.2). We write $\xi^l = c$ and use Eq. (17.23) to obtain (with $D_l = D$)

$$\rho\frac{d}{dt}(\chi c) + \boldsymbol{\nabla}.[\rho(1-\chi)c\mathbf{u}] = \boldsymbol{\nabla}.[\rho\chi D\boldsymbol{\nabla}c]. \tag{17.24}$$

Darcy's law (17.14) can be written

$$\rho\mathbf{u} = -\frac{\Pi}{\nu}[\boldsymbol{\nabla}p - \rho_l\mathbf{g}], \tag{17.25}$$

where $\nu = \mu/\rho_l$ is the liquid phase kinematic viscosity.

The energy equation (17.15) may be reduced, using Eqs. (17.17), (17.18), and (17.5), to

$$-Lm^s + \rho[\chi c_l + (1-\chi)c_s]\frac{\partial T}{\partial t} + \rho c_l\mathbf{u}.\boldsymbol{\nabla}T = \boldsymbol{\nabla}.[k\boldsymbol{\nabla}T], \tag{17.26}$$

and m^s is defined through Eq. (17.5a):

$$m^s = \frac{\partial}{\partial t}[\rho(1-\chi)]. \tag{17.27}$$

Together with the liquidus relation (17.19), Eqs. (17.21), (17.23), (17.24), (17.25), (17.26), and (17.27) serve to determine the variables ρ, χ, $\mathbf{u}$, c, p, T, and m^s; we thus have seven equations for seven unknowns. Of these, m^s, ρ, and T can be easily eliminated to reduce the model to four partial differential equations. In essence, as we shall later see, this model is a generalization of that describing convection in a porous medium, together with an additional evolution equation for the liquid mass fraction.

Liquid and solid regions

Equations in the liquid and solid regions are much more easy to write down. We denote the mush–liquid boundary as z_l and the mush–solid boundary as z_s. Then in $z > z_l$, the Boussinesq equations for the fluid motion are

$$\begin{aligned} \boldsymbol{\nabla}.\mathbf{u} &= 0, \\ \frac{dT}{dt} &= \kappa_l \nabla^2 T, \\ \frac{dc}{dt} &= D\nabla^2 c, \\ \rho\frac{d\mathbf{u}}{dt} &= -\boldsymbol{\nabla} p + \mu\nabla^2\mathbf{u} + \rho_l\mathbf{g}, \end{aligned} \tag{17.28}$$

where κ is the thermal diffusivity, $\kappa = k_l/\rho_l c_l$. In $z < z_s$, heat transfer is determined by

$$T_t = \kappa_s \nabla^2 T, \tag{17.29}$$

with $\kappa_s = k_s/\rho_s c_s$.

Boundary conditions

Associated with the conservation laws, there are flux continuity conditions. On the liquid–mush interface, these are

$$V_n[\rho]_-^+ = [\rho u_n]_-^+, \tag{17.30a}$$

$$V_n[\rho\chi c]_-^+ = \left[\rho c u_n - \rho\chi D\frac{\partial c}{\partial n}\right]_-^+, \tag{17.30b}$$

$$V_n[\rho L\chi]_-^+ = \left[\rho L u_n - k\frac{\partial T}{\partial n}\right]_-^+, \tag{17.30c}$$

where n represents the normal component and V_n is the normal velocity of the interface. In addition, we propose continuity of temperature, concentration, pressure, and liquid fraction; thus

$$[T]_-^+ = [c]_-^+ = [p]_-^+ = 0, \qquad \chi = 1. \tag{17.31}$$

Here $[\ ]_-^+$ denotes the jump across z_l; these conditions are standard, except for the prescription that $\chi = 1$, for which there is no *a priori* justification, and which we shall reconsider in Section 17.6.

At the solid–mush boundary z_s, we have flux conditions

$$V_n[\rho]_-^+ = [\rho u_n]_-^+,$$
$$V_n[\rho L\chi]_-^+ = \left[\rho L u_n - k\frac{\partial T}{\partial n}\right]_-^+, \tag{17.32}$$

and if we suppose that the basal temperature is subeutectic, then we prescribe also

$$c = c_E, \qquad [T]_-^+ = 0. \tag{17.33}$$

(If the basal temperature is not less than T_E, then a solute flux condition identical to Eq. (17.30b) would be appropriate, but for the remainder of this chapter, we ignore this possibility.)

Finally, we prescribe

$$T = T_b \quad (< T_E) \tag{17.34}$$

at the base of the vessel, and, for example,

$$T \to T_\infty, \qquad c \to c_\infty \tag{17.35}$$

as $z \to \infty$ in the liquid. More realistically, in a liquid of finite depth d, we could prescribe $T = T_\infty$ on $z = d$ and $\partial c/\partial z = 0$ there, so that the liquid composition is determined through global composition of solute. Additionally, we could prescribe $u_n = 0$, $p = p_\infty$ at $z = d$. In practice, such niceties can be ignored, because the liquid is effectively well mixed and can be considered as being of infinite extent, except, perhaps, for the temperature field.

17.3 A reduced model

We now turn to the problem of nondimensionalizing the equations. We define the liquidus temperature (at infinity) in the liquid to be

$$T_L^\infty = T_0 - \Gamma c_\infty; \tag{17.36}$$

thus $T_\infty - T_L^\infty$ is the degree of superheat. Then $T_L^\infty - T_E$ is a possible choice of temperature scale. The compositional range is $c_E - c_\infty$, and to account for the possibility that this is small, we choose $c_E - c_\infty$ as the composition scale. Finally, a natural length scale is d, the initial liquid depth. In fact, an independent (and more relevant) depth is that of the mush when convection begins there. If this is z_m, then z_m can be much less than d. Thus, although we denote d as our length scale, we will not treat it too religiously as the liquid depth, and will, for example, allow ourselves to consider the far field conditions (17.35) as occurring 'at ∞,' if convenient.

As we shall see, the mush thickness and evolution is controlled by thermal conduction; therefore we choose a thermal time and velocity scale. In summary, then, define scales

$$
\begin{gathered}
\mathbf{x} \sim d, \qquad \mathbf{u} \sim \kappa_l/d, \qquad t \sim d^2/\kappa_l, \qquad \rho \sim \rho_l, \\
c - c_\infty \sim c_E - c_\infty, \qquad T - T_L^\infty \sim T_L^\infty - T_E, \\
p - p_\infty - \rho_l^0 g z \sim \mu\kappa_l/\Pi_0,
\end{gathered} \tag{17.37}
$$

where we suppose

$$\Pi = \Pi_0 P(\chi), \tag{17.38}$$

with $\lim_{\chi \to 1} P(\chi) = 1$. Typical examples that have been used are $P = (1-\chi)^2$ or $P = (1-\chi)^3$. We have subtracted off a hydrostatic pressure $\rho_l^0 g z$, where z is the vertically upward coordinate, and in addition we suppose that the liquid density is given by

$$\rho_l = \rho_l^0\left[1 - \alpha_l\left(T - T_L^\infty\right) - \beta_l(c - c_\infty)\right], \tag{17.39}$$

where $\alpha_l, \beta_l > 0$ (c is the concentration of the light component), so that ρ_l^0 is the density of the liquid at its far field liquidus. Following the Boussinesq approximation, we take $\rho_l = \rho_l^0$ except in Darcy's law (and the Navier–Stokes equation in the liquid).

The nondimensional equations in the mush are then

$$T = -c, \tag{17.40a}$$

$$\rho = [1 - r(1-\chi)]^{-1}, \tag{17.40b}$$

$$\rho_t + \nabla.[\rho\mathbf{u}] = 0, \tag{17.40c}$$

$$\rho\frac{d}{dt}[\chi(1+\beta c)] + \nabla.[\rho(1-\chi)(1+\beta c)\mathbf{u}] = \frac{\beta}{Le}\nabla.[\rho\chi\nabla c], \tag{17.40d}$$

$$-S\frac{\partial}{\partial t}[\rho(1-\chi)] + \rho\bar{c}\frac{\partial T}{\partial t} + \rho\mathbf{u}.\nabla T = \nabla.[\bar{k}\nabla T], \tag{17.40e}$$

$$\rho\mathbf{u} = -P(\chi)[\nabla p - Rc\mathbf{k}], \tag{17.40f}$$

where $\mathbf{k}$ is a unit upward vector, and

$$
\begin{gathered}
\beta = (c_E - c_\infty)/c_\infty, \qquad Le = \kappa_l/D, \qquad S = L/\left\{c_l\left(T_L^\infty - T_E\right)\right\}, \\
\bar{c} = \chi + (1-\chi)(c_s/c_l), \\
\bar{k} = k/k_l = \alpha + (1-\alpha)(k_s/k_l), \\
R = \frac{\Delta\rho g d \Pi_0}{\mu\kappa_l},
\end{gathered} \tag{17.41}
$$

where $\alpha = \rho\chi$ (ρ being dimensionless), and

$$\Delta\rho = \rho_l^0\left[\beta_l(c_E - c_\infty) - \alpha_l\left(T_L^\infty - T_E\right)\right]. \tag{17.42}$$

This density difference $\Delta\rho$ may be written as $\rho_l^\infty - \rho_l^E$, being the difference between the far field liquidus density and that of the eutectic liquid.

In these equations, *Le* is the Lewis number, *S* is a Stefan number, and *R* is the relevant compositional Rayleigh number, which will act to drive buoyancy-induced convection in the mush if it is large enough.

Boundary conditions

On the liquid–mush boundary $z = z_l$, the conditions in (17.30) become

$$\begin{aligned} V_n[\rho]_-^+ &= [\rho u_n]_-^+, \\ V_n[\rho\chi(1+\beta c)]_-^+ &= \left[\rho(1+\beta c)u_n - \frac{\beta}{Le}\rho\chi\frac{\partial c}{\partial n}\right]_-^+, \\ V_n[\rho S\chi]_-^+ &= \left[\rho S u_n - \bar{k}\frac{\partial T}{\partial n}\right]_-^+, \end{aligned} \tag{17.43}$$

together with Eq. (17.31). On the solid–mush boundary, $z = z_s$,

$$\begin{aligned} V_n[\rho] &= [\rho u_n]_-^+, \\ V_n[\rho S\chi]_-^+ &= \left[\rho S u_n - \bar{k}\frac{\partial T}{\partial n}\right]_-^+, \\ c &= 1, \end{aligned} \tag{17.44}$$

and in the far-field liquid as $z \to \infty$,

$$T \to \Delta_\infty, \qquad c \to 0, \tag{17.45}$$

where Δ_∞ is the dimensionless superheat,

$$\Delta_\infty = \frac{T_\infty - T_L^\infty}{T_L^\infty - T_E}. \tag{17.46}$$

Notice that an alternative definition of Stefan number in terms of the applied temperature difference $T_\infty - T_E$ would be $St = L/\{c_p(T_\infty - T_E)\}$, and this is related to S in Eq. (17.41) by $St = S/(1 + \Delta_\infty)$.

Simplification

For typical values $\kappa \sim 1.47 \times 10^{-3}$ cm^2 s^{-1} and $D \sim 1.3 \times 10^{-5}$ cm^2 s^{-1}, we have $Le \sim 10^2 \gg 1$, which suggests that we neglect the compositional diffusion term in Eq. (17.40d). This is apparently a singular approximation as $1/Le$ multiplies the highest derivative, but in fact this seems not to be the case, because (with $T = -c$) the thermal conductive term is also diffusive for c.

We take $L \sim 3.14 \times 10^2$ kJ kg^{-1} and $c_l \sim 3.25$ kJ kg^{-1} K^{-1}, so that $L/c_p \sim 100$ K. Typical laboratory values of S are then $\lesssim O(1)$. For example, NH_4Cl has $\Gamma \sim 490$ K, so that with a composition range (as a mass fraction) $c_E - c_\infty \sim 0.1$, we have $S \sim 2$.

Next, consider the density mismatch $r = (\rho_s - \rho_l)/\rho_s$. For NH_4Cl, $\rho_s \sim 1.5 \times 10^3$ kg m^{-3}, $\rho_l \sim 1.1$ kg m^{-3}; thus $r \sim 0.3$. Though obviously not very small, r is smaller for some other alloys (e.g., $r \approx 0.03$ for lead–tin), and the advantage of putting it to zero is so dramatic that we do so without further apology. This is a kind of generalized Boussinesq approximation and has the effect of neglecting liquid flow induced by solidification shrinkage. An intermediate, and less drastic, approximation could be obtained by putting $r = 0$ except in the continuity equation; thus

$$\nabla.\mathbf{u} = \frac{r}{\rho}\frac{d\chi}{dt} \approx r\frac{d\chi}{dt}. \tag{17.47}$$

However, here we put $r = 0$ in Eq. (17.47) as well.

Our final approximation is also strictly motivated more by the ensuing simplifications that occur than (necessarily) by its accuracy, and this is to suppose that c_∞ is close to c_E, specifically that $\beta = (c_E - c_\infty)/c_\infty \ll 1$. In this case, it is necessary to rescale χ by introducing a scaled solid fraction ϕ by

$$\chi = 1 - \beta\phi. \tag{17.48}$$

Making these simplifications, we then obtain the reduced set of equations (noting that $\bar{c} = 1 + O(\beta), \bar{k} = 1 + O(\beta)$)

$$\begin{aligned} \nabla.\mathbf{u} &= O(r), && (17.49a)\\ -\frac{d\phi}{dt} + \frac{dc}{dt} + \nabla.(\phi\mathbf{u}) &= O(\beta, r), && (17.49b)\\ \frac{dc}{dt} &= \nabla^2 c - \beta S\phi_t + O(\beta, r), && (17.49c)\\ \mathbf{u} &= -\nabla p + Rc\mathbf{k} + O(\beta, r), && (17.49d) \end{aligned}$$

where we express the permeability coefficient $P(\chi) = P(1 - \beta\phi) = P(1) + O(\beta)$, and $P(1) = 1$.

Neglecting $O(\beta, r)$, Eqs. (17.49b) and (17.49c) can be written as

$$\begin{aligned} \phi_t &\approx \frac{dc}{dt}, \\ (1 + \beta S)\frac{dc}{dt} &= \nabla^2 c, \end{aligned} \tag{17.50}$$

and we retain the term in βS in case S is significant.

On $z = z_l$, we have

$$\begin{aligned} [u_n]_-^+ = [c]_-^+ = [p]_-^+ = 0, \qquad \phi = 0, \\ \left[\frac{\partial c}{\partial n}\right]_-^+ = \left[\frac{\partial T}{\partial n}\right]_-^+ = 0. \end{aligned} \tag{17.51}$$

On $z = z_s$,

$$c = 1, \qquad SV_n = -\left[\frac{\partial T}{\partial n}\right]_-^+, \qquad u_n = 0. \tag{17.52}$$

We see that the equations in the mush are those for convection in a porous medium, together with an equation for ϕ, which, however, uncouples from those for c, $\mathbf{u}$, and p. We then require two conditions on c and p (or u_n) at z_l and z_s, together with two relations to determine these moving boundaries. For z_s, these are given in Eq. (17.51). At z_l, we see that determination of the relevant two conditions requires a description of the liquid region, and we turn to this in the next section. Nondimensionalization of the fluid equations is left as an exercise.

17.4 No convection, similarity solution

The simplest situation occurs for small times, when there is no fluid motion, $\mathbf{u} = \mathbf{0}$. Convection in the liquid is initiated later, when the compositional flux becomes significant. If $\mathbf{u} = \mathbf{0}$, then

$$\begin{aligned} sc_t &= \nabla^2 c \quad \text{in } z_s < z < z_l, \\ T_t &= \nabla^2 T \quad \text{in } z > z_l, \end{aligned} \tag{17.53}$$

where $s = 1 + \beta S$. In addition, we have $c_t = (1/Le)\nabla^2 c$ in $z > z_l$; but because $Le \gg 1$, this implies c is constant to leading order, that is, $c = 0$ (only the gradient is $O(1)$ near z_l, but $c \approx 0$ uniformly). Thus the boundary conditions for c are

$$c = 1 \quad \text{on } z_s, \qquad c = 0 \quad \text{on } z_l, \tag{17.54}$$

and the free boundary z_l can be determined from the extra condition

$$\left.\frac{\partial c}{\partial n}\right|_{z_l-} = -\left.\frac{\partial T}{\partial n}\right|_{z_l+}. \tag{17.55}$$

The solid–mush boundary is given by $\dot{z}_s = O(1/S)$. Notice that

$$\beta S = \frac{L}{c_l \Gamma c_\infty} \tag{17.56}$$

is in fact independent of $c_E - c_\infty$, so that in a formal sense $\beta S = O(1)$ as $\beta \to 0$. This motivates our retention of βS in Eq. (17.50) and also suggests that $\dot{z}_s = O(1/S)$ $= O(\beta).O(1/\beta S) \sim \beta$ as $\beta \to 0$. We therefore have grounds for considering z_s approximately constant, and we therefore take $z_s = 0$.

These equations admit a simple similarity solution in terms of a similarity variable

$$\eta = \frac{z}{2\sqrt{t}}. \tag{17.57}$$

The solutions are

$$\begin{aligned} T &= \Delta_\infty \left[1 - \frac{\operatorname{erfc}\eta}{\operatorname{erfc}\alpha}\right], \quad \eta \geq \alpha, \\ c &= 1 - \frac{\operatorname{erf} s^{1/2}\eta}{\operatorname{erf} s^{1/2}\alpha}, \quad \eta \leq \alpha, \end{aligned} \tag{17.58}$$

where the free boundary location $\alpha = z_l/2\sqrt{t}$ is determined by

$$\frac{\Delta_\infty e^{-\alpha^2}}{\operatorname{erfc}\alpha} = \frac{s^{1/2} e^{-\alpha^2}}{\operatorname{erf} s^{1/2}\alpha}. \tag{17.59}$$

For a value $s = 1.25$ representative of NH_4Cl (with $c_\infty \approx c_E \approx 0.8$), a graph of α versus Δ_∞ is shown in Fig. 17.5. Asymptotic limits are given by

$$\begin{aligned} \alpha &\sim \frac{\sqrt{\pi}}{2\Delta_\infty} \quad \text{as } \Delta_\infty \to \infty, \\ \alpha &\sim \left(\frac{1}{s} \ln \frac{1}{\Delta_\infty}\right)^{1/2} \quad \text{as } \Delta_\infty \to 0. \end{aligned} \tag{17.60}$$

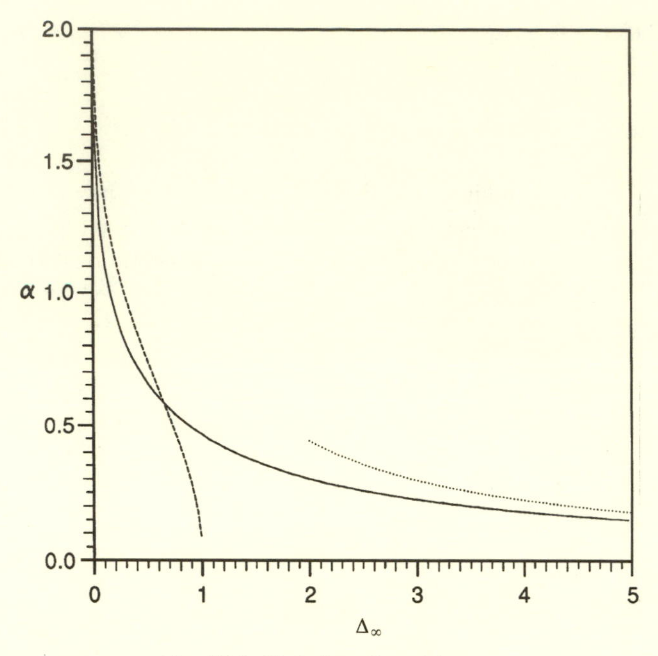

Fig. 17.5. Variation of α with Δ_∞.

17.5 Convection

When a suitable solution of ammonium chloride is chilled from below in the laboratory, a spreading mat of dendritic crystals grows upward from the base. At first the fluid above appears undisturbed. A little later, thin tendrils of fluid can be seen rising from the tip of the mushy region. These convecting plumes are reminiscent of the finger regime of double diffusive convection and suggest that the liquid above the mush has become unstable due to the compositional flux of light fluid delivered from below. The finger mode of convection is consistent with this flux being delivered to a stable thermally stratified environment. At first, there is no discernible effect on the mush. Indeed, it is consistent to view the mush as, effectively, an impermeable solid so far as the liquid is concerned, because its resistance to flow is much larger.

When the mush becomes several centimeters thick, undulations can be detected in the mush surface. Also the convective upwellings begin to become localized in pseudolinear chains. This is consistent with the onset of a second convective mode characterized by flow through the porous mush. We now seek to understand how this can occur.

Homogenization

In order to calculate a stability criterion for the secondary mush convection, we must describe the finger convection in the liquid. We adopt the assumption (based on observation) that the relevant width scale is small, of order $\varepsilon \ll 1$, and the resulting velocity field is large ($\sim 1/\varepsilon$, to balance heat conduction with advection). Specifically, we suppose that finger convection has a velocity field

$$\mathbf{u} = \frac{1}{\varepsilon}\mathbf{U}(x/\varepsilon) + v(x,t)\mathbf{k}, \tag{17.61}$$

where v accommodates a nonzero velocity in the mush, and we can then suppose $\mathbf{U}$ is periodic in $X = x/\varepsilon$. Near z_l, $\mathbf{U}$ will vary with z, but for $z - z_l = O(1)$, we take $\mathbf{U}$ to be vertical; thus $\mathbf{U} = w\mathbf{k}$.

Now consider the temperature field in $z > z_l$. It satisfies

$$T_t + \left\{\frac{1}{\varepsilon}w(x/\varepsilon) + v(x,t)\right\}T_z = \nabla^2 T, \tag{17.62}$$

and suitable boundary conditions are

$$\begin{aligned} T &\to \Delta_\infty, & z &\to \infty, \\ T &= 0, & z &= z_l. \end{aligned} \tag{17.63}$$

We can solve this problem using a multiple-scale approach. Define

$$x = \varepsilon X, \qquad z = \varepsilon Z, \tag{17.64}$$

and put

$$T = T(x, z, X, Z, t) \sim T_0 + \varepsilon T_1 + \ldots; \tag{17.65}$$

we obtain

$$\begin{aligned} L(T_0) &= w(X)T_{0Z} - (T_{0XX} + T_{0ZZ}) = 0, \\ L(T_1) &= -wT_{0z} - vT_{0Z} + 2(T_{0xX} + T_{0zZ}), \\ L(T_2) &= -T_{0t} + T_{0zz} + T_{0xx} + 2(T_{1xX} + T_{1zZ}) - wT_{1z} - vT_{1Z} - vT_{0z}, \end{aligned} \tag{17.66}$$

and we seek solutions T_i that are periodic with zero mean, so that $\partial T_i/\partial Z \to 0$ as $Z \to \infty$. We anticipate this last condition in view of the continuity flux condition that determines z_l: Specifically, if $\partial T/\partial z = O(1)$ on z_l, then balance of heat flux with the far field gives $\partial T/\partial Z = O(\varepsilon)$, whence $\partial T_0/\partial Z \to 0$ as $Z \to \infty$.

By inspection, solutions are

$$T_0 = T_0(x, z, t); \tag{17.67}$$

suppose $V(X)$ is defined by

$$V_{XX} = -w, \tag{17.68}$$

and V has zero mean. Specifically, if $w = \sum_1^\infty w_n \cos nX$, then $V = \sum_1^\infty (w_n/n^2)$ $\cos nx$. Then T_1 (suppressing secular terms $\propto Z$ as $Z \to \infty$) is given by

$$T_1 = -T_{0z}V, \tag{17.69}$$

so that

$$L(T_2) = -T_{0t} + T_{0zz} + T_{xx} - vT_{0z} + wVT_{0zz} - 2T_{0xz}V_X. \tag{17.70}$$

We now choose T_0 so that $L(T_2)$ has zero average (otherwise $T_2 \sim Z$ as $Z \to \infty$), and this requires that T_0 satisfy

$$T_{0t} + vT_{0z} = T_{0xx} + \left(1 + \frac{1}{2}\overline{\psi^2}\right) T_{0zz}, \tag{17.71}$$

where $w = \psi_X$, $V_X = -\psi$, so $\overline{wV}$ (average) $= \frac{1}{2}\overline{\psi^2}$. The enhanced longitudinal dispersion is the only effect of the fluid convection on the heat transport. In fact, $\frac{1}{2}\overline{\psi^2}$ is likely to be small, so that we can legitimately ignore it altogether.

To summarize, we wish to solve

$$\begin{aligned} s\frac{dc}{dt} &= \nabla^2 c \quad \text{in } 0 < z < z_l, \\ c &= 1 \quad \text{on } z = 0, \\ c &= 0 \quad \text{on } z = z_l, \end{aligned} \tag{17.72}$$

and

$$\begin{aligned} T_t + vT_z &= \nabla^2 T \quad \text{in } z > z_l, \\ T &= 0 \quad \text{on } z = z_l, \\ T &\to \Delta_\infty \quad \text{as } z \to \infty, \\ \left.\frac{\partial T}{\partial n}\right|_+ &= -\left.\frac{\partial c}{\partial n}\right|_-. \end{aligned} \tag{17.73}$$

In the mush, we also have to solve

$$\begin{aligned} \boldsymbol{\nabla}.\mathbf{u} &= 0, \\ \mathbf{u} &= -\boldsymbol{\nabla} p + Rc\mathbf{k}, \\ p &= 0 \quad \text{at } z = z_l, \\ u_n &= 0 \quad \text{at } z = 0; \end{aligned} \tag{17.74}$$

v is given by $v = \mathbf{u}.\mathbf{k}$ on $z = z_l$.

Linear stability

In the normal way, stability analysis of the similarity solution involves a linearization about it. In the situation where the basic state is time dependent, the procedure is less clear. The discussion at the beginning of Chapter 14, for example, does not apply, because the linearized operator is no longer autonomous. Indeed, exactly what one *means* by linear stability is not clear.

If an evolution equation

$$u_t = N(u) \tag{17.75}$$

has a basic time-dependent solution $u = u_0(\mathbf{x}, t)$, then on writing $u = u_0 + v$ and linearizing, we have the system

$$v_t = \mathcal{L}[u_0(t)]v. \tag{17.76}$$

If u_0 is periodic, then Floquet theory can be used. In the present situation, other methods must be resorted to.

Frozen-time hypothesis

For example, if one simply supposes that the basic solution is frozen at a particular instant, and examines the stability on this basis, we have the frozen-time hypothesis: put $t = \tau$, consider $u_0(\tau)$ as fixed, and seek solutions to

$$v_t = \mathcal{L}[u_0(\tau)]v, \tag{17.77}$$

which will then be of the form $v \propto \exp[\sigma t]\psi(\mathbf{x})$, and the eigenvalues σ can be calculated in the usual way. Then σ depends on the parameter τ, and we ascertain stability by finding $\tau = \tau_c$ when $\max\{Re(\sigma)\}$ first passes through zero. A variant in thinking about this supposes that u_0 varies slowly, or equivalently, that $Re\,\sigma$ is 'large' (which is manifestly not true when $Re\,\sigma = 0$). A different version, requiring a direct numerical simulation, computes the time evolution of some norm of v, for example, the L_2-norm

$$\|v\| = \left\{ \int |v|^2 \, d\mathbf{x} \right\}^{1/2}. \tag{17.78}$$

Separable solutions proportional to $\exp(\sigma t)$ have

$$Re\,\sigma = \frac{1}{\|v\|}\frac{\partial \|v\|}{\partial t}, \tag{17.79}$$

and we can therefore use Eq. (17.79) as a measure of local growth or decay. Instability is then determined locally by the criterion $\|v\|^{-1}\partial\|v\|/\partial t > 0$. In practice, this is rather subjective, and in particular there is no unique criterion for the 'onset' of convection, because the appearance of an $O(1)$ convective mode depends on the size and time of the initial perturbation.

To illustrate this point, consider $v^{-1}\dot{v} = -1 + t$, where a frozen-time hypothesis would identify instability at $t = 1$. If the initial perturbation is $v = \varepsilon$ at $t = t_0$, then

$$v = \varepsilon \exp\left[(t - t_0) \left\{ \frac{1}{2}(t - t_0) - 1 \right\} \right]. \tag{17.80}$$

A different criterion might be to identify instability with the time when $v = \varepsilon$ again, that is, $t = t_0 + 2$. Evidently, 'instability' is necessarily subjective and in particular may not correspond to observations.

With this proviso, we proceed with a frozen-time analysis. Put

$$\mathbf{u} = (\psi_z, -\psi_x) \tag{17.81}$$

for a two-dimensional solution; thus

$$\begin{aligned} \nabla^2\psi &= Rc_x, \\ s\frac{dc}{dt} &= \nabla^2 c, \end{aligned} \tag{17.82}$$

in $0 < z < z_l$, with

$$\begin{aligned} \psi &= 0, \qquad c = 1 \quad \text{on } z = 0, \\ \frac{\partial\psi}{\partial n} &= 0, \qquad c = 0 \quad \text{on } z = z_l. \end{aligned} \tag{17.83}$$

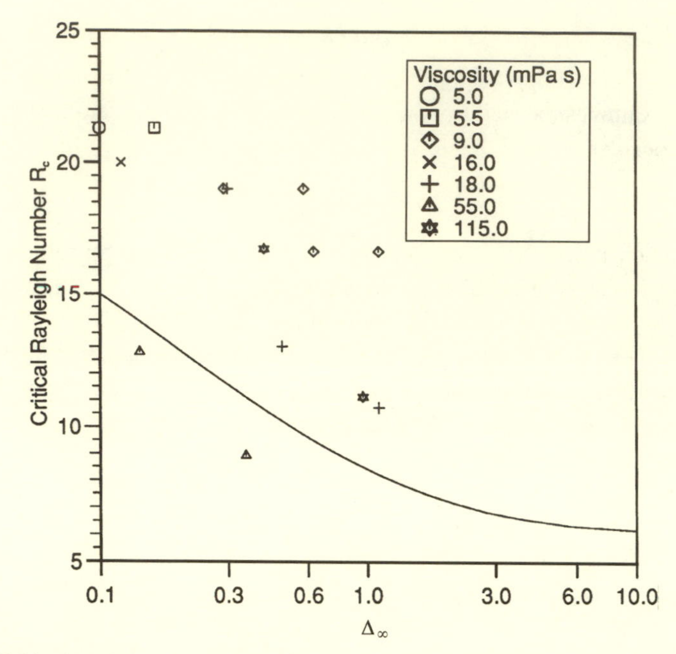

Fig. 17.6. Critical Rayleigh number for the onset of convection, as a function of dimensionless superheat Δ_∞.

The temperature field in $z > z_l$ is given as before. The stability analysis can be done by rescaling as

$$x = z_l x^*, \qquad z = z_l z^*, \qquad t = z_l^2 t^*, \tag{17.84}$$

where z_l is taken as *fixed*. The mush then lies in $0 < z^* < 1$, and the stability analysis is framed in terms of a rescaled Rayleigh number (cf. Eq. (17.41))

$$R_m = z_l R = \frac{\Delta\rho g z_m \Pi_0}{\mu\kappa_l}, \tag{17.85}$$

where z_m is the dimensional mush thickness. A numerical calculation then gives $R_c(s, \Delta_\infty)$, where for $R_m > R_c$, ψ grows (exponentially). For fixed $s = 1.25$, $R_c(\Delta_\infty)$ is plotted in Fig. 17.6, together with some experimental points representing the onset of convection. As might be expected from the remarks above, there is a good deal of scatter.

Boundary layer theory and chimney formation

If we have $z_l = z_l(t)$, and we define a moving coordinate $\zeta = z - z_l(t)$, then the temperature equation in the liquid is, from Eq. (17.73),

$$T_t - (\dot{z}_l - u_l)T_\zeta = T_{\zeta\zeta} + T_{xx}, \tag{17.86}$$

where $u_l = u|_{z_l}$. Because we require exponential decay of T as $\zeta \to \infty$, this suggests that we should have $u_l < \dot{z}_l$. More convincingly, Eq. (17.50) gives $\phi_t = \dot{c} \equiv dc/dt$.

Now because the scaled solid fraction is taken to be zero on z_l, then $\phi_t \geq 0$ there; thus $\dot{c} \geq 0$. Further, if $\mathbf{V}_l$ is the velocity of z_l, then, because $c \equiv 0$ on $z = z_l$,

$$c_t + \mathbf{V}_l.\boldsymbol{\nabla} c = 0 \quad \text{on } z_l. \tag{17.87}$$

Combining this with the criterion $\dot{c} \geq 0$ implies that for a solution to exist, we must have

$$(\mathbf{u} - \mathbf{V}_l).\boldsymbol{\nabla} c \geq 0 \quad \text{on } z = z_l, \tag{17.88}$$

or equivalently,

$$u_n \leq V_{ln}, \tag{17.89}$$

where $V_{ln} = \mathbf{V}_l.\mathbf{n}$, $u_n = \mathbf{u}.\mathbf{n}$, because $\partial c/\partial n < 0$.

This criterion implies that convection through the porous mush will only occur in a uniform way if R_m is low enough, because we expect $\|\mathbf{u}\|$ to grow as R_m increases. In fact, it is observed that, after the onset of convection in the mush, chimneys in the mush begin to appear. This is associated with the transgression of Eq. (17.89) as $\|\mathbf{u}\|$ increases, and the consequent formation of solid free channels. The existence of such channels, illustrated in Fig. 17.7, has a dramatic effect on the flow, because the upwelling plumes are now concentrated in the open channels, whereas the return flow takes place through the porous mush.

We will consider an analytic approximation to the flow akin to that used in Chapter 14 to describe convection at high Rayleigh number. Following the methodology of Chapter 14, we will anticipate that $|\mathbf{u}| \gg 1$, so that a quasi-static approximation in which $z_l \approx$ constant is justified. The geometry we consider is shown in Fig. 17.8. We choose the simplest such configuration, namely, a two-dimensional flow, and we suppose that the channel is thin, of thickness $2a \equiv 2\varepsilon d$, where d is here taken as half the channel spacing. The Poiseuille flux in the channel is given by $2\hat{G}a^3/3\mu$, where

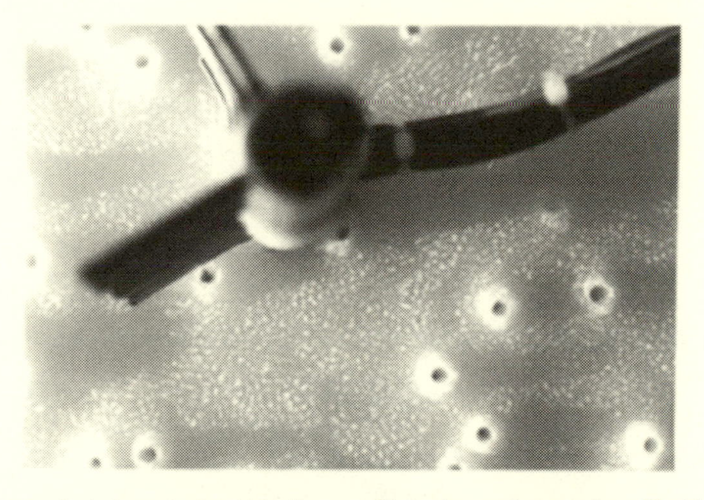

Fig. 17.7. A vertical view of a dendritically crystallizing layer of ammonium chloride crystals. The holes are channels within the mush. Photograph courtesy of Claude Jaupart.

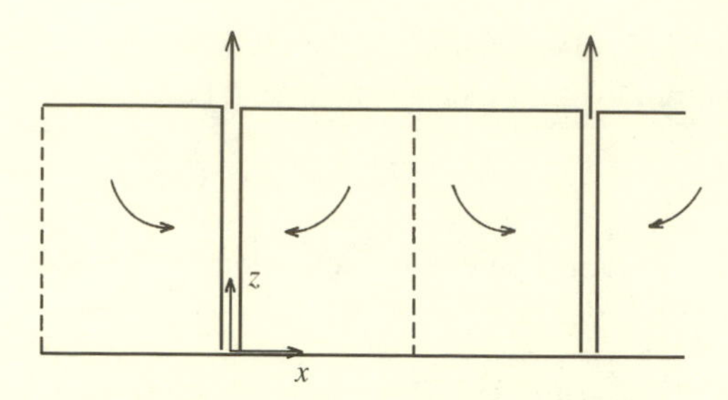

Fig. 17.8. Geometry for channelized flow

$\hat{G}$ is the driving pressure gradient. Using the nondimensionalized variables, and with the stream function given by Eq. (17.81), this implies that

$$\psi = -\frac{1}{3}\varepsilon^3 G/\gamma \tag{17.90}$$

on the tube wall, where G is the dimensionless pressure gradient,

$$G = -p_z + Rc, \tag{17.91}$$

and the dimensionless parameter

$$\gamma = \Pi_0/d^2 \tag{17.92}$$

is a relative measure of the dendritic spacing l_d (because $\Pi_0 \sim l_d^2$); typically $\gamma \ll 1$. We suppose that the solute concentration is uniform across the channel, $c = c^* : c^*$ will not necessarily be equal to the mush value at the wall, and the channel fluid may thus be superheated above the liquidus. We assume that the channel fluid temperature is also cross-sectionally uniform and that it equilibrates with the mush temperature; thus $T = -c$, where $c = c|_{\varepsilon+}$ is the mush concentration at the wall. The upward volume flux is $Q = -2\psi$, and a heat balance then implies $Q\partial T/\partial z = 2\partial T/\partial x$; thus T in the channel satisfies

$$-\psi\frac{\partial c}{\partial z} = \frac{\partial c}{\partial x}, \tag{17.93}$$

where $T = -c$ in the mush, $\partial c/\partial x$ being evaluated at $x = \varepsilon+$ in the mush. In addition, we take p to be continuous at the tube wall; thus with $G = -p_z + Rc^*$ ($c = c^*$ *in the channel*) and $\mathbf{u}.\mathbf{k} = w = -\psi_x = -p_z + Rc$ in the mush, we find from Eq. (17.90) that

$$\psi = -\frac{\varepsilon^3}{3\gamma}\{-\psi_x + R(c^* - c)\} \tag{17.94}$$

on the tube wall.

In order to determine the channel concentration c^*, we consider the solute flux from the channel. The actual solute equation in the channel is essentially

$$W\frac{\partial c}{\partial z} = \frac{1}{\varepsilon^2 Le}\frac{\partial^2 c}{\partial X^2}, \tag{17.95}$$

where we put $x = \varepsilon X$. The Poiseuille flow in the channel is

$$W = \frac{\varepsilon^2 G}{2\gamma}(1 - X^2), \tag{17.96}$$

with $G = -\psi_x + R(c^* - c)$. There are two particular end members. If $\varepsilon^4 GLe/\gamma \ll 1$, then $c_{XX} \sim 0$ and the channel concentration is thermodynamically equilibrated:

$$c^* \approx c \quad (\varepsilon^4 GLe/\gamma \ll 1), \tag{17.97}$$

where c is the mush value at $x = 0+$. If $\varepsilon^4 GLe/\gamma \gg 1$, then there is a boundary layer of thickness $X \sim (\gamma/\varepsilon^4 LeG)^{1/3}$ that delivers a solute gradient of order $-\partial c/\partial x \sim (\varepsilon LeG/\gamma)^{1/3}$ to the mush. More specifically, a similarity solution for the boundary layer equation gives a solute gradient in the mush

$$\frac{\partial c}{\partial x} = -\left(\frac{\varepsilon Le}{\gamma}\right)^{1/3} \left\{\frac{3}{\int_0^z dz/G}\right\}^{1/3} \frac{(c^* - c)}{\Gamma(1/3)} \quad (\varepsilon^4 GLe/\gamma \gg 1), \tag{17.98}$$

and lastly, a solute conservation law in the channel equivalent to that for T (in Eq. (17.93)) gives

$$-\psi \frac{\partial c^*}{\partial z} = \frac{1}{Le} \frac{\partial c}{\partial x}. \tag{17.99}$$

We now have sufficient conditions for the flow in the mush. We anticipate that $\varepsilon \ll 1$, so that the tube wall is at $x \approx 0$, and then (in a quasi-steady state)

$$\begin{aligned} \nabla^2 \psi &= Rc_x, \\ s[\psi_z c_x - \psi_x c_z] &= \nabla^2 c, \end{aligned} \tag{17.100}$$

with

$$\begin{aligned}
&\psi = 0, \qquad c = 1 \quad \text{on } z = 0, \\
&\frac{\partial \psi}{\partial n} = 0, \qquad c = 0 \quad \text{on } z = z_l, \\
&\psi = 0, \qquad c_x = 0 \quad \text{on } x = 1, \\
-\psi \frac{\partial c}{\partial z} = \frac{\partial c}{\partial x}, \qquad &\psi = -\frac{\varepsilon^3}{3\gamma}\{-\psi_x + R(c^* - c)\}, \qquad -\psi \frac{\partial c^*}{\partial z} = \frac{1}{Le}\frac{\partial c}{\partial x}, \\
&c^* = c \quad (\varepsilon^4 GLe/\gamma \ll 1), \quad \text{or} \\
\frac{\partial c}{\partial x} = -\left(\frac{\varepsilon Le}{\gamma}\right)^{1/3} &\left\{\frac{3}{\int_0^z dz/G}\right\}^{1/3} \frac{(c^* - c)}{\Gamma(1/3)} \quad (\varepsilon^4 GLe/\gamma \gg 1), \quad \text{on } x = 0.
\end{aligned} \tag{17.101}$$

The last four conditions on $x = 0$ give ψ and c as well as the channel concentration c^* at $x = 0$ and also serve to determine the unknown channel radius ε. A similar problem in determining z_l was circumvented earlier by choosing $\phi = 0$ there, equivalent to prescribing a continuous heat flux. For the channel, this is implicit in the heat flux condition (17.93), which itself follows from the assumption that the channels are (quasi-) steady, that is, $\dot{\varepsilon} = 0$.

Choice of scales

Firstly, observations in ammonium chloride experiments suggest $d \sim 3$ cm, $l_d \sim .1$ mm, and $a \sim 1$ mm, which (with tortuosity $\sim 10^2$) gives $\gamma \sim 10^{-5}$ and $\varepsilon \sim 3 \times 10^{-2}$. Also, $Le \sim 10^2$, so that *observations* suggest $\varepsilon^4 Le/\gamma \sim 10$ and is not small. In any event, the choice $c^* \approx c$ is problematic, as it then requires $\partial c/\partial z = \partial c/\partial x = 0$, and thus $c =$ constant throughout the mush (by a Cauchy–Kowalewski Taylor expansion). This suggests the requirement that $\varepsilon^4 Le/\gamma \gtrsim 1$.

Now observations suggest $\varepsilon^3/3\gamma \sim 1$ and $(\varepsilon Le/\gamma)^{1/3} \sim 10^2$, of which the latter implies $c \approx c^*$ anyway, if $\partial c/\partial x \sim O(1)$. Then $\partial c/\partial x = \partial c/\partial z = 0$, leading to the same problem as before. An alternative is to suppose that the chimneys induce a flow much larger than $O(1)$ (which is quite plausible anyway). Specifically, if $\psi \sim [\varepsilon LeG/\gamma]^{1/3}$, then the two terms in Eq. (17.98) balance. Because, if $R \sim O(1)$ at any rate, $G \sim \psi$, this corresponds to $\psi \sim G \sim (\varepsilon Le/\gamma)^{1/2}$. Thus we define

$$\varepsilon = \gamma^{1/3} h(z), \tag{17.102}$$

and $\delta \sim (\gamma/\varepsilon Le)^{1/2}$, whence

$$\delta = \gamma^{1/3}/Le^{1/2} \sim 10^{-3}; \tag{17.103}$$

we rescale ψ and G via

$$\psi, G \sim 1/\delta. \tag{17.104}$$

Then the rescaled problem is to solve

$$\begin{aligned} \nabla^2\psi &= \delta R c_x, \\ s[\psi_z c_x - \psi_x c_z] &= \delta \nabla^2 c, \end{aligned} \tag{17.105}$$

with

$$\begin{aligned} \psi = 0, &\qquad c = 1 \quad \text{on } z = 0, \\ \frac{\partial\psi}{\partial n}, &\qquad c = 0 \quad \text{on } z = z_l, \\ \psi = 0, &\qquad c_x = 0 \quad \text{on } x = 1, \end{aligned} \tag{17.106}$$

and on $x = 0$, we have $c^* \approx 1$, so that also

$$\begin{aligned} -\psi\frac{\partial c}{\partial z} &= \delta\frac{\partial c}{\partial x}, \\ \psi &= -\frac{1}{3}h^3\{-\psi_x + \delta R(1-c)\}, \\ \frac{\partial c}{\partial x} &= -\frac{1}{\delta\Gamma(1/3)}\left\{\frac{3h}{\int_0^z dz/G}\right\}^{1/3}(1-c), \end{aligned} \tag{17.107}$$

where $G = -\psi_x + \delta R(1-c)$.

Boundary layer solution

Putting $\delta = 0$, the stream function satisfies (uniformly) the outer problem

$$\nabla^2\psi = 0, \tag{17.108}$$

together with

$$\begin{aligned} \psi &= 0 \quad \text{on } z = 0, \\ \psi_n &= 0 \quad \text{on } z = z_l, \\ \psi &= 0 \quad \text{on } x = 1, \\ \psi &= \psi_0(z) \quad \text{(say) on } x = 0, \end{aligned} \tag{17.109}$$

whose solution can be obtained via Fourier series if z_l = constant (as is observed away from chimneys). The outer solution for c is $c = 0$ (to all orders of δ).

Near $x = 0$, we put $x = \delta\xi$; and to leading order, c satisfies

$$s\psi_0' c_\xi = c_{\xi\xi}, \tag{17.110}$$

with

$$-\psi_0 \frac{\partial c}{\partial z} = \frac{\partial c}{\partial \xi}, \tag{17.111a}$$

$$h = \{3\psi/\psi_x\}^{1/3}, \tag{17.111b}$$

$$\frac{\partial c}{\partial \xi} = -\frac{(1-c)}{\Gamma(1/3)} \left\{ \frac{9h}{-\int_0^z h^3\,dz/\psi_0} \right\}^{1/3} \quad \text{on } \xi = 0, \tag{17.111c}$$

and

$$c \to 0 \quad \text{as } \xi \to \infty. \tag{17.112}$$

The solution (because $\psi_0' < 0$) is

$$c = c_0 e^{s\psi_0'\xi}, \tag{17.113}$$

where $c = c_0(z)$ on $\xi = 0$, and c_0 satisfies

$$-\psi_0 c_0' = s c_0 \psi_0' = -\frac{(1-c_0)}{\Gamma(1/3)} \left\{ \frac{9h}{-\int_0^z h^3\,dz/\psi_0} \right\}^{1/3}. \tag{17.114}$$

Thus

$$c_0 = 1/k|\psi_0|^s, \tag{17.115}$$

where k must be determined by matching to the basal boundary layer near $z = 0$, $x = 0$, and then $\psi_0 = -\Psi$, where Ψ (> 0) satisfies

$$s\Psi' = \frac{(k\Psi^s - 1)}{\Gamma(1/3)} \left\{ \frac{9h}{\int_0^z (h^3/\Psi)\,dz} \right\}^{1/3}. \tag{17.116}$$

Thus, numerically, given Ψ, we solve for ψ in the mush, and then determine $h[\Psi]$ via Eq. (17.111b). Ψ can then be determined by integrating Eq. (17.116). Some care is needed in matching the solutions to the basal boundary layer at $z = 0$.

Summary

We have found a boundary layer formulation for channelized flow based on the limit $\delta = \gamma^{1/3}/Le^{1/2} \ll 1$. The consequent analysis is then valid provided $\delta R \ll 1$, that is, $R \ll Le^{1/2}/\gamma^{1/3}$. Once channels are initiated, the excess buoyancy in them dominates any such effect in the mush and controls the flow, which is passively dragged through the mush. A compositional boundary layer near the channel causes an increased solid fraction there. The dimensionless flux from the channel is $|\psi| \sim 1/\delta$, corresponding to a dimensional velocity of order $(\kappa/d).(1/\delta\varepsilon) = \kappa/\delta a$. Values of $\kappa \sim 10^{-3}$ cm^2 s^{-1}, $\delta \sim 10^{-3}$, and $a \sim 1$ mm, correspond to an order of magnitude of 10 cm s^{-1}, roughly comparable to that observed.

17.6 Modeling queries

In determining the location of the mush–liquid interface z_l, and also the channel boundary ε, we have used a condition of zero freezing. That is, at z_l we prescribe zero solid fraction ϕ, whereas at the channel boundary we implicitly assume continuity of heat flux. There is no *a priori* reason to assume such a condition. For example, a reasonable alternative would be to suppose that, because constitutional supercooling is the real reason why a dendritic mush forms, then the mush front advances just so as to eliminate the supercooling. This means that the temperature gradient in the liquid of the front is equal to the liquidus temperature gradient. In the analysis given in this chapter, this is tantamount to taking zero solid fraction at z_l, but this need not be necessarily the case, and in certain cases, one *cannot* prescribe zero solid fraction at z_l.

The question of interest is to understand how one could derive such an additional boundary condition. A similar problem arises in the study of the propagation of single dendrites and in the related problem of the Saffman–Taylor instability in a Hele–Shaw cell. Here one finds an infinity of possible solutions, parameterized by their speed (for example), although in practice only one appears to be selected. The mathematical situation seems analogous to the solutions of the Fisher equation

$$\begin{aligned} u_t &= u(1-u) + u_{xx}, \\ u &\to 1 \quad \text{as } x \to -\infty, \\ u &\to 0 \quad \text{as } x \to +\infty, \end{aligned} \tag{17.117}$$

which represents the propagation of a diffusing, resource-limited population. It can be shown that any initial data that decays fast enough at $\pm\infty$ tends to a traveling-wave solution with speed $c = 2$. Yet there are traveling-wave solutions for all wave speeds $c \geq 2$, and they are all linearly stable! The control of the final state is intimately bound up with the decay rate of u at $\pm\infty$. It has been suggested that the growth of a single dendrite might be controlled by the same kind of 'marginal equilibrium,' or 'marginal stability' criterion, and this idea has been carried across to mush evolution, via the concept of marginal supercooling. Nevertheless, it has no real justification.

Dendrite spacing

One possible source of a boundary condition lies in prescription of a further evolution equation. One such equation is certainly needed, because the dendrite spacing l_d is not known *a priori*. As a self-evolving system, the mush selects its own microstructure, and moreover this is manifested in the permeability because $\Pi \sim l_d^2$.

Two specific microstructural quantities related to the notion of a mean dendrite spacing are the (average) curvature H and the specific interfacial surface area s. If we make particular assumptions about the dendrite geometry, then expressions for H and s can be derived in terms of a length (l_d) and the solid volume fraction α_s. For a dendritic array consisting of cylinders of constant radius r and distance l_d apart, we find

$$\begin{aligned} H &\approx \left\{ \frac{\pi^{1/2} - 2\alpha_s^{1/2}}{\alpha_s^{1/2}} \right\} l_d^{-1}, \\ s &\approx \left\{ 2\alpha_s^{1/2}\left(\pi^{1/2} - 2\alpha_s^{1/2}\right) \right\} l_d^{-1}, \end{aligned} \tag{17.118}$$

which follow from the relation $r/l_d \approx \alpha_s^{1/2}/(\pi^{1/2} - 2\alpha_s^{1/2})$, and one might expect, in general, that $\Pi = \Pi(H, s)$. Drew (1990) has considered evolution equations for s and H, but no application to mushy zones has been made; and in particular, such equations can, for example, describe ripening of dendrites, but these require initial values of H and s at z_l. Although H/s could be determined by considering the mush front as a collection of individual dendrites, prescription of l_d there is still necessary.

It would seem axiomatic that the internal dendrite spacing is controlled by the local growth rate of dendrites. For such growth to occur, the interstitial fluid must be out of thermodynamic equilibrium. Thus, although it is reasonable to assume that the average temperature T of the pore fluid is equal to that of the interface T_i, we would *not* suppose that the pore fluid concentration c is equal to its interfacial value; rather (because solute is rejected) we suppose $c < c_i$, and thus (as we would prescribe thermodynamic equilibrium $T_i = T_0 - \Gamma c_i$ at an interface) the pore fluid is supercooled (by $\Gamma(c_i - c)$). In fact, we might expect interfacial curvature effects to be important; thus the temperature is given by the liquidus,

$$T = T_i = T_0 - \Gamma c_i - \frac{2\sigma H T_0}{\rho_s L}, \tag{17.119}$$

where σ is surface energy and H is mean curvature.

Because solute transport to the interface should be rate limiting, we have, in a quasi-steady state, that the normal outward velocity v of the interface is approximately given (in order of magnitude) by

$$v c_i \approx D[c_i - c]/l_d, \tag{17.120}$$

and the normal velocity is related to the interfacial area by

$$v = -\alpha_t/s, \tag{17.121}$$

where α is the liquid volume fraction. If, in addition, we posit the interfacial kinetic growth condition

$$v = m\Delta T_{uc} = m\left[\Gamma(c_i - c) + \frac{2\sigma H T_0}{\rho_s L}\right], \tag{17.122}$$

then we can use Eqs. (17.120)–(17.122) to calculate v, c_i, and l_d in terms of c and s, which we suppose are determined from the macroscopic equations. It is, however, a debatable procedure as to whether the extra condition (17.122) can properly be applied. Some justification might be made by analogy with the freezing rate of a single dendrite, but this simply pushes the problem back to a different one.

If we scale the variables as before (but v with κ/l_d and s with l_d^{-1}), then we obtain

$$\begin{aligned} v c_i &= \frac{1}{Le}(c_i - c), \\ v &= -\nu^2 \alpha_t/s, \\ v &= \delta_1(c_i - c) + \delta_2 H, \end{aligned} \tag{17.123}$$

where

$$\delta_1 = \frac{m\Gamma c_\infty l_d}{\kappa}\beta, \qquad \delta_2 = \frac{2m\sigma T_0}{\rho_s L\kappa}, \qquad \nu = \frac{l_d}{d}. \tag{17.124}$$

If we use values $\Gamma \sim 500$ K, $c_\infty \sim 0.8$, $l_d \sim 0.1$ mm, $\kappa \sim 10^{-3}$ cm^2 s^{-1}, $\sigma \sim 3 \times 10^{-2}$ J m^{-2} (for water), $T_0 \sim 400$ K, $\rho_s \sim 10^3$ kg m^{-3}, $L \sim 3 \times 10^5$ J kg^{-1}, $d \sim 10$ cm, and a value $m \sim 10^{-4}$ cm s^{-1} K^{-1}, then we find

$$\delta_1 \sim 0.4\beta, \qquad \delta_2 \sim 0.8 \times 10^{-6}, \qquad \nu^2 \sim 10^{-6}. \tag{17.125}$$

Apparently we can neglect the Gibbs–Thomson parameter δ_2. Eliminating v and c_i, we obtain the equation

$$1 + \nu^2 Le(\alpha_t/s) = Le\delta_1 c, \tag{17.126}$$

and if we ignore $O(\nu^2 Le) \sim 10^{-4}$, then

$$\delta_1 \approx 1/Lec. \tag{17.127}$$

Because $\delta_1 \propto l_d$, Eq. (17.127) can be construed as providing a local choice for the dendrite spacing:

$$l_d \approx \frac{D}{m\Gamma\Delta c}, \tag{17.128}$$

where $\Delta c = c - c_\infty$ (in unscaled variables) is the excess solute concentration.

The dubious part of this discussion is the prescription of Eq. (17.122). It does, however, indicate that some extra detail concerning a thermodynamic balance may hold the key to future progress.

Thermodynamic equilibrium

As has been alluded to above, it can be the case that thermodynamic equilibrium is not a good assumption, and hence a simple kinetic undercooling model such as that in Eq. (17.122) may be appropriate. A simple version of this at $z = z_l$ is

$$\dot{z}_l = m\Delta T_{uc}, \tag{17.129}$$

where ΔT_{uc} is the undercooling, that is, $T_L - T$, where T_L is the local liquidus temperature. This is taken to replace either the zero solid fraction condition or the marginal stability condition, whichever is applied.

A different version, applicable to the internal mush, would be to prescribe the condition (cf. Eqs. (17.121) and (17.122))

$$\phi_t \approx sm[\Delta T_{uc}]_+ = sm[T_L - T]_+ \tag{17.130}$$

in the mush. This equation requires prescription of a boundary condition for ϕ, which we might naturally suppose to be $\phi = 0$ at z_l. Strictly, this requires also $T = T_L$ at $z = z_l$, which is thus inconsistent with Eq. (17.129). Other possibilities can be suggested. If we consider a uniform (in temperature and composition) dendritic array with a gradient in mean curvature ∇H, then there is a corresponding gradient in the liquidus temperature; and thus we obtain liquidus-induced melting and refreezing, which results in the phenomenon of *Ostwald ripening*, whereby large dendrite arms grow at the expense of smaller ones. In this sense, H moves up ∇H gradients, whence we might expect an elliptical problem for H, and thus boundary conditions for H at z_l, for example. What these should be, however, is less clear. More sophisticated models of this type await future development.

17.7 Notes and references

The instability of a planar interface in a solidifying alloy was studied, in particular, by Mullins and Sekerka (1964). Since then, a lot of work has been done on nonlinear stability and pattern selection. In regard to the modeling of dendrites, a separate body of work concerns itself with the propagation of a single dendrite. Two relevant workshop proceedings are those edited by Davis et al. (1992) and Loper (1987). A remarkable solution owing to Ivantsov (1947) gives an exact solution for the growth of a dendrite in the form of a paraboloid of revolution, in the absence of surface tension. A whole body of research has risen around this theory, because if the surface tension γ is included as a small term, the problem is singularly perturbed, but it appears that the introduction of this term causes ripples of exponentially small (in γ) magnitude to occur (and Ivantsov's 'needle crystal' solution fails to exist). The problem is exemplified by the model (Hakim, 1991)

$$\varepsilon\dddot{\theta} + \dot{\theta} = \sin\theta, \tag{17.131}$$

which represents growth of an interface at a rate proportional to curvature; the angle θ is the angle of the surface to the direction of growth: That is, $\theta \to 0$ as $s \to -\infty$ and $\theta \to \pi$ as $s \to \infty$, where s is arc length along the surface (and $\dot{\theta} = d\theta/ds$). Hence ε is the surface tension term. The zero surface tension solution with $\varepsilon = 0$ is just

$$\theta = 2\tan^{-1} e^s, \tag{17.132}$$

and for small ε a regular perturbation expansion can be shown to exist, with the property that $\ddot{\theta}(0) = 0$ to all orders of ε, as is required for a symmetric solution. However, it can be shown that in fact $\ddot{\theta}(0)$ is nonzero (and is exponentially small in ε), which implies that a steady propagating symmetric dendrite does not exist. The study of such 'exponential asymptotics' is a subject of much current interest.

Mushy layers A lot of work has been done in the last ten years on the modeling of mushy layers, much of it by Grae Worster, whose review (Worster, 1992) is a useful source. Much of the present chapter is based on an M.S. thesis at Oxford written by Paul Emms, which he subsequently developed into a Ph.D. thesis. Some of this work is reported in Emms and Fowler (1994). The boundary layer solution in Section 17.5 is rather schematic and has not been reported elsewhere. There is clearly some further work to do, as, for example, the solution does not connect in an obvious way to the base at $z = 0$. On the face of it, we should expect a shear layer at the base of the chimney, where Ψ in Eq. (17.116) jumps rapidly to $k^{-1/s}$. More detailed work is necessary to understand the solution properly.

A rather different boundary layer theory has been presented by Worster (1991) based on the limit $R \gg 1$. He also assumes $\psi \sim 1$ rather than $\psi \gg 1$, as here, and the relation of his theory to the present description is therefore opaque. Evidently, this is a problem awaiting further research.

18

Ice sheet dynamics

The Earth has two major ice sheets, those covering Greenland and Antarctica, as well as a number of smaller ice caps, and numerous glaciers in Alaska, the Alps, and many other parts of the world. Ice sheets and glaciers are of awesome magnitude, and although relatively slowly moving, are capable of potent and significant effects. In former times, other parts of the Earth have been covered by ice, particularly during ice ages and in the last ice age, which finished about 10,000 years ago, two major ice sheets covered large parts of North America and northern Europe. The existence of these former ice sheets is a major cause of various geomorphologic features such as moraines (e.g., Cape Cod), drumlins, eskers, and proglacial lakes (the former Lake Bonneville, whose remnant is the Great Salt Lake). The understanding of the Earth's surficial features is thus partly determined by an understanding of the dynamics of how the ice sheets changed with time.

Ice sheets also influence, and are influenced by, climate. The existence of the Pleistocene ice sheets affected the oceanic circulation pattern and hence precipitation patterns. In turn, this affected the ice sheets themselves through the resultant snowfall. In addition, continental ice sheets affect the oceanic circulation via the freshwater flux released by meltwater and via their energy balance fluxes. On a smaller scale, glaciers both respond to, and are an indicator of, climatic change. It is well known that the 'Little Ice Age' in Europe between 1500 and 1900 is charted by the advance of the Alpine glaciers. In view of the current interest in the effect of CO_2 emissions on the Earth's climate, an understanding of the ways in which glaciers and ice sheets interact with the oceans and the atmosphere is of some significance.

Glaciers and ice sheets are vast and slow-moving edifices of solid ice. They flow under their own weight by solid-state creep processes such as the creep of dislocations in the crystalline lattice structure, and in this resemble rivers, except that they move more slowly (and are consequently much thicker). Glaciers typically have thicknesses on the order of 50–500 meters and move at velocities of 10–100 meters per year. Ice sheets move at comparable rates, but their thickness is several kilometers (up to five for the Antarctic). Their comparably lower relative velocity is due to a smaller surface slope (1/1000, as opposed to 1/100–1/10 for valley glaciers). Despite their slow movement and apparent changelessness, ice sheets and glaciers exhibit various interesting dynamic phenomena.

Various wavelike phenomena occur on glaciers, the most dramatic of which is the glacier *surge*, a periodically recurring advance that occurs at intervals of 10–100 years on a small number of glaciers, during which velocities can increase to hundreds of meters *per day*, and the ice surface can concomitantly lower by a hundred meters over several months.

A similar phenomenon (to some extent) occurs in ice sheets, where one sees drainage of the ice toward the coast occurring through a series of *ice streams*, which are highly crevassed rapid flows on the order of 50 km wide, bounded by regions of more stagnant ice. It is still a matter of debate as to what causes the distinction between the active ice stream and its more passive boundaries, although there is a suggestive analogy with the surging process. Moreover, these ice streams themselves can behave in a time-dependent manner. In Antarctica, ice flows over the trans-Antarctic mountains into the Ross Ice Shelf and drains to the sea through five streams, A–E. There is near-surficial evidence (buried crevasses) to suggest that ice stream C flowed rapidly until about 200 years ago, but suddenly switched off and is now largely dormant. An understanding of such features is of some interest.

Mostly, one expects the response of ice sheets to be slow, because their natural timescale is on the order of thousands of years. Nevertheless, this is a kind of geologically conditioned thinking, and one should not be complacent. Imminent collapse of the West Antarctic ice sheet, and a resulting rise of sea level by six meters (and thus disaster) is one possible scenario that has been considered. A much more dramatic example may be that which describes hypothetical events in the North Atlantic during the last ice age.

Evidence from ice cores in Greenland (where the oxygen-18 isotope ratio in bubbles in the ice gives an indirect measure of prevailing atmospheric temperature) indicate that during the last ice age, oceanic circulation in the North Atlantic may have repeatedly shut down, leading to sudden climatic temperature shifts of up to 10°C in as little as ten years. For obvious reasons, the mechanism of these *Dansgaard–Oeschger events* are of interest.

Possibly related *Heinrich events* evidenced in deep-sea sediment cores from the North Atlantic reveal periods where deposition of soft cretaceous mudstones took place. These mudstones come from Hudson Bay, and a mechanism that can account for both sets of events, when they occur at the same time, is that subglacial evacuation of the basal mudstones is associated with an increased ice flux and iceberg discharge through the Hudson Strait into the Labrador Sea. The resultant cold water flux then causes the North Atlantic Deep Water circulation to switch on. The periodic recurrence of these events is associated with 'megasurges' of an enormous glacier in the Hudson Strait, which evacuates the ice dome over Hudson Bay. We will come back to this concept in Section 18.5.

18.1 Basic equations and the shallow ice approximation

We consider the geometry indicated in Fig. 18.1. Ice is usually taken to be incompressible, and the flow is very slow; thus suitable equations of mass, momentum and

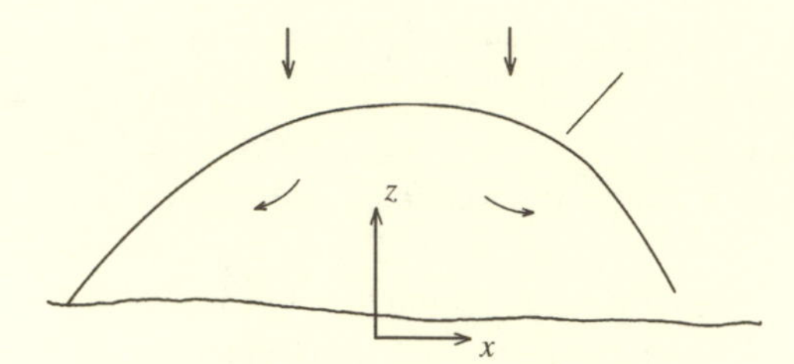

Fig. 18.1. Ice sheet geometry

energy are

$$
\begin{aligned}
u_x + v_y + w_z &= 0,\\
p_x &= \tau_{11,x} + \tau_{12,y} + \tau_{13,z},\\
p_y &= \tau_{21,x} + \tau_{22,y} + \tau_{23,z},\\
p_z &= \tau_{31,x} + \tau_{32,y} + \tau_{33,z} - \rho g,\\
\rho c_p \frac{dT}{dt} &= k\nabla^2 T + \tau_{ij}\dot{\epsilon}_{ij},
\end{aligned}
\tag{18.1}
$$

where $\tau_{11,x} = \partial\tau_{11}/\partial x$, etc., τ_{ij} is the deviatoric stress tensor, and $\dot{\epsilon}_{ij}$ is the strain rate tensor. Thus $\dot{\epsilon}_{ij}$ is $\frac{1}{2}(\frac{\partial u_i}{\partial x_j} + \frac{\partial u_j}{\partial x_i})$, and we assume a rheological flow law of the form

$$\dot{\epsilon}_{ij} = A(T)g(\tau)\tau_{ij}/\tau, \tag{18.2}$$

where the second stress invariant τ is defined by

$$2\tau^2 = \tau_{ij}\tau_{ij}. \tag{18.3}$$

Note that $\tau_{ij} = \tau_{ji}$, τ_{ii} $(= \tau_{11} + \tau_{22} + \tau_{33}) = 0$, and we apply the summation convention. For a constant viscosity fluid, A and g would be constant, but it is more normal to take a power law rheology for ice given by *Glen's law*:

$$g(\tau) = \tau^n, \quad n = 3, \tag{18.4}$$

and a thermally activated rate coefficient,

$$A = A_0 \exp[-Q/RT] \tag{18.5}$$

(although other forms can be used). The temperature variation of the effective viscosity is significant, encompassing three orders of magnitude over a temperature range of 50°K appropriate to the Antarctic, for example.

Shallow ice

The basis for solving the equations lies in the fact that whereas the lateral extent of the Antarctic, for example, is ~3,000 kilometers, its thickness is typically ~3 kilometers, and thus the aspect ratio is extremely large. This enables us to use the lubrication approximation to simplify the equations. Consistently, $x, y \gg z$, and thus the dominant stresses are the shear stresses τ_{13}, τ_{23}, also $u, v \gg w$, and the pressure is close to hydrostatic (or more appropriately, cryostatic).

To be specific, we scale the variables as follows:

$$
\begin{gathered}
x, y \sim l; \qquad z, \eta, h \sim d; \\
\tau_{13}, \tau_{23} \sim [\tau]; \qquad \tau_{11}, \tau_{22}, \tau_{12} \sim \epsilon[\tau]; \\
u, v \sim [u]; \qquad w \sim \epsilon[u]; \\
A \sim [A], \qquad g \sim [g], \qquad t \sim l/[u], \\
T - T_m \sim \Delta T, \qquad p - \rho g(\eta - z) \sim \epsilon[\tau],
\end{gathered}
\tag{18.6}
$$

in which T_m is the melting temperature (at ambient pressure, 273°K), $z = \eta$ is the top surface, $z = h$ is the base, and $\epsilon = d/l$. Of the scales l, d, $[\tau]$, ΔT, $[A]$, $[u]$, and $[g]$, we take l and ΔT as given from boundary data. We also suppose that the surface accumulation a is prescribed and of order $[a]$; hence we choose

$$[a] = \epsilon[u]. \tag{18.7}$$

The choice of $[A]$ and $[g]$ is so that the corresponding scaled functions are $O(1)$. For example, for Eqs. (18.4) and (18.5), we choose

$$[g] = [\tau]^n, \qquad [A] = A_0 \exp[-Q/RT_m]. \tag{18.8}$$

We thus have two relations to choose. One is a balance of shear stress and cryostatic pressure gradient, leading to

$$[\tau] = \rho g\, d\epsilon, \tag{18.9}$$

and the other is a balance of the shear rate in the flow law

$$\frac{[u]}{\nu d} = 2[A][g], \tag{18.10}$$

where we would normally take $\nu = 1$, but it will be useful to keep for Section 18.4, where we find that most shearing takes place in a basal region of thickness $\nu d \ll d$ when the flow is strongly thermally activated.

Evaluating the unknowns $[\tau]$, d, and $[u]$ gives the values

$$[\tau] = \left\{ \frac{\rho g[a]}{2\nu[A]} \right\}^{\frac{1}{n+1}}, \tag{18.11a}$$

$$d = \left\{ \frac{[\tau]l}{\rho g} \right\}^{1/2}, \tag{18.11b}$$

$$[u] = 2\nu[A]d[\tau]^n. \tag{18.11c}$$

A basic measure of consistency is then that these values correspond roughly to observations. In particular, the depth scale is

$$d = \left\{ \frac{l^{n+1}[a]}{2\nu(\rho g)^n[A]} \right\}^{\frac{1}{2(n+1)}} = \left(\frac{l}{\rho g} \right)^{1/2} \left\{ \frac{[a]\rho g}{2\nu[A]} \right\}^{\frac{1}{2(n+1)}}. \tag{18.12}$$

Choosing values $l = 3000$ km, $\rho g = 0.1$ bar m^{-1}, $[a] = 0.1$ m y^{-1}, and $[A] = 0.2$ bar^{-n} y^{-1}, we have, if $n = 3$ and $\nu = 1$, $d \approx 3.5$ km, which accords well with the observed order of magnitude and suggests the consistency of the suggested scales.

Scaling the variables as in Eq. (18.6), we obtain the equations in the following form:

$$
\begin{aligned}
&u_x + v_y + w_z = 0, \\
&\qquad 0 = -\eta_x + \tau_{13,z} + \epsilon^2[-p_x + \tau_{11,x} + \tau_{12,y}], \\
&\qquad 0 = -\eta_y + \tau_{23,z} + \epsilon^2[-p_y + \tau_{12,x} + \tau_{22,y}], \\
&\qquad 0 = -p_z + \tau_{13,x} + \tau_{23,y} - \tau_{11,z} - \tau_{22,z}, \\
&\qquad \frac{dT}{dt} = (\alpha/\nu)\tau A(T) g(\tau) + \beta(T_{zz} + \epsilon^2 T_{xx} + \epsilon^2 T_{yy}), \\
&\qquad \tau^2 = \tau_{13}^2 + \tau_{23}^2 + \epsilon^2\left[\tau_{12}^2 + \tfrac{1}{2}\{\tau_{11}^2 + \tau_{22}^2 + (\tau_{11} + \tau_{22})^2\}\right],
\end{aligned}
\tag{18.13}
$$

and the strain rate components are given by

$$
\begin{aligned}
\tau_{11} &= 2\mu u_x, \\
\tau_{22} &= 2\mu v_y, \\
\tau_{12} &= \mu(u_y + v_x), \\
\tau_{13} &= \mu(u_z + \epsilon^2 w_x), \\
\tau_{23} &= \mu(v_z + \epsilon^2 w_y),
\end{aligned}
\tag{18.14}
$$

where

$$
\mu = \nu\tau/Ag. \tag{18.15}
$$

Other than ϵ, the parameters in Eq. (18.13) are given by

$$
\alpha = \frac{gd}{c_p \Delta T}, \qquad \beta = \frac{\kappa}{d[a]}, \tag{18.16}
$$

where κ is the thermal diffusivity $k/\rho c_p$. Using values $g \sim 10$ m s^{-2}, $d \sim 3$ km, $c_p \sim 2 \times 10^3$ J kg^{-1} K^{-1}, $\Delta T = 50$ K, $\kappa = 38$ m^2 y^{-1}, and $[a] = 0.1$ m y^{-1}, we compute $\alpha = 0.3$, $\beta = 1/8$. We see that viscous heating is significant in the flow, whereas heat conduction is moderate to small for this choice of parameters.

Boundary conditions

At the surface $z = \eta(x, y, t)$, we prescribe zero stress; thus $\sigma_{ij} n_j = 0$ for $i = 1, 2, 3$. This leads to

$$
\begin{aligned}
\tau_{13} &= \epsilon^2[(-p + \tau_{11})\eta_x + \tau_{12}\eta_y], \\
\tau_{23} &= \epsilon^2[\tau_{12}\eta_x + (-p + \tau_{22})\eta_y], \\
p + \tau_{11} &+ \tau_{22} + \tau_{13}\eta_x + \tau_{23}\eta_y = 0;
\end{aligned}
\tag{18.17}
$$

in addition, the temperature may be prescribed,

$$
T = T_A, \tag{18.18}
$$

and there is a surface kinematic condition

$$
\eta_t + u\eta_x + v\eta_y - w = a, \tag{18.19}
$$

where a is the scaled accumulation rate ($a < 0$ signifies ablation).

At the base $z = h$, a prescribed geothermal heat flux G leads to a scaled boundary condition

$$\frac{\partial T}{\partial z} - \epsilon^2 h_x \frac{\partial T}{\partial x} - \epsilon^2 h_y \frac{\partial T}{\partial y} = -\left[1 + \epsilon^2 \left(h_x^2 + h_y^2\right)\right] \Gamma, \tag{18.20}$$

where

$$\Gamma = Gd/k\Delta T. \tag{18.21}$$

For a value $G \sim 5 \times 10^{-2}$ W m^{-2}, $k \sim 2.1$ W m^{-1} K^{-1}, we have $\Gamma \approx 1.5$. Equation (18.20) only applies if $T < 0$ at the base. If T reaches the melting temperature, then we fix $T = 0$, and the geothermal heat flux contributes to a production of water at the base. The resultant velocity of the ice downward is usually small, and can be neglected (see below).

Sliding

When $T < 0$, it is normal to assume a no-slip condition,

$$u = v = 0, \qquad T < 0 \tag{18.22}$$

at the base. However, when T reaches zero, the water produced enables the base to become partly lubricated, and in that case, some sliding can occur. Sliding is driven by the basal shear stress, and we write

$$u = u_b = \phi\tau_{13}, \qquad v = v_b = \phi\tau_{23} \tag{18.23}$$

in that case, where ϕ is a function of basal stress and water pressure that must be determined by a local analysis of conditions at the base. This is discussed further in Section 18.5.

The limit $\epsilon \to 0$

With $d = 3$ km, $l = 3000$ km, we have $\epsilon \sim 10^{-3}$, and in all practical circumstances one uses the limit $\epsilon \to 0$. It then follows that, to leading order,

$$\begin{aligned} \tau_{13} &\approx -(\eta - z)\eta_x, \\ \tau_{23} &\approx -(\eta - z)\eta_y, \\ \tau &\approx (\eta - z)|\nabla\eta|, \\ \tau_{13} \approx \mu u_z,& \qquad \tau_{23} \approx \mu v_z, \end{aligned} \tag{18.24}$$

where $\nabla = (\partial_x, \partial_y)$ is the horizontal gradient operator. Two integrations of the last two equations give (in principle) an expression for the horizontal flux vector

$$\mathbf{Q} = \int_h^\eta (u, v)\, dz, \tag{18.25}$$

and then the continuity equation together with the kinematic boundary condition gives the integrated mass conservation equation

$$\eta_t + \nabla.\mathbf{Q} = a. \tag{18.26}$$

18.2 Isothermal flow

If we take $A = 1$, then these integrations can be done explicitly. We put $\nu = A = 1$, $g = \tau^n$; thus

$$\begin{aligned} u_z &= \tau^{n-1}\tau_{13} = -(\eta - z)^n |\nabla\eta|^{n-1}\eta_x, \\ v_z &= \tau^{n-1}\tau_{23} = -(\eta - z)^n |\nabla\eta|^{n-1}\eta_y. \end{aligned} \tag{18.27}$$

Thus

$$\frac{\partial \mathbf{u}_H}{\partial z} = -(\eta - z)^{n-1}|\nabla\eta|^{n-1}\nabla\eta, \tag{18.28}$$

where $\mathbf{u}_H$ is the horizontal velocity (u, v); then

$$\mathbf{u}_H = \mathbf{u}_b - \frac{1}{n}[(\eta - h)^n - (\eta - z)^n]|\nabla\eta|^{n-1}\nabla\eta, \tag{18.29}$$

where $\mathbf{u}_b = (u_b, v_b)$ is the basal sliding velocity. The ice flux is

$$\mathbf{Q} = \int_h^\eta \mathbf{u}_H\,dz = H\mathbf{u}_b - \frac{H^{n+1}}{n+1}|\nabla\eta|^{n-1}\nabla\eta, \tag{18.30}$$

where $H = \eta - h$ is the ice thickness. Thus η satisfies the nonlinear diffusion equation

$$\eta_t = \nabla.\left[\left\{\phi H^2 + \frac{H^{n+1}}{n+1}|\nabla\eta|^{n-1}\right\}\nabla\eta\right] + a. \tag{18.31}$$

In general, this equation must be solved numerically, but its principal features can be derived semianalytically.

Steady state

The accumulation rate a is an essential part of Eq. (18.31), as without accumulation, the ice sheet would not exist. The simplest solution is in one space dimension (x), with no slip $(\phi = 0)$ and a level base ($h = 0$, whence $\eta = H$): Then the steady state ice flux is given by

$$Q = -|H_x|^{n-1}H_x H^{n+2}/(n+2) = \int_0^x a\,dx \triangleq s(x), \tag{18.32}$$

if we assume $a = a(x)$ and measure x from a central divide. Then

$$H^{2(n+1)/n} = \frac{2(n+1)}{n}\int_x^{x_m}[(n+2)s(x)]^{1/n}dx \tag{18.33}$$

gives H, where x_m denotes the margin, where H is prescribed to be zero: x_m is defined from $\int_0^{x_m} a(x)\,dx = 0$, in order to conserve mass. A useful approximation follows from assuming that accumulation is constant in the continental interior and all the ablation occurs at the margin. In this case, $a = 1$, $s = x$, and so (with $x_m = 1$)

$$\frac{H^{2(n+1)/n}}{2(n+2)^{1/n}} + x^{\frac{n+1}{n}} = 1, \tag{18.34}$$

giving the standard 'hyper-elliptical' shape.

Multiple steady states

A fundamental principle in the growth of the Pleistocene ice sheets is that accumulation varies with altitude. Two effects are important. Atmospheric temperature decreases at higher elevations, which will tend to increase accumulation. On the other hand, less moisture will be available. Particularly, thick ice is inland ice: Antarctica, for example, is essentially a desert. Therefore it is reasonable to take $a = a(\eta, x, t)$, and in certain circumstances this feedback can lead to multiple steady states. The simplest such situation is where $a = a(H)$ is a monotone increasing function. For simplicity, take $n = 1$ and define $\Phi = H^4/12$, so that

$$\Phi_{xx} + a(\Phi) = 0, \tag{18.35}$$

where we consider a as a function of Φ. If the maximum of Φ at $x = 0$ is Φ_m, then two integrations determine Φ_m implicitly via

$$\int_0^{\Phi_m} \frac{d\rho}{\left\{\int_\rho^{\Phi_m} a(\psi)\,d\psi\right\}^{1/2}} = \sqrt{2}x_m, \tag{18.36}$$

where x_m is the margin position. Considering x_m as a function of Φ_m defined by Eq. (18.36), we see that, if $a \to a_0$ when $\Phi \to 0$, then

$$x_m \sim (\Phi_m/2a_0)^{1/2} \quad \text{as } \Phi_m \to 0, \tag{18.37}$$

whereas if $a(\Phi) > c\Phi^\alpha$, $\alpha > 0$, then

$$\left(\frac{2c}{\alpha+1}\right)^{1/2} x_m < \Phi_m^{-(\alpha-1)/2} \int_0^1 \frac{dz}{[1 - z^{\alpha+1}]^{1/2}}, \tag{18.38}$$

and if $\alpha > 1$, then $x_m \to 0$ as $\Phi \to \infty$. This also applies only if $a > O(\Phi^\alpha)$, $\alpha > 1$ for $\Phi \to \infty$, and we see that essentially any convex function $a(\Phi)$ will lead to multiple steady states.

In reality, $a > H^4$ as $H \to \infty$ is unlikely, but if a increases sufficiently rapidly with H over a limited range of H, then we can expect there to exist a portion of the response curve Φ_m versus x_m of negative slope (and which is therefore unstable, in a manner analogous to that which occurs in thermal runaway).

Margins

At the margin $x = x_m$, it follows from Eq. (18.33) that in the general case where $s \sim x_m - x$ as $x \to x_m$, $H \sim (x_m - x)^{1/2}$, and the margin slope is singular. This is a common feature of degenerate diffusion equations such as Eq. (18.31) and suggests that a local rescaling of the problem is necessary there. Indeed, if $x_m - x \sim \delta \ll 1$, then (with $\mathbf{u}_b = \mathbf{0}$) $s \sim \delta$, $H \sim z \sim \delta^{1/2}$, $\tau_{13} \sim 1$, $\tau_{11} \sim \delta^{-1/2}$, $u \sim \delta^{1/2}$, $w \sim 1$, and $p \sim \delta^{-1/2}$, whence a distinguished limit occurs if $\delta = \epsilon^2$, and the neglected terms of $O(\epsilon^2)$ in Eq. (18.13) become comparable to the others. Thus with $x_m - x \sim \epsilon^2$, the full Stokes flow problem is recovered. This suggests that the margin is passive (it does not affect the behavior of the rest of the ice sheet), and that $H_x \sim 1/\epsilon$ there. In practice, computation of such details is an unnecessary complication.

Divides

As $x \to 0$ in Eq. (18.33), we have

$$|H_x| \sim x^{1/n}, \qquad u \sim x, \qquad \tau_{13} \sim x^{1/n}, \qquad p \sim \tau_{11} \sim x^{(1/n)-1}, \tag{18.39}$$

and the reduced model becomes invalid when $x \sim \epsilon$, when $\tau_{11} \sim p \sim \epsilon^{(1/n)-1}$, $\tau_{13} \sim \epsilon^{1/n}$, and $H_x \sim \epsilon^{1/n}$, and a local rescaling of the equations again leads to the full Stokes problem. Hence we find that the slope is continuous, although the curvature is singular, and that in reality, $H_{xx} \sim \epsilon^{(1/n)-1}$.

Plastic flow

The limit $n \to \infty$ corresponds to plastic flow, in the sense that no deformation occurs for τ less than a critical value, but motion occurs when τ reaches (and is held at) this value. In this case, we ignore Eq. (18.11c) and instead choose $[\tau]$ to be the critical shear stress. Then for flowing ice, all the deformation occurs at the base, where $\tau = H|\nabla\eta|$, and the plastic flow equations simply reduce to

$$H|\nabla\eta| = 1; \tag{18.40}$$

this can also be deduced from Eq. (18.31) by letting $n \to \infty$ in an appropriate way. The simplest case with $h = 0$ gives the eikonal equation for $\phi = \eta^2/2$,

$$|\nabla\phi| = 1 \tag{18.41}$$

with $\phi = 0$ on the boundary. Equation (18.41) can be solved for an arbitrary external boundary by solving Charpit's equations along characteristics. That is, define $\phi_x = p$, $\phi_y = q$; then we solve

$$\dot{x} = p, \qquad \dot{y} = q, \qquad \dot{\phi} = 1, \qquad \dot{p} = \dot{q} = 0, \tag{18.42}$$

where $\dot{x} = dx/d\tau$, etc., along characteristics that originate on the boundary given by

$$x = x_0(s), \qquad y = y_0(s), \tag{18.43}$$

on which also

$$\phi = 0, \qquad p_0^2 + q_0^2 = 1, \qquad p_0 x_0' + q_0 y_0' = 0 \tag{18.44}$$

(and $x_0' = dx_0/ds$, $y_0' = dy_0/ds$). If s measures arc length, then $x_0' = \cos\psi$, $y_0' = \sin\psi$, where ψ is the angle of the boundary to the x-axis. Then $p = p_0 = -\sin\psi$, $q = q_0 = \cos\psi$; the characteristics are straight lines

$$x = x_0(s) - \tau\sin\psi, \qquad y = y_0(s) + \tau\cos\psi, \tag{18.45}$$

on which $\psi = \tau$. We thus obtain a parametric solution in terms of s and τ (because $\psi = \psi(s)$ on the boundary).

Typically, characteristics intersect to form *shocks* across which ϕ (but not $\nabla\phi$) is continuous. In terms of the flow, the shocks correspond to ridges or divides (or catchment boundaries), and for a plastic flow, $\nabla\eta$ is discontinuous across these. For a viscous flow law such as Glen's law, the curvature is infinite at a ridge, whereas the slope is continuous. Figure 18.2 shows a ridge pattern calculated from a numerical solution of Charpit's equations for Eq. (18.40), with an irregular border.

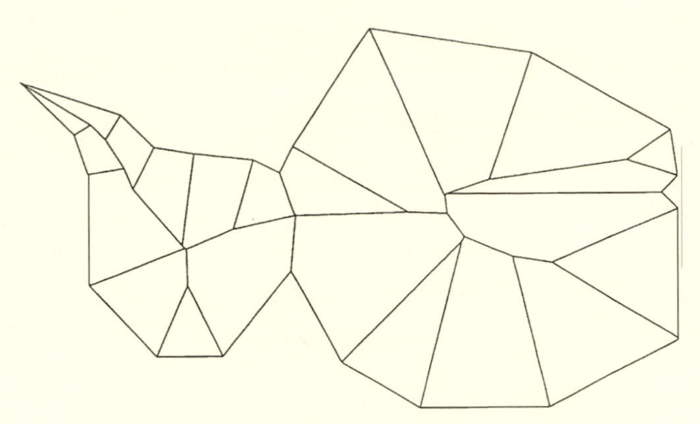

Fig. 18.2. Ridge pattern for plastic flow, irregular border. Reproduced courtesy of Clare Johnson.

Stability

Perturbations to the steady state decay diffusively, as one expects (at least when a is independent of η); to show this formally, however, requires some care owing to the singularity at the margin. The best way to treat the problem is to use the method of strained coordinates, as follows. Again we consider a one-dimensional situation, and take $n = 1$ for simplicity; the steady state is given by Eq. (18.33) and denoted by H_0. Writing $H = H_0 + H_1 \dots$, where $H_1 \ll H_0$, yields

$$H_{1t} = \frac{1}{3}\left(H_0^3 H_1\right)_{xx}; \tag{18.46}$$

putting

$$H_1 = e^{-\lambda t} w(x)/H_0^3 \tag{18.47}$$

gives

$$w'' + \left(3\lambda/H_0^3\right)w = 0, \tag{18.48}$$

together with suitable conditions at the margins, for example, $w(\pm 1) = 0$.

However, there is a problem. The two independent solutions of Eq. (18.48) give limiting behavior $H_1 \sim (x_m - x)^{-1/2}$, $H_1 \sim (x_m - x)^{-3/2}$ as $x \to x_m$, so that neither is bounded. Hence, the expansion is not uniformly valid as $x \to x_m$. In reality, x_m will move, so that the singularity in H is shifted slightly. The ordinary expansion method does not allow for this and compensates by creating a worse singularity in the perturbation. Thus we strain the x coordinate by putting

$$x = \xi + e^{-\lambda t}\sigma(\xi), \tag{18.49}$$

and put $H = H_0(\xi) + e^{-\lambda t}h(\xi) + \dots$. H_0 is again given by Eq. (18.33), namely,

$$H_0 = \left[12\int_{\xi}^{\xi_m} s(\xi)\, d\xi\right]^{1/4}, \tag{18.50}$$

where $s(\xi_m) = 0$. After careful simplification, the linearized equation for h can be written in the form (18.48) but where now $w = w(\xi) = H_0^3[h - \sigma H_0']$. We now *choose* $\sigma(\xi)$ so that h is no more singular than H_0 as $\zeta = \xi_m - \xi \to 0$. Because $H_0 \sim c\zeta^{1/2}$ as $\zeta \to 0$, it turns out that this can be ensured by choosing

$$\sigma = 2\delta(\alpha_0 + \alpha_1\zeta^{1/2})/c^4, \tag{18.51}$$

where α_0 and α_1 are known constants and $\delta \ll 1$ is a measure of the perturbation. We solve Eq. (18.48) subject to $w(\pm\xi_m) = 0$, obtaining a sequence of positive eigenvalues λ that give the decay rate of the various eigenmodes. Of interest is the corresponding motion of the margin. Because this is when $\xi = \xi_m$, we have

$$x_m \approx \xi_m + (2\delta\alpha_0/c^4)e^{-\lambda t}, \tag{18.52}$$

as we might have expected.

18.3 Steady, nonisothermal flow

Isothermal ice sheets are mathematically nice, but rather tame. Suppose now we consider the thermally activated flow law with $A(T)$ given by Eq. (18.5). From Eqs. (18.6) and (18.8), the dimensionless activation function is

$$A(T) = \exp\left[\frac{\gamma T}{1+\delta T}\right], \tag{18.53}$$

where

$$\gamma = \frac{Q\Delta T}{RT_m^2}, \qquad \delta = \frac{\Delta T}{T_m}. \tag{18.54}$$

Using values $\Delta T = 50$ K, $R = 8.3$ J mole^{-1} K^{-1}, $T_m = 273$ K, and $Q = 140$ kJ mole^{-1}, we have $\gamma \approx 11$, $\delta \approx 0.2$. The Frank-Kamenetskii approximation puts $\delta = 0$ and is reasonable here. We see that $\gamma \gg 1$, and base our approach on $\gamma \gg 1$; we will also take the limit $\beta \ll 1$, of qualitative rather than very great quantitative accuracy.

We revert to Eq. (18.24), with $\mu^{-1} \approx \tau^{n-1}e^{\gamma T}/\nu$, and we now retain $\nu \neq 1$, as we anticipate a shear layer where the temperature is largest (at the base). With $\mathbf{u}_H = (u, v) = \mathbf{U}$,

$$\begin{aligned} \frac{\partial \mathbf{U}}{\partial z} &= -\frac{(\eta - z)^n}{\nu}|\nabla\eta|^{n-1}e^{\gamma T}\nabla\eta, \\ \frac{dT}{dt} &= \frac{\alpha}{\nu}(\eta - z)^{n+1}|\nabla\eta|^{n+1}e^{\gamma T} + \beta T_{zz}. \end{aligned} \tag{18.55}$$

For simplicity, we suppose that the base is at the melting temperature everywhere – thus $T = 0$ at $z = h$ – but that no sliding occurs, and thus $\mathbf{U} = \mathbf{0}$ there. At $z = \eta$, $T = T_A < 0$, and η satisfies

$$\eta_t + \nabla.\left[\int_h^\eta \mathbf{U}\,dz\right] = a. \tag{18.56}$$

With $\gamma \gg 1$, the exponential terms are negligible except near the base, so that we have an outer solution determined by (assuming also $\beta \ll 1$)

$$\mathbf{U} = \mathbf{U}_0(x, y), \qquad dT/dt = 0. \tag{18.57}$$

Suppose T_A is spatially uniform, that is, $T_A = -1$; then $T = -1$ is the outer solution. There is a thermal boundary layer of thickness $O(\beta^{1/2})$ wherein T jumps from -1 to 0. In addition, there is a shear layer in which $T \sim 1/\gamma$ and $\mathbf{U}$ jumps to zero. Because (or if) $1/\gamma \ll \beta^{1/2}$, this shear layer lies within the thermal boundary layer, and because then $T \sim 1/\gamma$ for $h \sim \beta^{1/2}/\gamma$, we put, in the shear layer,

$$T = \theta/\gamma, \qquad z - h = \beta^{1/2} Z/\gamma, \tag{18.58}$$

to find

$$\begin{aligned} \frac{\partial \mathbf{U}}{\partial z} &\approx -\left(\frac{\beta^{1/2}}{\nu\gamma}\right) H^n |\nabla\eta|^{n-1} e^{\theta} \nabla\eta, \\ \frac{1}{\gamma^2}\frac{d\theta}{dt} &\approx \left(\frac{\alpha}{\nu\gamma}\right) H^{n+1} |\nabla\eta|^{n+1} e^{\theta} + \theta_{ZZ}. \end{aligned} \tag{18.59}$$

The relevant timescale is $t \sim 1/\gamma^2 \ll 1$, so that to leading order we neglect advection. We choose

$$\nu = \beta^{1/2}/\gamma \tag{18.60}$$

(so $\nu \sim 1/33$; note this multiplies the estimate of d in Eq. (18.12) by 1.5). Now $\alpha/\nu\gamma = \alpha/\beta^{1/2} \approx 0.85$, which we take as $O(1)$ (formally, therefore $\alpha = O(\beta^{1/2})$).

Put $g = -\theta_Z|_{Z=0}$, $G = -\theta_Z|_{Z\to\infty}$; then integrating Eq. (18.59), we have the far field velocity field

$$\mathbf{U}_0 = -\frac{\beta^{1/2}(G-g)}{\alpha H |\nabla\eta|^2} \nabla\eta, \tag{18.61}$$

and the flux is

$$\mathbf{Q} \approx \mathbf{U}_0 H = -\frac{\beta^{1/2}(G-g)}{\alpha |\nabla\eta|^2} \nabla\eta, \tag{18.62}$$

whence

$$\eta_t = (\beta^{1/2}/\alpha)\nabla.\left[\frac{(G-g)}{|\nabla\eta|^2}\nabla\eta\right] + a, \tag{18.63}$$

and another integration of Eq. (18.59) yields

$$G^2 - g^2 = (2\alpha/\beta^{1/2}) H^{n+1} |\nabla\eta|^{n+1}. \tag{18.64}$$

The final relation to determine G comes from solving the thermal boundary layer equation, because $-G$ is the boundary limit of the temperature gradient in the thermal boundary layer. Specifically, with $z - h = \beta^{1/2}\zeta$ and the outer vertical velocity $w = u_0 h_x + v_0 h_y - (z-h)(u_{0x} + v_{0y})$, the boundary layer equation for T is

$$\begin{gathered} u_0 T_x + v_0 T_y - \zeta(u_{0x} + v_{0y}) T_\zeta = T_{\zeta\zeta}, \\ T = 0, \qquad T_\zeta = -G \quad \text{on } \zeta = 0, \qquad T \to -1 \quad \text{as } \zeta \to \infty; \end{gathered} \tag{18.65}$$

the extra condition determines G in principle. There is a similarity solution $T = -\mathrm{erf}[\zeta/\sqrt{p(x,y)}]$, where p satisfies

$$u p_x + v p_y + 2p(u_x + v_y) = 4, \tag{18.66}$$

and then G is given by

$$G = \frac{2}{\sqrt{\pi p}}. \tag{18.67}$$

We integrate Eq. (18.66) along streamlines from ridges. In view of the degeneracy of Eq. (18.66) when $u = v = 0$, no initial condition for p at a ridge is necessary (because $p = 2/(u_x + v_y)$ there). For example, in one dimension, the solution is

$$p = \frac{4}{u^2}\int_0^x u\,dx, \tag{18.68}$$

where $x = 0$ denotes the ridge where $u = 0$. The solution is only consistent if the heat flux into the ice is less than the geothermal flux Γ, otherwise freezing occurs. The requisite condition (with no sliding) is that $g < \Gamma\beta^{1/2}$. With g given by Eq. (18.64), Eqs. (18.63), and (18.66) give a coupled pair of parabolic/hyperbolic equations for p (hence G) and η, whose solution must be sought numerically. Complications occur in the more general case, where frozen parts of the base occur, or if the surface temperature varies in space or time. On the other hand, this set represents a considerable simplification over the original coupled set of equations.

18.4 Drainage, sliding and ice-till coupling

When basal ice reaches the melting point, there is then a net heat flux arriving at the bed of the ice sheet, and consequently basal melt water is produced. The existence of subglacial water is well documented, both from the apparent existence of sub-Antarctic lakes, and from the existence of outlet streams in ice caps and glaciers. The existence of water has two main effects. Firstly, it can lubricate the bed sufficiently that the ice slides over the bed. Bedrock protuberances brake the flow, and the corresponding boundary condition is known as the sliding law, $\tau_b = f(u_b)$, where τ_b is basal shear stress and u_b is basal velocity (in the direction of the principal stress). More generally, Eq. (18.23) is, to leading order,

$$\mathbf{u}_b = -\phi H \nabla\eta, \tag{18.69}$$

where $\phi = u_b/f(u_b)$. Now in addition, sliding over a rough bed depends on the basal water pressure p_w through the (positive) *effective pressure* $N = p_i - p_w$, where p_i is the ice overburden pressure. The lower N is, the more nearly the ice is floating, and the lower the resistance. Sliding laws of the general form $\tau_b = cu_b^p N^q$ have been proposed, both on experimental and theoretical grounds. There is then the issue of how to determine N.

The determination of N depends on the constitution of the bed. When ice can slide over its bed, it is also able to erode it, and the resultant eroded debris forms a subglacial till that, when water saturated, is itself deformable. In this case, sliding at the base may be entirely due to deformation of the till, and in this case, an appropriate choice of ϕ in Eq. (18.69) is $\phi = s/\mu_t$, where s is the till thickness and μ_t is the till viscosity. In complete generality, s will itself evolve through an evolution equation of the form

$$s_t + \frac{1}{2}\nabla.(s\mathbf{u}_b) = E - m_s, \tag{18.70}$$

where E is the erosion rate ($E = E(u_b, \tau_b, N)$) and m_s is the sediment removal rate via subglacial streams. Till viscosity is expected to be nonlinear and to depend also on effective pressure N, $\mu_t = \mu_t(\tau_b, N)$, becoming smaller as N decreases; again the problem of determining N is raised.

Drainage

Water is observed to emerge from the front of glaciers through channels that may be cut upward into the ice, and a theory exists for the hydraulics of water flow through these channels. If $N > 0$, then they tend to close up (by viscous closure of the ice), but can be kept open by meltback of the channel walls due to viscous dissipation of the (generally turbulent) water flow in the channel. The theory leads to an equation for N in terms of the water flux Q_w, and at its simplest, this is an expression of the form

$$N = c_1 Q_w^a, \tag{18.71}$$

where (depending on choice of friction factor in the channel), $a = 1/5n$, where n is the exponent in Glen's law.

If ice flows over deformable till, it is not known what form the drainage system takes. One suggestion is that water flows in canals incised into the till, with the closure of the canals balanced by the transport of sediments downstream. Another is that a film of variable thickness (~1 mm) carries the flow in a thin sheet. In either case, the relation of N to Q_w may be of the form

$$N = c_2 Q_w^{-b}. \tag{18.72}$$

For the case of channels, with $\partial\tau_b/\partial N > 0$, we then have $\partial\tau_b/\partial Q_w > 0$. For the case of films or canals, $\partial\tau_b/\partial\mu_t > 0$, $\partial\mu_t/\partial N > 0$, $\partial N/\partial Q_w < 0$, so that $\partial\tau_b/\partial Q_w < 0$. In this case a general form of sliding law for ice sheets of the form

$$\tau_b = c u_b^r Q_w^{-s}, \tag{18.73}$$

with c, r, and s dependent on basal conditions, may be appropriate.

If $T < 0$, then the bed is frozen and we take $u_b = 0$. If $T = 0$, then the net heat flux to the bed is (with $\epsilon \to 0$) $G + \tau u_b + k\partial T/\partial z$. When scaled, this leads to a net production of water at the base (as a vertical velocity, i.e., water volume flux per unit area) of

$$v_w = \frac{\beta}{St}\left(\frac{\rho_i}{\rho_w}\right)\left[\Gamma + (\alpha/\beta)\tau u_b + \frac{\partial T}{\partial z}\right], \tag{18.74}$$

where

$$St = \frac{L}{c_p \Delta T}. \tag{18.75}$$

With $L = 3.3 \times 10^5$ J kg^{-1}, $St \sim 3.4$, so that with $\beta \sim 1/8$, v is insignificant from the point of view of the ice motion. The dimensional flux per unit area is $[a]v_w$ and of order 3 mm y^{-1}.

How the hydraulic system is configured now becomes important, because the spacing of canals will lead to a different relation of Q_w to v_w than if it is distributed in a film. At any rate, we can write $\partial Q_w/\partial s \propto v_w$, where s is a coordinate in the direction of water flow.

Basal water flow is directed down gradients of the hydraulic head, $p_w + \rho_w g h$. Scaling pressures and effective pressure with $\rho_i g d$, this is proportional to $\eta + \delta h - N^*$, where $\delta = (\rho_w - \rho_i)/\rho_i \sim 0.1$, $N^* = N/\rho_i g d$. In ice sheets, typically $N^* \ll 1$; thus water flows approximately down gradients of η, and we can write

$$-\nabla\eta.\nabla Q_w = \lambda|\nabla\eta|\left(\Gamma + (\alpha/\beta)\tau u_b + \frac{\partial T}{\partial z}\right) \tag{18.76}$$

to determine Q_w (if $T = 0$ and the right-hand side is positive), where λ depends on the basal drainage geometry. Together, Eqs. (18.73) and (18.76) give a nonlocal basal sliding condition that couples the ice flow to the temperature field.

A deformable till mechanism for the Hudson Strait megasurges

The basic idea, similar to that behind the notion of thermal regulation of glacier surges, is very simple. If the ice is thin over Hudson Bay, then the heat flux from the base is large, the base is below freezing, and no sliding occurs. In these circumstances, with low stress, flow may be negligible, so that the ice begins to thicken. As it does so, the basal temperature starts to rise, due both to increasing viscous dissipation and increasing ice depth. If the ice reaches melting point before the 'subfreezing' type steady state, and if sliding is sufficiently rapid, then velocity may increase rapidly, leading to a surge and reduction in depth, until the base freezes again. The resulting hypothetical cyclic pattern is similar to oscillations in many systems (the Belousov reaction is one) where different dynamics in two different regimes try to reach inaccessible equilibria, and in so doing, switch back and forth between the regimes. Because the model necessarily incorporates coupling between flow and thermal fields, the boundary layer theory proposed here is perhaps the simplest vaguely realistic model. This leads to the following problem.

To be simple, consider a two-dimensional ice sheet in $x > 0$, with a divide at $x = 0$, lying on a flat base $z = 0$. If we include basal sliding into the thermal boundary layer model, then (with $\eta_x < 0$) Eqs. (18.61), (18.63), (18.64), (18.67), (18.68), (18.73), and (18.76) lead to

$$\begin{aligned}
u &= u_b - \frac{\beta^{1/2}(G-g)}{\alpha\eta\eta_x},\\
\eta_t &= -\frac{\partial}{\partial x}(\eta u) + a,\\
G^2 - g^2 &= \frac{2\alpha}{\beta^{1/2}}\eta^{n+1}|\eta_x|^{n+1},\\
G &= \frac{u}{\left\{\pi\int_0^x u\,dx\right\}^{1/2}},\\
\eta|\eta_x| &= c u_b^r Q_w^{-s},\\
\frac{\partial Q_w}{\partial x} &= \Gamma - g + (\alpha/\beta)u_b\eta|\eta_x|,
\end{aligned} \tag{18.77}$$

for the six variables u, u_b, η, g, G, and Q_w, together with $Q_w = 0 = \eta_x$ at $x = 0$. These equations apply so long as $g < \Gamma$ (and also $g > 0$: if g reaches zero, then a temperate zone $T \equiv 0$ forms at the base, requiring further modeling considerations, which we ignore). If, on evolution of this set of equations, η decreases until g reaches Γ, we switch to frozen mode, where $T < 0$ at the base.

A simpler version of this model follows if we put $u_b = u$, $g = G$ and ignore the third equation in Eq. (18.77). We then have

$$\eta_t = -\frac{\partial}{\partial x}(\eta u) + a, \tag{18.78a}$$

$$u = Q_w^S(-\eta\eta_x)^R, \tag{18.78b}$$

$$\frac{\partial Q_w}{\partial x} = \gamma - \sigma u \eta \eta_x - \frac{\mu u}{\left\{\int_0^x u\,dx\right\}^{1/2}}, \tag{18.78c}$$

where R, S, γ, σ, and μ are positive parameters determined in terms of α, β, c, and Γ, after rescaling Q_w in Eq. (18.77). The basic structure of Eq. (18.78) is that of a nonlinear diffusion equation

$$\eta_t = \frac{\partial}{\partial x}\left[Q_w^S \eta^{R+1}|\eta_x|^{R-1}\frac{\partial \eta}{\partial x}\right] + a, \tag{18.79}$$

with Q_w determined by the quadrature in Eq. (18.78).

A toy model

The basic effect of the positive feedback term (the frictional heating term $-\sigma u \eta \eta_x$ in Eq. (18.78c)) can be understood simply by 'lumping' the derivatives. Specifically, we replace $\partial Q_w/\partial x$, $-\partial\eta/\partial x$, $\partial(\eta u)/\partial x$, and $\int_0^x u\,dx$ by the corresponding terms Q_w, η, ηu, and u; we obtain the simplified model

$$\dot{\eta} = a - \eta u, \tag{18.80a}$$

$$\gamma + \sigma u \eta^2 - \mu u^{1/2} = \left(\frac{u}{\eta^{2R}}\right)^{1/S}. \tag{18.80b}$$

The left-hand side of Eq. (18.80b) is a quadratic in $u^{1/2}$, which reaches zero if η is small enough. It is therefore clear that if μ is small enough and $S < 1$, then there will be a range of η for which there are three possible values of u, and hence also $Q = \eta u$. A typical example is shown in Fig. 18.3. If a has the intermediate value indicated, then cyclic oscillations will occur between the upper and lower branches.

The relation (18.80b) is valid approximately for $u > 0$, but if we allow $u \neq u_b$ and $G \neq g$, this will shift the validity of Eq. (18.80b) to a region $Q > Q_c$, below which the base will be frozen. This toy model indicates that cyclic surging can occur even if no basal switching occurs, and the base remains temperate throughout.

Toy models such as this are useful in gaining qualitative insight into mechanisms, but they are only suggestive. There are, of course, other possibilities; for example, even in the one-dimensional case, one could have a partly cold, partly temperate based solution. And where the flow is horizontally two dimensional, another possibility is a lateral distribution of fast- and slow-moving ice; and this is just what the flow in the ice streams of the Ross Ice Shelf consists of.

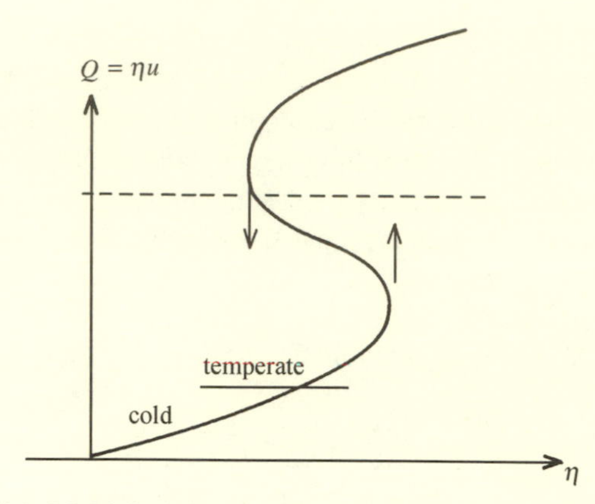

Fig. 18.3. Multiple-valued ice flux as a function of ice thickness.

18.5 Notes and references

The dynamics of ice sheets and glaciers is described by Paterson (1994). Another useful source is the book edited by Colbeck (1980). Glaciology has its origins in the nineteenth century observations of people such as Louis Agassiz, and many dynamic features of valley glaciers (e.g., surface waves) were observed and understood at that time. However, it was only after the second world war when John Nye, John Glen, and others brought the approach of physics to bear, that glaciology turned from its former observation-based descriptive nature, to being a more quantitative science. The recent rapid increase of observational data, for example owing to satellite observations, has led to a rise in stature of glaciology and has promoted an associated plethora of research, both theoretical and observational, into the behavior of ice.

Heinrich events The interpretation of Heinrich events, and the possibly related Dansgaard–Oeschger events, is a research topic of current interest, and the story presented here is only one possibility. Discussion of this can be found in the papers of Bond et al. (1992), Clark (1994), and MacAyeal (1993). The toy model in Section 18.5 is discussed by Fowler and Johnson (1995).

Ice sheet modeling The basis of the present chapter is an M.S. dissertation written by Peter Gillott at Oxford in 1985. The nonisothermal analysis was done later and was published by Fowler (1992), based on earlier work by Hutter, Yakowitz, and Szidarovsky (1986) and, Morland (1984). The 'shallow-ice' approximation is a terminology championed by Kolumban Hutter, whose book (Hutter, 1983) is a voluminous theoretical study based on a classical continuum mechanical approach.

Glen's flow law (after John Glen), given by Eq. (18.4), $\dot{\varepsilon} \propto \tau^n$, is not the only fit that can be made to the data, and it has the apparent drawback of having infinite viscosity at zero stress. Other polynomial forms for $g(\tau)$ have been suggested, particularly by Morland, but it is unlikely that these have much effect on the dynamics (though they may alleviate singularities at divides and margins).

More important, from the point of view of ice sheet modeling, may be the fact that as ice grains are deformed, their *fabric* is altered, and this leads to anisotropy of polycrystalline ice, particularly at the base of ice sheets, where the ice may be ten or twenty times softer in shear than would be otherwise expected (Shoji and Langway, 1988).

The first attempts at ice sheet modeling used the rigid-plastic flow described by Eq. (18.40); this was done by Nye (1951) in two dimensions. Reeh (1982) applied the present three-dimensional theory to the Greenland ice sheet.

Drainage and sliding Although it is well known that basal sliding occurs ubiquitously at the bed of glaciers and ice sheets if the ice is temperate, the sliding law that relates velocity and stress is much less certain, and the style of drainage under ice sheets is a matter of debate almost unconstrained by facts. Recent papers on these topics are by Walder and Fowler (1994) and Alley (1989). Sliding theory has a long and involved history. An interesting application of some of this theory is given by Bentley (1987), in an issue of the *Journal of Geophysical Research* (Vol. 92, No. B9) devoted to the topic of fast ice flow.

19

Chemosensory respiratory control

19.1 Respiratory physiology

Put most simply, we breathe because we need the energy available from the oxidative breakdown of foodstuffs to maintain our vital body processes and perform our normal activities. To this end, all forms of life have some mechanism of gas exchange. In microorganisms, the available cell surface is sufficiently large for adequate diffusion to take place. Higher animals, however, have specialized respiratory systems for this function. An adult man, for instance, has only 1.5–2.0 square meters of body surface area, but the 300 million alveolar air sacs in his lungs provide a total of over 150 square meters of area for gaseous exchange.

The respiratory system involves the lungs, heart, blood vessels, brain, and peripheral tissues. Deoxygenated ('venous') blood enters the lungs from the right side of the heart, is oxygenated there (carbon dioxide being released into the lungs in the process), travels to the left ('arterial') side of the heart, and is pumped into the peripheral tissues, where oxygen is extracted and carbon dioxide released from oxidative metabolism, before reentering the right side of the heart for the next cycle. During this process, the respiratory system keeps in control the values of over a dozen 'key' variables (gas partial pressures, pH of body fluids, ventilatory rate, etc.), and it is often conceptualized as a complex control system.

The mechanics of respiration

Inspired air travels down the windpipe to the lungs, ultimately reaching the terminal air sacs (the *alveoli*). We normally breathe about 12–15 times a minute, each breath drawing in approximately 500 ml of fresh air (the tidal volume), to join approximately 3 liters of alveolar air already present in the lung. During each expiration, 500 ml of air is breathed out, giving a total turnover of about 7 liters in a minute – the *minute volume*. During severe exercise, this figure may increase fifteen-fold. Not all the minute volume is available for gas exchange at the alveoli. About 150 ml of air occupying the air passages is pushed in and out with each breath. Hence the 'alveolar ventilation' is only about 5 liters a minute. Respiratory physiologists commonly (and confusingly) measure the rate of breathing in terms of the *ventilation* $\dot{V}$. The notation suggests that this is the rate of change of lung volume with time, but in fact it is no more than the average rate of breathing, usually expressed in liters per minute, and is thus properly an averaged measurement. For example, if one measures (at time t) the minute volume inspired in the preceding minute, this could be a definition of $\dot{V}$ (in liters per minute). If v is the lung volume, then this definition of $\dot{V}$ can be written mathematically as

$$\dot{V} = \frac{1}{\tau} \int_{t-\tau}^{t} \dot{v} H(\dot{v})\, dt, \tag{19.1}$$

where $\tau = 1$ minute. Equally, one could select other values of τ, and the definitions are only equivalent if (i) respiration is steady and (ii) $\tau \gg$ breath time. An alternative version of Eq. (19.1) is

$$\dot{V} = \frac{1}{2\tau} \int_{t-\tau}^{t} |\dot{v}|\, dt. \tag{19.2}$$

Ventilation is often used in laboratory studies, where (for example) the ventilation as a function of inspired CO_2 concentration might be studied. In such studies, a steady response is sought, and $\dot{V}$ is an appropriate variable. However, certain types of breathing (described more fully below) are time dependent in nature; and in this case, $\dot{V}$ becomes an irrelevant variable, unless the time dependence is much slower than the breathing frequency. Despite this, irregular breathing patterns are often portrayed in terms of the derived variable $\dot{V}$, rather than $\dot{v}$ (which would in fact be the measured variable). This practice is in the inscrutable domain of physiologists. In fact, many time-dependent patterns do exhibit such longer-term fluctuations, but breath-to-breath variability also occurs, and this is effectively filtered by plotting $\dot{V}$ versus time.

Gas exchange with blood flow

The heart pumps about 5 liters of blood (the 'cardiac output') past the lungs every minute. At any instant, approximately 60–140 ml of blood is present in the capillaries, and gases diffuse across the thin (0.2–1 μm) membrane barrier of the capillaries and alveoli.

The movement of gases is determined by the differences in partial pressures across the barrier. Normal inspired air has an oxygen partial pressure (pO_2) of 150 mm Hg, whereas the pO_2 of capillary blood is about 40 mm Hg.[1] Thus oxygen diffuses across the membrane to the blood. Similarly, a 40 mm Hg partial pressure of blood CO_2 enables loss of CO_2 to the lung (because inspired air contains virtually no CO_2). Moreover, this gas exchange is efficient in healthy individuals, so that arterial blood leaving the lungs typically reaches the gas partial pressures in the lungs.

Respiratory control

The basic oscillatory mechanism that determines breathing – the muscular expansion and contraction of the lung – is due to the *neurogenic drive*: a central neural mechanism controlled by various mechanical reflexes. The basic ventilation rate is controlled on a longer timescale by blood gas concentrations of CO_2 and O_2 through *chemoreceptors* situated in the brain (the central chemoreceptors) and the neck (the peripheral chemoreceptors), and it is this longer-term control that we seek to study in this chapter. It then makes sense to use $\dot{V}$ as a variable, and the 'respiratory controller' then parametrizes $\dot{V}$ as a function of blood gas concentrations.

[1] Millimeters of mercury Hg. Here, pressure is measured in terms of that exerted by the quoted height of a column of mercury. Atmospheric pressure ($\approx$1 bar $= 10^5$ Pa) is 760 mm Hg. Thus 150 mm Hg $\approx$ 0.2 bar.

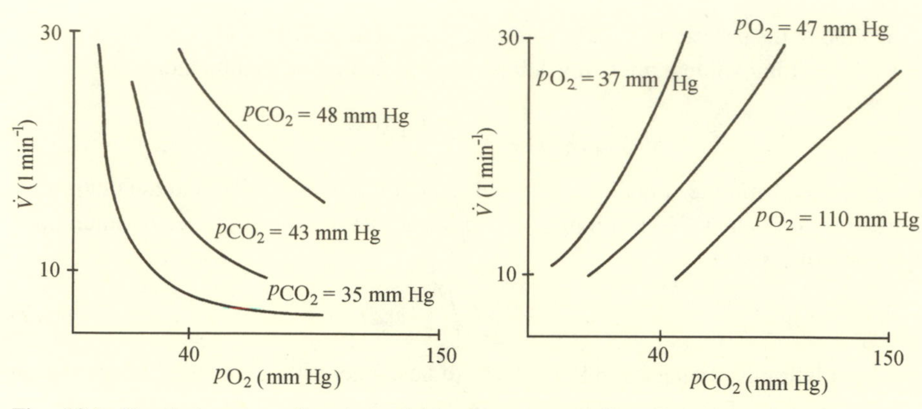

Fig. 19.1. Ventilation as a function of blood oxygen (left) and carbon dioxide (right) concentrations.

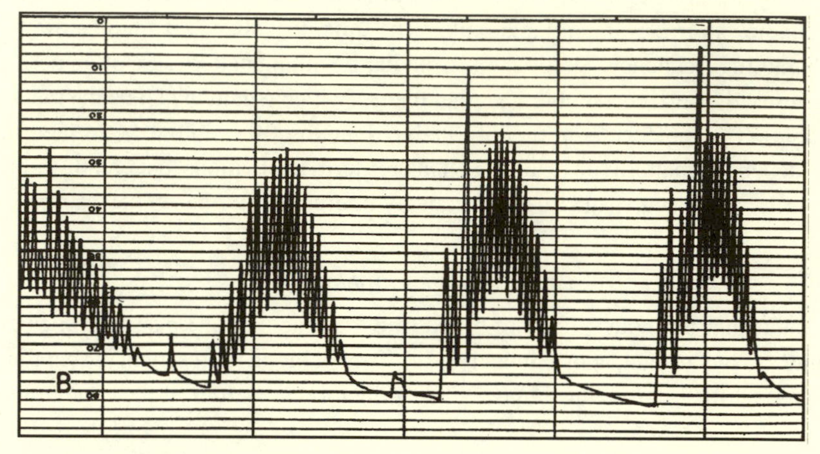

Fig. 19.2. Cheyne–Stokes breathing. The trace represents lung volume, with the downstroke representing expiration. Reproduced from Greene (1933) by permission. Copyright 1933, American Medical Association.

Broadly speaking, $\dot{V}$ decreases with increasing pO_2 in the arterial blood (at the neck), principally through the action of the peripheral chemoreceptors, whereas $\dot{V}$ increases with increasing pCO_2 (in the brainstem and cerebro-spinal fluid) through the action of the central chemoreceptors. Moreover, the two effects are coupled, as can be seen in Fig. 19.1.

Periodic breathing

One of the interests in modeling the respiratory system by a dynamic model is to understand various kinds of oscillatory ventilatory patterns, collectively termed *periodic* breathing. Normal periodic breathing is a slow waxing and waning of the breathing amplitude (i.e., of tidal volume). In certain circumstances, for example, when there is a nervous system disorder, or increased blood circulatory time, etc., these oscillations can become more pronounced, leading to *Cheyne–Stokes* breathing, where the ventilation can oscillate between periods of irregular breathing, and quiescent periods of little or no breathing. An example is shown in Fig. 19.2. Different forms of periodic breathing in such pathologies, depending on the characteristics of the irregular phases, go under the name of *ataxic* or *Biot* breathing. In addition, other varieties of periodic breathing are readily identified in infants (normal as well as those born preterm) and in healthy adults taken to high altitude. One aim of modeling respiration is to try and understand the mechanisms of such irregular breathing patterns.

19.2 The Grodins model

The Grodins model (actually enunciated in a paper by Grodins, Buell, and Bart (1967)) is a *compartment* model (see Fig. 19.3). Lungs, brain, cerebrospinal fluid (CSF), and tissues form four separate compartments, in each of which the concentrations of the blood gases O_2, N_2, and CO_2 are supposed to be functions of time only. Transmission between these compartments is effected by the blood flow through arteries and veins, which thus introduces delays that are themselves functions of blood flow rate Q. Because the blood flow is itself affected by blood gas concentrations, the heart effectively acts as a fifth compartment. Finally, the ventilation rate $\dot{V}$ is given as a prescribed function of blood gas concentrations.

In its entirety, the Grodins model is very complicated. It considers three blood gas concentrations: N_2, O_2, and CO_2, designated n, x, and c, respectively, in four compartments: lungs, brain, CSF, and tissues, designated with suffixes l, b, s, and t, respectively, together with arterial (a) and venous (v) concentrations of each gas; additionally, there are two blood flow variables Q (cardiac output) and Q_b (blood flow to the brain), numerous (variable) delay times τ_{ab}, τ_{at}, τ_{vb}, and τ_{vt}, representing the blood flow time from lung to brain or tissues via arteries or veins, etc., and finally, $\dot{V}$. In addition, there are a number of other variables, which arise in the description of various buffering relationships: for example, concentrations of hydrogen ions H^+ and of hemoglobin HbO_2 at various locations.

In view of the complexity, we make reference to the original Grodins, Buell, and Bart (1967) paper for the full model. Below we present a nondimensional model,

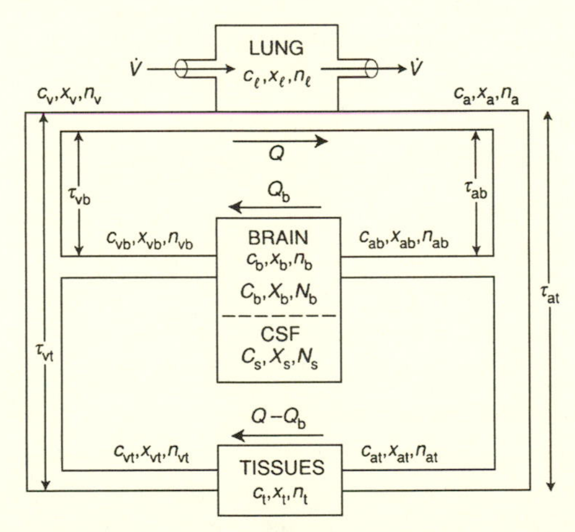

Fig. 19.3. Schematic diagram of the respiratory system. Reproduced from Fowler, Kalamangalam, and Kember (1993), by permission of Oxford University Press.

where every variable has been scaled as

$$x_a = [x_a]x_a^*, \tag{19.3}$$

and the asterisks dropped. We also ignore N_2, as it is a passive variable that uncouples from the determination of the other variables.

Conservation relations

Corresponding to each compartment, we have two material balance equations for CO_2 and O_2. With suffixes as described above, these are as follows:

In the lung:

$$\dot{c}_l = \dot{V}(p_0 - p_1 c_l) + Q(\Lambda_1 c_v - \Lambda_2 c_a), \tag{19.4a}$$
$$\dot{x}_l = \dot{V}(p_2 - p_1 x_l) + Q(p_3 x_v - p_4 x_a); \tag{19.4b}$$

here, p_j and Λ_k are various dimensionless parameters. The term in $\dot{V}$ arises through gas interchange with the external atmosphere (CO_2, O_2 concentrations proportional to p_0, p_2, respectively), whereas the term in Q represents gas interchange at the alveoli: Qc_v is the CO_2 arriving per unit time, and Qc_a is that leaving, so this whole Q term is the source of CO_2 for c_l. (The arterial c_a is not independent of c_l; it is determined by a chemical equilibrium constitutive relationship, see below. Most simply, we would put $c_a = c_l$; the correct relationship is more complicated, but has the same basic sense.)

In the brain:

$$\begin{aligned} \dot{c}_b &= \epsilon_2 + p_6 Q_b(c_{ab} - c_{vb}) - \epsilon_3(C_b - C_s), \\ \dot{x}_b &= -\Lambda_3 + Q_b(\Lambda_4 x_{ab} - \Lambda_5 x_{vb}) - \epsilon_4(X_b - X_s). \end{aligned} \tag{19.5}$$

Here, ϵ_2 and Λ_3 represent CO_2 production and O_2 consumption by metabolic processes in the brain; the Q_b term (Q_b is the blood flow to the brain) represents the net source of the gases for the brain, just as for the lungs; c_{vb} will be constituted in terms of c_b, whereas c_{ab} will be related to (in fact equal to) c_a at an earlier time (the delay being due to blood transit time between lung and brain). The capital letter symbols also represent amounts of the indicated variables, but whereas c_b, etc., are volume fractions, C_b, etc., are dimensionless partial pressures. These are of course related, so that C_b will be constituted in terms of c_b, etc. Finally, the last terms in each equation represent the transport to the CSF.

In the CSF:

$$\begin{aligned} \dot{C}_s &= p_{13}(C_b - C_s), \\ \dot{X}_s &= p_{14}(X_b - X_s); \end{aligned} \tag{19.6}$$

only transport from and loss to the brain occurs.

In the tissues:

$$\begin{aligned} \dot{c}_t &= \epsilon_6 + p_8 Q(c_{at} - c_{vt}) - Q_b(\epsilon_7 c_{at} - \epsilon_8 c_{vt}), \\ \dot{x}_t &= -p_9 + Q(\Lambda_6 x_{at} - \Lambda_7 x_{vt}) - Q_b(p_{10} x_{at} - p_{11} x_{vt}). \end{aligned} \tag{19.7}$$

As for the brain, ϵ_6 and p_9 are CO_2 production and O_2 consumption rates, and the two source terms are due to the blood flow to the tissues, $Q - Q_b$, bearing in mind that Q and Q_b are scaled differently.

Constitutive relationships

The primary variables in the eight equations above are $c_l, x_l, c_b, x_b, C_s, X_s, c_t$, and x_t; constitutive relations need to be posed for the fourteen subsidiary variables $c_v, c_a, x_v, x_a, c_{ab}, c_{vb}, C_b, x_{ab}, x_{vb}, X_b, c_{at}, c_{vt}, x_{at}$, and x_{vt}, as well as the blood flows Q and Q_b and the ventilation $\dot{V}$. These constitutive relations are determined compartment by compartment.

(a) Lung

Here c_a and x_a are determined in terms of c_l and x_l via the hemoglobin concentration. In full:

$$c_a = p_{15} - p_{16}h_a - p_{17}\left[\log\left(\frac{p_{18}c_a}{c_l} - \epsilon_{10}\right) - p_{14}\right] + \epsilon_{11}c_l, \tag{19.8a}$$

$$h_a = p_{20}[1 - \exp(-p_{21}S_1x_l)]^2, \tag{19.8b}$$

$$S_1 = \Lambda_8 p_a - \Lambda_9 p_a^2 + \Lambda_{10} p_a^3 - \Lambda_{11}, \tag{19.8c}$$

$$p_a = 1 - p_{14}\log(H_a), \tag{19.8d}$$

$$H_a = p_{22}\left[\frac{c_l}{p_{23}c_a - \epsilon_{13}c_l}\right], \tag{19.8e}$$

determines $c_a(c_l, x_l)$ via intermediaries p_a (pH of arterial blood) and H_a (H^+ concentration), as well as the HbO_2 (oxygenated hemoglobin) concentration h_a.

More simply, x_a is given by

$$x_a = \epsilon_{12}x_l + p_{19}h_a. \tag{19.9}$$

Equation (19.8) gives an algebraic set and poses little numerical difficulty. For analytic purposes, it is useful to derive an approximate relation from this. We can do this bearing in mind the rough distinction that $\epsilon_i \ll 1$, $p_j \sim O(1)$, $\Lambda_k \gg 1$ (hence the labeling of these parameters). Then typically $S_1 \gg 1$, $h_a \sim$ constant (p_{20}), and c_a is determined from Eq. (19.8a) as a monotone increasing function of c_l. For simple purposes, it suffices to take $c_a \sim c_l$.

In the remaining compartments, we do not detail the similar algebraic relations that apply.

(b) Brain

Here we find c_{vb}, x_{vb}, C_b, and X_b in terms of c_b and x_b. In particular, $c_{vb} \sim c_b$.

(c) Tissues

We obtain c_{vt}, x_{vt} in terms of c_t, x_t, and in particular $c_{vt} \sim c_t$.

The buffering relations above give eight relations for $c_a, x_a, c_{vb}, x_{vb}, C_b, X_b, c_{vt}$, and x_{vt}. There remain $c_v, x_v, c_{ab}, x_{ab}, c_{at}, x_{at}$, and the ventilation and blood flow.

Delays

Let τ_{ab}, τ_{at}, τ_{vb}, and τ_{vt} denote the blood flow transit times indicated in Fig. 19.3 from lung to brain, lung to tissues, brain to lung, and tissues to lung, respectively. Then it is evident that

$$\begin{aligned} c_{ab} &= c_a(t - \tau_{ab}), \\ x_{ab} &= x_a(t - \tau_{ab}), \\ c_{at} &= c_a(t - \tau_{at}), \\ x_{at} &= x_a(t - \tau_{at}), \end{aligned} \tag{19.10}$$

where the delays are not necessarily constant (if Q or Q_b vary), and satisfy

$$p_{55} = \int_{t-\tau_{ab}}^{t-\delta_0} Q\,dt, \qquad \epsilon_{23} = \int_{t-\delta_0}^{t} Q_b\,dt \tag{19.11}$$

(defining both τ_{ab} and δ_0) and

$$p_{55} = \int_{t-\tau_{at}}^{t-\delta_1} Q\,dt, \qquad p_{13} = \int_{t-\delta_1}^{t} (Q - p_{26}Q_b)\,dt. \tag{19.12}$$

On the venous side, we evidently have

$$Qc_v = p_{13}Q_b c_{vb}(t - \tau_{vb}) + (p_{24}Q - p_{13}Q_b)c_{vt}(t - \tau_{vt}), \tag{19.13a}$$

$$Qx_v = p_{32}Q_b x_{vb}(t - \tau_{vb}) + (p_{25}Q - p_{26}Q_b)x_{vt}(t - \tau_{vt}), \tag{19.13b}$$

and

$$\begin{aligned} \epsilon_{24} &= \int_{t-\tau_{vb}}^{t-\delta_2} Q_b\,dt, \qquad \epsilon_{25} = \int_{t-\delta_2}^{t} Q\,dt, \\ p_{56} &= \int_{t-\tau_{vt}}^{t-\delta_3} (Q - p_{27}Q_b)\,dt, \qquad \epsilon_{25} = \int_{t-\delta_3}^{t} Q\,dt, \end{aligned} \tag{19.14}$$

and Eqs. (19.10) and (19.13) give relations for the remaining six concentrations. We need, finally, equations for Q, Q_b, and $\dot{V}$.

Blood flow

Q and Q_b satisfy

$$\begin{aligned} \dot{Q} &= \Lambda_{12}U - \Lambda_{13}Q, \\ \dot{Q}_b &= \Lambda_{16}W - \Lambda_{13}Q_b, \end{aligned} \tag{19.15}$$

where U and W are complicated functions of x_l and c_l. In the Grodins model,

$$\begin{aligned} U &= U_1(c_l) + U_2(x_l) + p_{44}, \\ W &= W_1(c_l) + W_2(x_l) + p_{49}, \end{aligned} \tag{19.16}$$

with $U_1, W_1 = 0$ for $c_l < 1$, $U_2, W_2 = 0$ for $x_l > 1$. That is to say, for low c_l, blood flow is O_2 controlled, whereas for high x_l, it is CO_2 controlled. Blood flow will be constant if c_l is low *and* x_l is high.

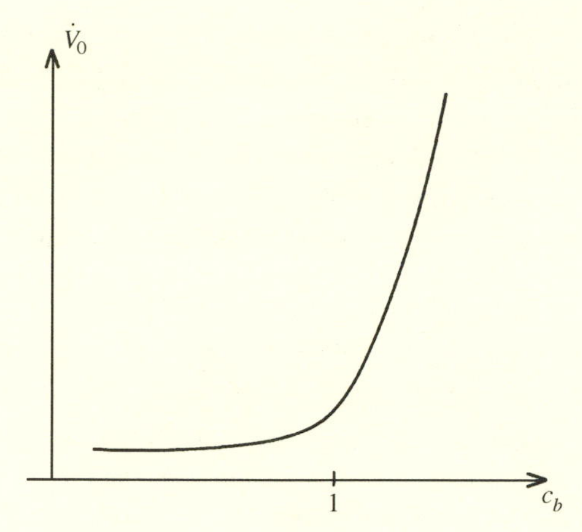

Fig. 19.4. Typical form of ventilation as a function of brain carbon dioxide concentration

Ventilation

A variety of controller functions might be chosen. One such is

$$\dot{V} = \dot{V}_0(c_b) + p_{57}\Lambda_{29}[1 - x_l(t - \tau_{a0})]_+^{4.9}, \tag{19.17}$$

where τ_{a0} is the lung to peripheral chemoreceptor transport delay. The brain CO_2 response curve $\dot{V}_0(c_b)$ is constant for $c_b < 1$ but increases roughly linearly thereafter (and is rather large for $c_b > 1$, see Fig. 19.4). Note that this formulation does not incorporate O_2–CO_2 interaction (Fig. 19.1), being based on the outdated "multiple factor" theory of respiratory control (Gray, 1946); in view of the succeeding analysis, however, this issue is of little significance.

19.3 Reducing the model

In nondimensionalizing the model, we have chosen typical observed values of the variables, together with a rather arbitrary choice of $[t] = 1$ minute. That is, we have not balanced terms in the equations, and this leads to a certain redundancy. For example, in Eq. (19.4) the definitions of Λ_1 and Λ_2 are

$$\Lambda_1 = \frac{[t]k_1\beta_2[Q][c_v]}{[c_l]}, \qquad \Lambda_2 = \frac{[t]k_1\beta_2[Q][c_a]}{[c_l]}, \tag{19.18}$$

and differ only in the scaling of c_a and c_v. Now because we choose $[c_a] = 0.5748$, $[c_v] = 0.613$, we find $\Lambda_1 = 25.37$, $\Lambda_2 = 24.72$, and we might as well have chosen $[c_a] = [c_v]$. While such reasoning can be applied *a posteriori*, it is probably quicker here, in view of the complexity of the model, to proceed without such anticipation.

The large number of parameters have been divided into three classes: ϵ_i, which are less than 0.1; p_j, which are between 0.1 and 10; and Λ_k, which are larger than 10. Thus we treat ϵ_i as small, p_j as $O(1)$, and Λ_k as large. This enables us to lay out an asymptotic path through the model, which we now propound. Of course, parameters will vary from case to case; moreover, the orders of magnitude get frayed at the edges (for example, $p_{39} = 0.1$, whereas $\epsilon_{24} = 0.08$; and $p_{47} = 8.99$, whereas $\Lambda_{13} = 10$) but mostly they serve as a useful indicator. In all, there are 26 ϵ_is, 58 p_js, and 29 Λ_ks (not all included here, however), and thus 113 parameters altogether. By choosing scales to balance terms, we could reduce this somewhat, but the model is still fearsome. We now show how it can be reduced dramatically.

Asymptotic reduction

Taking $\dot{V} \sim \Lambda$ (i.e., $\gg 1$) and $Q \sim 1$, we see from Eq. (19.4) that $x_l \to p_2/p_1$ on a (fast) timescale $t \sim 1/\Lambda$. From Eq. (19.15), the blood flow variables Q and Q_b also relax on this fast timescale, as does c_l, whereas c_b and c_t vary on the slower timescale $t \sim 1$. *If* we assume that no irregular behavior is associated with the fast time behavior of Q, Q_b, and c_l, then we can suppose these variables to be in (quasi-) equilibrium, and thus Eq. (19.4a) becomes, using the expression (19.13a) for Qc_v,

$$0 \approx (p_0 - p_1 c_l)\dot{V} - \Lambda_2 Q c_a + \Lambda_1 [p_{13} Q_b c_{vb}(t - \tau_{vb}) + (p_{24} Q - p_{13} Q_b) c_{vt}(t - \tau_{vt})]; \qquad (19.19)$$

for simplicity, we take the constitutive relations in the form

$$c_a \approx c_l, \qquad c_{vb} \approx c_b, \qquad c_{vt} \approx c_t. \qquad (19.20)$$

Thus

$$c_l(t) \approx \frac{p_0 \dot{V} + \Lambda_1 [p_{13} Q_b c_b(t - \tau_{vb}) + (p_{24} Q - p_{13} Q_b) c_t(t - \tau_{vt})]}{p_1 \dot{V} + \Lambda_2 Q}, \qquad (19.21)$$

and if $x_l = p_2/p_1 > 1$ (as it is for our estimates $p_1 = 1.65$, $p_2 = 2.34$), then $\dot{V} \approx \dot{V}_0(c_b)$ and we are concerned only with CO_2 control. In particular, Eq. (19.16) implies $Q \approx Q(c_l)$, $Q_b \approx Q_b(c_l)$, so that Eq. (19.21) determines c_l implicitly.

At leading order (neglecting $O(\epsilon_j)$ terms) the equations for c_t and c_b are, using Eq. (19.10),

$$\begin{aligned} \dot{c}_t &= p_8 Q [c_l(t - \tau_{at}) - c_t], \\ \dot{c}_b &= p_6 Q_b [c_l(t - \tau_{ab}) - c_b], \end{aligned} \qquad (19.22)$$

which are a pair of coupled delay recruitment equations, $c_l(t)$ being given as a function of $c_b(t)$ (through $\dot{V}$), $c_b(t - \tau_{vb})$, and $c_t(t - \tau_{vt})$ through Eq. (19.21).

A further reduction

In its present form the reduced model indicates that O_2 exists in equilibrium in the lung and throughout the body, due to the external source (p_2). The metabolic sources of CO_2 (ϵ_6 and ϵ_2), however, are very small ($\epsilon_6 \sim .008$, $\epsilon_2 \sim .078$) so that, although a primary function of ventilation is to remove CO_2 from the body, this is a secondary purpose from the *dynamic* point of view. The dynamics of CO_2 adjustment, as embodied in Eqs. (19.22) and (19.21), are concerned with distribution of CO_2 between the tissues, lungs, and brain. In the absence of an external source (p_0) we might expect a longer-term decline in CO_2 levels due to the loss term p_1 in Eq. (19.4a): this is not manifested in this model.

An even simpler model can be found by separating the timescales associated with c_t and c_b. Estimates of p_8 and p_6 are $p_8 \sim 0.15$ and $p_6 \sim 0.67$, which suggest that the limit $p_8 \ll p_6$ may be a useful starting point for analysis. This suggests that c_t is slowly varying as compared with c_b.

To estimate the delays, we have, from Eqs. (19.11), (19.12), and (19.14),

$$\begin{gathered} p_{55} \approx \int_{t-\tau_{ab}}^{t} Q\,dt, \\ p_{55} \approx \int_{t-\tau_{at}}^{t-\delta_1} Q\,dt, \qquad p_{13} = \int_{t-\delta_1}^{t} (Q - p_{26}Q_b)\,dt, \\ \tau_{vb} \approx 0, \\ p_{56} \approx \int_{t-\tau_{vt}}^{t} (Q - p_{27}Q_b)\,dt. \end{gathered} \tag{19.23}$$

Now rescale t by putting $t = p_{55}t^*$, and rescale all the delays similarly, $\tau_i \sim p_{55}$ (so the time unit now corresponds to p_{55} minutes, $\sim$10 seconds for $p_{55} \sim 0.18$); then (dropping the asterisks)

$$\begin{gathered} 1 \approx \int_{t-\tau}^{t} Q\,dt, \qquad 1 = \int_{t-\tau_{at}}^{t-\delta_1} Q\,dt, \qquad p_{13}/p_{55} \approx \int_{t-\delta_1}^{t} (Q - p_{26}Q_b)\,dt, \\ \tau_{vb} \approx 0, \qquad p_{56}/p_{55} = \int_{t-\tau_{vt}}^{t} (Q - p_{27}Q_b)\,dt, \end{gathered} \tag{19.24}$$

where $\tau = \tau_{ab}$. Equation (19.22) now becomes

$$\dot{c}_t = p_{55}p_8Q[c_l(t-\tau_{at}) - c_t], \tag{19.25a}$$

$$\dot{c}_b = p_{55}p_6Q_b[c_l(t-\tau) - c_b], \tag{19.25b}$$

and Eq. (19.21) is

$$c_l(t) \approx \frac{p_0\dot{V}_0 + \Lambda_1[p_{13}Q_bc_b(t) + (p_{24}Q - p_{13}Q_b)\,c_t(t-\tau_{vt})]}{p_1\dot{V}_0 + \Lambda_2 Q}. \tag{19.26}$$

We can take $c_t(t-\tau_{vt}) \approx c_t(t) \approx$ constant if c_t is considered to be slowly varying and $c_t \approx \bar{c}_l$ is the time average of c_l. Let us suppose no CO_2 is present in the atmosphere, and thus $p_0 = 0$; because $Q = Q(c_l)$ and $Q_b = Q_b(c_l)$, we have

$$c_l = \frac{\Lambda_1[p_{13}Q_b(c_l)c_b + \{p_{24}Q(c_l) - p_{13}Q_b(c_l)\}\bar{c}_l]}{p_1\dot{V}_0(c_b) + \Lambda_2 Q(c_l)}, \tag{19.27}$$

which defines (implicitly) c_l as a function of c_b. Equation (19.25b) is then a delay recruitment equation for c_b.

The simplest case is if Q and Q_b are constant (= 1, say). Then we write $c_b = c$, and

$$p_1\dot{V}_0(c_b)/\Lambda_2 Q = \delta v(c), \tag{19.28}$$

and the equation for c can be written in the form

$$\epsilon \dot{c} = -c + f(c_1), \tag{19.29}$$

where $c_1 = c(t-1)$, the delay $\tau = 1/Q = 1$, and

$$\epsilon = 1/p_{55}p_6Q_b,$$
$$f(c) = \frac{\beta c + \gamma}{1 + \delta v(c)}, \tag{19.30}$$
$$\beta = \frac{\Lambda_1 p_{13} Q_b}{\Lambda_2 Q}, \qquad \gamma = \frac{\Lambda_1(p_{24}Q - p_{13}Q_b)\bar{c}_l}{\Lambda_2 Q}.$$

Despite the enormous simplifications involved in writing Eq. (19.30), it is a model that contains all the seeds necessary to explain at least some aspects of oscillatory and chaotic behavior, and as such provides a starting point for understanding some of this behavior.

19.4 Oscillations and chaos

Typical values of the parameters in Eq. (19.30) are $\epsilon \sim 8$, $\beta \sim 0.12$, $\gamma \sim 1$, and $\delta = 1$, where a corresponding typical CO_2 ventilation response $v(c)$ is shown in Fig. 19.5. The corresponding form of $f(c)$ is shown in Fig. 19.6, and we see that Eq. (19.29) has a unique steady state $c = c^*$. To examine its stability, we linearize about c^*. Small disturbances proportional to $\exp(\sigma t)$ exist if

$$\epsilon\sigma = -1 + \lambda e^{-\sigma}, \qquad \lambda = f'(c^*), \tag{19.31}$$

and if we put $\sigma = \mu + i\omega$, then

$$\begin{aligned} \epsilon\mu &= -1 + \lambda e^{-\mu}\cos\omega, \\ \epsilon\omega &= -\lambda e^{-\mu}\sin\omega. \end{aligned} \tag{19.32}$$

Suppose $\mu > 0$. Then $e^{-\mu} < 1$, and if $|\lambda| < 1$, then $|\lambda e^{-\mu}\cos\omega| < 1$, leading to a contradiction. Thus $Re\,\sigma < 0$ for $|\lambda| < 1$. We treat λ as a bifurcation parameter. If

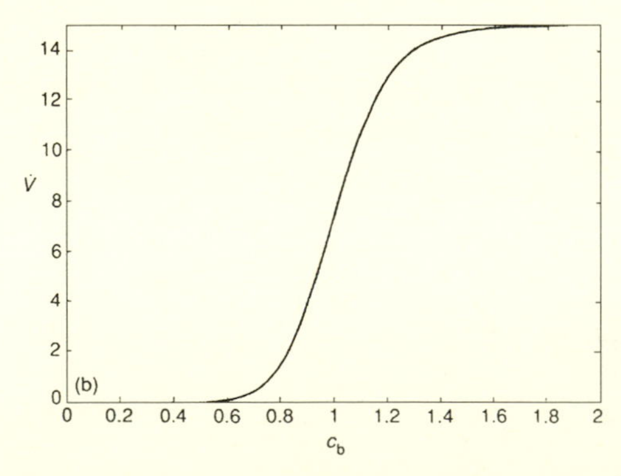

Fig. 19.5. Mackey–Glass (1977) type ventilation function. Reproduced from Fowler, Kalamangalam, and Kember (1993), by permission of Oxford University Press.

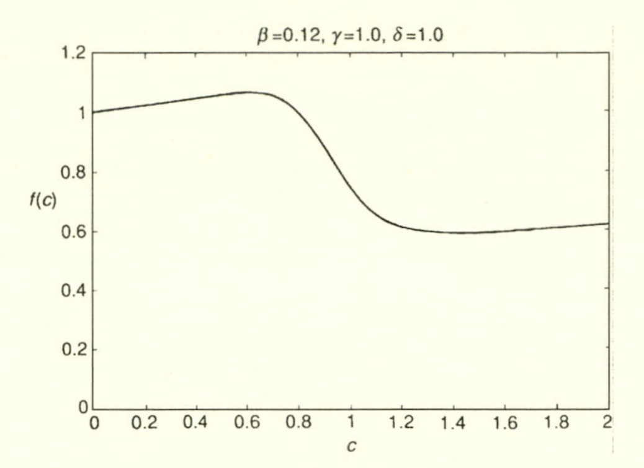

Fig. 19.6. Typical function $f(c)$ in (19.29). Reproduced from Fowler, Kalamangalam, and Kember (1993), by permission of Oxford University Press.

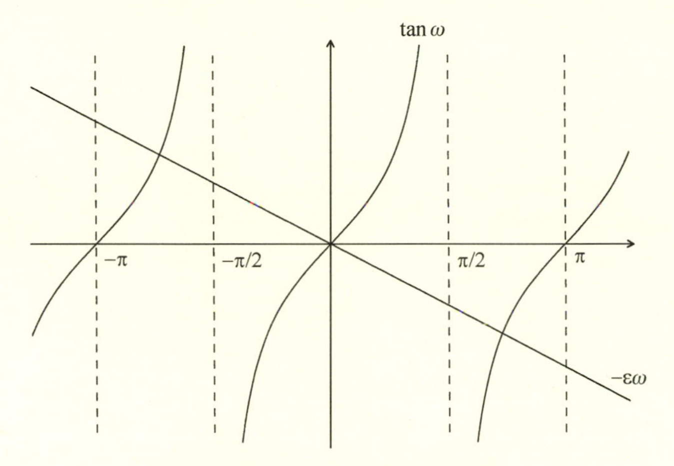

Fig. 19.7. Solution of Eq. (19.35)

an instability occurs for increasing $|\lambda|$, then $\mu = 0$ at the critical value, so that

$$
\begin{aligned}
1 &= \lambda \cos \omega, \\
\epsilon \omega &= -\lambda \sin \omega,
\end{aligned} \tag{19.33}
$$

which determines λ through

$$
\lambda = \sec \omega, \tag{19.34}
$$

and ω is a solution of the transcendental equation

$$
\tan \omega = -\epsilon \omega. \tag{19.35}
$$

It is easiest to locate the infinite number of roots of Eq. (19.35) graphically, as shown in Fig. 19.7. We discount the value $\omega = 0$, corresponding to $\lambda = 1$, as $f' < 0$ (and this would correspond to a saddle-node bifurcation, whereas we are specifically interested in oscillatory behavior).

Restricting our attention to values for which $\lambda < 0$, we see that there are an infinite number of values of λ for which $Re\, \sigma = 0$. Moreover, it is clear that the minimum value of $-\lambda$ occurs for $\omega = \omega_1$, which yields a value $\lambda = \lambda_1(\epsilon)$ at which a Hopf

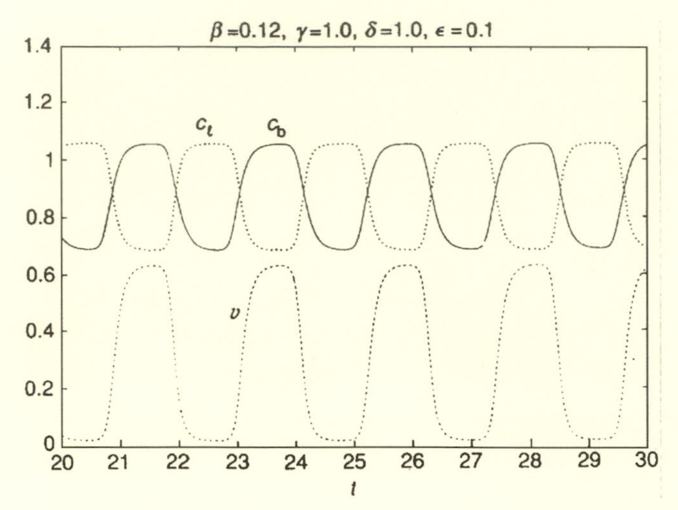

Fig. 19.8. An example of periodic breathing. Reproduced from Fowler, Kalamangalam, and Kember (1993), by permission of Oxford University Press.

bifurcation occurs to a periodic solution. Approximations to λ_1 can be obtained for $\epsilon \ll 1$ and $\epsilon \gg 1$, giving

$$\begin{aligned} \lambda_1 &\sim -(1+\epsilon^2\pi^2/2+\ldots) \quad \text{as } \epsilon \to 0, \\ \lambda_1 &\sim -\epsilon\pi/2 \quad \text{as } \epsilon \to \infty, \end{aligned} \tag{19.36}$$

and $-\lambda_1$ is monotonically increasing as ϵ increases.

We see that in comparison with the one-dimensional map $c \to f(c)$ (which is formally obtained from Eq. (19.29) as $\epsilon \to 0$), the differential term increases the stability, because the map undergoes a period-doubling bifurcation to a period-two cycle when $\lambda = -1$. Figure 19.8 shows an example of a periodic solution of the delay differential equation (dde) when the parameters are chosen as before, $\beta = 0.12$, $\gamma = 1$, $\delta = 1$, but $\epsilon = 0.1$. The solution begins to resemble the typical square wave form corresponding to a singular perturbation as $\epsilon \to 0$.

However, if $f(c)$ is steeper (and ϵ is smaller still) then this 'period-two' solution itself loses stability (as does the period-two orbit of the map) and chaotic behavior is obtained. An example is shown in Fig. 19.9(a) for $\beta = 1$, $\gamma = 0.4$, $\delta = 5$, and $\epsilon = 0.01$. The brain CO_2 oscillates chaotically in each of two states (a high level and a low level) with the period of alternation being equal to the delay time. A striking view of this chaos is seen in Fig. 19.9(b), where the corresponding ventilation $v(c)$ is plotted. We see that the ventilation oscillates between low, tranquil respiration, and highly variable increased states. Such chaotic bursts are extremely reminiscent of Cheyne–Stokes oscillations, although it must be remembered that Cheyne–Stokes breathing involves irregular oscillations from breath to breath, whereas the present model is for ventilation. Nevertheless, although our model is of the simplest possible form consistent with the respiratory control mechanism, this demonstrates that it has sufficient complexity to bear some relevance to clinically observed breathing patterns.

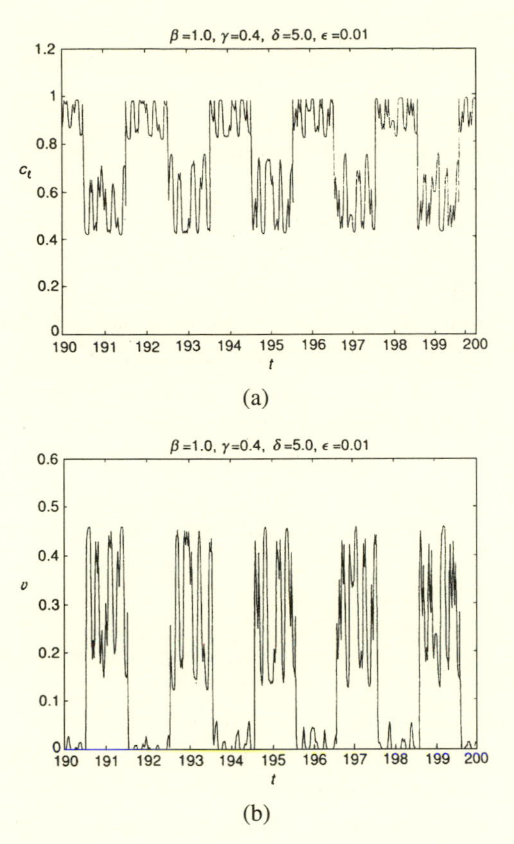

Fig. 19.9. Cheyne–Stokes-like chaotic oscillations in (a) blood CO_2 and (b) ventilation. Reproduced from Fowler, Kalamangalam, and Kember (1993), by permission of Oxford University Press.

19.5 Notes and references

Basic respiratory physiology is discussed in many textbooks, for example Mines (1986), Widdicombe and Davies (1991), Slonim and Hamilton (1987), Comroe (1974), and Bouhuys (1977). Two books that give some idea of modeling of respiration are by Whipp and Wiberg (1983) and Cramp and Carson (1988). It will be seen from these that the modeling concepts used are relatively primitive.

The Grodins model The model analyzed in this chapter was presented by Grodins et al. (1967). The analysis is based on an M.S. dissertation by Giri Kalamangalam, some of which appeared subsequently in the paper by Fowler et al. (1993). Giri Kalamangalam completed his doctoral thesis on further work, in particular, identifying important models for oxygen control and also identifying models where two or three delays are significant. A variety of more complicated dynamic behavior is possible.

Delay recruitment equations Equations of the form (19.29) arise in a variety of applications, for example, blood cell production (Mackey and Glass, 1977), lasers (Ikeda and Matsumoto, 1987) and population biology (Gurney et al. 1980). If the function $f(x)$ in the equation

$$\varepsilon \dot{x} = -x + f(x_1) \tag{19.37}$$

is unimodal (one humped) then an interesting question is what the relationship is between the chaos observed in the differential delay equation for small ε and that in the map ($\varepsilon = 0$), when $|f'|$ is large enough. One's initial thought might be that $\varepsilon \ll 1$ causes a singular perturbation such that (for example) the periodic solution generated by the Hopf bifurcation in $\varepsilon > 0$ becomes like a square wave for small ε, and indeed this can be observed numerically and has been studied by Chow and Mallet-Paret (1983). Furthermore, this basic ($T-$) periodic solution undergoes period-doubling bifurcations ($2T, 4T$, etc.) as ε is reduced or $|f'|$ is increased, and square-wave versions of these (joined by boundary layers at unit intervals) do exist. It is thus plausible to expect chaotic solutions of Eq. (19.37) to consist of square-wave type solutions consisting of values $x \approx x_n$ in intervals (t_n, t_{n+1}), where $t_{n+1} - t_n \approx 1$ and $x_{n+1} = f(x_n)$, joined by transition boundary layers, where (for large enough $|f'|$), $\{x_n\}$ is a chaotic sequence of iterates of the map $x \to f(x)$. So what happens? Numerically, we pose initial data

$$x = x_0 \text{ (constant)} \quad \text{for } t \in [-1, 0], \tag{19.38}$$

and if the map is chaotic, one finds instead that the boundary layers between successive piecewise constant regions grow out into the plateaus (see Fig. 19.10), and the long-time attractor is a chaotic motion that oscillates on the timescale $t \sim \varepsilon$.

A different way of thinking about Eq. (19.37) is as an infinite-dimensional map on the space of continuous functions $C[0, 1]$. Given a function $x_n(t)$, $t \in [0, 1]$, the solution $x_{n+1}(t)$ of Eq. (19.37) in $t \geq 0$ with $x_{n+1}(t) = x_n(t+1)$, $t \in [-1, 0]$, is simply

$$x_{n+1}(t) = x_n(1)e^{-t/\varepsilon} + \frac{1}{\varepsilon}e^{-t/\varepsilon} \int_0^t e^{s/\varepsilon} f[x_n(s)]\, ds, \tag{19.39}$$

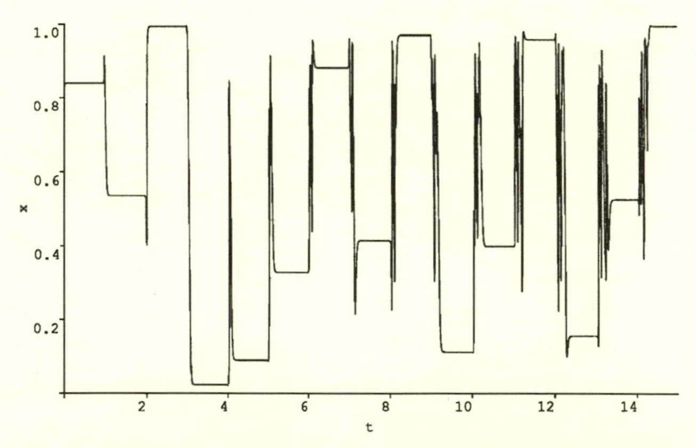

Fig. 19.10. Numerical solution of Eq. (19.37) with $f(c) = 4c(1-c)$, $\epsilon = 0.01$. The piecewise constant chaotic iterates of the map are eroded by chaotic oscillations on the timescale ϵ that spread from the transition layers. Reproduced courtesy of Jonathan Wattis.

and this defines a map $x_n \to x_{n+1}$ on $C[0, 1]$. Equivalently, this can be written as

$$x_{n+1}(t) = x_n(1)e^{-t/\varepsilon} + \int_0^{t/\varepsilon} f[x_n(t - \varepsilon\theta)]e^{-\theta}\, d\theta, \tag{19.40}$$

and if we define

$$t = \varepsilon\tau, \qquad x_n(t) = u_n(\tau), \tag{19.41}$$

then this is

$$u_{n+1}(\tau) = u_n(1/\varepsilon)e^{-\tau} + \int_0^{\tau} f[u_n(\tau - \theta)]e^{-\theta}\, d\theta. \tag{19.42}$$

If x is chaotic, then Eq. (19.42) suggests that the dynamics are ε-independent and essentially those of the map

$$u_{n+1}(\tau) = z_n e^{-\tau} + \int_0^{\tau} f[u_n(\tau - \theta)]e^{-\theta}\, d\theta, \tag{19.43}$$

where z_n is a random variable. Indeed, the distribution of x appears to be independent of ε as $\varepsilon \to 0$ (Ershov, 1992).

If access to chaos is not (directly) via that in the map, a different viewpoint may be necessary. We can write the solution as a trajectory in the infinite dimensional phase space $C[0, 1]$. Define $x_t(s) \in C[0, 1]$ for $t \geq -1$ by $x_t(s) = x(t + s)$. Then Eq. (19.39) can be written

$$x_t(s) = x_{-1}(1)e^{-t/\varepsilon}e^{-s/\varepsilon} + \int_0^{(t+s)/\varepsilon} e^{-\theta} f[x_{t-1}(s - \varepsilon\theta)]\, d\theta, \tag{19.44}$$

where we allow the extension of $x_t(s)$ to negative values of s, and the asymptotic limit of this trajectory for $\varepsilon \to 0$ is then

$$x_t(s) \sim \int_0^{\infty} e^{-\theta} f[x_{t-1}(s - \varepsilon\theta)]\, d\theta. \tag{19.45}$$

Hale and Sternberg (1988) suggest (and present numerical evidence) that the onset of chaos is associated with the occurrence of a homoclinic connection of the basic T-periodic orbit in the phase space $C[0, 1]$, similar to the situation in ordinary differential equations (see, for example, Sparrow (1982) or Wiggins (1988)).

20

Frost heave in freezing soils

20.1 Introduction

In areas that experience prolonged ground freezing in winter (Arctic and sub-Arctic regions of Canada and Russia, for example, but also many parts of North America, China, Scandinavia, etc.), it is common for the ground surface to rise or become otherwise distorted. This phenomenon is called *frost heave* and is responsible for an enormous amount of damage to roads and pavements, and also to buildings and foundations. It also gives rise to various natural features, as for example, in hummocks and stone circles and the dramatic pingos (see Figs. 20.1–20.3). All of these features are a result of frost heave, which occurs when ground is frozen due to a migration of groundwater toward the freezing front (or frost line). The heave is due not so much to expansion on freezing (although that helps, but is incidental) as to the fact that the water sucked toward the frost line freezes in a dramatic series of discrete ice lenses. Examples of these are seen in Figs. 20.4–20.5.

Soils range widely in particle size distribution. Soils with large grains (particles 60μ (microns) to 2 mm) are designated as *sand* (or, for larger particles still, *gravel*); finer soils with particle sizes 2–60μ are called *silts*, and the finest soils (particle size $< 2\mu$) are called *clays*. In general, soils have a distributed size fraction, and so one talks of 'clayey silt,' 'sandy silt,' and so on. The quantity and rate of heave vary with the soil type.

Generally it is found that soils composed of sand and gravel do not heave at all, whereas finer soils can produce considerable heave. Within these finer soils, it is found that soils in the fine silt to coarse clay range can produce spectacular heave, providing the overburden pressure (i.e., the superimposed load) is not too great, whereas the finest soils produce only small amounts of surface heave but are capable of overcoming substantial loads.

Particularly because of its importance in the construction of roads, pavements, and the foundation of buildings, and more recently because of the construction of oil and gas pipelines across permafrost regions in Alaska, frost heave has been a subject of engineering research since the early part of the century. As we shall see, it is a phenomenon whose explanation is still something of a mystery.

Early experiments into frost heave were conducted using closed systems.[1] These experiments seemed to confirm the theory of the time that heave was a result of the

[1] Unlike *open systems, closed systems* are unable to draw in or expel moisture.

Fig. 20.2. Stone circles in Spitzbergen. Photograph by Bernard Hallet, reproduced with permission, and kindly provided by Bill Krantz.

Fig. 20.3. Ice core in an eroded pingo. Photograph courtesy of Ross Mackay.

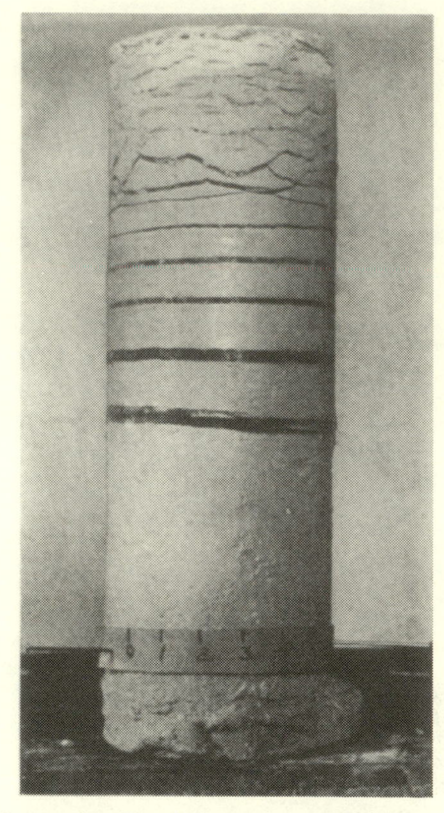

Fig. 20.4. Cylinder of frozen clay. The dark streaks are ice lenses. Reproduced from Taber (1930), by permission of the University of Chicago Press. Copyright 1930, the University of Chicago.

Fig. 20.5. Clay containing ice lenses from a street in Minnesota. Reproduced from Taber (1930), by permission of the University of Chicago Press. Copyright 1930, the University of Chicago.

increase in volume that occurs when water changes phase to ice. However, the theory was unable to explain such factors as the rhythmic banding of segregated seams of ice (ice lenses) that commonly form in heaved ground, nor could it account for excessive heaves of up to two feet, which would require frozen front depths of some twenty feet in order to be explained by this theory.

With the advent of the electric refrigerator, Taber (1930) was able to experiment extensively with open systems.[2] He found that he could not explain many of his observations using the above theory; most strikingly, he found that if water is replaced with a liquid that solidifies with a *decrease* in volume, frost heave will still occur.

Taber's experiments revealed many other characteristics of the frost heave process. He found that in relatively incompressible soils, ice lenses usually form and that their total thickness was equal to the amount of surface heave. Their formation seemed to depend on several factors, the most important being the size and shape of the soil particles, the rate of cooling, and the surface load.

The fundamental problem of frost heaving is to understand how an ice lens can grow in thickness by freezing on water from below and yet not incorporate the underlying soil particles. To explain this, Taber suggested that each soil particle would preferentially adsorb a thin film of 'water,' perhaps of only several molecules' thickness, similar to the adsorbed layers that form on many other solids in contact with water. Heaving and lens formation could then occur, providing such films could be maintained by water flow from below to balance the freezing from above. Physico-chemical explanations for the origins of such films include the electrical double layer, a model favored by Miller (1980) and that is consistent with descriptions of clay chemistry in particular. In this model, negative surface charges cause a thin film with excess positive charge to occur. This leads to an adsorption force on the film that binds it to the surface. One rationalization of this force is that the film pressure becomes anisotropic (Vignes-Adler, 1977), consisting of a *normal* pressure $p_\perp$ (to the surface) and a lateral pressure $p_\parallel$ (parallel to it). The difference

$$p_\perp - p_\parallel = p_d \tag{20.1}$$

is known as the *disjoining pressure*, and for a thin film of thickness h, can be taken as a function $g(h)$, dependent also on soil particle chemistry. Ultimately, it is the disjoining pressure that is responsible for maintaining lens formation, as if the film is not present, ice will freeze onto soil particles and grow past them (which is then called frost penetration).

Taber's hypothesis seemed to explain many of the mechanisms of frost heave and was supported by Beskow (1935) who also undertook extensive experimental investigations. Beskow substantiated many of Taber's findings and proposed empirical relationships for heave as a function of variables such as overburden pressure and particle size.

The complexity of the frost heave process led to no significant advances being made for the next twenty years. Gold (1957) introduced the concept of surface tension at the ice–water interface. He applied basic thermodynamic laws (including surface tension) at the ice front in order to relate freezing point depression to the curvature of the

[2] Soil usually behaves as an open system.

ice–water interface. From these equations he was able to conclude, as had been thought by Taber and Beskow, that because the narrow channels connecting the pores imposed constraints on the ice curvature, the water in the pores below the ice lens would not freeze until the temperature had dropped sufficiently to allow the ice to penetrate these channels. Gold's theory predicted that the advancing ice front would first penetrate into the pores with the largest interconnecting channels. He proposed that once ice had penetrated these pores the lens would become 'anchored' to the soil and cease to grow. If this assumption were true, it would be possible from knowledge of the distribution and size of the soil particles to calculate an upper limit on the heaving pressure attainable. His theory was developed by Everett (1961) and corroborative experiments were done by Penner (1959).

20.2 Primary frost heave models

In the 1970s, a number of models of frost heave were advanced, based roughly on the frost line description of Taber and on Gold's description of capillary forces (Outcalt, 1980; Dudek and Holden, 1979; Berg, Guymon, and Johnson, 1980). Before considering such a model, we need to consider the ice–water interface in more detail.

Frost-line suction

In Fig. 20.6, we present a schematic illustration of the ice–water interface. There are two types of water present: (free) pore water and bound or adsorbed film water. We are concerned with finding a relation between the *average* pressures p_i and p_w in ice and water.

The concept of stress, or normal stress, in a solid is well accepted. The concept of surface tension is less familiar, but nevertheless we adopt it. Denoting the normal ice stress as σ_n and the free pore water pressure as p, we would write

$$\sigma_n - p = \frac{2\sigma_{iw}}{r}, \tag{20.2}$$

where $2/r$ is the (mean) curvature of the interface. In an air–water system, this interface would be spherical, but in the ice–water system, the ice creeps too slowly to respond to the much more rapid evolution of the interface through freezing. For the case of *frost penetration*, where no lenses form, it is then natural to identify r with some mean pore size r_p, and specifically we would expect $r_p \sim d$, where d is the soil grain size. Identifying σ_n with p_i and p with p_w by a suitable average, we would

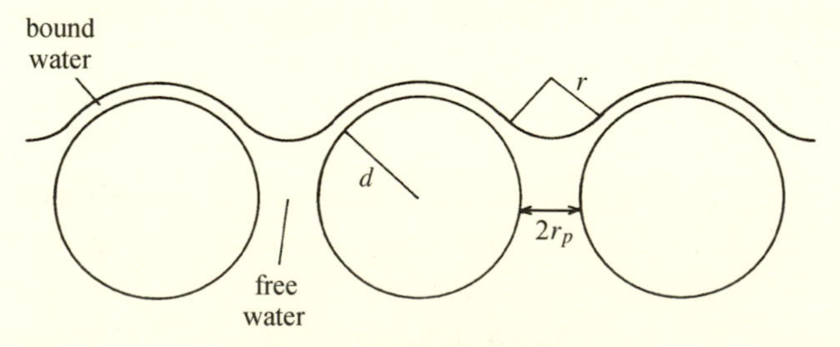

Fig. 20.6. Ice water interface in frozen soil. The circles are soil grains.

have the relation

$$p_i - p_w = \sigma_{iw}\bar{\kappa} \quad \text{(frost penetration)}, \tag{20.3}$$

where $\bar{\kappa} \sim 2/d$ is the mean curvature.

A transition from frost penetration to lens formation and thus heaving would then naturally be associated with soils for which $r > r_p$, that is to say, finer-grained soils. This broadly conforms to experience. When heaving occurs, the capillary suction $p_i - p_w$ must then be determined from the physics of the heaving process.

Adsorbed film dynamics

Consideration of the dynamics of thin films is addressed in the context of ice–water interactions by Vignes-Adler (1977) and Gilpin (1979). More generally, Hahn, Chen, and Slattery (1985) and Chen and Slattery (1982) consider the dynamics of thin films, based on a model of a viscous fluid, even though the layer may be only of several molecular thicknesses. With the geometry of Fig. 20.6, a suitable model for the film dynamics based on lubrication theory is

$$\begin{aligned} \frac{\partial h}{\partial t} + \frac{\partial q}{\partial s} &= 0, \\ q &= -\frac{h^3}{12\mu}\frac{\partial p_\parallel}{\partial s}, \end{aligned} \tag{20.4}$$

where μ is the effective film viscosity and s is circumferential distance from the pole of a grain. We take

$$p_\perp - p_\parallel = \frac{c}{h^\alpha} + \sigma_{iw}\left(\frac{\partial^2 h}{\partial s^2} - \kappa\right), \tag{20.5}$$

where the first term represents the disjoining pressure, and the second is an approximation for small h of the capillary suction (κ being the curvature of the grain, assumed constant). We take $p_\perp = p_i$ as constant, which is plausible because the ice lens is being accreted but not deformed. We expect c to be 'very small,' because the disjoining pressure should only be significant for $h \lll d$. Eliminating q, we have

$$h_t = \frac{\alpha c}{12\mu}\frac{\partial}{\partial s}\left[h^{2-\alpha}\frac{\partial h}{\partial s}\right] - \frac{\sigma_{iw}}{12\mu}\frac{\partial}{\partial s}\left[h^3 h_{sss}\right]. \tag{20.6}$$

If $s = 0$ represents the top of a grain, we expect to prescribe boundary conditions $h_s = h_{sss} = 0$ at $s = 0$, but conditions at the macroscopic 'triple junction' of ice/water/soil are less clear. Chen and Slattery (1982) prescribe $h_s = 0$, $h_{ss} =$ constant in their model, but such conditions are less obviously applicable for (20.6).

As a preliminary scaling, we choose

$$s, h \sim d, \qquad t \sim d/v_f, \tag{20.7}$$

where v_f is a typical freezing rate. With typical values of $c = 2.5 \times 10^{-22}$ J, $\alpha = 3$, $\mu = 10^{-3}$ kg m^{-1} s^{-1}, $v_f = 10^{-6}$ m s^{-1}, $d = 10^{-3}$ m, and $\sigma_{iw} = 33$ mN m^{-1}, we find the dimensionless version of Eq. (20.6) is

$$\delta h_t = \varepsilon\frac{\partial}{\partial s}\left[h^{2-\alpha}\frac{\partial h}{\partial s}\right] - \frac{\partial}{\partial s}[h^3 h_{sss}], \tag{20.8}$$

with

$$\delta = \frac{12\mu v_f}{\sigma_{iw}}, \qquad \varepsilon = \frac{\alpha c d^{1-\alpha}}{\sigma_{iw}}, \tag{20.9}$$

and $\delta \sim 3.6 \times 10^{-7}$, $\varepsilon \sim 2.3 \times 10^{-14}$. Thus $\varepsilon \sim \delta^2 \ll 1$.

As scaled, it is clear that surface tension dominates, and the disjoining pressure only becomes of significance when $h \ll 1$. The disjoining pressure balances the capillary term when $h \sim \varepsilon^{1/4}$ (if $\alpha = 3$), corresponding to a film thickness of 4×10^{-7} m, or 400 nm, rather thicker than the 10 nm considered appropriate by Gilpin (1979). However, the corresponding timescale is then $t \sim \delta/\varepsilon^{3/4} \sim 0.6 \times 10^4 \gg 1$, and is inappropriate.

On the face of it, one might hope that the presence of both the capillary and disjoining pressure terms would allow for a smooth transition between the film and ice–water interface. In fact, a balance is possible with $h \sim s - s_0 \sim \varepsilon^{1/2}$, but then one finds (if $h \sim s - s_0$ as $s/\varepsilon^{1/2} \to \infty$) that $h \to 0$ at finite s. In reality, the transition from finite slope interface to thin film induces large stresses in the ice, and this transition cannot be adequately modeled using Eq. (20.8). Thin film dynamics is described by a rescaling $h \sim (\varepsilon/\delta)^{1/(\alpha-2)}$, which is $\ll 1$ providing $\alpha > 2$, as we assume. The capillary term is then irrelevant, and its presence is only of consequence (see Hahn et al. (1985)), when its neglect would lead to an ill-posed problem. The rescaled equation is then

$$h_t = \frac{\partial}{\partial s}\left[\frac{1}{h^{\alpha-2}}\frac{\partial h}{\partial s}\right]; \tag{20.10}$$

and it follows by an application of the maximum principle (see Protter and Weinberger (1984)) that if h is initially positive everywhere, it remains so: pinchout cannot occur, and any film of finite extent will spread to cover the soil particle. In fact, if $h \sim c(s - Vt)^{\beta}$, $s > Vt$, with $h = 0$ for $s < Vt$, then we find

$$\beta = 1/(2 - \alpha), \qquad V = -c^{2-\alpha}\beta. \tag{20.11}$$

Thus fronts can only exist ($\beta > 0$) if $\alpha < 2$, but if they do, they move backward ($V < 0$) and the film eventually covers the particle surface. Also of note for a moving ice–water interface such that the film lies in $(-t, t)$ is a similarity solution

$$h = \frac{1}{t^{\alpha-2}}\phi(\eta), \qquad \eta = s/t; \tag{20.12}$$

when $\alpha = 3$, ϕ is given explicitly by

$$\phi = \frac{a}{1 + \frac{1}{2}a\eta^2}, \tag{20.13}$$

whence

$$h = \frac{at}{t^2 + \frac{1}{2}as^2}. \tag{20.14}$$

This solution is mass-conserving, and a is then determined implicitly by $\int_{-t}^{t} h\,ds = 2\sqrt{2a}\tan^{-1}[(a/2)^{1/2}]$. This magnitude of the disjoining pressure is instrumental in determining the film dynamics, but it has little direct effect on the frost heaving process itself.

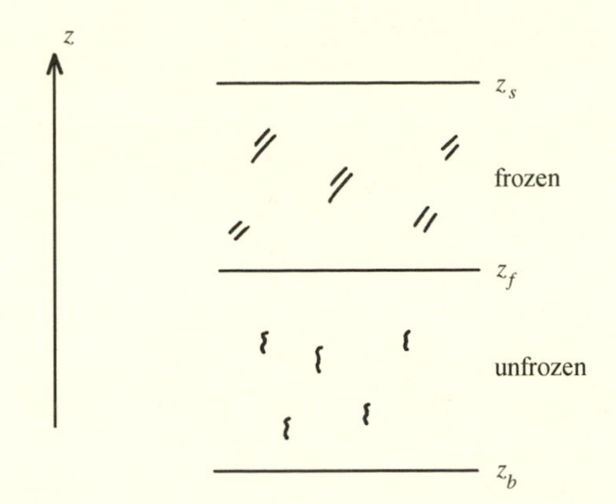

Fig. 20.7. Geometry of propagation of a freezing front into unfrozen soil

A mathematical model of primary frost heave

We consider only the one-dimensional situation here. Let z measure distance vertically upward from the initial surface location. The soil is confined to the region $z_b < z < z_s$, where z_b is the fixed base and z_s is the heaving surface. The freezing front is at z_f, so that $z < z_f$ is unfrozen, $z > z_f$ is frozen (see Fig. 20.7).

We assume the frozen region moves rigidly, that is, the soil is assumed rigid, except that it can separate at the frost line (if, for example, a lens is growing). Then

$$\dot{z}_s = v_i, \tag{20.15}$$

where v_i is the ice velocity at $z = z_f$. Suitable equations of heat and mass transfer are

$$\rho_f c_f [T_t + v_i T_z] = k_f T_{zz}, \quad z > z_f, \tag{20.16}$$

and

$$\rho_u c_u T_t + \rho_w c_w \frac{\partial}{\partial z}(vT) = k_u T_{zz}, \tag{20.17a}$$

$$v = -\frac{K}{\rho_w g}\left[\frac{\partial p_w}{\partial z} + \rho_w g\right], \tag{20.17b}$$

$$\frac{\partial(\phi \rho_w)}{\partial t} + \frac{\partial(\rho_w v)}{\partial z} = 0, \quad z < z_f, \tag{20.17c}$$

where ρ, c, and k represent density, specific heat, and thermal conductivity (f: frozen; u: unfrozen; w: water); T is temperature, v is the water flux in $z < z_f$, K is the hydraulic conductivity, and ϕ is the porosity, that is, the pore volume fraction. The second and third equations in Eq. (20.17) are Darcy's law and mass conservation.

The heat equations require T to be prescribed on z_s, z_f and z_b. If ϕ is given as $\phi(p_w)$ (cf. Chapter 13) then Eqs. (20.17b) and (20.17c) form a parabolic pair that require (for example) one condition for p_w on each of z_f and z_s. The heave rate is given by Eq. (20.15), and we then finally require two conditions to determine z_f and v_i.

The conditions we apply are the following:

$$\begin{aligned} T &= T_s \quad \text{on } z = z_s, \\ T &= T_f \quad \text{on } z = z_f, \\ T = T_b, &\qquad p_w = p_b \quad \text{on } z = z_s, \end{aligned} \tag{20.18}$$

where the last condition corresponds (for example) to the open laboratory system where the sample resides in a bath of water. The other condition for p_w at z_f comes from our discussion of suction, and we apply

$$p_w = p_f. \tag{20.19}$$

We consider the definition of p_f further below. The final two conditions for $\dot{z}_f$ and v_i come from a Stefan condition and a mass balance. We discuss these below.

Freezing temperature

In a medium where the pressure can vary, the freezing temperature changes by ΔT for a pressure change of Δp according to the Clapeyron relation

$$L\frac{\Delta T}{T} = \Delta p \Delta v, \tag{20.20}$$

where Δv is the specific volume change on melting (negative for water) and L is the latent heat. For the case where ice and water have two distinct pressures p_i and p_w, the appropriate generalization of this is

$$T_f = T_0\left[1 + \frac{p_w}{\rho_w L} - \frac{p_i}{\rho_i L}\right] = T_0\left[1 - \frac{p_w}{L}\left(\frac{1}{\rho_i} - \frac{1}{\rho_w}\right) - \frac{(p_i - p_w)}{\rho_i L}\right]. \tag{20.21}$$

In the latter form, we see that the second expression in brackets is the normal Clapeyron deficit, whereas the third expression represents exactly the Gibbs–Thomson effect, insofar as $p_i - p_w$ represents surface tension at the interface. This is the freezing temperature we prescribe at z_f.

Stefan condition

The rate of heat loss per unit area from the freezing front is $-[k\partial T/\partial z]_-^+$. This heat loss is balanced by the latent heat released on freezing water, either that which flows to z_f, a (heat) flux per unit area of $\rho_w L v$, or that frozen in situ by motion of z_f (downward), $-\rho_w L\phi\dot{z}_f$. Thus we have the Stefan condition

$$\rho_w L(v - \phi\dot{z}_f) = \left[-k\frac{\partial T}{\partial z}\right]_-^+. \tag{20.22}$$

Mass balance

Finally, mass balance of water substance can be determined using a pillbox argument. The net mass flux per unit area of water substance away from the interface is $\rho_i v_i - \rho_w v$. Here, specifically, it is assumed a lens lies in z_f+, because the heave

Fig. 20.1. A field of hummocks in Iceland. Photograph courtesy of Bill Krantz

rate (20.15) relates to the velocity. For frost penetration, the corresponding flux is $\rho_i(1-\phi)v_i - \rho_w v$. On the other hand, the rate of mass 'captured' by the interface is $(\rho_w - \rho_i)\phi(-\dot{z}_f)$; thus

$$-\phi\dot{z}_f(\rho_w - \rho_i) = \begin{cases} \rho_i v_i - \rho_w v & \text{(lens formation)} \\ \rho_i(1-\phi)v_i - \rho_w v & \text{(frost penetration).} \end{cases} \qquad (20.23)$$

A simplified model

On the basis of experimental conditions and observations, we expect typical pressure, length, and temperature scales to be

$$[p] \sim 1\,\text{bar} = 10^5\,\text{Pa}, \qquad d \sim 1\,\text{m}, \qquad \Delta T \sim 10\,\text{K}. \qquad (20.24)$$

The Stefan condition defines a timescale

$$t_h = \frac{\rho_w L d^2}{k\Delta T} \sim 10^7\text{s}, \qquad (20.25)$$

with $\rho_w \sim 10^3$ kg m^{-3}, $L \sim 3 \times 10^5$ J kg^{-1}, and $k \sim 2$ W m^{-1} K^{-1}. (We suppose $k_u = k_f = k$.) Also we can define a conductive timescale

$$t_c = d^2/\kappa \sim 10^6\,\text{s}, \qquad (20.26)$$

with $\kappa = k/\rho_w c_p$ and $c_p = 2$ kJ kg^{-1} K^{-1}. We thus expect the Stefan condition to be rate controlling. With $g \sim 10$ m s^{-2},

$$\rho_w g d/[p] \sim 0.1 \qquad (20.27)$$

is small; so also is the departure from isothermal freezing in the Clapeyron equation,

$$\frac{T - T_0}{\Delta T} \sim \frac{T_0}{\Delta T} \cdot \frac{[p]}{\rho_w L} \sim 10^{-2} \qquad (20.28)$$

with $T_0 = 273$ K. The liquid velocity scale is

$$v \sim K/\{\rho_w g d/[p]\}, \qquad (20.29)$$

and with typical values of $K \sim 10^{-11}$ m s^{-1} (clay), 10^{-8} m s^{-1} (silt), and 10^{-4} m s^{-1} (gravel), we find an equivalent range of water velocities:

$$\begin{aligned} v &\sim 10^{-10}\,\text{m s}^{-1}\ \text{(clay)}, \\ v &\sim 10^{-7}\,\text{m s}^{-1}\ \text{(silt)}, \\ v &\sim 10^{-3}\,\text{m s}^{-1}\ \text{(sand)}. \end{aligned} \qquad (20.30)$$

Of these, we will focus on silt, because sands do not generally heave, and clay is so impermeable that its water uptake is from within rather than from an external source. Then the Peclet number is $vd/\kappa \sim 10^{-1}$, and we do not expect advection to be significant.

Suppose we take ϕ, ρ_w as constant. Then $v = v(t)$, and

$$v = \frac{K(p_b - p_f)}{\rho_w g(z_f - z_b)}. \qquad (20.31)$$

To go further, we nondimensionalize the equations, by putting

$$\begin{aligned} T &= T_f + (\Delta T)T^*, \qquad z = dz^*, \qquad t = t_h t^*, \qquad v = (d/t_h)v^*, \\ v_i &= (d/t_h)v_i^*, \qquad p_w = p_f + (p_b - p_f)p^*, \end{aligned} \tag{20.32}$$

and dropping asterisks. Making the approximations alluded to above, we find (note that we suppose $k_u = k_f$, etc.)

$$\epsilon[T_t + v_i T_z] = T_{zz} \quad \text{in } z > z_f, \tag{20.33}$$

and

$$\begin{aligned} \epsilon[T_t + \alpha v T_z] &= T_{zz}, \\ v &= \frac{\Gamma}{z_f - z_b} \quad \text{in } z < z_f, \end{aligned} \tag{20.34}$$

where

$$\begin{aligned} \epsilon = t_c/t_h = c_p \Delta T/L, \qquad \alpha = \rho_w c_w/\rho_u c_u, \\ \Gamma = LK(p_b - p_f)/k\Delta T, \end{aligned} \tag{20.35}$$

and the boundary conditions are

$$\begin{aligned} T = T_s < 0 \quad &\text{on } z = z_s, \\ T = 0, \qquad v - \phi \dot{z}_f &= -\left[\frac{\partial T}{\partial z}\right]_-^+, \\ \dot{z}_s = v_i = (1+\delta)v - \delta\phi\dot{z}_f \quad &\text{on } z_f \text{ (lens formation)}, \\ \dot{z}_s = v_i = [(1+\delta)v - \delta\phi\dot{z}_f]/(1-\phi) \quad &\text{on } z_f \text{ (frost penetration)}, \\ T = T_b \quad &\text{on } z = z_b, \end{aligned} \tag{20.36}$$

where

$$\delta = (\rho_w - \rho_i)/\rho_i. \tag{20.37}$$

Typical values may be $\epsilon < 0.1$, $\delta \sim 0.1$, and $\Gamma \sim 1$, although the latter may certainly vary.

The simplest solution comes from taking $\epsilon \ll 1$, $\delta \ll 1$, and $\Gamma = O(1)$, as for silts. Then

$$\begin{aligned} T &\approx T_s\left(\frac{z - z_f}{z_s - z_f}\right), \quad z > z_f, \\ T &\approx T_b\left(\frac{z - z_f}{z_b - z_f}\right), \quad z < z_f, \end{aligned} \tag{20.38}$$

and we quickly find (for example, for lens formation)

$$\begin{aligned} \dot{z}_s = v &= \frac{\Gamma}{z_f - z_b}, \\ -\phi\dot{z}_f &= -v + \frac{|T_s|}{z_s - z_f} - \frac{T_b}{z_f - z_b}, \end{aligned} \tag{20.39}$$

two nonlinear equations for z_f and z_s. By writing

$$h_f = z_s - z_f, \qquad h_u = z_f - z_b, \tag{20.40}$$

the frozen and unfrozen thicknesses, respectively, then in terms of a nonlinear time-scale τ satisfying

$$\frac{dt}{d\tau} = \phi h_f h_u, \tag{20.41}$$

h_f and h_u satisfy (with $h' = dh/d\tau$)

$$\begin{aligned} h_f' &= |T_s|h_u - \{(1-\phi)\Gamma + T_b\}h_f, \\ h_u' &= -[|T_s|h_u - (\Gamma + T_b)]h_u, \end{aligned} \tag{20.42}$$

whose solution can be written down *exactly*. Appropriate initial conditions for a step-freezing experiment would be $t = 0$, $h_f = 0$, and $h_u = 1$ at $\tau = 0$. The heave rate is just Γ/h_u.

If heaving occurs via accumulation of water at an ice lens, then we add to the model (20.39) the condition $\dot{z}_f = 0$, which reflects the fact that the unfrozen soil is assumed to be rigid. This extra relation is facilitated by the criterion that the ice is unable to penetrate the pore space, which we associate with values of the suction pressure $< 2\sigma_{iw}/r_p$ (see Eq. (20.3)), where r_p is pore radius. If we define the basal effective pressure to be

$$p_e = P - p_b, \tag{20.43}$$

where P is the overburden pressure, and define the separation pressure to be

$$p_s = \frac{2\sigma_{iw}}{r_p}, \tag{20.44}$$

then the extra condition during lens formation is

$$\dot{z}_f = 0 \quad \text{if } \Gamma > \Gamma_c, \tag{20.45}$$

where

$$\Gamma_c = \frac{LK(p_e - p_s)}{k\Delta T}, \tag{20.46}$$

and in this case Γ is determined from Eq. (20.39). If Γ reaches Γ_c, then frost penetration occurs, and Eq. (20.45) is replaced by

$$\Gamma = \Gamma_c \quad \text{if } \dot{z}_f < 0, \tag{20.47}$$

and Eq. (20.39) is modified using Eq. (20.36). In this way the model can conceivably represent the formation of a sequence of ice lenses, although in practice it will not.

20.3 Secondary frost heave

Primary frost heave models suffer a serious drawback: they do not correspond to observations. In particular, they do not offer any mechanism whereby lens formation should occur periodically, or, indeed, at all. Insofar as they do predict heave, they are quantitatively inaccurate (see O'Neill (1983) for a review of this). There is also a damning mathematical fact. The freezing temperature for the pore water below the freezing front is given by the Clapeyron relation (20.20), that is,

$$T_f^u = T_0\left[1 - \frac{p_w}{L}\left(\frac{1}{\rho_i} - \frac{1}{\rho_w}\right)\right]. \tag{20.48}$$

Now the generalized Clapeyron relation predicts that at z_f,

$$T = T_f^u - (p_c/\rho_i L), \tag{20.49}$$

where p_c is the suction pressure. Therefore $T < T_f^u$ for some distance *below* the frost line, so that the pore water is supercooled. Indeed, both Taber and Beskow suggested that ice nucleating in the supercooled region would eventually block pore water and form a new lens. When we consider the result of supercooling in the solidification of alloys (see Chapter 17), it is natural to suppose that (unless we introduce solidification kinetics) thermodynamic equilibrium can be restored by allowing a 'mush' of partially frozen soil to exist below the lowest ice lens. This in fact is exactly what appears to happen.

It is observed (see the discussion by Miller (1978)) that the formation of new ice lenses actually occurs some distance *behind* the freezing front and not *at* the freezing front. The region below an ice lens that contains both water and ice is termed the *frozen fringe*. Miller (1972, 1977, 1978, 1980) developed a model of frost heave that allows the formation of ice lenses to occur within the frozen fringe. This mode of heave is known as *secondary heave* and is capable of attaining higher heaving pressures than those generated in the primary mode. Miller's model is acknowledged to be the most complete available and succeeds in predicting many of the characteristics of the frost heave process including the formation of new lenses within the frozen fringe.

20.4 Miller model of secondary frost heave

In the sections that follow we discuss the equations that are applicable in the frozen fringe. They are essentially those of O'Neill and Miller (1985) although some of them have been modified slightly.

Governing equations in the fringe

Throughout the following work we consider the soil matrix to be incompressible and thus only consider conservation of mass for the water and ice phases. Suppose that a soil has porosity ϕ, with water and ice volume fractions given by $W(\mathbf{x}, t)$ and $I(\mathbf{x}, t)$, respectively. Consideration of mass conservation for each phase gives

$$\frac{\partial W}{\partial t} + \nabla.\mathbf{U} = -\frac{S}{\rho_w}, \tag{20.50}$$

$$\frac{\partial I}{\partial t} + \nabla.\mathbf{V} = \frac{S}{\rho_i}, \tag{20.51}$$

where ρ_w and ρ_i are the densities of water and ice, $\mathbf{U}$ and $\mathbf{V}$ are the volume fluxes of water and ice, respectively, and S is the rate of freezing.

Conservation of energy is expressed by

$$-LS + \rho c_p \frac{dT}{dt} = k\nabla^2 T, \tag{20.52}$$

where L is latent heat, ρ is the overall density, and d/dt is a suitable (barycentric) material derivative. Because this term will be later ignored, we do not concern ourselves with its precise form here.

The sum of the pore constituent fractions is the porosity,

$$I + W = \phi, \tag{20.53}$$

and in Miller's model, the soil is taken as incompressible (and, in fact, undeformable); thus ϕ is constant.

In the fringe, we assume thermodynamic equilibrium, so that the generalized Clapeyron equation (20.21) gives T as

$$T = T_0\left[1 - \frac{p_w}{\rho_i L}\left(1 - \frac{\rho_i}{\rho_w}\right) - \frac{(p_i - p_w)}{\rho_i L}\right]. \tag{20.54}$$

The last term in brackets is the Gibbs–Thomson effect due to capillary suction, and in a two-phase medium, we can expect the mean curvature to be related to W; thus

$$p_i - p_w = f(W), \tag{20.55}$$

with f a monotone decreasing function of W. We shall expect f to be determined by the interfacial geometry, and Koopmans and Miller (1966) argue, and demonstrate experimentally, that f is the same curve as can be found in the wetting and drying of unsaturated, unfrozen soil, multiplied by the ratio of the surface tensions σ_{iw}/σ_{aw}. An example is shown in Fig. 20.8.

The simplest model for the water flow in the fringe is Darcy's law

$$\mathbf{U} = -\frac{K}{\rho_w g}[\boldsymbol{\nabla} p_w + \rho_w g\mathbf{k}], \tag{20.56}$$

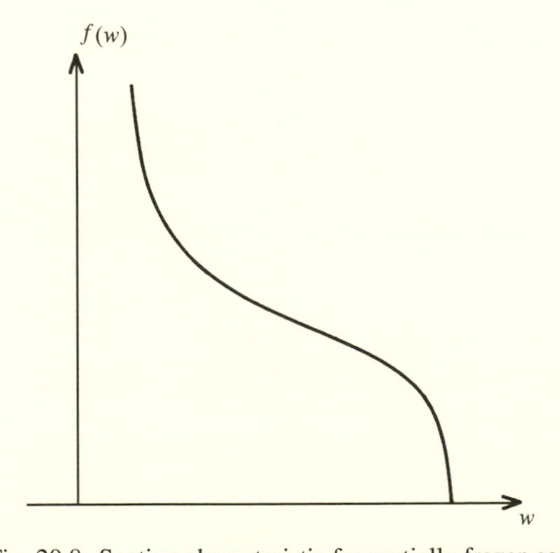

Fig. 20.8. Suction characteristic for partially frozen soil

where $\mathbf{k}$ is a unit vector directed vertically upward, and K is the hydraulic conductivity. Experiments indicate that K is a strong function of water content, and O'Neill and Miller (1985) take

$$K = K_0 (W/\phi)^\gamma, \tag{20.57}$$

with a value of γ of 7 or 9. The ice flux is simply

$$\mathbf{V} = I\mathbf{v}_i, \tag{20.58}$$

where $\mathbf{v}_i$ is the ice velocity. We come back to a prescription for this below.

Just as in an unfrozen soil, the confining pressure is distributed between the effective pressure p_e transmitted through the soil grains and the neutral stress p_n transmitted through the pores. This neutral stress is itself partitioned between the ice and water phases; thus

$$P = p_e + \chi p_w + (1 - \chi) p_i, \tag{20.59}$$

where the stress partition factor $\chi(W)$ varies between 0 and 1. Most simply, $\chi = W/\phi$, although O'Neill and Miller suggest other forms, for example $\chi = (W/\phi)^{3/2}$.

The total momentum equation would then, for a geostatic stress, be

$$P = P_0 + g \int_z^{z_s} \rho \, dz, \tag{20.60}$$

where ρ is the total soil density and P_0 is the surface load.

Regelation

The separation of an ice lens causes a conceptual problem. If a fringe exists, and if both ice and soil grains behave rigidly, then the ice must move past the soil grains if it forms a connected continuum. The alternative, suggested by Taber, is that ice crystals nucleate below the lowest lens and are not connected. In this case, we should choose $\mathbf{v}_i = \mathbf{0}$. O'Neill and Miller propose that, if the ice forms a connected medium, then it can in fact move past soil grains by *regelation*, that is, by melting and refreezing. Römkens and Miller (1973) have done experiments that show that particles move through ice under an imposed temperature gradient. On this basis, O'Neill and Miller suggest

$$\mathbf{v}_i = \mathbf{v}_i(t) = \mathbf{v}_i^+, \tag{20.61}$$

where $\mathbf{v}_i^+$ is the ice lens velocity.

However, there is a flaw in the prescription. For one thing, if differential frost heave occurs, then $\mathbf{v}_i^+$ will in general vary in space and so would $\mathbf{v}_i$; this is not consistent with the concept of rigidly moving ice. Secondly, and more subtly, the *averaged* velocity $\mathbf{v}_i$ used in the averaged equations is not necessarily the same as the instantaneous velocity. There is in addition the regelative flux, which though not instantaneously registered in the ice, does contribute over time. Following Gilpin's (1979) experiments, a natural modification is

$$\mathbf{v}_i = -\lambda \nabla T, \tag{20.62}$$

and we will adopt this law in the present exposition.

Initiation of an ice lens

When an ice lens forms, the soil particles on either side separate. This can only happen if the normal effective stress to the incipient lens becomes zero, because then there is nothing to keep the particles together. Miller's criterion for the initiation of a new ice lens (within the frozen fringe) is thus analogous to that of crack formation, and in geostatic stress, it is simply

$$p_e = 0. \tag{20.63}$$

This is a mechanical criterion, and in particular we can expect that the interlens pore water will remain unfrozen for a short while.

Boundary conditions

The equations in the frozen and unfrozen regions are just as those for primary heave. As boundary conditions, we take

$$T = T_s \quad \text{on } z = z_s, \tag{20.64}$$

for the top surface;

$$T = T_f, \tag{20.65a}$$

$$\dot{z}_l = 0, \tag{20.65b}$$

$$[\rho_w U_n + \rho_i V_n]_-^+ = 0, \tag{20.65c}$$

$$\rho_w L U_n^- = -\left[k\frac{\partial T}{\partial n}\right]_-^+, \tag{20.65d}$$

$$p_i = P, \quad \text{on } z = z_l, \tag{20.65e}$$

for the base of the lowest lens;

$$W = \phi, \tag{20.66a}$$

$$[U_n]_-^+ = 0, \tag{20.66b}$$

$$\left[\frac{\partial T}{\partial n}\right]_-^+ = 0, \tag{20.66c}$$

$$T = T_f, \tag{20.66d}$$

$$[p]_-^+ = 0 \quad \text{on } z = z_f, \tag{20.66e}$$

for the freezing front, and

$$T = T_b, \tag{20.67a}$$

$$p_w = p_b \quad \text{at } z = z_b, \tag{20.67b}$$

for the base. Here T_f is the Clapeyron freezing temperature. The condition $\dot{z}_l = 0$ is in fact a kinematic condition for soil particles at the lowest lens, that is, that they remain there. Then $\dot{z}_l = 0$ because we assume that the unfrozen soil is undeformable. The following two conditions, (20.65c) and (20.65d) are mass and energy balances, and the last states that the ice lens supports the load.

The first condition in (20.66) can be justified (only) by the requirement that the pore water beneath the fringe not be supercooled. It is analogous to the condition $\chi = 1$

used in Chapter 17 (Eq. (17.31)). The second and third are then mass and energy balances, and the fifth is a force balance. These conditions suffice to determine the solution, together with the heaving relation:

$$\dot{z}_s = v_{in}^+ = \mathbf{v}_i^+.\mathbf{n}, \tag{20.68}$$

which really comes from the assumption that the frozen soil behaves rigidly and is only strictly appropriate in one dimension.

Suppose, for the moment, we know the values of p_w and W in the fringe. Then Eqs. (20.64) and (20.65a) are suitable conditions for T in $z > z_l$, as T_f is given through (20.54). Equations (20.66e) and (20.67b) are the conditions for p_w in $z < z_f$, whereas Eqs. (20.66d) and (20.67a) are the corresponding temperature conditions. This leaves Eqs. (20.65b)–(20.65e) and (20.66a)–(20.66c) to determine the fringe variables, the ice velocity $\mathbf{v}_i$, and the unknown boundaries z_s, z_f, and z_l. By elimination of S (using Eq. (20.52)), T (using Eq. (20.54)), I (using Eq. (20.53)), $\mathbf{V}$ (using Eqs. (20.58) and (20.68)), and $\mathbf{U}$ (using Eq. (20.56)), we see that Eqs. (20.50) and (20.51) are a coupled pair of (one hopes) elliptic-parabolic equations for W and p_w. We then expect two conditions for these variables on z_l and z_f, and these are Eqs. (20.65d) and (20.65e) and (20.66a) and (20.66b). The extra conditions that determine z_f, z_l, and z_s are then Eqs. (20.66c), (20.65b), and (20.65c), respectively (together with Eq. (20.68)).

20.5 Simplifications

The computation of the model described above is extremely demanding and not really practical. What is needed is a simpler model that retains its accuracy but is computationally feasible. We now show how this can be done.

Thin frozen fringe

The thickness of the fringe can be estimated from the temperature drop across it. Denoting this by $[T]$, then

$$\frac{d_f}{d} \sim \frac{[T]}{\Delta T}, \tag{20.69}$$

where d_f is the fringe thickness, d is the overall depth scale, and ΔT is the temperature scale (for example, the applied temperature difference); from the Clapeyron equation, we expect $[T] \sim \sigma T_0/\rho_i L$ if $p_i - p_w = f(W) \sim \sigma$. Therefore

$$\epsilon = d_f/d \sim \sigma T_0/\rho_i L \Delta T, \tag{20.70}$$

and using typical values $T_0 \sim 300$ K, $\sigma \sim 1$ bar, and $\rho_i L \sim 3 \times 10^3$ bars, we find $[T] \sim 0.1$ K; thus $\epsilon \sim 10^{-2}$ if $\Delta T = 10$ K. On this basis, the fringe is thin; and our hope is to solve the equations in the fringe, allowing us to consider it as a surface across which various jump conditions can be applied.

We mentioned above that there is still unfrozen water in $z > z_l$. If this extends to $z = z_i$, say, similar reasons show that $z_i - z_l \sim \epsilon$. Eventually, we will include the whole region $z_f < z < z_i$ as our 'surface' of discontinuity. It should be noted that this really assumes bounded $f(W)$, for otherwise (as is practically the case), the soil will retain water down to some very low temperature – in that case, we should have to consider freezing to occur throughout the 'frozen' zone. We skip such peculiarities here.

Negligible gravitational acceleration

We will neglect gravity on the basis that $\rho g d/\sigma \sim 0.1$ is small.

Negligible heat advection

Typical heave and thus water velocities are of order $U \sim 10^{-7}$ m s^{-1}. Then the Peclet number is $Ud/\kappa \sim 10^{-1}$ for $\kappa \sim 10^{-6}$ m^2 s^{-1}. This suggests neglecting the heat advection term in the energy equation outside the fringe. It is also easy to show that heat advection can be neglected in the fringe.

Quasi-steady heat transport outside fringe

The thermal conductive timescale is $t_c = d^2/\kappa \sim 10^6$ s, whereas that for frost penetration is $t_f = d/|V_f|$, where $V_f = \dot{z}_f$. A Stefan condition such as Eq. (20.65d) (but not that particular equation) suggests $V_f \sim k[T]/d_f \rho_w L$. Then we find

$$t_f = \frac{d^2 St}{\kappa}, \tag{20.71}$$

where St is the Stefan number defined by

$$St = \frac{L}{c_p \Delta T}, \tag{20.72}$$

and L is the latent heat, c_p the specific heat. With $L \sim 3 \times 10^5$ J kg^{-1}, $c_p \sim 2 \times 10^3$ J kg^{-1} K^{-1}, and $\Delta T \sim 10$ K, then $St \sim 15$, so $t_f \gg t_c$, and we can neglect unsteady terms in the energy equation.

Permeability boundary layer

We now use the fact that the exponent γ in Eq. (20.57) is relatively large. This involves the idea of large activation energy asymptotics, and we can expect that the lowest value of K, that is, near the lens, is rate controlling. Essentially, the water flow is unimpeded except in a thin boundary layer near the lens, within which the pore water pressure jumps rapidly. The water flux is relatively constant in this layer, however.

Ignoring gravity, we have, for the normal water flux near the lens,

$$U_n = -\frac{K_0}{\rho_w g}(W_l/\phi)^\gamma \frac{\partial p_w}{\partial n}. \tag{20.73}$$

We write $n = z_l - \nu$; thus ν points downward, and we expand W in a local Taylor series for small ν, $W = W_l + W_l'\nu \ldots$, where $W_l = W(z_l)$, $W_l' = -\partial W(z_l)/\partial z$.

Then

$$U_n \approx \frac{K_l}{\rho_w g}\left[1+\frac{W_l'\nu}{W_l}+\dots\right]^{\gamma}\frac{\partial p_w}{\partial \nu}, \tag{20.74}$$

where

$$K_l = K_0(W_l/\phi)^{\gamma}, \tag{20.75}$$

and thus

$$\frac{\partial p_w}{\partial \nu} \sim \frac{\rho_w g U_n}{K_l}\exp\left[-\frac{\gamma W_l'\nu}{W_l}\right]; \tag{20.76}$$

the exponential deviates from K at values of $\nu \gg 1/\gamma$, but by then it does not matter, as p_w has reached its far field value. Integrating this, we thus obtain

$$p_w \sim p_b - \frac{\rho_w g U_n W_l}{\gamma W_l' K_l}\exp\left[-\frac{\gamma W_l'\nu}{W_l}\right], \tag{20.77}$$

where p_b matches the pressure to that in the far field (which is evidently simply $p_w = p_b$). Thus, putting $\nu = 0$, we have

$$\frac{\partial p_w}{\partial \nu} \sim \frac{\rho_w g U_n}{K_l}, \tag{20.78}$$

$$U_n \sim \frac{\gamma(f_l - P + p_b)W_l' K_l}{\rho_w g W_l}, \tag{20.79}$$

where n points upward, and each is evaluated on z_l. We have used Eq. (20.65e) together with Eq. (20.55) in evaluating p_w on z_l.

Quasi-stationary profiles in the fringe

In order to solve the equations in the fringe, we now show why the moisture and pressure profiles can be considered to be quasi-stationary, despite the disruption in the formation of new lenses. The adjustment process after the formation of a new lens is illustrated in Fig. 20.9. The ice pressure instantaneously adjusts to the overburden load, and the water pressure at the lens rapidly drops to its thermodynamic equilibrium value. The water pressure profile then adjusts rapidly to its equilibrium profile (and the adjustment is instantaneous if the water is incompressible, and very fast if it is slightly compressible). One can show that the relaxation of the moisture profile toward a quasi-stationary state takes a time of order $(\epsilon/\gamma^2)t_f$, while the next lens forms when the fringe temperature profile moves downward till the curve $-(1-\chi)f$ is next tangent to $p_w - P$. Because of the boundary layer profile of p_w, this takes a time $t_{\text{lens}} \sim (\epsilon/\gamma)t_f$, and thus the fringe profiles relax on a timescale much faster than the interval between lens formations. Furthermore, we see that the lens formation criterion is approximated by

$$P - p_b = (1-\chi_l)f_l, \tag{20.80}$$

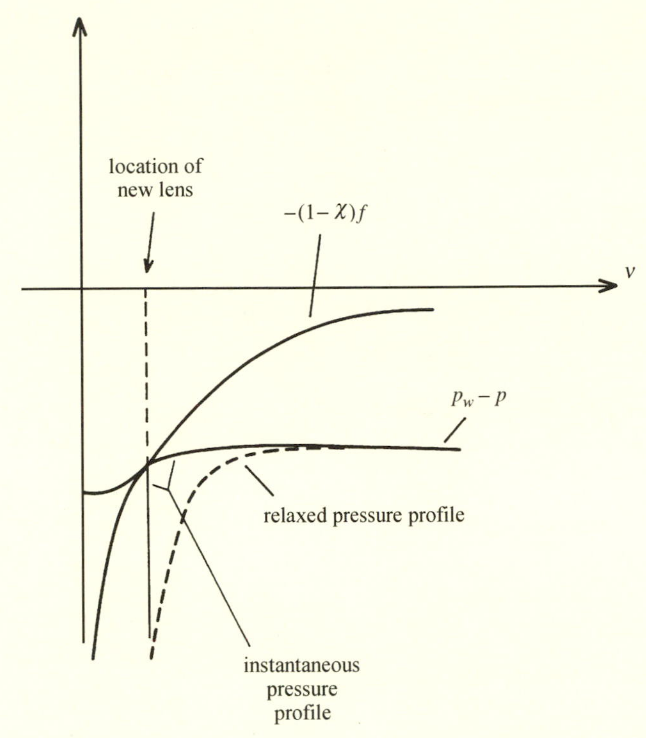

Fig. 20.9. Relaxation of the pore water pressure following the formation of an ice lens

where $f_l = f(W_l)$, $\chi_l = \chi(W_l)$. This then gives an expression for W_l as a function of the effective load $P - p_b$.

Quasi-continuous lens formation

In treating the fringe as a thin surface of discontinuity between the unfrozen ($z < z_f$) and frozen ($z > z_i \approx z_f$) regions, it is also consistent to consider the variation of z_f to be smooth. Because z_l moves downward in discrete jumps $\Delta z_l \sim \epsilon d/\gamma$ in times $\Delta t_l \sim t_{\text{lens}} \sim (\epsilon/\gamma)t_f$, we effectively approximate $\Delta z_l/\Delta t_l$ by its continuous limit, which will be $\dot{z}_f$. Despite this, we can still use the jump conditions at the lowest lens, as we have demonstrated that the transients during which these jump conditions are not satisfied are very short.

Integration of mass and energy equations across the fringe yield

$$
\begin{aligned}
[\rho_w W + \rho_i(\phi - W)]_{z_f}^{z_l^-} V_f &= [\rho_w U_n + \rho_i(\phi - W)v_{in}]_{z_f}^{z_l^-}, \\
\rho_w L[U_n - W V_f]_{z_f}^{z_l^-} &= k\left[\frac{\partial T}{\partial n}\right]_{z_f}^{z_l^-}.
\end{aligned}
\tag{20.81}
$$

We are now in a position to describe our reduction of the model.

20.6 A reduced model

We consider z_f as a moving surface separating the regions $z > z_f$ and $z < z_f$, in each of which heat transport is determined by

$$\nabla^2 T = 0. \tag{20.82}$$

The boundary conditions are

$$\begin{aligned} T &= T_s \quad \text{on } z = z_s, \\ T &= T_f \text{ (constant)} \quad \text{on } z = z_f, \\ T &= T_b \quad \text{on } z = z_b, \end{aligned} \tag{20.83}$$

and the moving boundaries are determined by

$$\dot{z}_s = v_{in}|_{z_l^+}, \tag{20.84}$$

(this really assumes one-dimensional heave), and v_{in}^+ and $\dot{z}_f$ are determined by mass balance across z_l:

$$[\rho_w U_n + \rho_i V_n]_{z_l-}^{z_l+} = 0 \tag{20.85}$$

and an energy balance from z_f to z_i:

$$\rho_w L[U_n - W v_f]_{z_f}^{z_i} = k\left[\frac{\partial T}{\partial n}\right]_-^+. \tag{20.86}$$

This completes the reduced model, except that certain quantities in Eqs. (20.85) and (20.86) need to be prescribed.

Equation (20.85) requires $U_n|_{z_l-}$ and W_l, whereas Eq. (20.86) requires $U_n|_{z_f}$. These are given by the permeability boundary layer solution (20.79):

$$U_n|_{z_l-} = \frac{\gamma W_l' K_l[p_b + f_l - P]}{\rho_w g W_l}; \tag{20.87}$$

the lensing criterion (20.80):

$$P - p_b = [1 - \chi(W_l)]f(W_l); \tag{20.88}$$

and the fringe mass balance (20.81) (giving $U_n|_{z_f}$):

$$[\rho_w W + \rho_i(\phi - W)]_{z_f}^{z_l-} V_f = [\rho_w U_n + \rho_i(\phi - W)v_{in}]_{z_f}^{z_l-}. \tag{20.89}$$

Of these, Eq. (20.87) requires W_l', which is found by differentiating the Clapeyron relation (20.54):

$$\left.\frac{\partial T}{\partial n}\right|_{z_l-} = \frac{T_0}{\rho_i L}\left[\delta \left.\frac{\partial p_w}{\partial n}\right|_{z_l-} + f_l' W_l'\right], \tag{20.90}$$

where $f_l' = \partial f/\partial W|_{z_l-}$; and using Eq. (20.78), we get

$$\left.\frac{\partial T}{\partial n}\right|_{z_l-} = \frac{T_0 W_l'}{\rho_i L}\left[\frac{\delta\gamma(f_l - P + p_b)}{W_l} + f_l'\right], \tag{20.91}$$

with

$$\delta = (\rho_w - \rho_i)/\rho_w. \tag{20.92}$$

This gives W_l', providing $\partial T/\partial n|_{z_l-}$ is known, and this is found from the fringe Stefan condition (20.81):

$$[\rho_w L(U_n - WV_f)]_{z_f}^{z_l^-} = \left[k\frac{\partial T}{\partial n}\right]_{z_f}^{z_l-}. \tag{20.93}$$

We thus see that the generalized Stefan condition (20.86) together with the heave rate can be given purely in terms of the heat fluxes at z_f and z_i, by algebraic elimination, using the subsidiary relationships (20.87), (20.88), (20.89), (20.91), and (20.93), while the heaving rate is then determined by Eq. (20.84) (in one dimension) together with Eq. (20.85).

Nondimensionalization

A little late in the day, we nondimensionalize the model by scaling

$$T - T_0 \sim \Delta T; \qquad z \sim d; \qquad V_f, v_i, U \sim \frac{\kappa}{dSt}, \qquad t \sim d^2 St/\kappa, \qquad f \sim \sigma. \tag{20.94}$$

Considering a one-dimensional solution only and adopting the Gilpin regelation assumption $v_i^- = -\lambda \partial T/\partial n|_{z_l^-}$, we find that the entire model collapses to two ordinary differential equations for z_f and z_s:

$$\dot{z}_f = \frac{[1 + \delta - (1 + \tilde{\beta})(\phi - W_l)\eta]G_i - (1 + \tilde{\beta})G_f}{\delta W_l + \phi + \tilde{\beta}(\phi - W_l) - \eta(1 + \tilde{\beta})(\phi - W_l)W_l}, \qquad \dot{z}_s = -\alpha[G_i - W_l \dot{z}_f], \tag{20.95}$$

where $G_i = \partial T/\partial z|_{z_i}$, $G_f = \partial T/\partial z|_{z_f}$, and the dimensionless parameters are defined by

$$\tilde{\beta} = \beta_l \left[\frac{f_l - N}{-W_l f_l' - \delta\gamma(f_l - N)}\right], \tag{20.96}$$

where N is the dimensionless effective load

$$N = \frac{P - p_b}{\sigma}, \tag{20.97}$$

β_l is given by

$$\beta_l = \frac{\gamma \rho_i L^2 K_l}{gkT_0}, \tag{20.98}$$

and for the thermal regelation model

$$\eta = \frac{\mu}{(1 + \tilde{\beta})}, \qquad \alpha = \frac{\tilde{\beta}(1 + \delta) + \mu(\phi - W_l)}{(1 + \tilde{\beta})}, \tag{20.99}$$

where

$$\mu = \frac{\lambda L \rho_w}{k}. \tag{20.100}$$

In one dimension, G_i and G_f can be written down explicitly, and in fact Eq. (20.95) have an exact solution: quite a reduction! The parameters μ and β_l are indicators of heave rate, and typical estimates using values $K_0 \sim 10^{-12} - 10^{-10}$ m s^{-1} for clay, $10^{-9} - 10^{-7}$ m s^{-1} for silt, $10^{-4} - 10^{-2}$ m s^{-1} for sand, and choosing $(W_l/\phi)^\gamma \sim 10^{-3}$ as representative, give

$$\begin{aligned} \beta_l &\sim 10^{-4} - 10^{-2} \quad \text{(clay)}, \\ \beta_l &\sim 10^{-1} - 10 \quad \text{(silt)}, \\ \beta_l &\sim 10^{4} - 10^{6} \quad \text{(sand)}, \end{aligned} \tag{20.101}$$

whereas if $\lambda \sim 10^{-10}$ m^2 s^{-1} K^{-1} (Gilpin (1979), then

$$\mu \sim 10^{-2}. \tag{20.102}$$

Because $1 - \chi$ and f are decreasing functions of W, we find that $W_l = g(N)$ is a decreasing function of N, and thus $\beta_l \sim [g(N)]^\gamma$ also decreases with N. Hence heave rate decreases with increasing load. Also, as soil becomes finer, β_l decreases Eq. (20.101); thus clays heave more slowly than silts owing to the effect of K_0. To get some idea of the effect of the load on $\tilde{\beta}$ in Eq. (20.96), choose simple (and not very realistic) expressions for f and χ:

$$f = 1 - (W/\phi), \qquad \chi = W/\phi. \tag{20.103}$$

We then obtain

$$W_l/\phi = 1 - \sqrt{N}, \tag{20.104}$$

and by substitution

$$\tilde{\beta} = \frac{\beta_0[1 - \sqrt{N}]^\gamma \sqrt{N}}{1 - \delta\gamma\sqrt{N}}, \tag{20.105}$$

where β_0 is β_l with W_l replaced by ϕ. The shutoff at $N = 1$ is rather artificial, as it corresponds to $W_l = 0$, which is due to the finite value of $f(0)$. For silts and particularly clays, it is more reasonable to have $f(W) \to \infty$ as $W \to 0$, in which case $W_l(N) \to 0$ as $N \to \infty$.

The behavior at $\sqrt{N} = 1/\delta\gamma$ is more interesting. It indicates a sudden increase in $\tilde{\beta}$ as the load approaches this value. We can see from Eq. (20.91) that this corresponds to W_l' becoming large. Considering Fig. 20.9, it is evident that the approximation that determines W_l through Eq. (20.88) breaks down as W_l' becomes large. Either $(1 - \chi_l) f_l$ increases beyond N, and lenses are more closely spaced, or else the occurrence of a point of tangency fails, and there is a terminal lens. When the exact tangency condition is included, it is found that the latter is what happens. Thus the model breaks down somewhere before this critical value of N is reached. To carry the model beyond the formation of the terminal lens, we lose the ability to prescribe W_l, but replace it by the condition that $\dot{z}_f = 0$ (and if this does not work, we must lose the fact that the fringe is thin). Figures 20.10–20.12 show typical computed heave rates, lens thicknesses, and lens initiation times before the formation of a terminal lens.

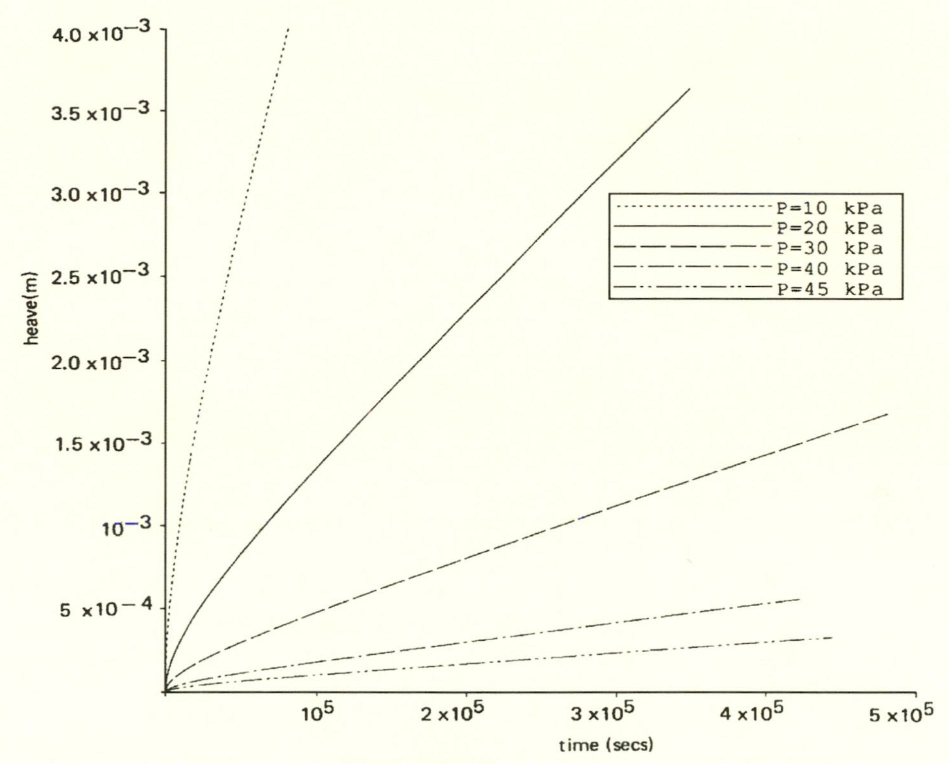

Fig. 20.10. Heave rate versus time for different overburdens. Reprinted from Fowler and Noon (1993) with kind permission of Elsevier Science–NL, Sara Burgerhartstraat 25, 1055 KV Amsterdam, The Netherlands.

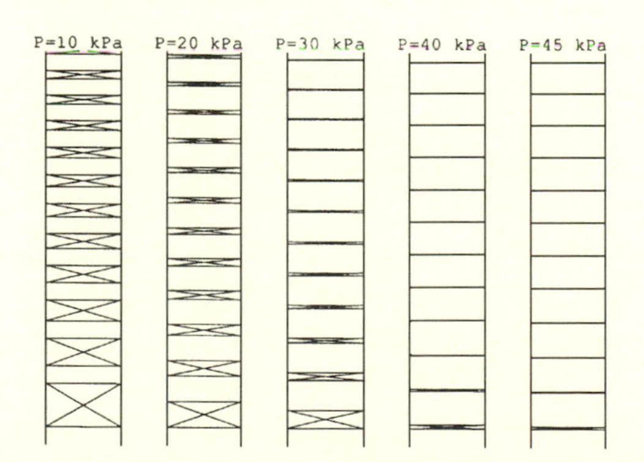

Fig. 20.11. Sequential ice lenses for different overburdens. Hatched regions indicate lenses. Reprinted from Fowler and Noon (1993) with kind permission of Elsevier Science–NL, Sara Burgerhartstraat 25, 1055 KV Amsterdam, The Netherlands.

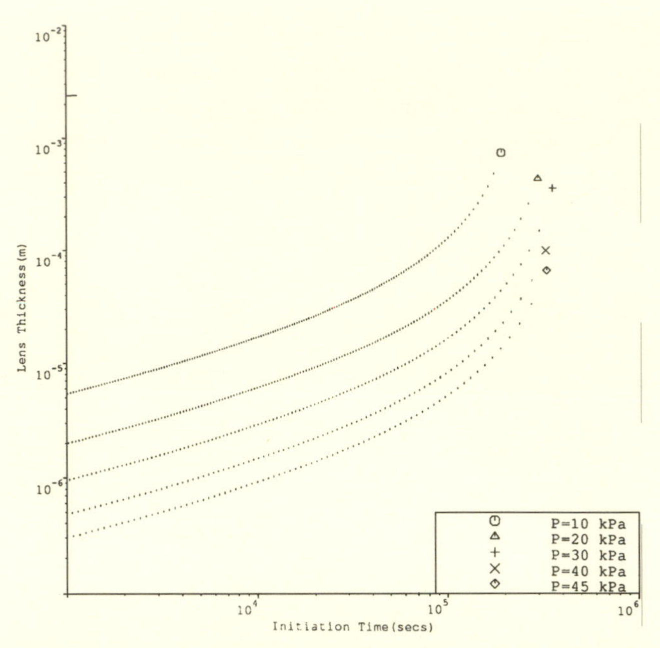

Fig. 20.12. Lens thickness as a function of time for different overburdens. Reprinted from Fowler and Noon (1993) with kind permission of Elsevier Science–NL, Sara Burgerhartstraat 25, 1055 KV Amsterdam, The Netherlands.

20.7 Notes and references

A useful review of the literature on frost heave is given by Black (1991) in a CRREL (Cold Regions Research and Engineering Laboratory) special report, which also includes seminal early papers by Beskow and Taber, including those by Beskow (1935) and Taber (1930).

The present chapter is based on a M.S. thesis (and later a doctoral thesis) by Chris Noon, which was motivated by O'Neill and Miller's (1985) numerical model, which was itself motivated by Miller's development of the concept of the frozen fringe. Although the existence of a frozen fringe is generally acknowledged, Miller's quantitative theory, and particularly the concept of a regelative ice flux is less universally accepted, and other theories have been proposed. A nice summary of the problem (as well as a generally useful introduction to frozen ground phenomenology) is given in the book by Williams and Smith (1989).

Much of the content of the present chapter appears in the paper by Fowler and Krantz (1994), and numerical solutions of the simplified model are given by Fowler and Noon (1993). Earlier attempts at simplifying the Miller model along similar lines are contained in the papers by Holden (1985), Piper, Holden, and Jones (1988), and Fowler (1989).

References

Acheson, D.J. 1990. *Elementary Fluid Dynamics*. Clarendon Press, Oxford.

Aldridge, C.J., and A.C. Fowler. 1996. Stability and instability in evaporating two-phase flows. *Surveys Math. Indus.*, **6**, 75–107.

Allen, J.R.L. 1985. *Principles of Physical Sedimentology*. Chapman and Hall, London.

Alley, R.B. 1989. Water-pressure coupling of sliding and bed deformation, I, Water system. *J. Glaciol.*, **35**, 108–118.

Anderson, O.L., and P.C. Grew. 1977. Stress corrosion theory of crack propagation with applications to geophysics. *Rev. Geophys. Space Phys.*, **15**, 77–104.

Aref, H., and E.P. Flinchem. 1984. Dynamics of a vortex filament in a shear flow. *J. Fluid Mech.*, **148**, 477–497.

Argoul, F., A. Arnéodo, and P. Richetti. 1991. Symbolic dynamics in the Belousov–Zhabotinskii reaction: from Rössler's intuition to experimental evidence for Shil'nikov's homoclinic chaos. In *A Chaotic Hierarchy*, G. Bauer and M. Klein, eds., p. 79, World Scientific, Singapore.

Aris, R. 1956. On the dispersion of a solute in a fluid flowing through a tube. *Proc. Roy. Soc.*, **A235**, 67–78.

Aris, R. 1975. *Mathematical Theory of Diffusion and Reaction in Permeable Catalysts.* 2 vols. Oxford University Press, Oxford.

Atkin, R.J., and N. Fox. 1980. *An Introduction to the Theory of Elasticity.* Longman, London.

Atkinson, B.K., ed. 1987. *Fracture Mechanics of Rock.* Academic Press, New York.

Atkinson, J.H., and P.L. Bransby. 1978. *The Mechanics of Soils: An Introduction to Critical State Soil Mechanics.* McGraw-Hill, New York.

Avrami, M. 1939. Kinetics of phase change, I. *J. Chem. Phys.*, **7**, 1103–1112.

Avrami, M. 1940. Kinetics of phase change, II. *J. Chem. Phys.*, **8**, 212–224.

Barenblatt, G.I., Iu. P. Zheltov, and I.N. Kochina. 1960. Basic concepts in the theory of seepage of homogeneous liquids in fissured rock [strata]. *J. Appl. Math. Mech.*, **24**, 1286–1303 (English translation of *Prikl. Mat. Mekh.*, **24**, 852–864).

Batchelor, G.K. 1967. *An Introduction to Fluid Dynamics.* Cambridge University Press, Cambridge.

Bear, J. 1972. *Dynamics of Fluids in Porous Media.* Elsevier, New York. (Dover reprint, Mineda, N.Y., 1988).

Bear, J., and Y. Bachmat. 1990. *Introduction to Modeling of Transport Phenomena in Porous Media.* Kluwer, Dordrecht.

Bear, J., and A. Verruijt. 1987. *Modeling Groundwater Flow and Pollution.* Reidel, Dordrecht.

Bebernes, J., A. Bressan, and D. Eberly. 1987. A description of blow-up for the solid fuel ignition model. *Indiana Univ. Math. J.*, **36**, 295–305.

Belyea, R.N., C.A. Miller, G.L. Baskerville, E.G. Kettela, K.B. Marshall, and I.W. Varty. 1975. The spruce budworm. (A series of eight articles.) *Forestry Chronicle*, **51**, 135–160.

Bender, C.M., and S.A. Orszag. 1978. *Advanced Mathematical Methods for Scientists and Engineers.* McGraw-Hill, New York.

Bentley, C.R. 1987. Antarctic ice streams: a review. *J. Geophys. Res.*, **92**, 8843–8858.

Berg, R.L., G.L. Guymon, and T.C. Johnson, 1980. Mathematical model to correlate frost heave of pavements with laboratory predictions. CRREL Report, 80-10, Hanover, N.H.

Bergles, A.E., J.G. Collier, J.M. Delhaye, G.F. Hewitt, and F. Mayinger. 1981. *Two-Phase Flow and Heat Transfer in the Power and Process Industries*. Hemisphere, Washington, D.C.

Beskow, G. 1935. Soil freezing and frost heaving with special applications to roads and railroads. *Swed. Geol. Soc.*, C, no. 375, Year Book no. 3, (Trans. J.O. Osterberg, Technological Institute, Northwestern University.)

Biot, M.A. 1941. General theory of three-dimensional consolidation. *J. Appl. Phys.*, **12**, 155–164.

Bird, R.B., R.C. Armstrong, and O. Hassager. 1977. *Dynamics of Polymeric Liquids: Vol. 1, Fluid Mechanics*. John Wiley, New York.

Black, P.B. 1991. Historical perspective of frost heave research. In: Historical perspectives in frost heave research. The early works of S. Taber and G. Beskow; P.B. Black and M.J. Hardenberg, eds., pp. 1–7. CRREL special report 91-23. CRREL, Hanover, N.H.

Black, P.B., and M.J. Hardenberg, eds., 1991. Historical perspectives in frost heave research. CRREL Spec. Rep. 91-23, Hanover, N.H.

Bond, G., H. Heinrich, W. Broecker, L. Labeyrie, J. McManus, J. Andrews, S. Huon, R. Jantschik, S. Clasen, C. Simet, K. Tedesco, M. Klas, G. Bonani, and S. Ivy. 1992. Evidence for massive discharges of icebergs into the North Atlantic ocean during the last glacial period. *Nature*, **360**, 245–249.

Booker, J.R. 1981. Large amplitude convection in porous media. In Report of the GFD summer program, 81–102, pp. 161–163. Woods Hole Oceanographic Institution, Woods Hole, Mass.

Bouhuys, A. 1977. *The Physiology of Breathing*. Grune and Stratton, New York.

Brandeis, G., C. Jaupart, and C.J. Allegre. 1984. Nucleation, crystal growth and the thermal regime of cooling magmas. *J. Geophys. Res.*, **89**, 10,161–10,177.

Bretherton, C.S., and E.A. Spiegel. 1983. Intermittency through modulational instability. *Phys. Letts.*, **96A**, 152–156.

Briggs, G.E., and J.B.S. Haldane. 1925. A note on the kinetics of enzyme action. *Biochem. J.*, **19**, 338–339.

Brooks, K., V. Balakotaiah, and D. Luss. 1988. Effect of natural convection on spontaneous combustion of coal stockpiles. *AIChE J.*, **34**, 353–365.

Buckmaster, J.D., and G.S.S. Ludford. 1982. *Theory of Laminar Flames*. Cambridge University Press, Cambridge.

Budyko, M.I. 1969. The effect of solar radiation variation on the climate of the earth. *Tellus*, **21**, 612–619.

Butterworth, D., and G.F. Hewitt, eds. 1977. *Two-Phase Flow and Heat Transfer*. Oxford University Press, Oxford.

Cao, G., A. Varma, and W. Strieder. 1993. Approximate solutions for nonlinear gas–solid noncatalytic reactions. *AIChE J.*, **39**, 913–917.

Carlson, J.M., and J.S. Langer. 1989. Properties of earthquakes generated by fault dynamics. *Phys. Rev. Lett.*, **62**, 2632–2635.

Carrier, G.F., and C.E. Pearson. 1976. *Partial Differential Equations*. Academic Press, New York.

Carrier, G.F., M. Krook, and C.E. Pearson. 1966. *Functions of a Complex Variable*. McGraw-Hill, New York.

Carslaw, H.S., and J.C. Jaeger. 1959. *Conduction of Heat in Solids*. Clarendon Press, Oxford.

Chan, Y.H., and D.L.S. McElwain. 1994. Conversion estimates for gas–solid reactions: reversible kinetics independent of solid concentrations. *Chem. Eng. Sci.*, **49**, 363–372.

Chandrasekhar, S. 1961. *Hydrodynamic and Hydromagnetic Stability*. Oxford University Press, Oxford.

Chen, J.-D., and J.C. Slattery. 1982. Effects of London–Van der Waals forces on the thinning of a dimpled liquid film as a small drop or bubble approaches a horizontal plane. *AIChE J.*, **28**, 955–963.

Chen, W.F., and E. Mizuno. 1990. *Nonlinear Analysis in Soil Mechanics: Theory and Implementation*. Elsevier, Amsterdam.

Chorley, R.J., ed. 1969. *Introduction to Physical Hydrology*. Methuen, London.

Chow, S.-N., and J. Mallet-Paret. 1983. Singularly perturbed delay-differential equations. In *Coupled Nonlinear Oscillators*, J. Chandra and A.C. Scott, eds., pp. 7–12, North–Holland Math. Studies, Vol. 80, North–Holland, New York.

Chow, V.T. 1959. *Open-Channel Hydraulics*. McGraw-Hill, New York.
Christensen, U.R. 1984. Heat transfer by variable viscosity convection and implications for the Earth's thermal evolution. *Phys. Earth. Planet. Int.*, **35**, 264–282.
Clark, P.U. 1994. Unstable behaviour of the Laurentide ice sheet over deforming sediment and its implications for climate change. *Quat. Res.*, **41**, 19–25.
Colbeck, S.C., ed. 1980. *Dynamics of Snow and Ice Masses*. Academic Press, New York.
Comroe, J.H. 1974. *Physiology of Respiration*, 2nd ed., Year Book Medical Publishers, Chicago.
Copley, S.M., A.F. Giamei, S.M. Johnson, and M.F. Hornbecker. 1970. The origins of freckles in unidirectionally solidified castings. *Metall. Trans.*, **1**, 2193–2204.
Courant, R., and D. Hilbert. 1953. *Methods of Mathematical Physics*, Vol. I. John Wiley, New York.
Coussy, O. 1995. *Mechanics of Porous Continua*. John Wiley, New York.
Cowling, T.G. 1957. *Magnetohydrodynamics*. Wiley-Interscience, New York.
Cramp, D.G., and E.R. Carson, eds. 1988. *The Respiratory System*. Croom Helm, London.
Crank, J. 1975. *The Mathematics of Diffusion*, 2nd ed. Clarendon Press, Oxford.
Crank, J. 1984. *Free and Moving Boundary Value Problems*. Clarendon Press, Oxford.
Cushman, J.H., ed. 1990. *Dynamics of Fluids in Hierarchical Porous Media*. Academic Press, London.
Dagan, G. 1984. Solute transport in heterogeneous porous formations. *J. Fluid Mech.*, **145**, 151–177.
Davis, S.H., H.E. Huppert, U. Müller, and M.G. Worster, eds. 1992. Interactive dynamics of convection and solidification. *NATO ASI Series E., Vol. 219*. Kluwer, Dordrecht.
Dibrov, B.F., M.A. Livshits, and M.V. Volkenstein. 1977a. Mathematical model of immune processes. *J. Theor. Biol.*, **65**, 609–631.
Dibrov, B.F., M.A. Livshits, and M.V. Volkenstein. 1977b. Mathematical model of immune processes, II, Kinetic features of antigen-antibody interrelations. *J. Theor. Biol.*, **69**, 23–39.
Do Carmo, M.P. 1976. *Differential Geometry of Curves and Surfaces*. Prentice-Hall, Englewood Cliffs, N.J.
Dodd, R.K., J.C. Eilbeck, J.D. Gibbon, and H.C. Morris. 1982. *Solitons and Nonlinear Wave Equations*. Academic Press, New York.
Dold, J.W. 1985. Analysis of the early stage of thermal runaway. *Quart. J. Mech. Appl. Math.*, **38**, 361–387.
Dorning, J. 1989. An introduction to chaotic dynamics in two-phase flow. *AIChE Symp. Ser.*, **85**(269), 241–248.
Drazin, P.G. 1983. Solitons. *L.M.S. Lect. Note Ser., Vol. 85*. Cambridge University Press, Cambridge.
Drazin, P.G. 1992. *Nonlinear Systems*. Cambridge University Press, Cambridge.
Drazin, P.G., and W.H. Reid. 1981. *Hydrodynamic Stability*. Cambridge University Press, Cambridge.
Drew, D.A. 1983. Mathematical modeling of two-phase flow. *Ann. Rev. Fluid Mech.*, **15**, 261–291.
Drew, D.A. 1990. Evolution of geometric statistics. *SIAM J. Appl. Math.*, **50**, 649–666.
Drew, D.A., and R.T. Wood. 1985. Overview and taxonomy of models and methods for workshop on two-phase flow fundamentals. Nat. Bureau of Standards, Gaithersburg, Md., Sept. 22–27.
Dudek, S.J.-M., and J.T. Holden, 1979. A theoretical model for frost heave. In *Numerical Methods in Thermal Problems*, R.W. Lewis and K. Morgan, eds., pp. 216–229, Pineridge Press, Swansea.
Duffin, W.J. 1968. *Advanced Electricity and Magnetism*. McGraw-Hill, London.
Duffin, W.J. 1980. *Electricity and Magnetism*, 3rd ed. McGraw-Hill, London.
Dullien, F.A.L. 1979. *Porous Media: Fluid Transport and Pore Structure*. Academic Press, New York.
Dussan V., E.B., and S.H. Davis. 1974. On the motion of a fluid–fluid interface along a solid surface. *J. Fluid Mech.*, **65**, 71–95.
Edelson, D., R.J. Field, and R.M. Noyes. 1975. Mechanistic details of the Belousov–Zhabotinskii oscillations. *Int. J. Chem. Kinet.*, **7**, 417–432.

Emerman, S.H., D.L. Turcotte, and D.A. Spence. 1986. Transport of magma and hydrothermal solutions by laminar and turbulent fluid fracture. *Phys. Earth Planet. Int.*, **41**, 248–259.

Emms, P.W., and A.C. Fowler. 1994. Compositional convection in the solidification of binary alloys. *J. Fluid Mech.*, **262**, 111–139.

Engelund, F. 1970. Instability of erodible beds. *J. Fluid Mech.*, **42**, 225–244.

England, A.H. 1971. *Complex Variable Methods in Elasticity*. John Wiley, New York.

Ershov, S.V. 1992. Asymptotic theory of multidimensional chaos. *J. Stat. Phys.*, **69**, 781–812.

Everett, D.H. 1961. The thermodynamics of frost damage to porous solids. *Trans. Faraday Soc.*, **57**, 1541–1551.

Field, R.J., and M. Burger, eds. 1985. *Oscillations and Travelling Waves in Chemical Systems*. John Wiley, New York.

Field, R.J., and H.D. Försterling. 1986. On the oxybromine chemistry rate constants with cerium ions in the Field-Körös-Noyes mechanism from the Belousov–Zhabotinskii reaction: the equilibrium $HBrO_2 + BrO_3^- + H^+ = 2BrO_2^\bullet + H_2O$. *J. Phys. Chem.*, **90**, 5400–5407.

Field, R.J., E. Körös, and R.M. Noyes. 1972. Oscillations in chemical systems, II, Thorough analysis of temporal oscillation in the bromate-cerium-malonic acid system. *J. Amer. Chem. Soc.*, **94**, 8649–8664.

Field, R.J., and R.M. Noyes. 1974. Oscillations in chemical systems, IV, Limit cycle behaviour in a model of a real chemical reaction. *J. Chem. Phys.*, **60**, 1877–1884.

Finn, R. 1986. *Equilibrium Capillary Surfaces*. Springer-Verlag, Berlin.

Fowler, A.C. 1978. Linear and nonlinear stability of heat exchangers. *J. Inst. Maths. Applics.*, **22**, 361–382.

Fowler, A.C. 1989. Secondary frost heave in freezing soils. *SIAM J. Appl. Math.*, **49**, 991–1008.

Fowler, A.C. 1992. Modeling ice sheet dynamics. *Geophys. Astrophys. Fluid Dyn.*, **63**, 29–65.

Fowler, A.C. 1993. Towards a description of convection with temperature and pressure dependent viscosity. *Stud. Appl. Math.*, **88**, 113–139.

Fowler, A.C., and C. Johnson. 1995. Hydraulic run-away: a mechanism for thermally regulated surges of ice sheets. *J. Glaciol.*, **41**, 554–561.

Fowler, A.C., and W.B. Krantz. 1994. On the O'Neill/Miller model for secondary frost heave. *SIAM J. Appl. Math.*, **54**, 1650–1675.

Fowler, A.C., and C.G. Noon. 1993. A simplified numerical solution of the Miller model for secondary frost heave. *Cold Reg. Sci. Technol.*, **21**, 327– 336.

Fowler, A.C., G.P. Kalamangalam, and G. Kember. 1993. A mathematical analysis of the Grodins model of respiratory control. *IMA J. Maths. Appl. Med. Biol.*, **10**, 249–280.

Frank-Kamenetskii, D.A. 1955. *Diffusion and Heat Exchange in Chemical Kinetics*. Princeton University Press, Princeton, N.J.

Freeze, R.A., and J.A. Cherry. 1979. *Groundwater*. Englewood Cliffs, N.J.

French, R.H. 1994. *Open-Channel Hydraulics*. McGraw-Hill, New York.

Frenzen, C.L., and P.K. Maini. 1988. Enzyme kinetics for a two-step enzymic reaction with comparable initial enzyme-substrate ratios. *J. Math. Biol.*, **26**, 689–703.

Freund, L.B. 1990. *Dynamic Fracture Mechanics*. Cambridge University Press, Cambridge.

Friedman, A. 1964. *Partial Differential Equations of Parabolic Type*. Prentice-Hall, Englewood Cliffs, N.J.

Gel'fand, I.M. 1963. Some problems in the theory of quasilinear equations. *AMS Transl. Ser. 2*, **29**, 295–381.

Gelhar, L.W. 1984. Stochastic analysis of flow in heterogeneous porous media. In Fundamentals of Transport Phenomena in Porous Media, M.Y. Corapcioglu and J. Bear, eds., pp. 673–717, *NATO ASI Series E, Vol. 82*, Martinus Nijhoff, Dordrecht.

Ghil, M., and S. Childress. 1987. *Topics in Geophysical Fluid Dynamics: Atmospheric Dynamics, Dynamo Theory, and Climate Dynamics*. Springer–Verlag, Berlin.

Gibbon, J.D., and M.J. McGuinness. 1981. Amplitude equations at the critical points of nonstable dispersive physical systems. *Proc. R. Soc. Lond.*, **A377**, 185–219.

Gilpin, R.R. 1979. A model of the 'liquid like' layer between ice and a substrate with applications to wire regelation and particle migration. *J. Colloid Interface Sci.*, **68**, 235–251.

Glass, L., and M.C. Mackey. 1988. *From Clocks to Chaos*. Princeton University Press, Princeton, N.J.

Gold, L.W. 1957. A possible force mechanism associated with freezing of water in porous materials. *High. Res. Board Bull.*, **168**, 65–72.

Goldstein, S., ed. 1938. *Modern Developments in Fluid Dynamics*. Clarendon Press, Oxford. (Reprinted by Dover Publications, Mineola, N.Y., 1965.)

Gould, P.L. 1994. *Introduction to Linear Elasticity*, 2nd ed. Springer–Verlag, Berlin.

Gray, J.S. 1946. The multiple factor theory of respiratory regulation. *Science*, **102**, 739–744.

Greene, J.A. 1933. Clinical studies on respiration. IV: Some observations on Cheyne-Stokes respiration. *Arch. Intern. Med.*, **52**, 545–563.

Grodins, F.S., J. Buell, and A.J. Bart. 1967. Mathematical analysis of digital simulation of the respiratory control system. *J. Appl. Physiol.*, **22**, 260–276.

Groves, C.G., and A.D. Howard. 1994. Minimum hydrochemical conditions allowing limestone cave development. *Water Resour. Res.*, **30**, 607–615.

Gurney, W.S.C., S.P. Blythe, and R.M. Nisbet. 1980. Nicholson's blowflies revisited. *Nature*, **287**, 17–21.

Györgi, L., and R.J. Field. 1991. Simple models of deterministic chaos in the Belousov–Zhabotinsky reaction. *J. Phys. Chem.*, **95**, 6594–6602.

Györgi, L., and R.J. Field. 1992. A three-variable model of deterministic chaos in the Belousov–Zhabotinsky reaction. *Nature*, **355**, 808–810.

Hagan, P.S., M. Hershowitz, and C. Pirkle. 1988. A simple approach to highly sensitive tubular reactors. *SIAM J. Appl. Math.*, **48**, 1083–1101.

Hager, B.H., and M. Gurnis. 1987. Mantle convection and the state of the Earth's interior. *Rev. Geophys.*, **25**, 1277–1285.

Hahn, P.-S., J.-D. Chen, and J.C. Slattery. 1985. Effect of London–Van der Waals forces on the thinning and rupture of a dimpled liquid film as a small drop or bubble approaches a fluid–fluid interface. *AIChE J.*, **31**, 2026–2038.

Hakim, V. 1991. Computation of transcendental effects in growth problems: linear solvability conditions and nonlinear methods in the example of the geometrical model. In Asymptotics Beyond all Orders. H. Segur, S. Tanveer, and H. Levine, eds., pp. 15–28, *NATO ASI Ser. 3, Vol. 284*. Plenum, New York.

Hale, J.K., and N. Sternberg. 1988. Onset of chaos in differential delay equations. *J. Comp. Phys.*, **77**, 221–239.

Hasimoto, H. 1972. A soliton on a vortex filament. *J. Fluid Mech.*, **51**, 477–485.

Heineken, F.G., H.M. Tsuchiya, and R. Aris. 1967. On the accuracy of determining rate constants in enzymatic reactions. *Math. Biosci.*, **1**, 115–141.

Herrero, M.A., and J.J.L. Velazquez. 1990. Stability analysis of a closed thermosyphon. *Euro. J. Appl. Math.*, **1**, 1–24.

Hetsroni, G., ed. 1982. *Handbook of Multiphase Systems*. Hemisphere, Washington, D.C.

Hewett, T.A. 1986. Fractal distributions of reservoir heterogeneity and their influence on fluid transport. *Soc. Petrol. Eng.*, paper no. 15386.

Hewitt, G.F., and D.N. Roberts. 1969. Studies of two-phase flow patterns by simultaneous flash and X-ray photography. Report no. AERE-M2159, Atomic Energy Research Establishment, Harwell, Oxfordshire, England.

Hill, R. 1950. *The Mathematical Theory of Plasticity*. Oxford University Press, Oxford.

Hinch, E.J., 1991. *Perturbation Methods*. Cambridge University Press, Cambridge.

Hocking, L.M., and K. Stewartson. 1971. On the non-linear response of a marginally unstable plane parallel flow to a three-dimensional disturbance. *Mathematika*, **18**, 219–239.

Hocking, L.M., and K. Stewartson, 1972. On the nonlinear response of a marginally unstable plane parallel flow to a two-dimensional disturbance. *Proc. R. Soc. Lond.*, **A326**, 289–313.

Hocking, L.M., K. Stewartson, and J.T. Stuart. 1972. A nonlinear instability burst in plane parallel flow. *J. Fluid Mech.*, **51**, 705–735.

Holden, J.T. 1985. Approximate solutions for Miller's theory of secondary heave. *Proc. Fourth Int. Symp. Ground Freezing*, pp. 498–503. A.A. Balkema, Rotterdam.

Holling, C.S., D.D. Jones, and W.C. Clark. 1979. Ecological policy design: a case study of forest past management. In *Pest Management: Proceedings of an International Conference*, G.A. Norton and C.S. Holling, eds. Pergamon Press, Tarrytown, N.Y.

Hoppensteadt, F.C. 1975. *Mathematical Theories of Populations: Demographics, Genetics and Epidemics*. Society for Industrial and Applied Mathematics, Philadelphia.
Horowitz, I., and A. Ioinovici. 1985. Budworm-forest system: application of quantitative feedback theory. *Int. J. Systems Sci.*, **16**, 209–225.
Howard, A.D., W.E. Dietrich, and M.A. Seidl. 1994. Modeling fluvial erosion on regional to continental scales. *J. Geophys. Res.*, **99**, 13,971–13,986.
Howard, L.N., and G. Veronis. 1987. The salt-finger zone. *J. Fluid Mech.*, **183**, 1–23.
Hutter, K. 1983. *Theoretical Glaciology*. D. Reidel, Dordrecht.
Hutter, K., S. Yakowitz, and F. Szidarovsky. 1986. A numerical study of plane ice sheet flow. *J. Glaciol.*, **32**, 139–160.
Ikeda, I., and K. Matsumoto. 1987. High-dimensional chaotic behaviour in systems with time-delayed feedback. *Physica*, **29D**, 223–235.
Ikeda, S., and G. Parker, eds. 1989. River meandering. A.G.U. Water resources monograph 12. American Geophysical Union, Washington, D.C.
Ishii, M., 1975. *Thermo-Fluid Dynamic Theory of Two-Phase Flow*. Eyrolles, Paris.
Ivantsov, G.P. 1947. Temperaturnoye pol'e vokrug sharoobraznovo tsilindricheskovo i igloobraznovo kristalla, rastushchevo v pereochlazhdennom raslav'e. *Dokl. Akad. Nauk. SSSR*, **58**, 567–569.
Jeffreys, H., and B. Jeffreys. 1953. *Methods of Mathematical Physics*. Cambridge University Press, Cambridge.
Jimenez, J., and J.A. Zufiria. 1987. A boundary layer analysis of Rayleigh–Bénard convection at large Rayleigh number. *J. Fluid Mech.*, **178**, 53–71.
Jones, D.D. 1979. The budworm site model. In *Pest Management: Proceedings of an International Conference*, G.A. Norton and C.S. Holling, eds., pp. 91–156. Pergamon Press, Tarrytown, N.Y.
Keener, J.P. 1988. *Principles of Applied Mathematics*. Addison-Wesley, Redwood City, Calif.
Keller, J.B. 1966. Periodic oscillations in a model of thermal convection. *J. Fluid Mech.*, **26**, 599–606.
Kennedy, J.F. 1963. The mechanics of dunes and anti-dunes in erodible-bed channels. *J. Fluid Mech.*, **16**, 521–544.
Kevorkian, J., and J.D. Cole. 1982. *Perturbation Methods in Applied Mathematics*. Springer–Verlag, Berlin.
Khan, A.S., and S. Huang. 1995. *Continuum Theory of Plasticity*. John Wiley, New York.
Kilpinen, A. 1988. An on-line model for estimating the melting zone in a blast furnace. *Chem. Eng. Sci.*, **43**, 1813–1818.
Kooi, H., and C. Beaumont. 1994. Escarpment evolution on high-elevation rifted margins: Insights derived from a surface processes model that combines diffusion, advection, and reaction. *J. Geophys. Res.*, **99**, 12,191–12,209.
Koopmans, R.W.R., and R.D. Miller, 1966. Soil freezing and soil water characteristic curves. *Soil Sci. Soc. Am. Proc.*, **30**, 680–685.
Kramer, S., and M. Marder. 1992. Evolution of river networks. *Phys. Rev. Lett.*, **68**, 205–208.
Lacey, A.A. 1983. Initial motion of the free boundary for a non-linear diffusion equation. *IMA J. Appl. Math.*, **31**, 113–119.
Lacey, A.A., J.R. Ockendon, and A.B. Tayler. 1982. "Waiting-time" solutions of a nonlinear diffusion equation. *SIAM J. Appl. Math.*, **42**, 1252–1264.
Lai, W.M., D. Rubin, and E. Krampl. 1993. *Introduction to Continuum Mechanics*. Pergamon Press, Oxford.
Lambe, T.W., and R.V. Whitman. 1979. *Soil Mechanics, SI Version*. John Wiley, New York.
Landau, L.D., and E.M. Lifshitz. 1959. *Fluid Mechanics*. Pergamon Press, Oxford.
Landau, L.D., and E.M. Lifshitz. 1986. *Theory of Elasticity*, 3rd ed. Pergamon Press, Oxford.
Lapwood, E.R. 1948. Convection of a fluid in a porous medium. *Proc. Camb. Phil. Soc.*, **44**, 508–521.
Lawn, B.R., and T.R. Wilshaw. 1975. *Fracture of Brittle Solids*. Cambridge University Press, Cambridge.
Ledinegg, M. 1938. Instabilität der Strömung bei Natürlichen und Zwangumlauf. Die Wärme, **61**, 891–898.
Le Mesurier, B.J., G.C. Papanicolaou, C. Sulem, and P. Sulem. 1988. Local structure of the self-focussing singularity of the nonlinear Schrödinger equation. *Physica D*, **32**, 210–226.

Lewis, G.C., W.B. Krantz, and N. Caine. 1993. A model for the initiation of patterned ground owing to differential secondary frost heave. *Proc. 6th Int. Conf. Permafr.*, Vol. 2, pp. 1044–1049.

Lighthill, M.J. 1958. *An Introduction to Fourier Analysis and Generalised Functions*. Cambridge University Press, Cambridge.

Lin, C.C., and L.A. Segel. 1974. *Mathematics Applied to Deterministic Problems in the Natural Sciences*. Macmillan, London.

Liouville, J. 1853. Sur l'equation aux differences partielles $\partial^2 \log \lambda/\partial u\, \partial v \pm \lambda/2a^2 = 0$. *J. Math. Pures Appliq.*, **18**, 71–72.

Lister, J.R. 1990. Buoyancy-driven fluid fracture: the effects of material toughness and of low-viscosity precursors. *J. Fluid Mech.*, **210**, 263–280.

Loper, D.E., ed. 1987. Structure and Dynamics of Partially Solidified Systems. *NATO ASI Series E*, Vol. 125. Martinus Nijhoff, Dordrecht.

Lotka, A.J. 1920. Undamped oscillations derived from the law of mass action. *J. Amer. Chem. Soc.*, **42**, 1595–1599.

Ludwig, D., D.D. Jones, and C.S. Holling. 1978. Qualitative analysis of insect outbreak systems: the spruce budworm and forest. *J. Anim. Ecol.*, **47**, 315–332.

Ludwig, D., D.G. Aronson, and H.F. Weinberger. 1979. Spatial patterning of the spruce budworm. *J. Math. Biol.*, **8**, 217–238.

Lunardini, V.J., and A. Aziz. 1993. Perturbation techniques in conduction-controlled freeze-thaw heat transfer. CRREL Monograph 93-1, Hanover, N.H.

MacAyeal, D. 1993. Binge/purge oscillations of the Laurentide ice sheet as a cause of the North Atlantic's Heinrich events. *Paleoceanography*, **8**, 775–784.

Mackey, M.C., and L. Glass. 1977. Oscillations and chaos in physiological control systems. *Science*, **197**, 287–289.

Male, D.H. 1980. The seasonal snowcover. In *Dynamics of Snow and Ice Masses*. S.C. Colbeck, ed., pp. 305–395. Academic Press, New York.

Manga, M. 1996. Waves of bubbles in basaltic magmas and lavas. *J. Geophys. Res.*, **101**, 17,457–17,465.

Masek, J.G., and D.L. Turcotte. 1993. A diffusion-limited aggregation model for the evolution of drainage networks. *Earth Planet. Sci. Lett.*, **199**, 379–386.

Massey, B.S. 1986. *Measures in Science and Engineering*. Ellis Horwood, Chichester, England.

May, R.M. 1980. Nonlinear phenomena in ecology and epidemiology. *Ann. N.Y. Acad. Sci.*, **357**, 267–281.

McNamee, P.J., J.M. McLeod, and C.S. Holling. 1981. The structure and behaviour of defoliating insect/forest systems. *Res. Popul. Ecol.*, **23**, 280–298.

Meinhardt, H. 1976. Morphogenesis of lines and nets. *Differentiation*, **6**, 117–123.

Meyer-Peter, E., and R. Müller. 1948. Formulas for bed-load transport. *Proc 3rd Ann. Conf. Int. Assoc. Hydraul. Res.*, pp. 39–64. Stockholm.

Michaelis, L., and M.I. Menten. 1913. Die Kinetik der Invertinwirkung. *Biochem. Z.*, **49**, 333–369.

Miller, R.D. 1972. Freezing and heaving of saturated and unsaturated soils. *High. Res. Rec.*, **393**, 1–11.

Miller, R.D. 1977. Lens initiation in secondary heaving. *Proc. Int. Symp. Frost Action in Soils*, pp. 68–74. University of Luleå.

Miller R.D. 1978. Frost heaving in non-colloidal soils. *Proc. Third Int. Conf. Permafrost*, pp. 708–713. National Research Council of Canada. Ottawa, Canada.

Miller, R.D. 1980. Freezing phenomena in soils. In *Applications of Soil Physics*, D. Hillel, ed., Chapter 10, pp. 254–299. Academic Press, New York.

Mines, A.H. 1986. *Respiratory Physiology*, 2nd ed. Raven Press, New York.

Moresi, L.-N., and V.S. Solomatov. 1995. Numerical investigation of 2D convection with extremely large viscosity variations. *Phys. Fluids*, **7**, 2154–2162.

Morland, L.W. 1984. Thermo-mechanical balances of ice sheet flow. *Geophys. Astrophys. Fluid Dyn.*, **29**, 237–266.

Morris, R.F., ed. 1963. The dynamics of epidemic spruce budworm populations. *Mem. Ent. Soc. Can.*, No. 21, 332 pp.

Mullins, W.W., and R.F. Sekerka. 1964. Stability of a planar interface during solidification of a dilute binary alloy. *J. Appl. Phys.*, **35**, 444–451.

Murray, J.D. 1977. *Lectures on Nonlinear Differential Equation Models in Biology*. Oxford University Press, Oxford.
Murray, J.D. 1989. *Mathematical Biology*. Springer–Verlag, Berlin.
Nayfeh, A.H. 1973. *Perturbation Methods*. John Wiley, New York.
Newell, A.C. 1985. *Solitons in Mathematics and Physics*. Society for Industrial and Applied Mathematics, Philadelphia.
Newell, A.C., and J.A. Whitehead. 1969. Finite bandwidth, finite amplitude convection. *J. Fluid Mech.*, **38**, 279–303.
North, G.R., R.F. Calahan, and J.A. Coakley, Jr. 1981. Energy balance climate models. *Rev. Geophys. Space Phys.*, **19**, 91–121.
Noyes, R.M., and J.-J. Jwo. 1975. Oscillations in chemical systems, X, Implications of cerium oxidation mechanisms for the Belousov–Zhabotinskii reaction. *J. Amer. Chem. Soc.*, **97**, 5431–5433.
Nye, J.F. 1951. The flow of glaciers and ice sheets as a problem in plasticity. *Proc. R. Soc. Lond.*, **A207**, 554–572.
Ockendon, H., and J.R. Ockendon. 1995. *Viscous Flow*. Cambridge University Press, Cambridge.
Ockendon, J.R., and W.R. Hodgkins, eds. 1974. *Moving Boundary Problems in Heat Flow and Diffusion*. Oxford University Press, Oxford.
Ó Mathúna, D. 1971. The differential equations of enzyme reaction kinetics – the regularisation of a singular perturbation. *Proc. R. Ir. Acad.*, **A71**, 27–51.
O'Neill, K. 1983. The physics of mathematical frost heave models: a review. *Cold Reg. Sci. Technol.*, **6**, 275–291.
O'Neill, K., and R.D. Miller. 1985. Exploration of a rigid ice model of frost heave. *Water Resour. Res.*, **21**, 281–296.
Orszag, S.A., and L.C. Kells. 1980. Transition to turbulence in plane Poiseuille and plane Couette flow. *J. Fluid Mech.*, **96**, 159–205.
Outcalt, S.I. 1980. A step function model of ice segregation. *Proc. Second Int. Symp. Ground Freezing*, pp. 515–524. Norwegian Institute of Technology, Trondheim.
Parker, G. 1978. Self-formed straight rivers with equilibrium banks and mobile bed, Part 1, The sand-silt river. *J. Fluid Mech.*, **89**, 109–125.
Paterson, A.R. 1983. *A First Course in Fluid Mechanics*. Cambridge University Press, Cambridge.
Paterson, W.S.B. 1994. *The Physics of Glaciers*, 3rd ed. Pergamon Press, Oxford.
Pedlosky, J. 1979. *Geophysical Fluid Dynamics*. Springer–Verlag, Berlin.
Penner, E. 1959. The mechanism of frost heaving in soils. *High. Res. Board Bull.*, **225**, 1–13.
Piper, D., J.T. Holden, and R.H. Jones. 1988. A mathematical model of frost heave in granular materials. *Proc. Fifth Int. Conf. Permafrost*, pp. 370–376. Tapir Publishers, Trondheim, Norway.
Poincaré, J.H. 1893. *Les Méthodes Nouvelles de la Mécanique Céleste*, Vol. II. Gauthiers-Villars, Paris. English translation: New methods of Celestial Mechanics, ed. D.L. Goroff. American Institute of Physics, 1993.
Polubarinova-Kochina, P. Ya. 1962. *Theory of Ground Water Movement*. Princeton University Press, Princeton, N.J.
Price, M. 1985. *Introducing Groundwater*. George Allen and Unwin, London.
Protter, M.H., and H.F. Weinberger. 1984. *Maximum Principles in Differential Equations*, Springer–Verlag, Berlin.
Quareni, F., and D.A. Yuen. 1988. Mean-field methods in mantle convection. In *Mathematical Geophysics*, N.J. Vlaar et al., eds., pp. 227–264. D. Reidel, Dordrecht.
Ramichandran, P.A., and L.K. Doraiswamy. 1982. Modeling of non-catalytic gas-solid reactions. *AIChE J.*, **28**, 881–900.
Reeh, N. 1982. A plasticity theory approach to the steady-state shape of a three-dimensional ice sheet. *J. Glaciol.*, **28**, 431–455.
Régnière, J., and M. You. 1991. A simulation model of spruce budworm (Lepidoptera: Tortricidae) feeding on balsam fir and white spruce. *Ecol. Model.*, **54**, 277–298.
Richards, K. 1982. *Rivers: Form and Process in Alluvial Channels*. Methuen, London.
Roberts, G.O. 1979. Fast viscous Bénard convection. *Geophys. Astrophys. Fluid Dyn.*, **12**, 235–272.

Roberts, P.H. 1967. *An Introduction to Magnetohydrodynamics*. Elsevier, New York.
Robinson, F.N.H. 1973. Electromagnetism. Oxford University Press, Oxford.
Robinson, J.L., and M.J. O'Sullivan. 1976. A boundary-layer model of flow in a porous medium at high Rayleigh number. *J. Fluid Mech.*, **75**, 459–467.
Römkens, M.J.M., and R.D. Miller, 1973. Migration of mineral particles in ice with a temperature gradient. *J. Coll. Interface Sci.*, **42**, 103–111.
Röthlisberger, H. 1972. Water pressure in intra- and sub-glacial channels. *J. Glaciol.*, **11**, 177–203.
Rubinow, S.I. 1975. *Introduction to Mathematical Biology*. John Wiley, New York.
Rudnicki, J.W. 1980. Fracture mechanics applied to the earth's crust. *Ann. Rev. Earth Planet. Sci.*, **8**, 489–525.
Saltzman, B. 1978. A survey of statistical-dynamical models of the terrestrial climate. *Adv. Geophys.*, **20**, 183–304.
Schlichting, H. 1979. *Boundary Layer Theory*. McGraw-Hill, New York.
Schofield, A.N., and C.P. Wroth. 1968. *Critical State Soil Mechanics*. McGraw-Hill, New York.
Segal, G. 1986. Geometry of surfaces. Mathematical Institute Lecture Notes, Oxford University.
Segel, L.A. 1984. *Modeling Dynamic Phenomena in Molecular and Cellular Biology*. Cambridge University Press, Cambridge.
Segel, L.A., and M. Slemrod. 1989. The quasi-steady state assumption: a case study in perturbation. *SIAM Rev.*, **31**, 446–477.
Sellers, W.D. 1969. A global climate model based on energy balance of the earth-atmosphere system. *J. Appl. Meteorol.*, **8**, 392–400.
Sellmeijer, J.B., and M.A. Koenders. 1991. A mathematical model for piping. *Appl. Math. Model.*, **15**, 646–651.
Shields, A. 1936. Anwendung der Aehnlichkeitsmechanik und der turbulenzforschung auf die geschiebebewegung. Mitteilung der Preussischen versuchsanstalt für Wasserbau und Schiffbau, Heft 26, Berlin.
Shoji, H., and C.C. Langway. 1988. Flow-law parameters of the Dye 3, Greenland deep ice core. *Ann. Glaciol.*, **10**, 146–150.
Silveston, P.L., R.R. Hudgins, S. Bogdashev, N. Vermjakovskaja, and Yu. Sh. Matros. 1994. Modeling of a periodically operating packed-bed SO_2 oxidation reactor at high conversion. *Chem. Eng. Sci.*, **49**, 335–341.
Skempton, A.W. 1960. Effective stress in soils, concrete and rocks. In *Pore Pressure and Suction in Soils*, pp. 4–16. Butterworths, London.
Slonim, N.B., and L.H. Hamilton. 1987. *Respiratory Physiology*, 5th ed. The C.V. Mosby Company, St Louis, Mo.
Smith, T.R., and F.P. Bretherton. 1972. Stability and the conservation of mass in drainage basin evolution. *Water Resour. Res.*, **8**, 1506–1529.
Solymar, L. 1984. *Lectures on Electromagnetic Theory*, 2nd ed. Oxford University Press, Oxford.
Sparrow, C. 1982. *The Lorenz Equations: Bifurcations, Chaos and Strange Attractors*. Springer–Verlag, Berlin.
Spence, D.A., P.A. Sharp, and D.L. Turcotte. 1987. Buoyancy-driven crack propagation: a mechanism for magma migration. *J. Fluid Mech.*, **174**, 135–153.
Stanshine, J.A. 1976. Asymptotic solutions of the Field-Noyes model for the Belousov reaction, II, Plane waves. *Stud. Appl. Math.*, **55**, 327–349.
Stanshine, J.A., and L.N. Howard. 1976. Asymptotic solutions of the Field-Noyes model for the Belousov reaction, I, Homogeneous oscillations. *Stud. Appl. Math.*, **55**, 129–165.
Stark, C. 1991. An invasion percolation model of drainage network evolution. *Nature*, **352**, 423–425.
Stedinger, J.R. 1984. A spruce budworm-forest model and its implications for suppression programs. *Forest Sci.*, **30**, 597–615.
Stefan, J. 1891. Ueber die Theorie der Eisbildung, insbesondere über die Eisbildung im Polarmeere. *Ann. Phys. Chem.*, **42**, 269–286.
Stewartson, K., and J.T. Stuart. 1971. A non-linear instability theory for a wave system in plane Poiseuille flow. *J. Fluid Mech.*, **48**, 529–545.

Stuart, J.T. 1960. On the non-linear mechanics of wave disturbances in stable and unstable parallel flows. *J. Fluid Mech.*, **9**, 353–370.

Stuart, J.T. 1967. On finite amplitude oscillations in laminar mixing layers. *J. Fluid Mech.*, **29**, 417–440.

Szekely, J., J.W. Evans, and H.Y. Sohn. 1976. *Gas–Solid Reactions*. Academic Press, New York.

Taber, S. 1930. The mechanics of frost heaving. *J. Geol.*, **38**, 303–317.

Takayasu, H. 1990. *Fractals in the Physical Sciences*. Manchester University Press, Manchester.

Tayler, A.B. 1986. *Mathematical Models in Applied Mechanics*. Clarendon Press, Oxford.

Taylor, G.I. 1953. Dispersion of soluble matter in a solvent flowing slowly through a tube. *Proc. Roy. Soc.*, **A219**, 186–203.

Timoshenko, S., and J.N. Goodier. 1970. *Theory of Elasticity*. McGraw-Hill, New York.

Titchmarsh, E.C. 1937. *Introduction to the Theory of Fourier Integrals*. Oxford University Press, Oxford.

Townsend, A.A. 1956. *The Structure of Turbulent Shear Flow*. Cambridge University Press, Cambridge.

Tritton, D.J. 1988. *Physical Fluid Dynamics*. Oxford University Press, Oxford.

Tucker, G.E., and R.L. Slingerland. 1994. Erosional dynamics, flexural isostasy, and long-lived escarpments: a numerical modeling study. *J. Geophys. Res.*, **99**, 12,229–12,243.

Turcotte, D.L. 1979. Convection. *Revs. Geophys. Space Phys.*, **17**, 1090–1098.

Turcotte, D.L. 1992. *Fractals and Chaos in Geology and Geophysics*. Cambridge University Press, Cambridge.

Turcotte, D.L., and G. Schubert. 1982. *Geodynamics*. John Wiley, New York.

Turner, J.S. 1973. *Buoyancy Effects in Fluids*. Cambridge University Press, Cambridge.

Tyson, J.J. 1976. *The Belousov–Zhabotinskii Reaction*. Springer–Verlag, Berlin.

Tyson, J.J. 1985. A quantitative account of oscillations, bistability, and travelling waves in the Belousov–Zhabotinskii reaction. In *Oscillations and Travelling Waves in Chemical Systems*. R.J. Field and M. Burger, eds., pp. 93–144. John Wiley, New York.

Tyson, J.J. 1994. What everyone should know about the Belousov-Zhabotinsky reaction. In Frontiers in Mathematical Biology, S.A. Levin, ed., *Lecture Notes in Biomathematics, Vol. 100*, pp. 567–587. Springer–Verlag, Berlin.

Van Dyke, M. 1975. *Perturbation Methods in Fluid Mechanics*, annotated ed. Parabolic Press, Stanford, Calif.

Van Dyke, M.D. 1982. *An Album of Fluid Motion*. Parabolic Press, Stanford, Calif.

Vidal, C., and A. Pacault. 1982. Spatial chemical structures, chemical waves, a review. In *Evolution of Order and Chaos*, H. Haken, ed., pp. 74–99. Springer–Verlag, Berlin.

Vignes-Adler, M. 1977. On the origin of the water aspiration in a freezing dispersed medium. *J. Colloid Interface Sci.*, **60**, 162–171.

Volterra, V. 1926. Variazionie fluttuazioni del numero d'individui in specie animali conviventi. *Mem. Acad Lincei.*, **2**, 31–113.

Wake, G.C., and M. Hood. 1993. Multiplicity of solutions of a quasilinear elliptic equation in spherical domains. *Math. Comput. Model.*, **18**, 157–162.

Walder, J.S., and A.C. Fowler. 1994. Channelised subglacial drainage over a deformable bed. *J. Glaciol.*, **40**, 3–15.

Wallis, G.B. 1969. *One-Dimensional Two-Phase Flow*. McGraw-Hill, New York.

Ward, R.C., and M. Robinson. 1990. *Principles of Hydrology*. McGraw-Hill, New York.

Weertman, J. 1976. Milankovitch solar radiation variations and ice age ice sheet sizes. *Nature*, **261**, 17–20.

Welander, P. 1967. On the oscillatory instability of a differentially heated loop. *J. Fluid Mech.*, **29**, 17–30.

Whalley, P.B. 1987. *Boiling, Condensation and Gas–Liquid Flow*. Oxford University Press, Oxford.

Whipp, B.J., and D.H. Wiberg, eds. 1983. *Modeling and Control of Breathing*. Elsevier Biomedical, New York.

Whitham, G.B. 1974. *Nonlinear Waves*. John Wiley, New York.

Widdicombe, J., and A. Davies. 1991. *Respiratory Physiology*, 2nd ed. Edward Arnold, London.

Wiggins, S. 1988. *Global Bifurcations and Chaos: Analytical Methods.* Springer–Verlag, Berlin.

Willgoose, G., R.L. Bras, and I. Rodriguez-Iturbe. 1989. A physically based channel network and catchment evolution model. Report No. 322, Dept. of Civil Engineering, Mass. Inst. of Technol., Cambridge, Mass.

Willgoose, G., R.L. Bras, and I. Rodriguez-Iturbe. 1991a. A coupled channel network growth and hillslope evolution model, 1, Theory. *Water Resour. Res.*, **27**, 1671–1684.

Willgoose, G., R.L. Bras, and I. Rodriguez-Iturbe. 1991b. A coupled channel network growth and hillslope evolution model, 2, Nondimensionalization and applications. *Water Resour. Res.*, **27**, 1685–1696.

Williams, P.J., and M.W. Smith. 1989. *The Frozen Earth.* Cambridge University Press, Cambridge.

Winfree, A.T. 1972. Spiral waves of chemical activity. *Science*, **175**, 634–636.

Worster, M.G. 1991. Natural convection in a mushy layer. *J. Fluid Mech.*, **224**, 335–359.

Worster, G. 1992. The dynamics of mushy layers. In Interactive Dynamics of Convection and Solidification. Davis et al., eds., pp. 113–138, *NATO ASI Series E, Vol. 219.* Kluwer, Dordrecht.

Yih, C.-S. 1979. *Fluid Mechanics.* West River Press, Ann Arbor, Mich.

Zuber, N., and J. Findlay. 1965. Average volumetric concentration in two-phase flow systems. *Trans. ASME, Ser. C., J. Heat Transf.*, **87**, 453–468.

Index